Managing Construction Projects

An Informatio

Managing Construction Projects

An Information Processing Approach

Second Edition

Graham M. Winch
Professor of Project Management
Centre for Research in the Management of Projects
Manchester Business School
The University of Manchester

WILEY-BLACKWELL

A John Wiley & Sons, Ltd., Publication

This edition first published 2010

Blackwell Publishing was acquired by John Wiley & Sons in February 2007. Blackwell's publishing programme has been merged with Wiley's global Scientific, Technical, and Medical business to form Wiley-Blackwell.

Registered office
John Wiley & Sons Ltd, The Atrium, Southern Gate, Chichester, West Sussex, PO19 8SQ, United Kingdom

Editorial office
2121 State Avenue, Ames, Iowa 50014-8300, USA

For details of our global editorial offices, for customer services and for information about how to apply for permission to reuse the copyright material in this book please see our website at www.wiley.com/wiley-blackwell.

Library of Congress Cataloging-in-Publication Data
Winch, Graham.
Managing construction projects : an information processing approach / Graham M. Winch. — 2nd ed.
p. cm.
Includes bibliographical references and index.
ISBN 978-1-4051-8457-1 (pbk. : alk. paper) 1. Building—Superintendence. 2. Building—Planning. 3. Communication of technical information. I. Title.
TH438.W56 2010
690.068--dc22

2009016226

A catalogue record for this book is available from the British Library.

Set in 10.5/12 pts Bembo by Macmillan Publishing Solutions
Printed in Singapore by Ho Printing Singapore Pte Ltd

2 2010

DEDICATION

This book is dedicated to the memory of Geoffrey Roy Winch,
engineer extraordinary, 1922–1999,
and to the love of Sandra Schmidt

It is a massive review of the art and science of the management of projects, which has the great virtue of being a good read wherever it is touched. It spills the dirt on things that went wrong, elucidates the history so that you can understand the industry's current stance, draws on other countries' experience and explains the latest management processes. Throughout it is liberally sprinkled with anecdotes and case histories which amply illustrate the do's and don'ts for practitioners wishing to deliver projects on time with expected quality and price. It is a valuable book for students and practitioners alike.

John D. Findlay
Director
Stent

This is a valuable source for practitioners and students. It covers the A–Z of project management in a confident contemporary manner and provides a powerful and much needed conceptual perspective in place of a purely prescriptive approach. The engaging presentation introduces a range of challenges to establishing thinking about project management, often by making comparisons between practices in the UK and those of other countries.

Peter Lansley
Professor of Construction Management
University of Reading

Contents

Preface to 1st Edition

The management of construction projects is a problem in information, or rather, a problem in the lack of information required for decision-making. In order to keep the project rolling, decisions have to be made before all the information required for the decision is available. Decision-making in construction is, therefore, about robust decisions, rather than optimal decisions. This paradox is at the heart of the book, which explores the high-grade project management skills required to manage under uncertainty. The book does not provide easy answers, but ways of thinking about challenging problems. Construction project management is not easy otherwise we would have solved the problems by now, but it can be done better. This book draws extensively from practice in other industries to show how it can be done better.

The book is intended for those practitioners – let us call them reflective practitioners – who wish to develop their capabilities to manage the whole rather than the parts, and for those students on masters' courses who are being trained in those capabilities. Drawing on a wide range of research, it does not summarise received wisdom, but proposes new ways of thinking about managing construction projects better. Its basic assertion is that we have to treat the management of construction projects as a holistic discipline, managing from inception to completion, rather than a set of fragmented professional domains. It is this vision of an integrated construction project management that the book attempts to define.

If construction project management is a problem in information, what is the role of information and communication technology (ICT)? The answer is simple – it is central. ICT pervades this book. Although only one chapter is explicitly devoted to the topic, there are continual references to the role of the new, distributed, generation of ICT in the management of construction projects.

The argument is supported throughout by vignettes, and case studies complement each chapter. These are not intended to show only good or bad practice, but to illustrate the argument and stimulate reflection. Those using the book for teaching will be able to use them as teaching cases. A full version of the Channel Fixed Link case study is available from the Blackwell website http://www.blackwellpublishing.com/winch, together with a complete set of lectures and associated graphics.

The book has been long in the making, and draws on a wide range of intellectual sources read over some 30 years. It might be helpful for the reader to identify some of the key influences here, for they are very much embedded in the text but without them this book could not have been written. The main ones are, in order of reading, are as follows:

- Peter Berger and Thomas Luckmann, *The Social Construction of Reality*. The first text – before the rise of Anthony Giddens – to articulate the dynamic dialectic between structure and process.
- Jay Galbraith, *Organization Design*. The source of the idea of organisations as information processing systems.
- Oliver Williamson, *Markets and Hierarchies*. The founding work of transaction cost economics, and hence the leading theory of inter-firm relations.
- Marian Bowley, *The British Building Industry*. Written about 40 years ago, much of the analysis of the construction industry as system is as relevant today as it ever was.
- Peter Morris, *The Management of Projects*. The first project management text to raise the discipline out of the tool box and into the boardroom as a strategic discipline.

In preparation for the writing of this text, a review of the project management literature (Winch, 2000a) identified five generic project management processes:

- Defining the project mission
- Mobilising the resource base
- Riding the project life cycle
- Leading the project coalition
- Maintaining the resource base.

The first four of these provide the overall structure of the book. Constraints of space and time mean that the fifth could not be addressed here. Readers seeking insights into this last process should refer to David Gann (2000) on construction innovation, and Jan Druker and Geoff White (1996) on human resource management in construction.

The range of acknowledgement that this book requires is vast, and I can only be selective here. First and foremost, thanks must go to Graham Ive who first encapsulated the approach taken here that project management is a problem in information, and who inspired the writing of the book in the first place. Second, thanks go to the students on the MSc Construction Economics and Management at The Bartlett, University College London, who have had these ideas tried out on them over the past 10 years, and have contributed some of the case material. In particular, the contribution of the 2000/1 cohort who participated in a feedback seminar on version 1 of the book is warmly thanked. The 2001/2 cohorts on the MSc Engineering Project Management and fourth year MEng Civil Engineering at the Manchester Centre for Civil and Construction Engineering at UMIST worked with version 3 and allowed the argument and graphics to be fine-tuned. The approach of this book is different from the UMIST tradition in project management – represented by Roy Pilcher, Nigel Smith, Peter Thompson, Stephen Wearne and others – but it is, I hope, complementary.

Third, thanks go to the bodies which have funded the research over the past 10 years, which has allowed particular aspects of the argument to be explored. These include the Economic and Social Research Council, The Leverhulme Trust, Plan Construction et Architecture and, most notably, the Engineering and Physical Sciences Research Council (EPSRC). As researchers, Aalia Usmani, Naomi Clifton, Andrew Edkins, Bríd Carr, Steve North and John Kelsey have contributed to this book more than they probably appreciate. In particular, the book draws in a variety of ways on material developed by the EPSRC funded VIRCON project, a four-university collaboration of Teesside (Nash Dawood), University College London (Alan Penn), UMIST and Wolverhampton (Lamine Mahdjoubi). I am especially grateful to Nash Dawood and the team from Teesside for the data on, and images of, the Centuria Building.

The first draft of Chapter 14 was prepared while the author was Velux Visiting Professor at the Department of Civil Engineering, The Technical University of Denmark. Thanks go to Sten Bonke and Axel Gaarslov for their hospitality, and to Rob Howard and Christian Koch for their help in the development of the chapter. Steve North read version 3 of the chapter in detail and mitigated some of my misunderstandings of the issues. He, together with John Kelsey, was also enormously helpful in the development of Cases 10 and 11. In particular, John Kelsey grappled with the details of the critical chain methodology, giving the argument a robustness that would otherwise be missing. He also read and commented on version 2 of Chapters 10 and 11 in their entirety. Ghassan Aouad and Ming Sun of Salford University willingly provided materials for figures in Chapter 14. I am especially grateful to Peter Morris, who read and commented on the whole of version 2. Pam Hyde, administrator of the Project Management Division of the Manchester Centre, helped with the final production of version 4. None of these, of course, bears any responsibility for the argument in the text of the book.

Finally, warmest thanks go to Shilpi Kawar who acted as editorial assistant on the book, handling permissions and processing the text for printing, as well as drawing all the diagrams and researching some of the vignettes, mainly for Chapter 4.

The images that illustrate this book are of the various stages in the project life cycle of one of the more remarkable millennium projects in the UK – the Millennium Bridge, which links St Paul's Cathedral in the City of London to the Tate Modern Museum in Southwark. The vision of a *Financial Times* journalist, the bridge opened to the public in 2002, after a false start during 2000 when it had to be closed on its day of opening due to excessive lateral movement. It provides a wonderful example of the excitement and challenges of managing construction projects where the project mission is both to push the technological envelope and to make a major contribution to urban culture. Despite these problems, the bridge has already won a place in the affections of Londoners, and will doubtless go on to become a major landmark for London and its people. Further information is available in Deyan Sudjic's book, *Blade of Light: the Story of the Millennium Bridge.*

Graham M. Winch
Manchester

Preface to 2nd Edition

> In the wide ocean upon which we venture, the possible ways and directions are many; and the same studies which have served for this work might easily, in other hands, not only receive a wholly different treatment and application, but also lead to essentially different conclusions. Such indeed is the importance of the subject that it still calls for fresh investigation, and may be studied with advantage from the most varied points of view. Meanwhile we are content if a patient hearing is granted us, and if this book be taken and judged as a whole.

Jacob Burckhardt (1990, p. 19) thus introduced his distinctive perspective on the Italian Renaissance in 1860, suggesting that complex phenomena are best investigated using multiple perspectives. This second edition develops the information processing perspective introduced in the first as a distinctive contribution to the available perspectives on managing construction projects. The information processing perspective cannot claim to be comprehensive, but we do suggest that it is a worthy way of venturing on that wide ocean. In particular, the information processing perspective deepens understanding of the dynamics of the construction project process through life from the value proposition inherent in the project mission to the functioning asset generating that value for its owners and users.

The information processing perspective has been developed through three main influences since 2002.

- A move to Manchester Business School (MBS) within the new University of Manchester formed in 2004 from UMIST and the old Victoria University of Manchester. This created opportunities and incentives to read different literatures, attend different conferences and teach students with a broader perspective on managing projects than would be found within an engineering or built environment school. This edition is more clearly about the business of managing construction projects than the first, reflecting the needs of students on MBS' MBA for Construction Executives.
- The criticisms made by a number of researchers of the first edition. These are identified explicitly in Chapter 1, and addressed – hopefully adequately –

throughout the text. Here we reiterate Burckhardt's plea that the book be read as a whole, rather than criticised piecemeal.

- Research developing a more cognitive understanding of managerial information processing drawing on the work of Karl Weick (1995) and Alfred Schutz (1967) in collaboration with Eunice Maytorena on managing risk and uncertainty, and Kristian Kreiner on future perfect organizing respectively.

The text has been updated throughout to refer to current standards and practice in the industry, as well as the latest research findings. I am particularly grateful to Ghassan Aouad of Salford University and Martin Riese of Ghery Technologies for their help with the revisions to Chapter 14. One important systematic change throughout is to refer to 'schedule' rather than 'programme' in construction project planning. The two reasons are that, first, 'programme' in some countries (e.g. USA and France) refers to the brief and not the project plan and, second, it avoids confusion in relation to the concept of 'programme management' introduced in Chapter 15.

Interactions are always very important in the development of ideas, and I would particularly like to thank the members of the Managing Projects group of the Business Systems Division of MBS (Nuno Gil, David Lowe, Eunice Maytorena, Cliff Mitchell, Mike Pryce and Mark Winter) as well as the broader membership of the Centre for Research in the Management of Projects at MBS. Three years as a Visiting Professor (2006–2008) at the *Center for Ledelse i Byggeriet* (Centre for Management Studies of the Building Process) at Copenhagen Business School working with Professor Kristian Kreiner and his team have broadened my theoretical perspectives as well as helped me to appreciate the contribution of high-quality ethnographic research to understanding the construction project process. Of course, none of the above bears responsibility for the content of this edition.

As Burckhardt would have wished, the first edition has already provoked debate; I very much hope that the second edition will continue that debate as well as meet the needs of future cohorts of students of *Managing Construction Projects*. The images on the cover of this edition are of remarkable value generation – the Eden Project in Cornwall, UK. They show the architects' section of the geodesic roof structure (courtesy Grimshaw Architects) and a view of the completed biome (photo Ben Foster: courtesy of Eden Project), capturing the creation of these remarkable buildings through the life cycle from conception to completion. Further information on the project is given in Case 17 and a visit (http://www.edenproject.com) is heartily recommended for a sense of how the built environment can and should be created. Finally, I would like to add my enormous gratitude to Sandra for her diligence in preparing the final version of this text.

Graham M. Winch
Manchester

Part I
Introduction

Chapter 1
The Management of Construction Projects

1.1 Introduction

'Between the idea
And the reality. . . .
Between the conception
And the creation. . . .
Falls the Shadow'

One of the principal ways in which modern societies generate new value is through projects which create physical assets that can then be exploited to achieve social and economic ends – factories for manufacturing goods, offices and shops for delivering services, hospitals for health care and tunnels for transport. Societies even create assets that are exploited for largely symbolic purposes, such as opera houses and cathedrals. In a typical modern society, around half of all physical asset creation (fixed capital formation) is the responsibility of the construction industry, thereby generating around 10% of national wealth (gross domestic product). These figures are much higher for rapidly developing countries. The creation of these assets is the principal force in the dynamics of cities and change in the built environment and, therefore, one of the major sources of social and economic change. This book is about how such assets are created effectively and efficiently so that they meet the needs of the clients which make the investments, thereby providing a net gain to the economy and society for which they are created.

The creation of new values is not an easy mission – as the liberties taken with T.S. Eliot's *The Hollow Men* in the epigraph above are intended to capture. Many problems have to be solved between the initial idea for a new asset, through its realisation on site, to the client starting to exploit it. This book covers the whole of this process conceived as a progressive reduction of uncertainty through time. In other words, it argues that the problem of managing construction projects is principally a problem in the management of information and its progressive embodiment in a physical asset. As a director of a leading European construction corporation puts it, 'HBG's core competence is the generation and management of information'[1]. The book will, thereby, shine a penetrating light into the shadow between the conception of a constructed asset and its physical creation.

The book is not aimed at any particular professional group within the construction industry; rather it is aimed at all those whose working lives are committed to the creation of constructed assets – at all professional groups. These include the representatives of the clients who provide the capital; the designers who turn ideas into specifications; the constructors who turn specifications into reality on site; as well as those who manage and regulate the overall process on behalf of the client and society. Creating new value through construction projects is an inherently collaborative process, and all have their specialist skills to deploy. The central premise of this book is that these specialisms can be deployed more effectively in the context of an understanding of the process as a whole. Thus, one of the most important measures of the success of this book will be the extent to which it helps in the creation of a common language for discussing the management of construction projects between different professional groups. The perspectives and terminology used in this book may be a little unfamiliar at times; this is because the book is deliberately written from a perspective of managing the entire project process, rather than the contribution of any one professional group to it.

More specifically, the objectives of this book remain unchanged for this edition:

- to provide a total project perspective on the management of construction projects from inception to completion;
- to apply business process analysis (BPA) to the management of projects;
- to define basic principles of construction project management which will allow readers to apply these principles to their particular management problems;
- to review and synthesise the large number of different tools and techniques proposed for improving construction performance, from risk management and value management, through to supply chain management and quality assurance;
- to place the use of information and communication technologies (ICTs) at the heart of the construction project management process.

In achieving these objectives, the book will provide a holistic perspective that will allow practitioners and more advanced students to place their particular specialisms – be it risk management, design management or site management – in the broader context of the project process as a whole. The sheer variety of proposed ways of improving the performance of the construction process can be daunting, even for the most enthusiastic practitioner. By placing all these different initiatives in the context of the entire project process, and by articulating basic principles of good management rather than the latest fads, this book will provide help in sorting good practice from fashionable practice. As such, it aims to facilitate the development of the evidence-based management of construction projects which 'first and foremost, is a way of seeing the world and thinking about the craft of management; it proceeds from the premise that using better, deeper logic and employing facts, to the extent possible, permits leaders to do their jobs more effectively'[2].

1.2 Projects as the creation of new value

All modern societies and economies are dynamic – the only certainty is change. Many of these changes are the result of unforeseen interactions of complex forces, but societies also change through deliberate action, and one of the most important forms of deliberate action is to invest in physical assets which can then be exploited to provide the goods, services and symbols that society needs. Governments invest in schools to provide education services and in bridges to provide transport services; firms invest in shops to provide retail services and in houses to provide homes. Investments are also made in redundant quarries to create an inspirational ecological experience as at the Eden Project in Cornwall (which we shall revisit in Case 17) or a five-star hotel as at 松江區 (Songjiang) near Shanghai or on a smaller scale as a theatre at Dalhalla, Rättvik, Sweden. Investments are made to transform coastlines such as the Delta and Zuidersee projects in The Netherlands which created millions of hectares of farmland and the extensive marine works to 'help solve Dubai's beach shortage' in Nakheel's three Palm and The World developments – The World alone adds 232 km to Dubai's coastline[3]. Cities change as shops are refurbished and new metros are built. Increasingly, these investments are made by partnerships of the public and private sectors. What all these investments have in common – whether directly for profit or not – is that they create something where there was nothing, create new assets to be exploited for private benefit and public good. It is in this sense that construction projects are about the creation of new value in society.

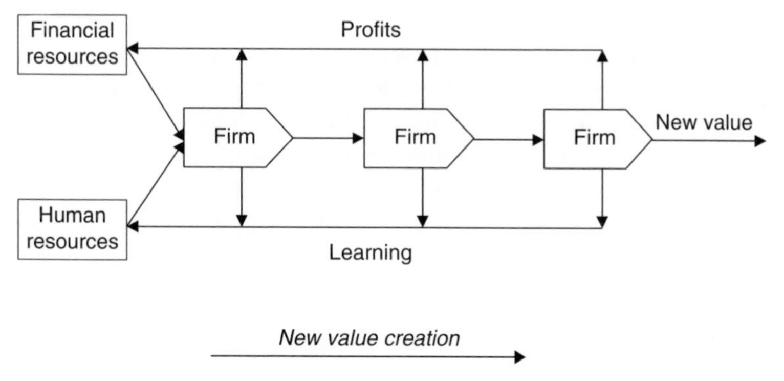

Fig. 1.1 Construction projects as the creation of new value.

This process forms a 'value system'[4] as illustrated in Fig. 1.1; how projects add value for clients through the value system will be explored in more detail in Chapter 3. The fundamental inputs to the process are capital and human resources – capital resources to cover the costs of investment; human resources to transform ideas into reality. The return on capital from the process is the profits taken out of the process by the participating firms. The return on human resources is the learning that takes place as problems are solved through the project life cycle. The effective achievement of both of these returns on the resources deployed in the creation of constructed assets is problematic – construction firms have low profitability compared to other sectors, and learning often stays with the individual, rather than being captured by the firm. As will be explored in Part IV, these two problems are linked.

1.3 The project as an information processing system

All organisations are, in essence, information processing systems[5]. In order to function they must monitor their environment, take decisions, communicate their intentions and ensure that what they intended to happen does happen. In manufacturing organisations, these information flows generate and control flows of materials as well, but many service organisations are purely devoted to managing flows of information. Information flows are the heart of the business process in all organisations. These information flows are directed and enabled by the structure of the organisation, and the problem of management is the problem of continually shaping processes by manipulating the structure – what has been called the tectonic approach to organisation[6].

The analogy of a river is useful here. What is of interest in a river is the flow of water, which irrigates crops, provides a transport route, enables the generation of hydroelectric power and is a source of leisure and repose. Yet it is through altering the banks that we shape the flow – dams and weirs create lakes and power; dykes and canals control direction; docks and locks facilitate transport; bridges and tunnels mitigate the downside of the river as a barrier. At the same time, the action of the water erodes banks, weakens riverine structures and silts navigation channels. The process – the flow of water – cannot be directly managed; we have to manage the context in which it flows, but those flows also change the ways in which we manage. The same, I suggest, applies to organisations and their flows of information, and much of this book will be about how we manage the project process through managing the organisational structure of projects, and how the project process in turn shapes those organisational structures.

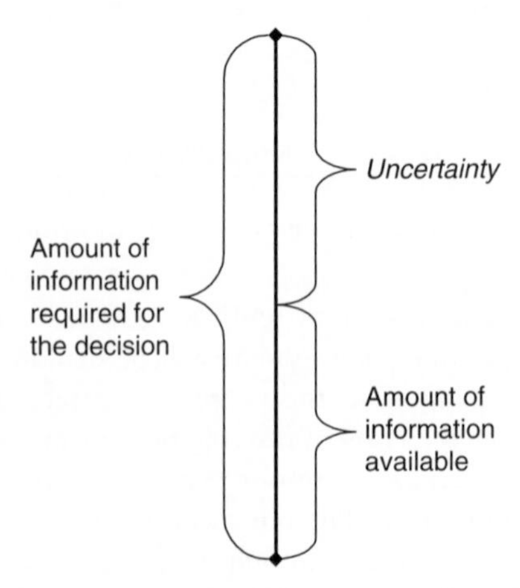

Fig. 1.2 The definition of uncertainty (source: developed from Galbraith, 1977, Fig. 3.1).

The fundamental problem in the management of information is uncertainty; in other words, the lack of all the information required to take a decision at a given time. Figure 1.2 illustrates Jay Galbraith's definition of uncertainty as the difference between the information required for a decision and the information available. This uncertainty has two sources:

- *Complexity*, or the condition where the information is, in principle, available, but it is too costly or time-consuming to collect and analyse;
- *Predictability*, or the condition where the past is not a reliable guide to the future – the future is, by definition, unknowable, but past experience is a valuable, if not infallible, guide to the future in many situations.

The challenge of managing projects in the context of uncertainty is the central theme of this book, while we will focus explicitly on the cognitive issues this poses in Chapter 13.

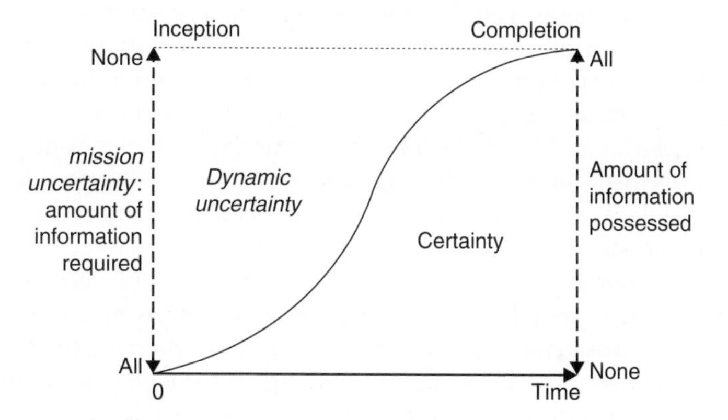

Fig. 1.3 The project process as the dynamic reduction of uncertainty through time (source: developed from Winch *et al.*, 1998).

At the inception stages of a construction project, uncertainty is very high – the asset of the future is little more than an idea and possibly a few sketches. How high depends upon a number of factors such as the extent to which the asset is a copy of the ones existing; the extent to which standardised components and solutions can be used; and the extent of the requirement for new technologies to solve the particular problems posed by the project. This may be thought of as the level of *mission uncertainty* inherent in the project. As the project moves through the life cycle, uncertainty is reduced as more information becomes available – ambiguities in design are resolved; geotechnic surveys are completed; regulatory approval is obtained; component suppliers provide their shop drawings; and contractors successfully complete their tasks. The level of uncertainty at a particular point in the project life cycle relative to earlier and later points in the project life cycle may be thought of as the level of *dynamic uncertainty* on the project. This framework is illustrated in Fig. 1.3, which shows how uncertainty is progressively reduced through time, and how certainty increases until all the information required for the project is available at completion and embodied in the asset created. The area to the left of the S-curve represents information still to be acquired, that is uncertainty; that to the right represents what is known, that is certainty.

1.4 Project management and the management of projects

Construction projects have been 'managed' since time immemorial. Traditionally, this was the responsibility of the 'master of the works' – a concept retained in the modern French *maître d'œuvre* – but the emergence of a concept of 'project management' is a phenomenon of the nineteenth century[7]. Project management emerged as industrial societies started to build complex systems such as rail and power networks. This concept was adopted by the US aircraft industry in the 1920s, came to maturity in the US defence programme in the 1950s and gained international attention with the space programme in the 1960s. Project management is essentially an organisational innovation – the identification of a team responsible for ensuring the effective delivery of the project mission for the client. However, it has become associated with a particular set of tools and techniques – most notably critical path analysis – which has stunted its development. As the concepts of project management diffused to the construction industry from the 1960s onwards, it was this toolbox, rather than the broader management concept, which was adopted[8].

Peter Morris (1994) argues strongly that project management is about the total process, not just about realising a specification to time, cost and quality. For this reason, he distinguishes the 'management of projects' as a strategic approach from 'project management' as a toolbox approach to delivering the project mission. This book adopts Morris' perspective and argues for a holistic approach to managing the construction project. Effective management tools are vital – and will be discussed in detail in Part IV – but they are no substitute for a strategic overview of the process of realising a constructed asset, and skills in managing the disparate stakeholders in the project. However, this book is not just about the activities of the designated project management team, but about all those who are responsible for ensuring that the project mission is achieved – including project architects, site supervisors and contracts managers as well as client representatives. To be effective, the principles of the management of projects need to infuse the project process – construction project managers cannot operate effectively as an external add-on harrying those responsible for actually adding value.

1.5 Projects and resource bases

Construction projects mobilise capital and human resources. The capital that finances the process comes from the client and its financiers. The human resources that enable the progressive reduction of uncertainty through time are supplied

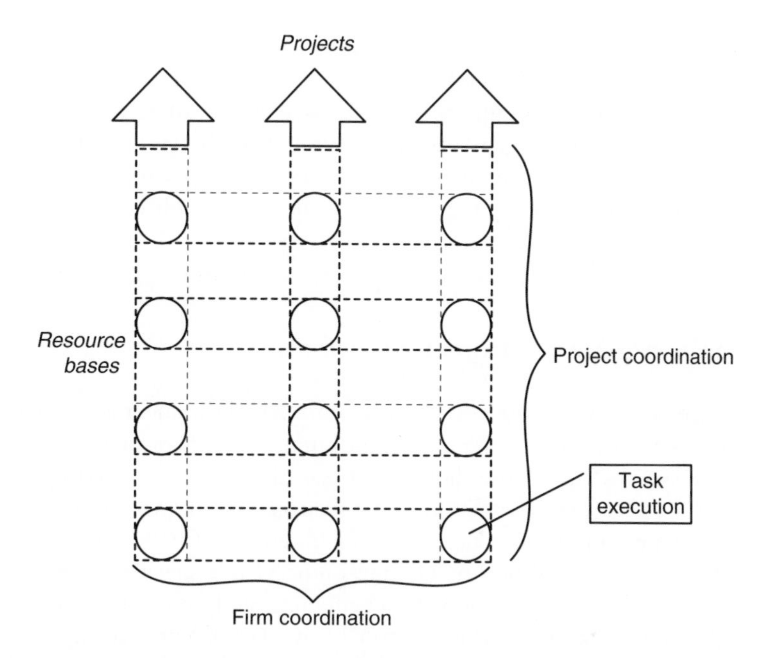

Fig. 1.4 Project organisation as a coalition of resource bases and a portfolio of projects (source: developed from Fellows *et al.*, 1983, Fig. 1.1).

by the firms on the supply side of the construction industry, which act as skill containers[9] for these resources. Resources of equipment are also typically supplied by firms in the construction industry. Components and materials are usually supplied by firms outside the construction industry, although some construction firms are vertically integrated backwards into frequently used sources of components such as prefabricated concrete elements and materials such as aggregates. Our focus here will be on the mobilisation of human resources and specialist equipment.

Firms are different from projects – projects are temporary organisations with no autonomous capability; they rely entirely on mobilising the resources supplied by clients and the firms in the construction industry for their existence. Each project requires a large number of different types of human and equipment resources which are held by the firms on the supply side; we can think of these as the *resource bases* of the construction industry. It is with these resource bases that the continuing capacity to create constructed assets lies. These groupings of resource bases are often called the project team. However, as will be explored in Part V, the number of people involved is, in practice, too large to be meaningfully called a team. Moreover, as will become clear – particularly in Parts II and III – all these different resource bases have different interests. We can more usefully think of these groupings of resource bases mobilised on the project as the *project coalition* which comes together around shared objectives so that each member can meet its individual objectives. One of the main reasons why interests differ is that most resource bases will be supplying resources to more than one project at once, and can find themselves juggling resources between projects. We can, therefore, most usefully think of projects as coalitions of resource bases co-ordinated by the project management team, indicated by the vertical dimension in Fig. 1.4, and firms as participating in portfolios of projects co-ordinated by the resource-base firm, indicated by the horizontal dimension, with project and firm meeting through task execution.

1.6 The five generic project processes

Business process analysis has become increasingly influential in a number of industries – both in the re-engineering of business processes to maximise the benefits of ICT systems and in the diffusion of lean thinking. Conceptually, there are important links between the notion of the management of projects as the management of the entire project life cycle and the development of BPA. This is clear from Thomas Davenport's formulation of a business process as 'a specific ordering of work activities across time and place, with a beginning, and end, and clearly identified inputs and outputs: a structure for action'[10], and James Womack and Dan Jones' argument[11] that the emergence of project management foreshadowed their own concepts of lean thinking. The concepts behind BPA and lean thinking are central to the agenda for change set out in the UK Construction Task Force's report, on *Rethinking Construction* – colloquially known as the Egan Report. We will revisit these themes in the conclusions, showing how they have evolved into the *revaluing construction* agenda.

The approach adopted here to identifying the principal project process is that of BT[12] which identified five first-order processes (Manage the Business; Manage People and Work; Serve the Customer; Run the Network; and Support the Business). Within these five, some 15 second-order business processes were identified. The structure of this book will draw upon a review of the body of empirical studies on the management of projects across the full range of project-orientated industries which identified five first-order project processes[13] – defining the project mission; mobilising the resource base; riding the project life cycle; leading the project coalition; and maintaining the resource base. Within these five, a larger number of more focused business processes such as risk management, supply chain management and quality management will then be explored.

1.7 Critiques of the first edition

The first edition of this text was generally well received – which is why you are reading the second one now – but it did attract a number of criticisms which we will try to address in this section.

Stuart Green has argued that the attempt to place the analysis of the process of managing projects in its institutional context is welcome, but also argue that the institutionalism deployed in the book is more 'old' than 'new' in that it is structurally deterministic. Green then goes on to suggest that 'there is seemingly little recognition of the role of discourse in the shaping of self-identities that lead to action, and how such streams of action combine over time to reshape context'[14]. Green's principal influences in this argument are Giddens, and Powell and DiMaggio[15]. Green is correct to point out that the argument in the book does not explicitly rely upon Gidden's structuration theory; however, the discussion of the 'tectonic approach' on page 6 shows that it is rooted in Gidden's work and articulates the same[16] dialectic of structure and process that Green advocates. The metaphor of the river in section 1.3 has been developed to make this point clearer and the overall approach is captured in the *tectonic approach* presented in Fig. 1.5. Green's advocacy of a discourse approach, we would suggest, is compatible with a tectonic approach, save in one crucial respect. This is the tendency, well displayed in the empirical section of Green's chapter, to focus only on process while ignoring outcomes, a weakness shared by much constructivist analysis[17].

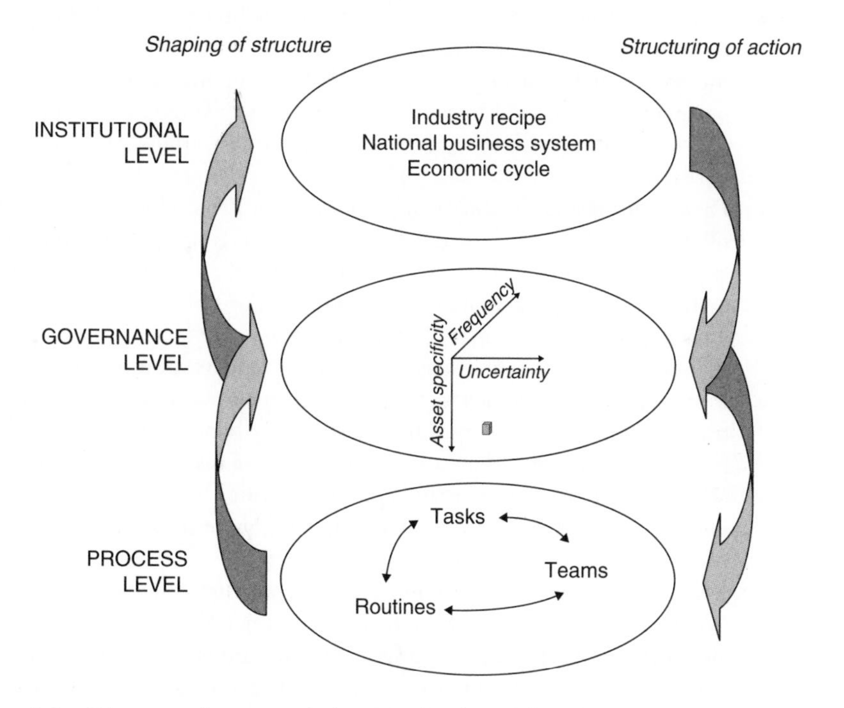

Fig. 1.5 The tectonic approach (source: developed from Winch, 2006a, Fig. 14.2).

Mark Winter and Tony Szczepanek[18] argued that the perspective on projects as the creation of new value is compromised by its reliance on Porter's concept of a value chain. Winter and Szczepanek prefer to draw on the work of Normann[19] who emphasises the co-creation of value between customer and supplier, and argue for a concept of a project as a 'value creation process'. This criticism would appear to be based on a misreading of Porter. The value chain concept does, indeed, focus on the single suppling firm, but as Porter emphasises, any value chain is part of a larger value system in which 'a firm's product eventually becomes part of its buyers value chain .... Gaining and sustaining competitive advantage depends on understanding not only a firm's value chain but how the firm fits into the overall value system'[20]. That said, Porter focuses on the value chain in his analysis and does not develop the value system concept. The work of Normann and his colleagues provides a valuable, but not incompatible, development of the value system concept, and the concept of the construction project as a value creation process will be developed further in Chapter 3.

Lauri Koskela and Glenn Ballard have argued that this book takes an economist's approach to managing construction projects, rather than a 'production' approach. Their arguments have already been discussed in a detailed response[21]; here we will review some of the broader points of difference. Koskela and Ballard argue that the tectonic approach advocated here:

- *Focuses on transactions rather than production.* While it is true that the section on mobilising the resource base does focus on transactions, this is only one of the four generic project processes explored in the book. We submit that the perspective developed in the book the merit of integrating both a production and a transaction cost perspective within one framework as is articulated in Part III.
- *Focuses on information flows rather than material flows.* This is perfectly true, but is inherent in the nature of the process of managing construction projects. As is explored in Chapter 15, task execution – be it a materials processing or information processing activity – is not the responsibility of the project manager. This responsibility is for co-ordination between tasks, not in executing the tasks themselves. This, we submit, is an inherently information processing activity.
- *Places uncertainty reduction at the heart of the project process.* Again, this is perfectly true, but the critique comes from a strangely backward view that all the information for the next decision is acquired as a result of the previous decision. We submit that although this contention might well hold in perfectly stable environments, this is hardly tenable in the dynamic, forward-looking environment of projects as we will see in Chapter 13.
- *Neglects the possibilities for improvement by direct intervention in the production process.* We agree that there is considerable scope for process improvement in materials flows on the project, but this is not the direct responsibility of the project manager, but of the managers responsible for task execution, and remain convinced of the need to mould information flows structurally rather than directly, although the implementation of ICT as discussed in Chapter 14 may provide a partial exception to this.

In sum, we share the assessment of Clegg *et al.*[22] that the lean construction approach advocated by Koskela, Ballard and their colleagues represents a contribution to the traditional systems analysis-derived approaches to managing projects that they purport to criticise and as will be seen later, their principal contribution is to add to the toolbox for riding the project life cycle, rather than at the stratetgic level of managing projects as a whole. The perspective is then neo-bureaucratic, rather than professional – a point to which we will return in section 17.6.

1.8 A theoretical perspective on managing construction projects

Peter Morris, conclusion to his keynote speech at the first Project Management Institute Research Conference that 'the challenge for research . . . is precisely the perceived weakness of the discipline's theoretical base'[23] echoes a widespread perception of researchers and reflective practitioners in both the project management field and the construction project management subfield. Disciplines – in both the academic and the professional senses – mature through the development of a coherent body of ideas that deepens understanding and enables predictive propositions, and so it might be useful to be more specific regarding the theoretical perspectives deployed here. We will here present them as assertions; elaborating them adequately to convince readers of their strength is the task of the following chapters:

- Projects are temporary organisations consisting of a coalition of firms chartered by a client; as such they have distinctive properties which no current theory of organisation can comprehend[24].
- Projects move through distinctive life cycles because of their determinate character as temporary organisations; the termination date for the temporary organisation is typically specified more or less accurately at its foundation[25].
- Project managers are intendedly rational decision-makers, satisficing in the face of uncertainty, whose rationality is both bounded and shaped by impulse[26]. This implies that moving through the project life cycle is essentially a process of *structured sensemaking*[27] in which project managers respond to cues in the situation and make sense of them through actions which yield further information – what Weick calls enactment. We call it structured because the sensemaking is facilitated through structured routines for search and action.
- Routines are an essential element of managerial activity, yet their implementation is contradictory in that they both constrain and enable managerial action[28].
- Projects are embedded in contexts that are both organisational and institutional, simultaneously shaping and being shaped by these contexts[29].

The overall tectonic approach to the argument in this book has been elaborated since the first edition and is shown in Fig. 1.5. In the tectonic approach, the institutional level of analysis shapes and is shaped by decisions made at the governance level. Decisions at the governance level select the organisational structures within which the project process flows, but these processes also shape governance-level decisions. The process level is where the project is performed through a flow of information which initiates and controls the flow of materials. In terms of the river analogy presented in section 1.3, the institutional level is the underlying geography and geology of the landscape through which the river flows; the governance level is the banks of the river (whether natural or artificial); and the process level is the flow of water to the termination of the project in the 'sea' of facility operation. The institutional level will be discussed in Chapter 2; the governance level in Part III and Chapter 15; and the process level in Parts II and IV.

1.9 A practical contribution to managing construction projects

As well as deploying a distinctive theoretical perspective, the text also aims to make a strong practical contribution to managing construction projects more effectively. To indicate the contribution we hope to make, we will use the (UK) Office of Government Commerce's leaflet *Common Causes of Project Failure*[30] to identify more precisely where this text can contribute:

(1) *Lack of clear links between the project and the organisation's key strategic priorities, including agreed measures of success*; this will be covered in Chapter 3.
(2) *Lack of clear senior management and Ministerial ownership and leadership*; these issues will be covered in Chapters 15 and 16.
(3) *Lack of effective engagement with stakeholders*; Chapter 4 tackles this in detail.
(4) *Lack of skills and proven approach to project management and risk management*; the whole of Part IV addresses these issues, with a focus on risk management in Chapter 13.
(5) *Too little attention to breaking development and implementation into manageable steps*; some of these issues are discussed in Chapter 8, with the scheduling issues covered in Chapter 11.
(6) *Evaluation of proposals driven by initial price rather than long-term value for money (especially securing delivery of business benefits)*; again this is the topic of Chapter 3 supported by Chapters 9 and 10.
(7) *Lack of understanding of, and contact with the supply industry at senior levels in the organisation*; this is covered in Chapter 5 with the more contextual issues implied here covered in Chapter 2.
(8) *Lack of effective project team integration between clients, the supplier team and the supply chain*; Chapters 6 and 7 address the issues here.

1.10 The plan of the book

Chapter 2 assesses the role of the socio-economic context of construction projects for their effective management. Different national construction industries are organised to solve common problems in different ways. These differences have evolved over centuries and have a profound effect on the ways in which projects are managed. While the principles explored in this book remain valid for all advanced societies, the details of their application will need to be adapted for specific national contexts. This chapter indicates some of the main points of variation. In conclusion, Chapter 17 explores the prospects for the development of the management of construction projects – suggesting how we might learn from other project-orientated sectors to mitigate our weaknesses, and how they might learn from our strengths.

The central chapters of the book follow the structure defined by the five generic project processes. Part II investigates the definition of the project mission – how do clients decide what they want, and how can members of the project coalition most effectively advise them on the full range of possibilities open to them? What tools are available for rapidly providing visualisations of the possibilities? How can all the different stakeholders be managed, some of which may be totally opposed to the project in principle? The outcome of this process defines the project mission, which allows the identification and mobilisation of the resource bases required for its realisation, discussed in Part III. How can such resource bases be selected and motivated, both those in direct contract with the client and those mobilised as subcontractors?

Once the resources are in place, they have to be managed through time as they deliver on their commitments to the project. Thus, Part IV covers the core tools and techniques of the management of construction projects, while placing them in a broader, strategic perspective. Part V switches attention to the more social aspects of the management of construction projects, exploring differences in the organisation of the project management function, and the importance of effective leadership and teamwork.

Readers may be puzzled as to why there is no explicit reference to ICT in this overview. This is because ICT is central to the information processing approach to organisations, not an optional extra. Discussions of the role of ICT are embedded in the discussions of the business processes on which it is deployed, although of course, at the present state of the art, ICT is of more use for a process such as information management than it is for stakeholder management, so the amount of discussion will vary. However, some specific issues around ICT are addressed in Chapter 14.

1.11 Summary

This chapter has laid out the information processing approach to the management of construction projects as the principal source of the creation of new value in modern societies that will be developed in this book. In order to give an early taste of how it fits together, Case 1 applies it to the construction of the Channel Fixed Link. However, before we move to developing the perspective in detail, Chapter 2 sets out the context of managing construction projects which influence the ways in which they are managed.

Case 1
The Channel Fixed Link

The fixed link under the Channel/La Manche is one of the most challenging construction projects completed in the twentieth century. The range of challenges its project managers faced well illustrate the importance of taking a holistic approach to the management of construction projects. While the performance of the project on the traditional criteria of schedule, budget and conformance to specification is superior to the majority of mega-projects, it was widely seen at the time of its opening in 1994 as a failure. An *ex post* re-evaluation of the cost-benefit case for the project in 2003 – 10 years after it opened – has argued that its net present value is negative by over £10m in 2004 prices and it was therefore a burden on the UK economy. However, this argument ignores the fact that the bulk of the capital came from outside the UK. While there might be a large disbenefit to the global economy, the economy of the Brussels–London–Paris triangle has surely gained significantly because it reaped most of the benefits and paid few of the costs. After a major financial restructuring and the opening of the High Speed 1 through to London, Eurotunnel finally moved into profit in 2008.

Defining the project mission was fraught and an egregious case of strategic misrepresentation. The completed project was the third attempt that had actually started tunnelling; the other two had been abandoned as key stakeholders lost commitment to the project because of economic and political pressures. The fear among the Eurotunnel project management team that this would happen again should Labour win the 1987 election led them to commence tunnelling – thereby sinking capital – before the design had been adequately developed, leading to some expensive design changes. Although the technical solution implemented had been developed in the 1950s, this focus on the technology led to serious errors in the definition of the project mission. Throughout the early phases, the mission was defined in terms of providing a tunnel as a challenging, but relatively well-defined, civil engineering problem. It was only around 1990 that it became clear that the true project mission was to provide an integrated transport system – a much more challenging systems engineering problem using many innovative technologies. This failure to define the mission properly led to inadequate attention being paid to the design of the mechanical and electrical services, procurement of the rolling stock and the commissioning of the system as a whole.

The *mobilisation of the resource base* also created serious – indeed showstopping – management problems. The main problem was that the constructors – who formed the Transmanche-Link (TML) consortium – were also the promoters of the project. As a result, the construction contract was signed when their representatives were also on the client side. This generated enormous suspicion on the part of other stakeholders – most notably among the global banking consortia that were providing the capital – that the contract was biased towards the interests of TML. As a result, Eurotunnel's project management team was obliged to play tough publicly with TML in a masterly display of scapegoating, and its chief executive gained a ferocious reputation among TML managers. A related problem was the use of inappropriate contracts for different parts of the works. Only the tunnelling contract was incentive based; the contract of the rolling stock was a cost-plus one, and the fit-out and termini were on a lump sum. As might have been predicted, the cost-plus contract witnessed by far the largest percentage cost overruns, while the lump-sum contract was the focus of most of the crippling arguments between the stakeholders, which diverted attention away from actually delivering the project mission.

Against this context, *riding the project life cycle* was extremely difficult and escalation inevitable. Although sophisticated schedule and budgetary management systems were in place, they could not be meaningfully used as management tools because of the continuing negotiations between TML and Eurotunnel. Everything was open to negotiation as the project coalition moved from one crisis to another. Schedules and budgets were typically set as the result of tense negotiations to justify outcomes, not to plan project realisation. Tools and techniques can only be effective for project management where appropriate organisational contexts exist for their implementation. Despite this, the project achieved outcomes that compare favourably with other major civil and petrochemical engineering projects around the world. Indeed, in one respect, the project performed better than the benchmarks – it worked. A high proportion of very large petrochemical facilities fail to meet their planned performance criteria, and the track record of the IT sector in delivering large systems is appalling. On the criteria of fitness for purpose and conformance to specification, the fixed link is a great success.

Leading the project coalition was extremely difficult and overwhelmed more than one senior executive. Senior executives lost their jobs, marriages and nerves. On site, there were particular management problems in the early stages of the tunnelling on the British side as the TML member firms responsible failed to work together in a co-ordinated manner. This breakdown of managerial control led to lost lives, as well as to problems with the schedule. Perhaps surprisingly, there were few intercultural problems between the British and the French. The relatively bureaucratic British approach with heavy reliance on systems and procedures contrasted with the more action-orientated French approach, but this did not appear to cause problems. What is most remarkable about the human resources deployed on the project is the extremely high level of commitment to the project, even as it entered its final commissioning stages.

The construction of the Channel Fixed Link was a remarkable adventure, mobilising massive resources and capturing the imagination of the world. On most criteria it was a very successful project, outperforming on budget and schedule most other projects of a similar scale, and working almost perfectly once opened, yet it represents a textbook example of project escalation derived from strategic misrepresentation. Many of the management problems encountered were generated very early on during the definition of the project mission – the lack of clarity regarding the roles of different stakeholders led to mistrust; the inappropriate definition of the mission as a civil engineering project rather than an integrated transport system project led to lack of management attention to key elements of the mission; these problems in definition were compounded by errors in the mobilisation of the resources bases, and in combination, these made riding the project life cycle very difficult. Leading the project mission in this context became intense – too intense for some.

Sources: Winch (1996b); Fetherston (1997); Winch (2000b); Winch *et al.* (2000); Anguera (2006).

Notes

1 Seminar, TU Delft, May 2000.
2 Pfeffer and Sutton (2006, p. 74).
3 http://www.nakheel.com/developments/ (accessed 07/07/08).

4 Porter (1985).
5 This is the central thrust of the major contributions to organisation theory of James March (March and Simon, 1993; Cyert and March, 1992), Herbert Simon (1976) and, more recently, Jay Galbraith (1977). See Mintzberg (1979) for the broader context of this body of organisation theory.
6 See Winch (1994a) which reports on the co-ordination of the engineering/manufacturing interface in 15 UK engineering firms and shows how information flows initiate and control material flows.
7 Pinney (2001) shows how the basic concepts of project management evolved during the nineteenth century, and how they started to become clearly articulated in contrast to the emergent theory of repetitive manufacturing associated with, for instance, scientific management. The railways were seminal in this development, although the lessons of the earlier canal-building period were not forgotten, and the construction of the great seaways of the later nineteenth century posed enormous managerial challenges.
8 This critique is developed in Morris (1994); see also Giard and Midler (1993).
9 The concept of 'skill container' is taken from Kristensen (1996).
10 Davenport (1993, p. 5).
11 Womack and Jones (1995, p. 156).
12 Cited in Davenport (1993, Chapter 2).
13 Winch (2000a).
14 Green (2006, p. 234).
15 For example, Giddens (1984); Powell and DiMaggio (1991).
16 See particularly Winch (1994a, p. 5).
17 See the critique of Weick's work in Winch and Maytorena (forthcoming).
18 Winter and Szczepanek (2008).
19 For example, Normann and Ramirez (1993).
20 Porter (1985, p. 34). The misreading is both understandable and widespread given the counter-intuitive use of the term 'chain' by Porter to denote one link in the overall system.
21 Koskela and Ballard (2006) and Winch (2006b); see also Koskela and Howell (2008).
22 Clegg *et al.* (2006).
23 Morris (2002, p. 53).
24 The original insight here comes from Cherns and Bryant (1984), followed by Bryman *et al.* (1987) and became a founding proposition of the Scandinavian school of project management research (Lundin and Söderholm, 1995). However, there has been little attempt to combine theorisation of the temporary organisation with theories of inter-firm organisation to provide a more encompassing theory of project organisation.
25 Morris (1994) and Lundin and Söderholm (1995) both conceptualise the life cycle and examples in the practice of managing projects come in forms as varied as the advocacy of value engineering and stage-gate processes.
26 This assertion adopts the Carnegie school's behavioural theory – see Simon (1955), Cyert and March (1992) and Shapira and Berndt (1997) for an application to construction project management; it also accepts the critique of the 'coolly cognitive' Carnegie approach (Adler and Obstfeld, 2007) developed from a reading of Dewey (2002). In this perspective, there is no contradiction between the notion of 'rationality' and the notion of 'impulse' because Carnegian rationality is about how things happen, not why.
27 The concept of sensemaking is very much associated with the work of Karl Weick (1979, 1995), and has been applied to managing projects by Thomas (2000) and Ivory *et al.* (2006) amongst others. See Walsh (1995) for an overview of the wider sensemaking literature, and Winch and Maytorena (2009) for a critique of the solipsistic tendency in sensemaking research. The contribution of sensemaking in project risk management is explored further in section 13.2.
28 The importance of routines for economic activity was first analysed by Nelson and Winter (1982), while Dewey (2002) argues for the profound importance of 'habit' in social interaction. From this perspective, project management practices as routines are both constraining 'disciplines' in the analogy of a prison as in Foucault (Burrell, 1988) and enabling prerequisites of action as in Dewey (2002).

29 Engwall (2003) shows how projects have history and context, while the particular inspiration for this conceptualisation of embedment is Giddens (1984). Applications in the project context have been made by Bresnen *et al.* (2004), Sydow (2006) and Manning (2008) amongst others.

30 Office of Government Commerce (2005b).

Chapter 2
The Context of Construction Project Management

2.1 Introduction

> 'Men make their own history, but they do not make it just as they please: they do not make it under circumstances chosen by themselves, but under circumstances directly encountered, given and transmitted from the past. The tradition of all the dead generations weighs like a nightmare on the brain of the living.'

Karl Marx[1] opens his analysis of the 'farce' that he considered the regime of Louis Bonaparte to be in mid-nineteenth-century France, with wise words on the ways in which the present is shaped by the past, and how it is necessary to understand the past to be able to form a vision of the future. This chapter will show how the practice of construction project management is embedded in the history and context of construction as a social and economic activity, thereby exploring the institutional level of Fig. 1.5. History will be explored through the lens of the industry recipe for construction, while context will be explored through the lens of national construction business systems. The history and context of one particular construction sector – the UK – are presented in Case 2.

Our thinking here is profoundly influenced by the work of Marian Bowley[2]. The idea of 'the system' in British building as a distinctive form of industrial organisation was first espoused by her. Bowley identified it as a highly structured set of relationships along lines of social class with architects at the top, followed in rank order by engineers, surveyors and builders. Within this system she identified 'the establishment' as the version of the system approved by architects. She then explored in some detail the evolution and malfunctions of the system. Alongside this system one can also identify the systems for civil engineering with the civil engineer in the dominant position, and speculative housing with the developer or developer/builder in the dominant position. However, it is perhaps indicative of the force of Bowley's argument that when one thinks of the construction industry in the UK, it is the architect-dominated establishment to which one reflexively turns.

Bowley's emphasis on institutionalised sets of interests was a profound insight, providing a subtle analysis of the interactions between the institutional and governance levels in what we call in Case 2 the professional system. The system allocated roles, defined responsibilities and specified liabilities. Effectively it defined some actors as proactive and others as reactive, dubbed some with the rank of profession and tarred others with the brush of commerce. In this system, legitimacy was provided by the principal clients, which increasingly became dominated by the state. Crucially, it established the reward and penalty structure for the actors in the British construction industry where a relatively stable set of rules of the game co-ordinates the actions of the players in the business system. Actors within such systems of action act rationally, but with a rationality that can only be understood within the logic of the system, as expressed in the rules of the game. The same conceptual lens can be effectively applied to other national construction industries[3].

In this perspective, the rules of the game for building and civil engineering provide the structure of incentives for the actors in the system, encouraging each actor into particular types of behaviour and tending to punish digressions from these rules of the game. Patterns of behaviour become institutionalised so that they act back upon the actors through the process of *structuration* – the rules of the game come to be seen as given, normal, the only way to do things. Careers and status become dependent upon certain rules; threats to those rules become personal attacks. The system has a powerful momentum, and planned change is difficult because no one actor can grasp the whole system. Yet such systems are also dynamic because of the inherent contradictions that they often contain. The rest of this chapter will deepen the insights generated by Bowley on the organisation of the construction sector by combing them with more recent theoretical developments from neo-institutional theory and international comparisons of business systems.

2.2 The industry recipe for construction

An *industry recipe* can be defined as 'the business-specific world view of a definable "tribe" of industry experts and is often visible articulated into its rituals, rites of professional passage, local jargon and dress'[4]. It forms the first element of the institutional context of the tectonic model in Fig. 1.5. The recipe thereby provides the cognitive dimension of the forces of institutional change within a sector, providing the language of the 'rules of the game'[5] in that sector. Industry recipes are typically analysed in terms of the 'constructs' that articulate the rules of the game in that organisational field, and fields can have competing while overlapping sets of constructs without losing coherence. We are not aware of any work using the lens of industry recipes so in the absence of such research, Table 2.1 is offered as a starting point for debate regarding the traditional industry recipe in construction, and its limitations in the twenty-first century. We will return to this table when we discuss the issues in the reform of the traditional industry recipe in the concluding chapter.

Table 2.1 Elements of the traditional industry recipe in construction (sources: developed from NAO, 2001, Fig. 13 and Tavistock, 1966).

Clients	Designers	Contractors	Specialist suppliers
Procurement			
• Contractors selected on lowest price rather than quality	• Underbidding to get work leading to poor design development which needs rework during execution • Failing to act professionally	• Underbidding to get work relying on poor specifications, client changes and cost variations to make a profit • Price-ringing and cover pricing to share out work between firms • Use of 'Dutch auctions' to drive down specialist suppliers prices	• Underbidding to get work relying on poor specifications, client changes and cost variations to make a profit • Price-ringing and cover pricing to share out work between firms
Briefing and specification			
• Poor briefing and definition of requirements with insufficient focus on user needs and the functionality of the facility • Lack of focus on the business case for the facility	• Insufficient weight given to users' needs and constructability • Use of prescriptive specifications which stifle innovation and restrict the scope for value management • Pursuing agenda not related to the needs of the project	• Reluctance to point out weaknesses in specifications so as to provide the basis for later claims	
Design and planning			
• Limited awareness of potential available solutions • Limited understanding of value management • Limited understanding of the benefits and uses of prefabrication and standardisation	• Little integration of design teams or of the design and execution processes • Limited use of value management • Reluctance to use prefabrication and standardisation • Reluctance to involve specialist suppliers in design	• Poor planning leading to wasteful process and accidents • Limited use of value management • Limited use of prefabrication and standardisation	

Table 2.1 Continued

Clients	Designers	Contractors	Specialist suppliers
• Appointing designers separately from the rest of the team • Making late variations to requirements	• Overdesign to reduce risk of litigation		
Project management			
• Poor project management skills • Tendency to pass risk on rather than identify it, allocate it appropriately and manage it • Reliance on contracts to resolve problems with adversarial relationships	• Resistance to the integration of the supply chain • Limited understanding of risk management • Limited understanding of the true cost of construction components and processes	• Limited project management skills with stronger emphasis on managing contracts rather than work flows • Reliance on contracts to resolve problems with adversarial relationships • Late payments to specialist suppliers generating cash flow problems • Limited understanding of the true cost of components and processes • The industry 'produces a climate of endemic crisis which becomes self-perpetuating. The type of man who can best handle this situation tends to have a crisis type of personality. He thrives on this situation and is unwilling to entertain the possibility or validity of any form or planning or control that is not short-term and completely flexible'[6]	• Limited project management skills with stronger emphasis on managing contracts rather than work flows • Orientation towards crisis management rather than effective planning • Reliance on contracts to resolve problems with adversarial relationships • Over-committing on workload so resources have to be rationed between sites

2.3 National business systems in construction

The second element at the institutional level is the national business system[7]. There are many different approaches to these issues, but what they all have in common is that the national context shapes the strategy and performance of construction firms in nationally distinctive ways. Although it is individual firms that compete in international markets, it is empirically observable that if a country has a world-class firm in a particular sector, it typically has more than one. Examples such as Japanese car manufacturers, British pharmaceutical companies, Belgian chocolate manufacturers, Italian fashion houses and Swiss banks come immediately to mind.

These nationally specific patterns of national strengths (and weaknesses) in international performance can only rarely be explained by natural resource endowments or other geographical advantages (i.e. comparative advantage). More typically they are the results of historical and institutional factors which have allowed particular strengths to be developed, and domestic firms to engage in intensive rivalry which hones them for international competition. Michael Porter[8] cites the example of auctioneering where the global networks developed under Empire, the wealth and cosmopolitan character of London, the strengths in arts of the British educational system, and a benign regulatory environment allowed four British firms to become dominant in the worldwide fine art auctioneering industry. This dominance has been sustained through intensive rivalry between these four firms, particularly between Christie's and Sotheby's.

Although analysis of business systems has tended to remain at the national level, the sectoral level also has a very strong influence on the structure and performance of the firm, and as discussed in section 2.2, industry recipes can be identified for each sector in terms of the taken-for-granted assumptions about how firms in that sector ought to be managed. The environmental influences on each firm will have, therefore, a sectoral component and a national component. Thus, a French construction firm and a British construction firm will share similarities in comparison to banks or car manufacturers from these countries, but these two firms will also display differences because of their membership of the French and British business systems respectively. Within this perspective, the construction business system can be defined as the nationally specific organisational field for construction.

The construction business system inevitably shares many of its characteristics with the broader national business system, while at the same time displaying its own sectoral characteristics. There have been many attempts to classify different types of national business system, but these are not always consistent and it is not our intention to review them here. However, most classifications distinguish between three basic types of advanced economy business system:

- *Anglo-Saxon* type business systems (e.g. the USA and the UK) with a greater reliance on liberal market values, relatively low levels of state regulation, greater reliance on the stock market for industrial finance and relatively low levels of worker protection.
- *Corporatist* type systems (e.g. Germany and The Netherlands) with more negotiated co-ordination between the 'social partners', greater willingness to

intervene in the market to protect social values, greater reliance on banks for industrial finance and relatively high levels of worker protection.

- *State-led* systems (e.g. France and Japan) with more extensive co-ordination of the economy by the state, relatively high levels of worker protection, greater reliance on the state for industrial finance and a desire to promote national champions in various industrial sectors.

The relationships between the actors in the system at the institutional level can be seen as one of competitive collaboration. They must all collaborate together in coalitions on particular projects mobilised by clients in order to achieve their aim as firms of staying in business; at the same time they compete with each other for influence over the system as a whole. These types of dynamics are found in a number of industrial sectors which rely on highly skilled professionals[9]. This competition is typically conducted by the different representative bodies, such as the American Institute of Architects, the Institution of Civil Engineers and the Ordre des Architectes, rather than by the individual firms themselves. The organisation of the construction project process at any particular time is the outcome of this competitive struggle. For instance, in France, the position of architects has recently been reinforced and contractors are being pushed back from involvement early in the process, while in Germany and the UK, it is being eroded as contractors seek earlier involvement and more control over the project process.

What might the factors be that allow some actors in the system to become relatively powerful compared to others? A number of different factors may be suggested, often working in combination:

- *Ability to solve complex problems for the client.* It can be suggested that it is not closeness to the client as such which generates power, but the unique ability to solve the client's more complex problems. The traditional role of the architect and the consultant engineer in the British system rests to an important extent on this complex problem-solving through the briefing process. In sharp distinction, the use of *concours* to decide which design is to be chosen ruptures the briefing process and thereby removes part of the problem-solving dynamic from the relationship between the client and the architect, encouraging a retreat into formalism in design. This could be part of the explanation for the relatively weak role of the architect in France.
- *The blessing of the state.* The most remarkable example of this is the civil engineer in France, with an elite trained in the École Nationale des Ponts et Chaussées and organised in the Corps des Ponts et Chaussées, acting as the instrument of state policy throughout the national territory[10]. This blessing more usually comes in the form of statutory protection, but this can generate a formalism in compliance with regulations. For instance, in many countries – such as Germany and Japan – only the architect can apply for building permission, yet in such countries 'signature' architects exist who will sign the application for a small fee, and in Italy, where the applicant must be an architect or engineer, the regulations are widely ignored.
- *Ability to manage risk for the client.* Within the dynamics of the contracting system, particular risks are posed for clients, which they typically manage by recruiting

specialist actors. For instance, the rise of the general contractor is very much associated with efficiencies gained through the superior co-ordination of the construction process. However, this leaves the general contractor in a powerful position in relation to the client, because of the post-contract asset specificities generated, as discussed in section 6.3. Clients, and their advisors for the conception process, typically attempt to redress this power through the complex contracts analysed in section 6.4. In the UK, where the general contractor first emerged, this problem has long been handled through the development of a control actor – the principal quantity surveyor – which, in effect, mediates the power of the general contractor on behalf of the client.

From this perspective, it can be suggested that one element of the industry recipe, from the point of view of the client as the actor who capitalises the project process, is to face a balanced project coalition, where no single type of actor wields too much power. In most markets such problems are solved through ensuring an adequate number of buyers and suppliers to allow competition. This is not easy in construction because of the difficulties, in most cases, of writing complete contracts, thanks to the high levels of mission uncertainty. This issue is discussed extensively in Part III. In upstream design, the process is inherently uncertain as design and regulatory issues are resolved; in downstream execution, site-related uncertainties remain and clients typically wish to retain the option of change.

2.4 The regulatory context

A particularly important aspect of the institutional level is the regulatory context shown in Fig. 2.1. The dynamic between the actors at the governance level is both influenced by, and influences, the regulatory context. This has five different aspects:

- The national *legal system*
- The national *zoning regulations* and procedures for the determination of where and what type of built facilities can be constructed
- The national *construction regulations* and procedures for ensuring integrity of the constructed product, and, in particular, the safety and comfort of building users and the public
- The national arrangements for *labour market regulation* – particularly the arrangements for employment, training and safety
- How the state as client chooses *procurement policies* for its own built assets.

2.4.1 The national legal system

The national legal system is one of the main dimensions of difference between national construction business systems, and its effect is pervasive throughout the

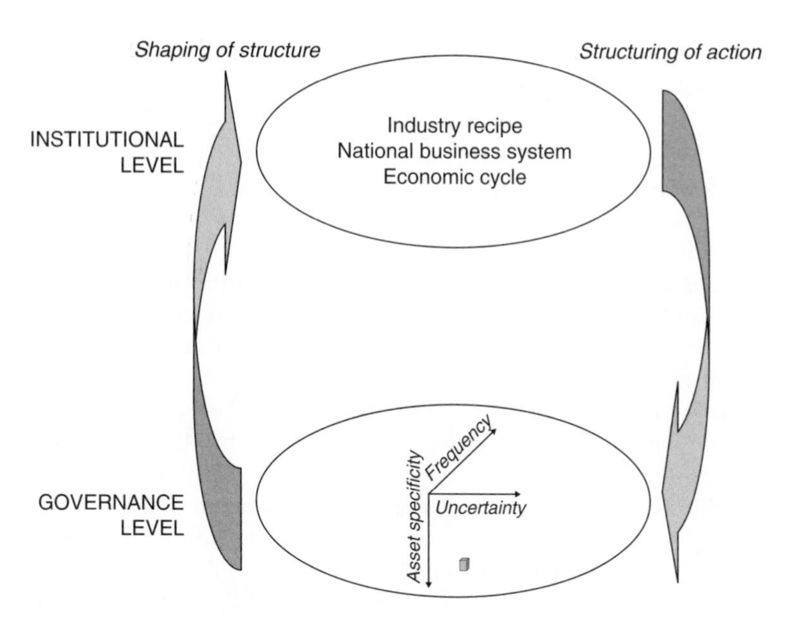

Fig. 2.1 National construction business systems: a conceptual framework.

regulatory context because of the extensive reliance on legal instruments. Broadly, two basic types of legal system can be identified[11]:

- The common law systems (e.g. the USA, Canada, Australia and England)
- The codified systems (e.g. France, Italy, Scotland and Germany).

Under common law systems there is no codified body of law. Judicial interpretation and precedent play a very important role, and statute is used as a policy instrument to influence judicial interpretation. In codified legal systems, the basic code, passed by the legislature, is the point of reference – for instance the Code Civil in France and the Bürgerliches Gesetzbuch in Germany – and the role of the judiciary is formally limited to interpreting that law. However, such codes cannot foresee all developments, and so in many codified law countries, there is a body of case law which, while not formally having the status of precedent, is used as an authoritative guide to interpretation and application of the codes. In the French system a distinction is drawn between civil law covering contracts between private actors and administrative law covering contracts made between public agencies and private actors.

Within the national legal framework, a number of sector-specific legal obligations significantly affect the incentives of actors in the system. For instance:

- Liabilities and remedies for defects to completed buildings vary greatly between countries.
- Opportunities for settling contractual disputes without litigation vary.
- The ability to pass on liability to sub-contractors and suppliers varies.

- In many countries design services are remunerated through standardised fee scales, while in the UK these are unlawful on competition grounds.

2.4.2 *Urban and rural zoning*

Although most nations require local government bodies to draw up plans for the urban and regional development of their area of jurisdiction, the legal status and interpretation of these plans vary greatly. In the *codified* systems, such plans have the force of law and any proposal for construction that is in accordance with the codes cannot be refused. An exception to this is Italy, where the planning system is underdeveloped, not systematically applied and undermined by corruption. The common law systems, on the other hand, are more flexible and open to considerable interpretation and negotiation. The local plan is, for instance, not legally binding and can be overridden if broader considerations so require. There is also the right of appeal in the case of dispute over interpretation of the plan at local level. Important differences also exist in the amount of information that is required for an application for planning permission – varying from relatively little in France to a high level of detail in Germany, with the UK, The Netherlands and Italy between the two. These differences result in large variations in the time it takes to reach approval for major infrastructure projects, ranging from an average of 5 years in France to 20 in the UK[12].

2.4.3 *Safety and comfort of building users*

The construction regulations covering the structural integrity of the built facility, the safety of its users and, increasingly, its energy performance also form an important element of the regulatory context. A review of the regulatory frameworks in a number of European countries identified two dimensions – the degree of responsibility of the actors themselves for the interpretation of the regulations compared to the degree of responsibility of the state, and the extent to which they are performance or prescription orientated. The results of this analysis are summarised in Fig. 2.2. In those countries where the actors themselves have high levels of responsibility for compliance with the regulations and the regulations are performance rather than prescription orientated – such as France and Belgium – a distinctive control actor emerges to help the client ensure compliance with the regulations. This is the *bureau de contrôle*, described in panel 2.1.

Panel 2.1 The *bureau de contrôle*

The *bureau de contrôle* (BdC) first emerged in France in response to quality problems in structural concrete during the 1920s. These *bureaux* are hired by clients to provide quality control – both during the design stage when they review the architect's drawings for compliance with the construction codes, and during execution on site where they check that the works conform to the specification. They work on a fee basis as part of the client's professional team. This is what distinguishes them from building control in the UK and the *Prüfstatiker* in Germany, which are regulatory agencies. Hiring a BdC usually brings the benefit of a reduction in the client's premium for project insurance (*décennal* in France). Under the loi Spinetta of 1977, the use of a BdC is obligatory for public works in France, and is widespread in the private sector. Although not obligatory, its use is also widespread in Belgium. The largest French BdC – SOCOTEC – is a major operation with its own research laboratories and is a major source of expertise on the technical performance of buildings in use.

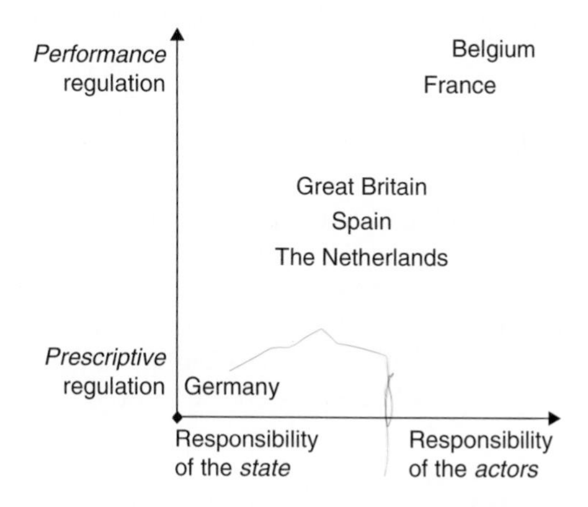

Fig. 2.2 Construction regulation systems for the integrity of the product (source: developed from Bazin, 1993).

2.4.4 *Labour market regulation*

The regulatory frameworks for labour markets – particularly in a labour-intensive process such as construction – play a profound role in shaping the possibilities for the management of the construction projects. This is also an area where the weight of the European Union is increasingly felt, particularly through the Social Chapter. Issues to be addressed here include working conditions, forms of employment, health and safety, and the organisation of education and training. Not all of these areas can be covered here, and so three of particular interest will be identified:

- Self-employment on a labour-only subcontracting basis is not lawful in France and Germany, and many firms retain their operative workforce on a permanent basis. In contrast, in the UK this practice is lawful and around 45% of the construction workforce is self-employed, and few building sites deploy employed operatives in the main trades in any numbers – see section 7.5 for a more detailed discussion of this issue.

- The social organisation of technical expertise varies enormously – for instance, the socio-economic basis of the formation and deployment of engineers is very different in France, Germany and the UK. In France the *corps* is the main organising institution, while in the UK it is the *professional institution*; in Germany it is *die Kammer*. The USA combines elements of the British and the French systems. These differences have profound implications for attempts at the mutual recognition of qualifications.
- The organisation of the training of operatives varies greatly. Comparative research has identified three models in national construction industries[13] – the craft-based *competitive model* (UK and Italy), which is employer led and based on apprenticeships in traditional crafts; the *dual model* (Germany), where the 'social partners' work together to provide skill-based training; and the *scholastic model* (France), where training is largely skill-based but takes place mainly outside the workplace.

2.4.5 The state as client

Finally, the state's administrative policies towards the procurement of the built product are a critical dimension of variation[14]. In a typical country, around 50% of construction output has the state or its agencies as a client, and many of the more inexperienced private sector clients tend to follow the state model. For civil engineering, the state is the predominant client in all countries. For these reasons, the European Commission has interested itself in those administrative policies which affect equity in the selection of suppliers.

The following are just a few of the ways in which the state as client influences the dynamics of the national construction business system:

- The decision of the state to use concessions for the private finance of public works has the potential to radically change incentives. The long-established use of the concession for public works and services in France – particularly for water – led to the sector becoming dominated by a few very large construction corporations which were principally utility companies.
- The demand for probity in public sector procurement tends to lead to relatively low-trust transaction governance with reliance on competitive tendering. This is clear in both Italy and Japan where, following major corruption scandals, the aim is to separate design from construction, and in the UK where the public sector is responding with difficulty to the diffusion of partnering.
- The state may respond to lobbying from particular actors who believe they are not being fairly treated and may change the power balance in the system in their favour, as in France where the loi MOP of 1985 was aimed at rolling back the power of the contractor.
- The extent to which designers – particularly architects – are appointed through *concours* or on the basis of reputation has a strong effect on the way the conception process is organised – see section 5.3.
- Other policy agendas to do with local economic development may also play a role – in Germany, competitive tendering has deliberately favoured local

contractors to enhance trust, while the Dutch public sector has effectively condoned price-ringing on public contracts on social policy grounds as part of the 'polder model' corporatist culture[15]. This is an area in which the European Commission has interested itself, making such policies difficult to pursue.

2.5 The construction cycle

The third element of the institutional level is the broader economy. Construction is an enormously important part of any advanced economy. Including all the supply chain, it typically accounts for around 10% of the wealth generated each year in advanced economies, and up to 20% in newly industrialising countries. Each year about 50% of the capital invested in assets in an advanced economy is spent on construction, and around 70% of the stock of assets is constructed facilities. The efficient and effective construction and maintenance of these assets is profoundly important for the overall success of all economies, and because constructed assets have such long lifespans, errors made in deciding which assets to build will have much longer-term implications than for other types of assets. Construction is by far the most important way in which societies create new value.

This new value creation takes place as part of the general activity of the economy – intimately dependent on the health or otherwise of that economy. Although there is considerable debate among economists regarding the amplitude of the cycles, it is obvious to any observer that this economic activity is cyclical – going through periods of expansions and contractions of economic activity within an overall upward trend. Because constructed assets are investment goods, the amplitude of the cycles facing construction firms is greater than for the economy in general because of the accelerator effect, defined in panel 2.2. Research by government economists shows that this is certainly the case for UK construction over the past 30 years[16].

Panel 2.2 The accelerator effect

A change in levels of demand for consumer goods and services does not translate directly into demand for the investment goods used in their supply, but it is magnified. This is because investment is lumpy – a new factory is intended to pay back over more than a single year, so the initial capital investment to meet a given consumer demand is front-loaded. Similarly, when consumer demand falls, existing assets are adequate for supply and no new ones need to be purchased.

So, the accelerator is given by:

$$\nu.\Delta O = \Delta K$$

where v is the ratio of value of capital equipment to its annual output (the accelerator coefficient), O the value of output and K the value of new investment.

Source: Ive and Gruneberg (2000).

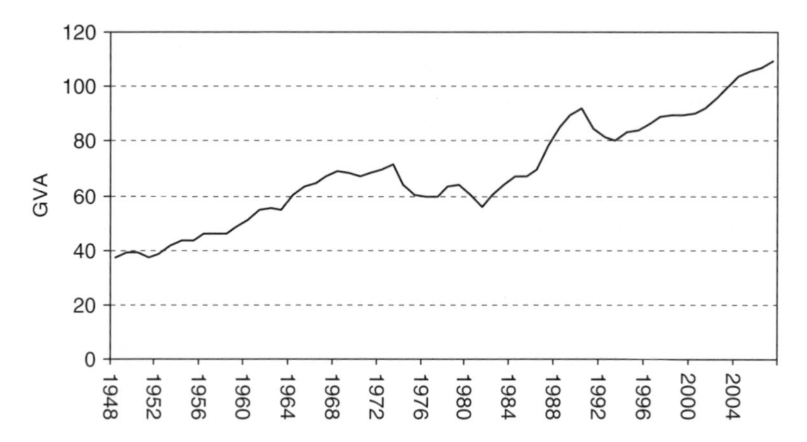

Fig. 2.3 The UK Construction Cycle 1948–2007 (source: Office for National Statistics).

Figure 2.3 shows the construction cycle in the UK from 1948 to 2007 measured in terms of gross value added (output less input bought from other sectors) in constant prices for the narrowly defined contracting sector which serves as a useful proxy for activity in the sector as a whole. It shows clear peaks of activity in 1973 and 1990, and troughs roughly 3 years later in 1976 and 1993. At the time of writing it is likely that the 2007 figure is another peak of activity, and that output will fall away for the next few years.

The problem this poses for construction project managers is that they need to know where they are in the cycle when planning projects. If the project goes on site during periods of upswing, then input prices may be higher than expected when budgets are set, and there may be difficulties in obtaining the inputs when required by the schedule. During downswings, the project may benefit from the opposite effects, but there is also a greater chance that the client may cancel the project as its investment appraisal looks less attractive than it did during the upswing. In other words, the construction cycle as part of the overall business cycle is another important factor in increasing the level of mission uncertainty facing decision-makers on the project. As illustrated in Case 1, the Channel Fixed Link project faced this problem.

A second problem is that the large size of the sector, combined with the role of the public sector as a client, tempts the state to attempt to regulate the overall level of economic activity through financing construction projects. Following this policy, in times of boom, the government cuts public spending to cool the economy down, while in times of slump, the government launches projects to stimulate economic activity. Such a policy was behind much infrastructure investment in Japan during the 1990s, for instance, as its government tried to stimulate its stalled economy[17]. The problem is that the lead time in any project compared to the wavelength of cycles means that it is difficult to ensure that the project is at the execution stage – which contains the largest proportion of the total investment – while the economy is still in a slump. The problem that this strategy poses for construction project managers is that there is a strong temptation under such circumstances to fail to define the project mission properly, and poor mission definition makes the successful completion of the project more difficult, as much of the rest of this book will illustrate.

2.6 The development of concession contracting

The state is typically a key actor in processes of change in industrial sectors, and construction is no exception. Case 2 provides some illustrations of this point from the UK. Here we focus on one state-driven force for change that is worldwide[18]. Throughout the developed nations, a fiscal crisis of the state became dramatically apparent following the economic crisis of 1973 as the unprecedented growth of the post-1945 period came to an end, with its strong commitment to state spending on welfare as a central plank of the post-war political settlement. The first response was to cut back dramatically on public sector investment in constructed facilities. Over the next 20 years, the ability of most Western nation-states to provide adequate amounts of capital to fund public investment needs, as a trend, deteriorated. Most recently, this trend has been particularly pronounced within the European Union as member states struggle to meet the Maastricht criteria on state debt. Elsewhere, the political commitment to reducing taxation levels, coupled with the growth in welfare budgets because of rising unemployment and an ageing population, has reinforced these trends.

This is the same kind of problem that faced the first nation-states – France and the UK – as they struggled to construct a national infrastructure during the seventeenth century. How could roads be constructed that would allow trade in goods and the movement of armies? The two solutions were strikingly different. The British relied on the market and allowed entrepreneurs to build turnpikes – the first was opened in 1654 under the Commonwealth. At the same time, the French took a very different route under Louis XIV and his minister Colbert, and granted concessions to rich bourgeois to construct roads. As the canals and railways were built in the eighteenth and nineteenth centuries, the two countries followed their national models. The British infrastructure was built in the manner shown in panel 2.7 – leading to the railway mania of the 1840s. Meanwhile in France, the *Corps des Ponts* rigorously monitored the construction of the railway concessions granted to a small number of companies[19]. As a result, the British had a complete rail network decades earlier than the French. Other rapidly industrialising countries such as the USA drew heavily on these two models to finance their infrastructure requirements, and developed their own distinctive land grant approach. The two models were then spread throughout the world by such colourful entrepreneurs as Thomas Brassey, Ferdinand de Lesseps and Henry Meiggs – 'a handful of bankers and contractors controlled nearly all the railway building in the world, outside the USA, between 1840 and 1870, and a large share of transport developments in the half-century thereafter'[20].

By the twentieth century, the state was increasingly taking over the finance of infrastructure construction. Financiers increasingly preferred to lend to established operators and the state itself, while many infrastructure operators were effectively nationalised in the era following 1918. However, the slow deterioration of the infrastructure acquired by direct public finance during the boom after

1945, coupled with demands for new infrastructure as the economy grew and patterns of economic activity changed, meant that the fiscal crisis could not be solved through a permanent reduction in capital investment in infrastructure. International research on competitiveness[21] also showed that one of the most important things that the state could and should do to support firms was to provide an efficient and effective infrastructure for economic activity. In many countries, one strategy has been privatisation – in sectors such as telecoms, water and rail, the problem was simply sold to the private sector with varying degrees of success so far as stimulating capital investment is concerned. However, in many areas, the state could not, or did not wish to, transfer responsibility to the private sector, and so new solutions had to be found; the answer was an old one – concession contracting on the French model. Known variously as Design Build Finance Operate (DBFO), Build Own Operate Transfer (BOOT) and so on, they all derive more from the French than from the British tradition.

The essence of concession contracting as an instrument of public policy is that the state invites potential concessionaires to bid for the concession to finance, build and operate the facility for a pre-defined period, the capital investment being repaid through the revenue stream generated by the operation of the facility. It is this state-led nature of project promotion which distinguishes concession from the more privately promoted model of nineteenth-century British infrastructure development. Prior to its diffusion in the 1980s, the concept of concession for the provision of public facilities was unknown in English law[22]. UK infrastructure development during the first half of the nineteenth century was almost entirely privately promoted, with the state merely facilitating and regulating its construction – entrepreneurs were free to choose where they built canals, docks and railways themselves. The concession, on the other hand, is very much an instrument of public policy, with the state determining what facilities are required, and inviting bids to supply it.

2.7 Summary

All construction projects are launched into a distinctive history and context – they both shape and are shaped by that institutional context. The actors in the project coalition work within the rules of the game articulated in the industry recipe, while these rules evolve in response to new challenges. A construction business system provides more or less space for innovation, but that innovation is obliged to start from where the system is, not where the actors would like it to be. Path dependence is at the heart of the evolution of construction business systems. Construction business systems, in turn, are embedded in the wider national business system and cannot escape the dynamics of that system. In a highly regulated industry such as construction where project execution is inherently site specific, globalisation will continue to have less of an impact than it has had in many other sectors.

Different construction business systems have developed different solutions to the common problems of creating the constructed assets upon which their economies depend. Here, lack of convergence is an advantage because benchmarking with foreign projects and firms can yield alternative solutions to these common problems[23]. Although solutions can rarely be borrowed directly from abroad, the demonstration effect of doing things differently can raise questions about existing ways of working much better than the type of theoretically derived speculation that has characterised much of the current debate about industry development, at least in the UK.

Moving on from the seminal work of Marian Bowley, this chapter has developed the concept of the construction business system as articulated through an industry recipe shaping project organisation and performance. It has analysed both how the interactions between the actors within the project coalition are shaped by the historical legacy of the construction business system, and in particular, the impact of the regulatory context on coalition actor behaviour. It has then identified the impact of the construction business cycle on decision-making by the project manager and discussed the important new changes in policy by public sector clients in a large number of countries associated with concession contracting. We are now ready to address the challenges of managing construction projects.

Case 2
The UK Construction Business System

Market relations in the UK construction sector emerged in the Middle Ages as the crown and church required large concentrations of labour to build their castles and cathedrals, particularly in the period of relative labour shortage after the Black Death in the mid-fourteenth century. These demands led to a labour market, particularly for masons, outside the traditional feudal ties of obligation which was how most building was accomplished during the period. The master craftsman predominated and the prospective owner of the building bought the materials directly and paid the labour by the day in what might be called the *craft system*. Two examples of this system in operation are provided in panels 2.3 and 2.4.

Panel 2.3 Refurbishing Canterbury Cathedral

In September 1174, the choir of Canterbury Cathedral was badly damaged in a fire. Various French and English masters were consulted, but the one who won the confidence of the monks was William of Sens. After a careful survey, he recommended the demolition of the remains of the choir and the construction of a new structure. He arranged the purchase of the stone from quarries in Caen, and devised the lifting tackle for the loading and unloading of the ships that were to transport it across the channel. He also prepared the templates for the masons who were doing the actual carving.

He supervised the works in detail for the next 5 years, until he was badly crippled in a fall when a scaffolding collapsed under him. After attempting to direct the works from his bed, he resigned his commission and returned to France, to be replaced by an English master who was also called William. As the works progressed, it became possible to place the relics of the saints rescued from the tombs in the old choir in their new resting places, and use parts of the structure for worship by 1180. Although no progress was made in 1183 because of lack of funds, 1184 saw the substantial completion of the works with the roofing of the structures.

Source: Harvey (1972, Appendix A).

Panel 2.4 A courtier's castle

Work commenced at Kirby Muxloe Castle in 1480 on a large rectangle with towers at each corner on the site of an earlier castle which was incorporated into the foundations. The client, Lord Hastings, appointed his steward as clerk of works who made all the payments to the craftsmen and labourers who were paid on a day-rate basis. Many of the workers were local, but the labourers came from Wales, while a number of bricklayers came from Flanders to execute patterned brickwork. The master-mason was not on site continually but came for a few weeks each year. Unfortunately, both the client and the project were cut short when the former was beheaded in June 1483 by 'a poisonous bunch-backed toad', but his widow carried on and completed the works that were already in hand.

Source: Emery (1989).

On larger projects such as cathedrals, considerable amounts of design activity were required in order to co-ordinate the works. This was usually carried out by master-masons who became increasingly specialised in design, as opposed to construction, activities, and were much sought after by bishops wishing to glorify God in gothic stone. However, these 'architects' grew from the ranks of masons and remained intimately involved with the work of the craftsmen they directed. The craft system passed on its distinctive organisation of construction around the materials used – carpenter, mason and so on – which is still prevalent today. In the craft system, conception and construction were the combined responsibility of the master craftsmen, while control was carried out directly by agents of the client such as its clerk of works. Clients were also very happy to involve themselves deeply in the design and construction processes.

The rise of a rich merchant class in fifteenth- and sixteenth-century Florence led to the emergence of a new actor – the architect – who was capable of articulating the merchants' desire for expression through building. The architect took on the task of co-ordinating the building crafts that had emerged from the medieval guilds, which now had a much reduced range of responsibilities in what might be called the *trade system*. This became widespread throughout Europe. It is distinguished from the craft system by the role of the architect independent of the crafts. Perhaps the most important legacy of the trade system is in the organisation of conception in the role of the architect – particularly as theorised by Alberti – as a unique combination of conception and control actor. Acting simultaneously as the artist of the built form, the client's advisor on cultural matters and co-ordinator of the construction process, the architect slowly developed as the principal actor in the system. For the first time, under the trade system, a project actor emerged who could *preconceive* the built form on behalf of the client independently of the construction process.

The trade system was slow to diffuse to England, but through the sixteenth-century houses became more explicitly designed. During the great rebuilding of country houses during the Tudor era, an architectural consciousness slowly emerged, with Robert Smythson, a master-mason, as its best known exponent. Inigo Jones, widely acknowledged as the first English architect in the Renaissance sense, practised as surveyor of the King's works during the first half of the seventeenth century. However, it was not until the even larger rebuilding of country houses after the restoration, and the rebuilding of London after the Great Fire of 1666, that the trade system became fully established, heralding the first golden age of English architecture, as illustrated in panel 2.5. Because the works were now conceived in advance, new forms of payment could emerge, and the tradesmen were increasingly paid on a measure and value basis according to the work done. This stimulated the development of surveying techniques.

Panel 2.5 The Queen Anne churches

In the early eighteenth century, a programme of church building was commissioned – a programme that became known as the Queen Anne churches. The surveyors appointed were responsible for designing the church, providing an estimate of its costs, selecting the trade contractors, measuring their work and supervising the works on site. These surveyors included some of the most illustrious names of English architecture, notably Hawksmoor. Contractors for each trade were selected on the basis of a competitive tender – known as a 'proposal' – organised at the appropriate point in the construction programme. Masonry was by far the most important trade, but bricklaying, carpentry, plumbing and plastering were also significant elements of the works. The tender was on a schedule of rates, and payment was on the basis of the weight or quantity of materials fixed, or the area of work completed against this schedule. Although the proposals were against a previously developed architectural design, they tended to include such details as timber sizes. The surveyors were also in the habit of changing the design as the works progressed. Trade contractors normally supplied their own materials and labour. In addition, dayworks would also be agreed with the surveyor. Cost and time overruns were endemic on this building programme, and there were continual problems with the quality of the bricks supplied to the works by the bricklaying trade contractors.

Source: Yeomans (1988).

Under the pressures generated by the French wars in particular, and the industrial revolution more generally, the trade system began to break down. The Barrack Office was established in 1793 in order to provide accommodation for the unprecedentedly large numbers of soldiers mobilised against France. At first this relied on the trade system, but the urgency of the building programme meant that a shift was made to 'contracting in gross', where a single contractor undertook financial responsibility for execution in a single contract. Thus, both pre-design and pre-measurement became essential parts of the new system, and the first important British general contractor – Copland – emerged. A government enquiry in 1828 pronounced in favour of the new system, despite complaints from architects. Although contracting 'by the great' was not unknown in previous centuries, it does not appear to have been in conjunction with a separate surveyor and was not typical. The dynamics behind the emergence of the professional system are well illustrated by Cubitt's London Institution project, outlined in panel 2.6.

Panel 2.6 The London Institution

The London Institution contract was undertaken by Cubitt in 1815. This building, now demolished, was let on a contracting in gross basis to a very tight programme. It was a very large contract for its time, and the institution was in a hurry for the building. Stiff penalty clauses for non-completion within the specified time were therefore attached to the contract. In order to reduce his risk in the face of this penalty clause, Cubitt decided to employ all the trades directly, rather than subcontracting them. The project ran into a number of problems that remain depressingly familiar: the haste of the work meant inadequate preparation; considerable modifications were required to the foundations of the building; the architect was slow to deliver the working drawings; and considerable cost overruns were experienced. However, the project was successfully completed, the blame for the problems fell largely on the architect, and the project formed the basis of Cubitt's subsequent career.

Source: Hobhouse (1971, Chapter 1).

During the latter part of the eighteenth century, the task of after-measurement had been increasingly delegated to the measurer by the architect. As general contracting emerged, the measurer began to take responsibility for measuring the quantities to be built in advance so as to facilitate the accuracy and fairness of the tendering process. This new task became institutionalised in the role of the quantity surveyor, around the distinctive competence of the bill of quantity as a control tool. The task of co-ordinating the separate trades was delegated to the new master builder who took on the entire works for a fixed price. The architect was left mainly with the tasks of conception and quality control; indeed those occupied with measuring or building were excluded from membership of the Institute of Architects. Thus, the architect's role became even more clearly focused on conception, with important control tasks delegated to the quantity surveyor, and all responsibilities for co-ordinating construction passed to the general contractor. These changes had the advantage for the architect of reinforcing the role as a professional rather than a craftsman, a gentleman rather than a tradesman.

The most important feature of this emerging system – which may be called the *professional system* – was the general contractor undertaking work conceived by others and subject to independent control. For the first time, a project actor emerged to whom the client could effectively transfer some of the risks inherent in the construction process. During the same period in the early nineteenth century, many of the institutions that later served to give the system its enormous momentum were founded – in particularly the Institution of Civil Engineers in 1818 and the (Royal) Institute of British Architects in 1834. By the 1860s the professional system was fully established with price-based competitive tendering for works in response to full bills of quantities which relied on fully detailed drawings for their production. Encouraged by John Ruskin, architects increasingly defined themselves around a distinctive competence based on creativity as artists rather than intimate participants in the construction process.

In civil engineering the trend had a different trajectory, but a similar outcome in the adoption of the professional system during the second half of the nineteenth century. The building of the infrastructure of the first industrial nation – turnpikes, canals and railways – was undertaken on the basis of private promotion. These promoters were sometimes landowners or other interested parties, but particularly with the advent of the railways, they were themselves engineers, such as the Stephensons and the Brunels. Initially, the actual works were divided into small lots and let to local contractors who were closely supervised by the engineers.

During the 1830s, Joseph Locke on the Grand Junction Railway developed the role of the general contractor to take over a broader responsibility for the works in partnership with the engineer. As the momentum of railway building grew, contractors such as Thomas Brassey increasingly took over the promotion task, and between 1844 and 1866, half the lines were promoted by contractors, often working in partnership with engineers. This shift suggests that the greatest risks lay with the construction process rather than the engineering design, and that the actor who could most effectively bear the greatest risks had the best chance of raising the capital required. In 1845, Samuel Peto began the practice of accepting payment in the shares of the line being built. The 1850s saw the emergence of project finance companies such as Crédit Mobilier working in close collaboration with the great contractors, and the railway contractors were increasingly vertically integrated operations, providing rolling stock as well as the tracks. The logic of action in what may be called the *charter system* is illustrated by the building of the Millwall Docks, in panel 2.7. A very important difference between the charter system and the contemporary system of private finance is that the latter is an instrument of public policy – it is how the state procures the assets it needs to deliver public services – while the charter system is merely a licence for purely private finance and exploitation.

Panel 2.7 The Millwall Docks

The way in which the charter system worked is, well illustrated, by the construction of the Millwall Docks, now the site of the Canary Wharf development. The story is complex, but the basic details of interest here are that the docks were promoted by a loose consortium of a railway engineer and two civil engineering contractors. Once the act of parliament had been obtained in 1864, the new company pressed ahead with a public subscription for capital. This was underwritten by the English branch of Crédit Mobilier. Immediately upon conclusion of the financement in March 1866, the construction contract was signed with the two promoting contractors, and it included in their contract sum the costs of providing cover for the shareholders' loans for the first 2 years of the project. However, this was already well in excess of the company's approved borrowing power. Lack of confidence in the management, generated by some rather dubious dealings, led to many shareholders not meeting the second call for funds, and loan capital had to be sought.

The financial collapse in May 1866 meant that although a further act of parliament raised the capitalisation ceiling, there was no chance of raising further funds. The sponsoring contractors therefore provided the working capital themselves. However, the crisis also hit shipbuilding on the Thames very severely and meant that the original market for the dock was now in question. Further capital had to be raised to fund investments in warehousing and other transport dock installations. The docks opened for business in March 1868 following much acrimony between the contractors and the company which completed the works using direct labour. The ensuing court case was dropped in favour of a negotiated settlement in which the promoting contractors were paid the outstanding sums they were owed in equity.

Source: Guillery (1990).

The financial crash of 1866 took away much of the competitive advantage of the promoter-contractors, and banks increasingly preferred to lend to governments and established firms rather than to finance projects directly. Clients were increasingly public authorities such as the Metropolitan Board of Works, and during the last quarter of the century, competitive tendering for civil engineering contracts became universal. The railway companies increasingly developed their own engineering expertise. The consulting engineer became more important, earning Brunei's jibe that the consulting engineer was a man who was prepared to sell his name but nothing more, while the enterprise increasingly took the form of the civil engineering contractor of today. The main difference from the professional system in building was that no equivalent of the quantity surveyor emerged – surveying remained a sub-discipline within engineering, and the engineer retained a strong control role in addition to the conception role. Although the crash of 1866 is undoubtedly the proximate cause of this shift to a professional system in civil engineering, it can also be located within a more general shift after 1860 from a society dominated by the entrepreneurial ideal and regulated by Adam Smith's hidden hand, towards the beginning of a society dominated by the professional ideal in which the role of the state was to regulate the free market in the interests of the wider community.

The manifestation of this more general societal development within the contracting system was the insulation of the activities facing the highest uncertainty in the design stages from market forces altogether, through the development of the professionally organised consultant engineer reimbursed on a fee basis, and the evolution of control actors responsible for regulating those activities that remained subject to market forces – principally construction – on behalf of the client and the wider community. How far this had come by the last quarter of the century is illustrated by the Tay Bridge disaster, in panel 2.8.

Panel 2.8 The Tay Bridge disaster

On 28 December 1879, a newly completed bridge across the Firth of Tay collapsed in a storm while a passenger train was on it. Although there were clear failures in quality control, particularly in the foundry, by the contractor during construction, and although the railway company failed to comply with speed restrictions on the bridge imposed by the Railway Inspectorate, and the engineer acted on Railway Inspectorate advice regarding wind loadings on structures of this kind, it was the consulting engineer Thomas Bouch who took the blame for the disaster. The Board of Trade enquiry concluded that 'the bridge was badly designed, badly constructed and badly maintained' and Bouch was ultimately responsible for overseeing the construction and supervising the maintenance as well as the engineering design.

Sources: Prebble (1979); Thomas (1972).

As the public sector became a client of greater and greater importance in the market, it opted for the professional system, seeking reassurance from appointed architects, quantity surveyors and consulting engineers for conception and control, and relying on competitive tendering for construction on fully detailed designs. This generated generally high standards in the constructed product and met public concern for transparency and accountability in the system, but led to relatively high costs and a deskilling of those responsible for execution on site.

A number of themes can be identified in these developments. Firstly, the changes are, to a very important extent, associated with changes in the nature of the client and its needs. The principal client, if that is not too much of an anachronism, for the craft system was the church and, to a lesser extent, the crown. The emergence of the architect/surveyor and the trade system is associated with the rise of rich merchants and an educated aristocracy influenced by the ideals of the Renaissance. The emergence of the professional system in the UK is associated with the emergence of new types of clients needing new types of buildings associated with the industrial revolution and the rise of municipal government, but particularly with the large-scale building programmes mounted by the crown during the French wars.

Secondly, although these succeeding systems replaced each other as the dominant model, the earlier ones survived to meet particular client needs. Particularly in the vernacular tradition, the craft system survived and has a place today in the repair and maintenance sector as well as in its more pervasive legacy of the division of labour on site. The trade system survived well into the twentieth century in Scotland and elsewhere in Europe, and left its profound legacy of the architectural role. The history is more one of successive layering than elimination.

Thirdly, the method of establishing the price for the work changed in important ways, with profound consequences for the motivation of project actors within each system. The medieval mason was paid on a time basis. These rates were subject to market forces, and the periodic attempts to regulate wages by law generally failed. However, there was little motivation to improve productivity or change methods, and the client had no way of passing risk on to others. A more sophisticated system of measure and value became associated with the trade system, where each master tradesman was paid a sum in proportion to the amount of work completed related to the cost of inputs plus a mark-up for profit. Although the award of the contract by competition did provide some incentive to reduce input costs, risks associated with the works remained with the client. Competitive tendering on a lump sum basis was first associated with the contracts for military works during the French wars, and rapidly became the norm. This, in turn, stimulated further developments in the surveying role and led to the emergence of the quantity surveyor. Competitive tendering for general contracts intensified competition, and many risks associated with budget and programme, particularly the latter, could now be more effectively transferred.

Each of the three systems, in their ideal-typical forms, have their own way of solving the problems of conception, construction and control. The craft system combines all three roles in the activities of the mason, while the trade system witnesses the definition of separate actors for conception and control on the one hand, and construction on the other. One of the professional system's most distinctive features is the separation of the conception and control functions, together with a reinforcement of the construction function with development of the general contractor.

For over a century the system was steadily reinforced – the surveyors acquired their charter in 1881 as the Surveyors Institution, and became the Royal Institution in 1946. Although the accuracy of the bill of quantities method had been proven with the rebuilding of the Palace of Westminster after the fire of 1834, it was not until 1922 that a Standard Method of Measurement was finally agreed, although the Scottish surveyors in Edinburgh had had one since 1773. The Institute of British Architects (IBA) received its charter in 1837, and in 1931 'architect' became a protected title. The first standard form of construction contract was issued by the Royal Institute of British Architects (RIBA) in 1903. Eventually the task of developing such standard forms was taken over by the Joint Contracts Tribunal (JCT) after its foundation in 1931, and the professional system became firmly institutionalised in its series of standard forms. Newer professions copied their older peers – the Institution of Structural Engineers received its Royal Charter in 1934, the Chartered Institute of Building Services Engineers in 1978 and the Chartered Institute of Building in 1980.

The reform of the professional system

It was not until the 1960s that the professional system began to change in Britain. Innovations in tendering procedures were made, particularly a shift towards selective rather than open competitive tendering, and the first applications of a new form of procurement imported from the US – management contracting – were made. During the 1980s the pace of change gathered speed, accelerating more during the 1990s. Again it was new types of clients with new requirements for constructed facilities which led the way, and for the first time in over a century, these clients came mainly from the private sector:

- The property boom of the 1980s favoured management contracting, and then construction management, in the drive to improve schedule performance in terms of both level and predictability.
- The growth of out-of-town shopping centres during the 1980s favoured the integration of design and execution to achieve efficiency benefits.
- The privatisation of the utilities during the 1980s and 1990s created a whole new set of private sector clients with programmes of complex works, and it was these clients which led in the development of partnering.
- The convergence of commitments to low taxation with growing requirements for investment in dilapidated public facilities encouraged the development during the 1990s by the state of a variety of different forms of public/private partnerships in which the private sector financed, constructed and operated the facility on the basis of a long-term concession.

During the post-war heyday of the professional system there had been a string of enquiries into the performance of the industry, which had articulated a number of recurring themes around procurement, relationships, performance and the role of the public sector client. Integration and collaboration were recurring themes. Thoughts of reform revived in the late 1980s, and since then the UK construction industry has been through a blizzard of reports and initiatives rethinking construction. Broad consensus on the inherent weaknesses of the professional system for twenty-first-century construction was not echoed in mutual understanding of what integration and collaboration really meant in the context of the exogenous drivers for change identified above.

Broadly, there have been two dynamics of endogenous change over the last 20 years in UK construction. The first is – remarkably for the immediate post-Thatcher era – essentially corporatist in that it attempts to bring all the stakeholders in the process together with government by giving them voice through their representative bodies. It is only distinguished from true corporatism by the absence of the trade unions from the table. This strand resulted in the Latham Report of 1994 and initiatives aimed at the institutional level and governance levels of Fig. 1.5 through legislative and regulatory reform. The second is identifiably new labour in that it prefers to deal directly with key opinion formers in the industry leading change by demonstration rather than regulation. This resulted in the Egan Report of 1998, and subsequent reform movement focused much more on the process level of Fig. 1.5. The present arrangement where a Strategic Forum for Construction takes a strategic view of the industry as a whole (and includes the trade unions) and Constructing Excellence works at performance improvement on projects contains elements of both these strands, but is arguably closer to Egan than to Latham.

In a largely parallel development, the UK government – driven by HM Treasury – began to take an active interest in how it bought construction services. Tentative experiments with private finance for infrastructure during the 1980s had matured by the early 2000s into a full-blown preference for private finance over crown building (i.e. finance from tax revenues) to achieve – in the words of one government minister – 'the end of the BAD old days – Build and Disappear' (cited *Financial Times* 04/04/96). Increasingly elaborate arrangements which attempted to combine private finance and therefore returns on that finance with public accountability for asset exploitation reached their zenith with large hospitals and the privatisation of London Transport – the total capital value of the 563 deals signed by April 2003 was £35.5bn.

According to Partnerships UK, there are now four basic types of privately financed procurement:

- *concession* – typically for infrastructure projects, where an asset is provided for which users pay directly, such as a tramway. The Second Severn Bridge in panel 2.9 is an example, as is the M6 Toll road.
- *private finance initiative (PFI)* – typically buildings for the delivery of public services, where an asset is provided and the public service provider pays a fee based on the availability of the asset for exploitation. This is the most common form of private finance of public assets and widespread for facilities such as hospitals and HM Treasury's own building.
- *public–private partnership* (PPP) – typically used to increase the exploitation of underused public assets where the public and private sectors share in the

returns from the sale, transfer or other exploitation of the publicly owned asset. London Transport is an example.

- *company limited by guarantee* – typically used where privatised companies are not viable without risk-sharing with the state, at least in the last resort. Network Rail is an example.

Growing awareness of the limitations of this approach prompted an HM Treasury review in 2003 which recommended the abandonment of PFI for smaller projects under £20m because of the transaction cost overhead of bidding and negotiation. Changes to accounting procedures associated with the switch to International Financial Reporting Standards also reduced the public accounting advantage of private finance as the debt now had to be carried on the public sector accounts. Growing concern about the costs of bidding and negotiation in a context of a boom in conventionally financed projects meant that the supply of willing bidders for those PFI projects that came forward reduced. Many projects – particularly in the health sector – were unable to reach close, and similar problems affected the schools programme. Growing concern was also expressed about the product integrity of some of the facilities provided through PFI procurement. The failure of Metronet, one of the two PPP contractors on London Underground in 2007, added to the gloom. The early 2000s would appear to have seen the peak of private finance at around 10% of government expenditure on construction.

Panel 2.9 The Second Severn Bridge concession

By the late 1980s, the Severn Bridge linking the Bristol region to South Wales was becoming overloaded, potentially hampering the economic development of South Wales. Following extensive survey work, the decision was taken in 1986 to locate a crossing further downstream, but the problem of finance was not resolved. Following the model of the Queen Elizabeth II Bridge at Dartford, an invitation to tender was launched in April 1989 to pre-qualified concessionaires. The successful bidder was Severn River Crossing plc, a joint venture (see section 7.7 for definition) consisting of the following partners shown with their respective shares of the equity of the joint venture (JV).

GTM Entrepose SA	35%
John Laing plc	35%
Barclays de Zoete Wedd	15%
Bank of America	15%

Finance was raised using the following instruments
Equity capital of £100 000
A loan from the European Bank for Investment of £150m, guaranteed by a letter of credit from the financial members of the JV
A loan of £190m from a syndicate of banks
An index linked debenture stock to the value of £131m at the retail price index plus 6%
Government subordinated debt, equal to the value of the existing bridge of £60m
The income from tolls on the existing bridge, predicted to amount to £150m over the period of construction.

The design was for a cable-stayed bridge 5125m in length, costed at £270m. The concession, signed in October 1990, was to last 30 years from April 1992 *or* until the concessionaires had received an income on tolls from the two bridges of £957m at 1988 prices.
The JV then made a construction contract with a 50:50 consortium of John Laing Construction and GTM Europe. These two, in turn, sub-contracted for design work to a consortium of Sir William Halcrow and Société d'Etudes et d'Equipement d'Entreprises (SEEE), the latter a subsidiary of GTM. The JV also contracted with a Laing subsidiary for maintenance and a GTM subsidiary for the operation of the toll booths. In order to represent its interests, the UK Department of Transport as concessor hired two firms of engineering consultants: Maunsell to supervise construction – they had done the original concept work – and Flint O'Neill to supervise operations and maintenance. A further firm – appointed jointly by the concessor and concessionaire – acted as checker of the design work. The project was successfully completed to both programme and budget in April 1996; however, its opening was pushed back to June because of delays in the construction of the approach roads under different, conventional procurement. As the Project Director put it, concession projects require 'a different culture in terms of project management, with very tight controls on cost particularly. There is no opportunity on a project like this for seeking reimbursement for additional costs'.

Source: Campagnac (1996).

HM Treasury was also very concerned about the larger bulk of projects that were financed by taxpayers who continually overran schedule and budget in both construction and IT. A series of procurement guidance notes started to be produced which began to address some of the long-standing weaknesses in government procurement. This policy was given a much higher profile by the launch of the Office of Government Commerce in 2000 as part of the Levene recommendations on government purchasing as an office of HM Treasury tasked with developing policy standards and guidelines, and supporting government departments in implementing them. In 2005, its remit was extended to the NHS and local government. One important result of its establishment was the start of cross-fertilisation from the IT sector, and concepts such as gateway reviews and programme management began to enter the language of construction project management. Its guidelines have fully engaged with the reform agenda, and the most recent set published in 2007 represents a sophisticated view of the role of the client in effective project management in creating process integrity in construction.

Around the same time – and apparently in reaction to the complete silence of the Egan report on product integrity (chapter 3) issues – HM Government established the Commission for Architecture and the Built Environment (CABE) in 1999 under the auspices of the Department for Culture, Media and Sport and published *Better Public Buildings: A Proud Legacy for the Future* in 2000. This initiative has stimulated a sea change in public sector attitudes to product integrity within government, legitimising the role of design champions. In 2002, CABE and the OGC published a joint report on *Improving Standards of Design in Public Buildings*.

So has all this activity resulted in a shift towards what might be called a *production system* where production is defined as the organisational integration of product and process? We will return to a more general assessment of the answer to this question in the concluding chapter, but here we will indicate some of the signs of what has, and has not, happened.

- Latham argued that the public sector should become a best practice client. If the publications of the OGC are a good guide, then this aim has arguably been met. However, implementation is patchy, and, particularly in local government, there are many clients which are still the industry recipe of the professional system.
- The ability of occasional public sector clients to manage design-led projects with architectural stars remains very weak as a string of failures including Bath Spa, Clissold Leisure Centre, Colchester Arts Centre and, most spectacularly, The Scottish Parliament demonstrates.
- Delivery performance on central government projects has improved significantly. The National Audit Office sample of 142 such projects suggests a 100% improvement in predictability between 1999 and 2004, which is near the 20% year-on-year Egan target. In particular, a number of public sector clients have been innovative in developing partnered framework agreements, both crown and privately financed.
- In 2008, 112 firms were indicted by the Office of Fair Trading for 'cover pricing' on local government contracts – that is finding out what other tenderers were thinking of offering and then pricing higher because they did not want the work on this occasion, but did not wish to offend a potential client. Nine of these firms were accused of the much more serious offence of bid rigging – that is offering compensation to bidders putting in higher prices. This followed fines of a total of £3.7m on 38 flat roofing contractors between 2004 and 2006 for bid rigging and market sharing.
- The adjudication provisions of the post-Latham legislation have worked well, speeding dispute settlement and minimising arbitration and litigation, but the fair payment provisions have been less successful, and are presently the subject of review and possible further legislation.
- The Egan targets as measured through the Key Performance Indicators, with the notable exception for the one on contractor's profitability which has nearly doubled, have not been met between 2000 and 2007, although there have been significant improvements in schedule predictability and client satisfaction. In terms of absolute performance, clients do not receive their facilities any quicker or cheaper in 2000 than they did in 2007.
- Although the rhetoric is of 'integration' in the sector, this would appear to be more rhetoric than reality. On the supply side, integration would appear to amount to little more than loose associations of firms that reconfigure for each project – there have certainly been few signs of moves towards vertical integration along the supply chain.

This is a mixed score card, and Sir John Egan himself gave it 4 out of 10 in a speech at the House of Commons in 2008. At the institutional and governance levels, considerable progress has been made over the past 20 years in the construction sector, although more would always be welcome. It is at the process level that change has been slow, yet it is at this level that projects are managed and facilities delivered. Many of the vignettes that follow will attest to the process innovations that have taken place, but they remain at the level of best practice – even advanced practice – rather than standard practice throughout the industry. Perhaps such a lag is inevitable because it was always argued that reform – particularly at the governance level – was a prerequisite for more collaborative and innovative relationships at the process level.

Sources: Adamson and Pollington (2006); Cacciatori and Jacobides (2005); McMeeken (2008); Murray and Langford (2003); National Audit Office (2001, 2005a); Winch (1996a, 2000c); http://www.building.com; http://www.bre.co.uk; http://www.cabe.org.uk and http://www.ogc.gov.uk (accessed on various dates).

Notes

1 Marx (1968, p. 97).
2 Her *The British Building Industry* (1966) remains the seminal analysis of the organisation of the UK construction industry. It is, perhaps, depressing that so little has changed since she wrote.
3 See the two special issues of *Building Research and Information*: Construction Business Systems in the European Union (**28**, 2000) and Global Construction Business Systems (**31**, 2003). See also Sha (2004) on China.
4 Spender (1989, p. 7).
5 The concept of rules of the game comes from North (1990).
6 Although old (Tavistock Institute, 1966, p. 50) this quotation has a depressingly contemporary feel to it even down to the assumed gender of site managers.
7 There are two main strands to this work – that of economists such as Porter (1990) who are concerned with how any competitive advantage differs between nations and that of sociologists and historians such as Whitley (1992, 1999) or Herrigel (1996) concerned with the social and historical roots of contemporary business performance.
8 See his *The Competitive Advantage of Nations* (1990).
9 Abbott (1988) analyses this same process of competitive collaboration among the US health care professions.
10 Bourdieu (1989) and Thoenig (1987) provide detailed analysis of this process.
11 See Zweigert and Kötz (1998) and Marsh (1994).
12 See Ponthier (1993) on architects and planning permission, and Reitsma (1995) on infrastructure development.
13 Campinos and Grando (1988).
14 DiMaggio and Powell (1983) stress the importance of the state as a force for changes in shaping what they call organisational fields.
15 Perceptions of the welfare benefits of this approach have changed rapidly in The Netherlands recently; see panel 5.9 for further details.
16 Unpublished structure conduct performance assessment of the UK construction industry by the Department of the Environment, Transport and the Regions (1998).
17 The UK government announced that it would adopt a similar strategy in October 2008.
18 Martinand (1993) provides an overview of the French experience and Miller (2000) the US experience.
19 For a more systematic comparison of these two parallel histories, see Campagnac and Winch (1997).
20 Middlemas (1963, p. 307).
21 The work of Porter (1990) is seminal here.
22 This point is made by Marcou *et al.* (1992). It is a strict definition. In Ireland, the British government was obliged to provide soft loans to make the railways viable, while in India the East India Company offered interest payment guarantees to encourage contractors to build the lines on the basis of a 99-year concession. This proved to be an expensive way to build railways, and

after the implementation of direct imperial rule, the government built the lines itself from 1870 onwards.

23 See, for instance, the study by Winch and Carr (2001a) of productivity on structural concrete works in France and the UK.

Further reading

Ive, G. and Gruneberg, S. (2000) *The Economics of the Modern Construction Sector.* Basingstoke, Macmillan.
An authoritative guide to the economics of the UK construction industry.

Linder, M. (1994) *Projecting Capitalism: A History of the Internationalization of the Construction Industry.* Westport, Greenwood Press.
Drawing extensively on *Engineering News Record* and other original sources, this is a remarkable history of the international construction industry from Brassey to Bechtel.

Winch, G. (ed.) (2000) Construction Business Systems in the European Union. *Building Research and Information* **28** 2 and (2002) Global Construction Business Systems. *Building Research and Information* **30** 6.
Provides analyses by national experts of recent developments in 11 different national construction industries from a business systems perspective.

Part II

Defining the Project Mission

The first step in the management of any construction project is to define what is wanted. A bridge or a building is a major capital investment. The term *facility* will be used to define all those capital assets that are created through processes conventionally associated with the construction industry – the outcomes of the management processes analysed in this book. Moreover, such assets have very long lives, typically much longer than other capital assets. Whether the client is a public authority, fulfilling its commitments to the electorate to provide public services; a private corporation creating the facilities required for its own business processes; or an individual purchasing a home, a facility is a major investment which will shape the quality of life and the competitiveness of the business for years to come. Such investment decisions are inherently *strategic*, and so it is the disciplines of strategic management that we turn to in this section to help us understand the process of defining the project mission.

There has been a lively debate regarding the precise character of strategic management[1], which there is no need to rehearse here – much of it relates to the corporate level, rather than the business level where project investment decisions are usually made. For our purposes, defining the project mission is a strategic decision-making process because:

- it is about the relationship of the client organisation to its economic and social environment;
- it is a proactive process of allocating scarce resources between alternative projects;
- it is related to the medium- and long-term future of the client;
- it is a prerequisite for the successful implementation of any strategy of expansion or diversification.

Any decision by the client organisation to expand existing capabilities or pursue new opportunities requires additional investment, and a high proportion of that investment will typically be on facilities. Even the most high-tech industries require buildings for staff, trenches for fibre-optic cables and the erection of radio masts. Acquisition of new facilities can take place either through purchase or rent from a property developer, or by direct procurement, depending on the functional specificity of the asset required. In the former case, the asset will often be acquired as a shell and core, and require fitting out to meet the client organisation's specific needs. Where second-hand assets are acquired, refurbishment of the existing facility will often be required. The remarkable capability of constructed assets to be adapted and readapted is one reason for their longevity[2]. In all cases, most strategic decisions to develop the capabilities of the organisation require, at some point, investment in constructed assets.

In line with the definition of the corporate mission used by strategic managers as the 'overriding purpose in line with the values or expectations of the stakeholders'[3], the term *project mission* is used in this book to refer to the overall strategic intent of the project – what is to be delivered to the client. This mission is then broken down into goals and quantifiable objectives during the project definition process. Intended strategy is not the same as realised strategy, so the project mission is in a continual process of reappraisal during the life cycle of the project as shown in Fig. II.1. Project missions are rarely fully realised because:

- they are formulated under high levels of uncertainty regarding the social and economic conditions in which the asset will be exploited;
- assumptions made as the basis for strategy formulation prove to be untenable as new information becomes available through the project life cycle;
- new opportunities present themselves to which the facility can be adapted;
- stakeholders change their minds.

It follows from this analysis that the key criterion for project success is not that the project mission is fully achieved, but that the realised asset fully matches the client's needs *at the time of realisation*. The management of this gap between intention as articulated in the project mission and realisation will be fully explored in Part IV.

Why, then, bother to define a project mission if it is going to change anyway? There are three very good reasons:

- The process of definition as the participants articulate their understandings tests the intuition and analysis upon which the strategy is based for consistency and viability.
- The defined mission allows the communication of strategic intent to the diverse project stakeholders – both those whose active participation is required to realise the facility and those who have the power to disrupt that delivery.
- The defined mission provides the baseline for the planning and control of the project process through the life cycle.

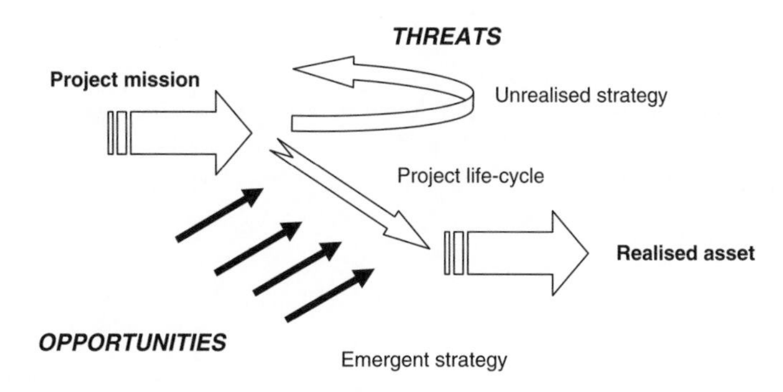

Fig. II.1 The project mission and the realised asset (source: developed from Mintzberg, 1987, Fig. 1).

Notes

1 Ansoff (1968) and Mintzberg (1994) are major contributions to this debate.
2 See Brand's (1994) fascinating account of the ways in which buildings evolve over time.
3 Johnson and Scholes (2002, exhibit 1.1).

Chapter 3
Deciding What the Client Wants

3.1 Introduction

> The project was "assembled round a hole like a Polo mint ... [there was] no client driving it forward with a vision of what the operator needed to have".

The organisation's overall corporate strategy will set the basic parameters of decision-making. Decisions regarding, for instance, how many hospitals should be built, or how many new stores are required to achieve market share and turnover objectives, are corporate strategy decisions. With those decisions will come basic parameters which define the bounds of the project mission – hurdle rates for project appraisal; policies on space standards; branding policies for customer facilities and the like. This chapter focuses on the strategic decisions associated with defining the mission for a single project, or closely related set of projects. As indicated by Sir Alastair Morton, Co-chairman, Eurotunnel[1], in the epigraph above, such strategic decisions require that clients understand their business processes and the ways in which these processes can be enhanced by investment in facilities.

The chapter will first address how clients understand their business so that they can articulate their requirements to those that will be designing and constructing their new facility by developing a balanced scorecard of asset value. It will then turn to the ways in which clients appraise the returns on, and benefits of, any particular investment project and hence choose between competing project proposals, and also look at the ways in which such appraisals can be manipulated for ulterior motives. These issues will be brought together in a summary project mission definition model. The role of negotiations between different stakeholders during this process will be the subject of Chapter 4.

3.2 From artefact to asset: facilities as new value

It is worth starting by examining what exactly is meant by the word 'value'. It has been in use in English at least since the fourteenth century in both of its intertwined uses of value as a measure of worth and value as an explicitly held belief[2]. One long-standing tradition in defining value as a measure of worth is to make a distinction between 'value in use' and 'value in exchange' first made by the Scottish economist Adam Smith[3]. He argued that although water has no value in exchange it has considerable value in use while a diamond has no value in use but considerable value in exchange. This distinction was associated with a belief that only physical things could have a value, and that labour was the source of all value. The development of privatised water companies and industrial uses for diamond over the past 200 years indicates some of the problems in making this distinction. The current debates within the construction sector[4] typically deploy this labour theory of value by arguing that exchange values for buildings do not fully reflect their use values to their occupiers and the two are seen as in tension.

The development of neoclassical economics with its emphasis on price formation through the marginal equalisation of supply and demand made the distinction between exchange and use value irrelevant because *utility* was determined by what the market was prepared to pay to satisfy its wants[5]. We will be critical of perspectives derived from neoclassical economics in section 3.6, but its understanding of utility and the role of information in decision-making are important planks in our argument. From this perspective, the argument for treating buildings as assets rather than artefacts is not merely an advocacy of 'good design' where use value should be preferred over exchange value, but an argument that we need to develop a much deeper understanding of the benefits of good design so as to enhance the returns flowing from the expected benefits to fund any additional costs that might be incurred in achieving them through 'value-added' investment to provide a net increase in utility. In this perspective, any inability to trade the completed building is a problem in information in that potential users who value the building as highly as the current users have not yet been identified. Deeper and more widely diffused understanding of how buildings add value for clients should, in principle, reduce these information problems.

Clients invest in facilities because these provide a utility which can be exploited by themselves or others to provide goods or services which generate benefits, thereby providing the returns on the investment. So, the most important aspect of project definition is to understand how this utility is generated. In essence, utility is generated through business processes, so the key to this understanding is business process analysis. When discussing the budget for a new facility, it is common to address the capital and running costs of the facility, but how the facility will add value to the client's business processes is often articulated only intuitively. Even the more sophisticated approaches of whole life costing – discussed in section 9.7.3 – limit themselves to the costs-in-use of the facility. However, the evidence for commercial buildings – and almost certainly for most other types of facility – is that these capital and whole life costs are relatively trivial in the total cost profile. Table 3.1 shows that capital and costs-in-use account for around 8% of overall costs per square foot for a commercial building over a 20-year span. The vast bulk of the costs are accounted for by the personnel costs (mainly salaries) of the people working in the building[6].

Table 3.1 The costs of buildings in use (source: calculated from Wright *et al.*, 1995, Table 1).

Personnel using building	92%
Annualised cost of construction	5%
Utilities and facility management	3%

The balanced scorecard concept starts with the observation that financial criteria should not be the only measure of a firm's performance because 'an overemphasis on achieving and maintaining short-term financial results can cause companies to overinvest in short-term fixes and to underinvest in long-term value creation'[7]. Value in both senses of the word is created through attention to customer needs, efficiency and effectiveness in internal business processes, and the ability to innovate, as identified above. The concept of a balanced scorecard has now been applied to the value created by facilities – buildings in particular – and is summarised in Fig. 3.1. The argument is that symbolic quality, spatial quality and indoor environmental quality (IEQ) are all potential areas for value-added investment – that is, investment in the facility over and above that required to meet minimum functional requirements and regulatory obligations.

The problem is that facilities have typically been perceived as artefacts rather than assets, where artefacts are perceived as things which have no capacity to create further value for their owners and users. While they might have a resale value as artefacts, they are seen by clients simply as a cost of doing business and hence something to be minimised. Assets, on the other hand, have the capacity to create further value because of their design. They are an investment rather than a cost, not simply because property values outperform other investment opportunities, but because greater investment can return greater benefits from the exploitation of the asset to provide services valued by users.

Financial Value	Indoor Environmental Quality
NPV CAPEX OPEX Market value	Ventilation Daylighting Acoustics Controllability
Spatial Quality	**Symbolic Quality**
Interaction Isolation Integration Security	Image Branding Public interest Power

Fig. 3.1 The balanced scorecard for constructed assets (source: developed from Spencer and Winch, 2002, Fig. 4.2).

3.3 Understanding spatial quality and business processes

As the citation from Davenport in section 1.6 indicates, the spatial aspects of business processes are central to their analysis. Broadly, client organisations have four types of business process:

- information flows;
- resource flows;
- material flows;
- people flows.

The first two are essentially non-spatial. Information flows through communication systems of either the traditional or the electronic type, or it is embodied in people and is therefore a type of people flow. Financial resources are similarly non-spatial in their flows. Material flows are demanding consumers of space, and hence buildings, and material flows between buildings require a transport infrastructure. Similarly, people flows require space for their own movement, or movement enabled through the transport infrastructure. Understanding flows of materials and people is central to the definition of the project mission, because it is by enabling such flows in the most efficient and effective manner that the value of the facility is generated.

One way of understanding the spatial aspects of business processes is to use space syntax analysis – presented in panel 3.1 – which analyses the effect that the spatial configuration of the building or urban space has on shaping movement patterns. Where different work groups need to physically interact – such as in organisations whose competitive advantage comes through innovation – their relative spatial locations are important enablers of that interaction, as shown in panel 3.2. Ensuring that an office encourages interaction between staff, while allowing quiet space for activities requiring focus or privacy, is an inherently spatial problem. Similarly, the spatial integration of a commercial centre plays a significant role in determining which retail units have the greater footfall past their door and hence more opportunities to tempt shoppers inside. The same principles apply at the urban level, with, for instance, the distribution of the incidence of crime on a housing estate being linked to how integrated the spaces on the estate are to the main thoroughfares as illustrated in panel 3.3. Similarly, it was argued at the planning enquiry for the new Heron Tower in the City of London that expensive tall buildings enhance business processes by clustering teams in close proximity so that they compete intensively with each other[8], while research on hospital design has stressed the importance of individual rooms rather than large wards in patient recovery[9].

Panel 3.1 Space syntax

The *integration* of a particular space with the other spaces linked to it is a major factor in explaining the amount of movement through that space – this is the fundamental insight of space syntax analysis. Integration is a measure of how many corners a person moving through from one part of the space to another must turn. Spaces that are highly integrated experience more movement than those less integrated. This can be explained by people following lines of sight as they choose the direction in which to move. Thus, space syntax analysis is particularly important where movement has elements of browsing – such as in shopping – or where the effective functioning of the organisation is facilitated through informal interactions between staff.

Source: Hillier and Hanson (1984).

Panel 3.2 Spatial configuration and people flows

Three R&D groups were dispersed on more than one floor in a converted textile mill which consisted of a number of interconnected buildings. When they moved together into a new building providing a square, open-plan office, interaction levels rose significantly. Research in an advertising agency shows how the level of interaction between staff is strongly influenced by the spatial configuration of the offices – open plan with movement spaces running through generates much more interaction than cellular offices. In both R&D and advertising, the business process is, in essence, the creation of new ideas, and people generate new ideas largely through random interactions with other people working on similar problems. However, in order to work up those ideas into usable forms, they also need relative isolation from interruption by others.

Sources: Allen (1977, Chapter 8); Penn *et al.* (1999).

Panel 3.3 Space and crime

The theories of 'defensible space' as a way of designing a safe urban environment have had a profound influence over the past 30 years in shaping our urban areas. However, the design solutions they encouraged of creating fragmented spaces which discourage through movement, particularly where pedestrians and vehicles are separated, can be shown to increase the risk of crime. A space syntax study of three English towns has shown that houses in cul-de-sacs have a greater risk of crime, and that those with the least risk are in conventional street patterns with houses looking out on to both sides of the street.

Source: Hillier and Shu (2000).

3.4 Indoor environmental quality and business processes

While the spatial quality of the building can provide utility through enabling efficient and effective business processes, utility can also be generated by providing an appropriate level of IEQ. Poor IEQ can lead to loss of productivity, increased absenteeism and illness among employees, and turnover among tenants. One US survey – reported in Table 3.2 – of the reasons why commercial tenants move offices showed that the worst problem cited related to the design of the tenanted building in over half the cases, with 30% citing heating, ventilation and air conditioning (HVAC) and indoor air quality as the worst problem. If a higher IEQ reduces tenant turnover, an owner will save the costs of voids, as well as be able to charge a higher rent because of the productivity factors. The quality of IEQ also directly affects personnel productivity, as a recent review of UK surveys of buildings in use shows – see panel 3.4 for information on the Post-occupancy Review of Buildings and their Engineering (PROBE) studies. These productivity benefits come from four distinctive variables[10]:

- personal control – staff like to be able to personally control their working environment;
- speedy response to reported discomfort by facilities managers;
- shallow plan forms that allow all workstations natural light;
- small workgroups allocated their own spaces.

Table 3.2 Reasons for moving office (source: Mudarri, 2000, Table 8.1).

Worst problem	Percentage of responses
HVAC and indoor air quality	30
Elevators	12
Building design	7
Loading docks	6
All others	45

HVAC, heating, ventilation and air conditioning.

Panel 3.4 Understanding IEQ: the PROBE studies

Supported by the UK government, detailed evaluations of new buildings in use have been made, covering the technical performance of HVAC installations; the energy efficiency of the buildings; and user satisfaction with the indoor environment (measuring on 49 variables) using a development of the well-established Building Use Studies instrument. Twenty-three buildings were evaluated and the reports published in the *Building Services Journal*. The principal conclusion is that performance is highly variable and frequently not as expected at the design stage, and that a commissioning period post-handover would be very beneficial. This latter point is presently being pursued through the Soft Landings initiative.

Sources: Cohen *et al.* (2001); Way and Bordass (2005); http://www.usablebuildings.co.uk/ (accessed 28/07/08).

3.5 Symbolic quality: beyond peer review

While there is a growing body of rigorous research on spatial quality and IEQ, the value added of symbolic quality remains obscured and occluded by the debates within the architectural profession on architectural quality. The tenor of the debates can be seen at work in the pages of the architectural press where projects are reviewed and criticised – not always constructively. This peer review process reaches its apogee in the awards ceremonies that have become a common feature of many areas of business. In the UK, the highest profile prize is the Stirling Prize which is awarded annually to the 'architects of the building which has been the most significant for the evolution of architecture in the past year'. The short-listed buildings are visited by the members of an expert review panel who make their final decision just before the award dinner which is televised nationally. Some indicative citations for prize-winning projects include[11]:

> 'It is a new icon. In the last few years there has been the London Eye, the Angel of the North and now this. It is the one new piece of architecture that will be remembered by people this year'. (Wilkinson Eyre's Gateshead Millennium Bridge)
>
> 'It hits you straight between the eyes as soon as you get there. It has the same movement, youth, agility, pizzazz, front to it that its students have – it's very seductive. The immediate impact on everyone as we arrived was to go wow'. (Herzog and de Meuron's Laban Dance Centre, London)
>
> 'The proof of the extraordinary architectural ambition and design vision is to be seen in every aspect and detail of the finished building. At the outset, Miralles made a major contribution in leading the clients towards a proper understanding of their needs and the final formulation of the role and function of the building. Further, through his awareness of the problems and knowledge of the subject, the architect has formulated the philosophy of the role of the Parliament and reflected it in his architectural interpretation'. (Enric Miralles' Scottish Parliament Building)

However, the prize does tend to ignore other aspects of the project and has been awarded for buildings that came in seriously over budget and schedule such as the Scottish Parliament Building and the Lord's Media Centre by Future Systems. As one critic puts it:

> 'Architects still yearn to see themselves as modernist pioneers creating a form-breaking architecture by pushing at the boundaries of technology. Combine a space-age image, the exotic use of boat-building technology and the feel-good story of a small, highly principled but embattled firm that had been on the brink of collapse, and you hit all the right notes'[12].

This is the view advocated – arguably to the point of parody – in King Vidor's 1948 film *The Fountainhead* of the architect as lone genius struggling against the conservatism of the aesthetically challenged committee members who fund public and private developments. There are projects which successfully steer between these two poles. Buildings such as Herzog and de Meuron's Tate Modern in London and Grimshaw Architect's Eden Centre in Cornwall presented in Case 17 can be architecturally feted as well as meet client aspirations and receive public acclaim. In other cases opinions can be strongly divided. The Scottish Parliament at Holyrood provoked much unhappiness amongst the Scottish people for both its design and the associated budget and schedule overruns, and was voted as one of the 12 worst buildings in Great Britain in a poll for the Channel 4 television series *Demolition* in which 10,000 people participated. However, it was lauded by the Presiding Officer of the Parliament when it was awarded the Stirling Prize on the grounds that 'The judges have decided that Holyrood is not just a working legislature but a work of art constructed on a world-heritage site where the history and land of Scotland fuse together'[13].

The fundamental problem with the somewhat fractious debate on architectural quality is that it is essentially about buildings as artefacts rather than assets, and therefore adds little to our understanding of how buildings add value for clients. In order to start an exploration of the contribution of symbolic quality to asset value we can identify four facets as shown in Fig. 3.1[14]:

- *Branding* is a central element in mass marketing and many clients have commissioned facilities that express some dimension of their market position. Often this is done through an association with the modernity of the day. From London's Michelin and Hoover buildings, through to New York's Seagram Building, consumer goods companies have tried to reinforce the contemporary image of their products with their facilities. A trend in wineries is to have what are essentially factories designed by leading architects to convert them into visitor attractions such as the Faustino Winery in Gumiel de Hizan, Spain (Foster Associates), the Bodegas Protos Winery in Peñafiel, Spain (Rogers Stirk Harbour) and the Hall Winery in Napa, California (Gehry Partners)[15]. The same principles are applied in any number of 'designer' hotels, restaurants and boutiques. Most recently, branding around sustainability themes such as Marks and Spencer's Plan A described in Case 5 has become an important market positioning criterion.
- *Image* can also be of central concern to clients. Building high is one of the classic statements of image as the competition between banks on the Hong Kong Waterfront demonstrates, and cities throughout the world compete with each other by constructing architecturally innovative public facilities. Other clients consider the image of the buildings they commission for their attractiveness to potential staff. For instance, the CEO of the McLaren Group argued that the rationale for the McLaren Technology Centre near Woking, UK, was that:

  > 'Our company is all about people, we all want the company to win – and I'm not just talking about winning grands prix. I'm talking about everything we are involved in. We need highly motivated, dedicated people, and such people can only exist if you provide them with an environment in which they can aspire to be the best. Quite simply, great facilities attract great people. That's where the story of this project really has its roots'[16].

- *Public benefit.* Any investment in constructed facilities has important externalities and one of the most important externalities is the contribution to the surrounding area. Attractively and interestingly designed buildings enhance surrounding areas; dull and ugly ones detract from it. It is not possible to regulate for symbolic quality, and so the public spirit of the client is vital here.

Clearly, the clients who ought to have this most at heart are those in the public sector, and there is a fine tradition of public building in most countries, even if it has rather been diminished in the period after 1945 driven by a myopic cost-of-artefact as opposed to public asset agenda[17]. Commercial developers can also benefit from attention to public benefit – developments that create attractive spaces around them tend to achieve higher rents than those that do not[18].

- *Power.* The symbolic use of buildings to express power has a long tradition. Religious and political leaders have often sought to express their power by commissioning fine buildings as glorifications of their aesthetic tastes, and the architectural profession – as shown in Case 2 – has its roots in the ostentation of rich merchants and landowners. Democracies are not immune to such pretentions, as desire for 'princely action' as expressed in the Parisian *grands projets* of recent French presidents demonstrates[19]. Similarly, the Channel Fixed Link presented in Case 1 was strongly supported by the political leaders of the UK and France of the day – on the French side in the tradition of *grands projets*; on the British by an ideological commitment to demonstrate what the free market could achieve that the public sector could not[20]. Location is also important in expressing power – Bill Hillier notes how buildings that symbolise political power rupture the natural, commercially led evolution of urban street patterns[21].

3.6 Justifying the investment

Once the utility of the asset has been identified, the benefits thereby generated need to be valued so that the resources required for the investment can be justified. The problem of the allocation of resources to projects is the capital budgeting problem – how does the client choose for investment those projects that will yield the greatest return? It is through solving the capital budgeting problem that the mission objectives for the project schedule and budget are set.

The good practice method[22] for solving this is to calculate the net present value (NPV) of the proposed investment, as shown in panel 3.5. The essence of NPV calculation is that the value of money today is greater than the value of money at a future point in time; therefore, benefits accruing in the future need to be *discounted* to their current value for an appraisal to be made. The rate of discount applied is a function of the *opportunity cost of capital*, in other words the return on capital that could be obtained if it were invested in the next best type of investment. One benchmark often taken is government bonds. NPV calculation is essentially a cash flow calculation – it compares the outflow of cash required for the investment with the discounted inflow of cash arising from the exploitation of the asset, and it is, therefore, frequently referred to as a *discounted cash flow* calculation. So long as the NPV of the income is greater than the investment outflow at the chosen rate of return, the project is worth pursuing in principle. When comparing a number of projects for scarce resources, those with the higher rates of return should be favoured. This rate of return is known as the return on capital employed (ROCE).

The calculation of the NPV requires both the cash inflows and outflows to be precisely measured, and herein lie three very important difficulties. Firstly, not all benefits can be precisely valued. While the cost of investment can be known relatively precisely, this is not so easy for many of the benefits, particularly with buildings. As discussed in section 3.2, many of the benefits of value-added design are poorly understood, with symbolic quality being particularly difficult to measure. It is not possible to value in a quantitative way all the benefits flowing from such investments, except by imputing essentially arbitrary figures, yet such investments play a vital role in the quality of life for those working in and around them.

Panel 3.5 Calculating net present value

The NPV of a project is given by:

$$NPV = C_{out} \frac{C_{in}}{1+r}$$

The discounted cash flow of a stream of future cash inflows is, therefore, given by:

$$NPV = C_{out} + \sum \frac{C_t}{(1+r_t)^t}$$

where C_{out} is the (negative) cash flow required for the investment, C_{in} the (positive) return on the investment, r the rate of return required by investors for their capital, and t the number of periods over which returns are expected.

A second problem is that market prices may be a poor guide to either or both of the costs of investment and the benefits flowing from exploiting the asset, and cost/benefit techniques may be more appropriate. There are, in essence, two reasons for this:

- Markets are distorted by various forms of market failure such as monopoly, the tax system or the availability of otherwise unused resources. This is a common problem in developing countries, which is why the World Bank pioneered the development of cost/benefit analysis in the 1960s for the appraisal of infrastructure development projects.
- Elements of either costs or benefits are not traded and so are intrinsically difficult to value. On the input side, externalities such as damage to the environment need to be added to the direct investment costs, while on the output side, benefits flowing from the investment such as reduced journey times, increased safety and generalised macroeconomic impacts also need to be assessed on many projects. For instance, the relatively known investment in the facilities for the London Olympics needs to be appraised against the stimulating effect on the

regeneration of a derelict area of London, and to the London economy generally from increased tourism in addition to the actual cash flow from receipts.

The technique which has been developed to allow such factors to be taken into account in NPV calculations is *shadow pricing*, described in panel 3.6. Shadow prices, once established and agreed, can be fed into the appraisal calculations in very much the same way as market prices. However, because shadow prices – by definition – cannot be directly observed, the methodologies used to measure them are difficult in practice and often questionable in theory.

Panel 3.6 Shadow pricing

The concept of a shadow price is simple – it is the net loss or gain associated with having one unit more or less of the asset. If one more kilometre of road costing £*x* will generate £*y* worth of leisure time of commuters, where $y>x$ at present values, then the investment provides a positive return. In practice, it is profoundly problematic: *y* is given by the amount that the appropriate decision-maker would be prepared to pay for an extra unit of whatever *y* is shadow pricing. For full cost benefit analysis, therefore, prices must be assigned to such emotive assets as how much society is prepared to pay to save an additional human life or to retain a particular view of a rural landscape. For instance, the value of a fatality is currently (2007) valued at £1.43m (made up of £0.5m in lost output; £0.9m in grief and loss; and a small balance of the costs of actually dealing with the accident) by the UK Department for Transport.

The third problem is central to the management of projects – both the cost and the benefits of the investment are uncertain. On the benefits side of the equation:

- The benefits arising from exploiting the asset may prove to have been optimistically valued.
- The operational costs of the facility may be higher than predicted.
- The facility may not be capable of being operated as planned.
- The facility may be delivered late, thereby pushing the income stream further into the future and possibly missing the market opportunity.

On the costs side, uncertainty arises for the following reasons:

- The investment required may turn out to be higher than expected.
- Late delivery of the facility may mean that existing, less efficient facilities have to be kept operating for longer.
- Late delivery of the facility may mean later commencement of the income stream.

All of these uncertainties can be managed within the project and this is the purpose of this book.

3.7 Strategic misrepresentation in investment appraisal

In line with our commitment to the Carnegie school of decision-making under bounded rationality articulated in section 1.8, we have presumed so far that decision-makers are acting in good faith even if their judgement turns out to be flawed in hindsight. However, recent research has confirmed what many have long suspected – decision-makers sometimes act in bad faith when making project investment appraisals. Since at least the construction of the world's first steam-hauled passenger railway – the Liverpool and Manchester Railway (LMR) – which overran its budget by 45% in the 1820s, infrastructure projects have been notorious for their inability to keep to budget and schedule, but at least the LMR was profitable. Many large infrastructure projects also fail to achieve their objective in terms of benefits for economy and society, and some become 'white elephants' – Table 3.3 gives some examples. Is this consistent inability to deliver infrastructure projects because of the incompetence of the supply side? While the supply side tends to get the blame for these failures, we suggest that other dynamics are also at work which not only explain why the business case consistently fails to stack up, but also explain why we are not learning to do it any better. The argument is that the promoters of projects are systematically biased towards the overestimation of the benefits of a project and an underestimation of the costs of a project – what has been dubbed *strategic misrepresentation*[23]. Although the literature is sometimes confused on this point, it is important to distinguish the strategic misrepresentation which has organisational drivers from the phenomenon of *optimism bias* which has psychological drivers and will be discussed in section 10.9.

Strategic misrepresentation is inherent in the principles of capital budgeting because the principal use of the NPV calculation, once it has been shown to be positive, is to rank order the projects so that the ones showing the highest returns relative to the others are preferred. If all the figures on each side of the equation were known with certainty, then this would not present a problem, but there are two linked areas of weakness. The first is that both benefits and costs are projected into

Table 3.3 Indicative cost overruns on major infrastructure projects (source: Flyvbjerg *et al.*, 2003, Table 4.2).

Project	Construction cost overrun (%)	Actual traffic as a percentage of forecast in opening year
Humber Bridge (UK)	175	25
Channel Fixed Link	80	18
Baltimore Metro (USA)	60	40
Tyne & Wear Metro (UK)	55	50
Portland Metro (USA)	55	45
Buffalo Metro (USA)	50	30
Miami Metro (USA)	35	15
TGV du Nord (France)	25	25

the future – sometimes a long time into the future – and are therefore matters of judgement, not fact. While there are very well-established techniques for addressing this uncertainty, they are rooted in the elicitation of subjective probabilities which are inherently fallible and open to bias as will be discussed in section 13.3. The second is that many of the benefits of investments are not estimates of real prices, but based on shadow pricing techniques measuring 'willingness to pay' and so introduce another area of subjectivity into the process. An important consequence of this considerable room for judgement in decision-making is that the persuasiveness of the project promoters becomes a major factor in making a particular NPV calculation stack up better than that of competing claims on the same resources.

The problems of strategic misrepresentation and optimism bias are now policy concerns and agencies responsible for the funding of infrastructure in countries such as Norway[24] and the UK are developing methods to tackle these issues in investment appraisal. A recent UK government report[25] focused particularly on urban rail projects. It found that the funding system encouraged 'optimism bias' in the same way that the 10 cent dollar created perverse incentives on the Boston Central Artery/Tunnel project described in Case 13. In the UK, cities promote projects which are funded by central government. Projects have to meet value-for-money criteria which are based on capital budgeting methods. The report found that it was in almost nobody's particular interest to mitigate strategic misrepresentation. Local interests obviously favour bringing investment to their city, and if doubts are expressed about the business case, elected representatives in Parliament are expected to lobby on behalf of the project. Supply side interests clearly have an interest in the project going ahead. In a more ambiguous position is the Department for Transport (DfT) trying to referee this game, but – not mentioned in the report – the DfT is arguably incentivised in the government resource allocation rounds by making the case for transportation projects vigorously as opposed to other claims on public funds such as defence and health expenditure. The only actor with a clear incentive not to strategically misrepresent is the finance ministry, HM Treasury, which probably explains why it has taken the lead in addressing this issue. Similarly, it is the Norwegian Finance Ministry which is taking the lead on the implementation of the Quality at Entry procedures for the evaluation of major projects in that country.

The Danish authors of the DfT report make the following recommendations in an attempt to address these issues:

- Ensure that those who benefit from the project share in the cost overruns. Thus, the DfT now requires 25% local funding on urban rail projects.
- Provide independent appraisal – the 'outside view' – of business cases, and, in particular, ignore special pleading that the project under consideration will be better managed than others by sticking rigorously to the historical data on percentage uplifts and using 'reference class forecasting'[26].
- Create a culture of deterring strategic misrepresentation by penalising project promoters who prepare poor cost estimates and rewarding those who prepare sound ones.
- Formalise requirements for business case development and risk management.

- The reports authors have elsewhere argued for the involvement of private finance on the grounds that such finance tends to perform more rigorous due diligence. The argument could also be made that such capital is internationally mobile and less prone to local suasion to invest in poor projects. However, this conclusion is tempered by rampant strategic misrepresentation on the Channel Fixed Link project despite 100% private sector funding[27].

The inherent uncertainties of investment appraisal mean that optimism bias in good faith and strategic misrepresentation in bad faith are pervasive in our predictions about the future. Together, they are important explanations of budget and schedule overruns in construction projects. Arguably, strategic misrepresentation is an adaptive response to decision-making under uncertainty – if you do not know the facts, you might as well give yourself the benefit of the doubt. Because it is pervasive, one could argue that it is not a problem – projects will inevitably overrun, but we simply allow for uplifts in total budget to take this into account. However, the existence of strategic misrepresentation seriously threatens the rationalistic basis of investment appraisal, particularly if some project promoters are better at playing the game of strategic representation better than others. Honest project promoters might produce more accurate estimates, but they are also more likely to have funding for their project rejected and so lose the possibility of demonstrating their wisdom. We do not know how pervasive strategic misrepresentation is in investment appraisal, but the inference from the data presented by Flyvbjerg and his colleagues is that it is widespread – much more research is needed in this area into its extent and dynamics.

3.8 Defining the project mission: a conceptual framework for product integrity

Our task is now to pull together all these disparate strands in the definition of the project mission – the new facility needs to meet a set of functional requirements, symbolic desires and investment criteria in the absence of strategic misrepresentation. Different missions will imply different sets of trade-offs between these criteria – a library to house a collection donated to a university by an affluent benefactor will have different symbolic and functional criteria from a local municipal lending library. The university library will need to express symbolically the importance of the collection, the ego of the benefactor, integrate sympathetically into the existing university buildings and pay particular attention to curatorial issues. The municipal library will need to be inviting to a local population, and to house and lend the collection efficiently and effectively. The university library is likely to require a higher budget and longer schedule than the municipal library.

These issues have been captured in the context of the car industry in the concept of *product integrity*, because 'a company can be fast and efficient, but unless it produces great products . . . it will not achieve competitive advantage'[28], as shown in panel 3.7. This concept is equally applicable to the constructed product, and Fig. 3.2 suggests how it might be applied. The three dimensions of the integrity of the constructed product are defined in terms of quality, because the quality of the asset resulting from the process is fundamental for the creation of new value:

- The value-added aspects discussed in sections 3.2–3.5 are captured by the quality of *conception* where value is added through design measured by benefits such as higher than average rents achieved, greater employee satisfaction and the like.
- The functional aspects defined as the minimum functional requirements for the facility are captured by the quality of *specification* measured by the fitness for purpose of the completed facility.

Panel 3.7 Product integrity in the car industry

Based on research with the majority of firms in the global car industry, a Harvard Business School research team developed a penetrating analysis of the role of design in competitive advantage. They argue that successful products have product integrity:

- external integrity, or the match of the product's performance to the customer's expectations;
- internal integrity, or the consistency of the engineering between the product's component parts.

They developed performance metrics for the new product development process:

- engineering hours, a measure of the cost of the process;
- development lead time, a measure of time to market;
- total product quality, a combination of measures of design quality (measured by expert review); conformance quality (measured by consumer surveys) and perceived total quality (measured by customer repurchase intentions and expert evaluations).

Source: Clark and Fujimoto (1991).

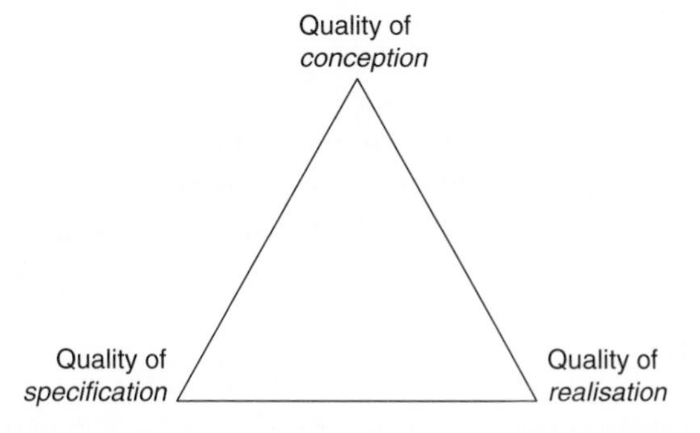

Fig. 3.2 Product integrity in construction: the quality of intention.

- The schedule and budget are captured by the quality of *realisation*, in terms of the objectives set for schedule and budget, and the service delivery experience for the client measured through process benchmarks for comparator buildings.

Trade-offs within the three criteria take many forms, for example:

- The type of open-plan spaces that generate the randomised interactions shown to be important for innovation and creativity have also been shown to demotivate staff because of the lack of control this implies over the working space.
- Stakeholders may have very different ideas regarding how the symbolic aspects of the building are best expressed – this is a frequent cause of clashes among architects, for instance.
- Schedule and budget often require to be traded off against each other – shorter schedules typically imply a higher budget for the construction.

Trade-offs between the criteria also have to be made – a high quality of conception typically implies a higher budget because of factors such as the greater proportion of the space given to circulation areas, and the higher quality materials used for finishes.

How are these trade-offs to be managed? The rationalist would argue that all elements can be reduced to an NPV calculation through either market or shadow pricing, but this poses serious problems in practice:

- PV calculations are themselves open to considerable uncertainties and hence debate – the future income streams are expectations, not certainties.
- There is widespread lack of awareness among clients regarding the impact on business process performance of spatial configuration and IEQ factors – arguably, there is a serious problem of market failure here.
- Stakeholders are rarely unanimous in their valuation of shadow prices.
- Disaggregating the PVs of the components of the facility is a treacherous process – is the higher rental stream realised for the upmarket development designed by a renowned architect because of the branding effect of the name, the play of light in the entrance lobby, or the prime location in the city? Successful project promoters and developers have a gut feel for these factors, not an Excel spreadsheet.

Judgement is essential in investment appraisal – experienced bankers would argue that only around 30% of the information required for a decision is in the figures[29]. The rest is down to judgement based on experience and the ability to play the political game of strategic misrepresentation.

3.9 Summary

Defining the project mission *is* difficult. The most intuitive way of doing it is to point to existing similar facilities as models of either what is wanted or what is not wanted. In the absence of good analysis about how constructed facilities add value for clients by enhancing the performance of their operational processes – be they the celebration of God or mammon – this is, perhaps, the best that can be done, but it opens the opportunity for the strategic misrepresentation of the business case. Most clients still treat buildings and infrastructure as artefacts that are costs rather than assets that yield returns. This chapter has indicated some of the recent research that is starting to offer a better insight into how buildings do create value for clients through providing spatial environments that enhance employee efficiency and effectiveness, attract customers and symbolise values, yet there is a long way to go before product integrity in the constructed product is understood as well as it is in other industries.

We have also stressed the importance of symbolic value, which is closer to the ethical dimension of that word. Constructed assets are much more than utilities – they are also expressions of our culture, and always have been. It is for good reason that high-tech companies such as Dyson and Linn commission avant-garde architects to build their factories, but those reasons cannot be captured in a balance sheet. They are about the intangibles of image and confidence. Building high – be it medieval Bologna or twenty-first century Dubai – has always been an expression of the competitive virility of an economy and society. Building grand is an expression of a common, or imposed, culture. This symbolism can never be fully captured analytically, only interpreted subjectively. This does not mean, however, that the expression of symbolic values is not a central element of the project mission. For some buildings – particularly religious and cultural buildings – the symbolism is all. Even for hard-headed business people, symbolism is very important when they wish to express their corporate values in the facilities where their staff work or customers visit.

This chapter has articulated the definition of the project mission as a problem in strategic management, explored how we can move from thinking in terms of artefacts to thinking in terms of assets in investment appraisal and identified the vexed issue of strategic misrepresentation in such appraisals. The concept of product integrity was introduced to capture the three dimensions of quality in the project mission. The role of the project manager is to facilitate the client in coming to an appropriate definition of the project mission that can then be championed through the project life cycle to deliver value for the client. At various points, references have been made to the differing interests of the stakeholders in the project and it is to this issue that we now turn to in Chapter 4.

Case 3
Defining the Mission at the University of York

The University of York – one of the UK's top 10 universities – is among the smaller of the universities founded in the great expansion of the British university system during the 1960s. Built on swampy ground on the outskirts of York, the campus is arranged in seven colleges around an artificial lake. Three categories of buildings were initially constructed:

- simple, robust buildings with little flexibility such as halls of residence organised around a basic 'pantry' of six study/bedrooms;
- simple, robust buildings with high flexibility to house teaching departments – single storey with raised floors so that services could be easily redirected and with continuous top daylighting to allow internal reconfiguration;
- symbolic buildings of architectural distinction – either new build such as the Central Hall and J. B. Morrell Library, or sensitive refurbishment of existing, listed buildings such as King's Manor and Heslington Hall.

The development of the built stock of the university has gone through three phases:

- 1962 to 1972 – rapid expansion following the master plan, mainly new build using the CLASP system of prefabricated construction for the simple, robust buildings, and a traditional 'one-off' approach for the symbolic buildings;
- 1973 to 1992 – consolidation, with a focus on adaptation of the facilities built during phase 1 to meet changing needs in a context of major funding constraints;
- 1993 to 1998 – renewed expansion in strategically chosen areas, with both new build and refurbishment using traditional procurement and materials.

The principal players in all three phases were the client, represented throughout by the same person, the architects RMJM, again with continuity of personnel, and the locally based contractor Shepherd. Overall the programme can be considered to have been successful – the phase 1 buildings were delivered to time and cost, thanks to the choice of prefabrication, and were favourably reviewed by the architectural community, even if there were a number of minor problems on handover. As a student at York in the 1970s, I can attest to their fitness for purpose.

Commitment to prefabrication was already waning even at the end of the first phase, as its costs rose compared to traditional methods, and construction switched from prefabricated panels to blockwork for internal partitions. This shift was reinforced for new build during the later phases as a major design criterion became that the new buildings should not look like the existing ones; also, cost-in-use considerations came to the fore with the poor thermal performance and relatively costly maintenance of the CLASP system, encouraging a shift to traditional brickwork for external cladding, with pitched and tiled roofs.

Despite the continuity of people and partners over the 35-year programme, there appears to have been almost no attempt at organisational learning to activate the facilities feedback loop shown in Fig. 9.2, by learning how the buildings were performing in use. Two exceptions were a study of laboratory use in 1967 and a student survey in 1983, both of which broadly indicated high levels of satisfaction. Otherwise, little has been done formally, except on energy costs. Only in 1995 was the original master plan of 1962 revisited and its relevance

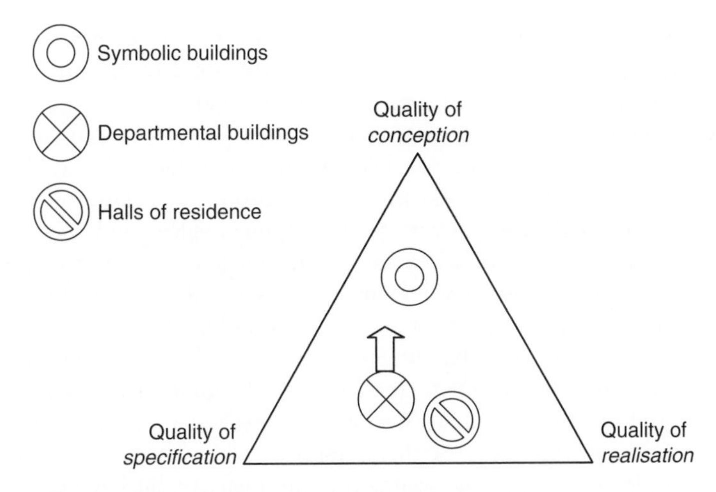

Fig. 3.3 The project mission at the University of York.

assessed, and then detailed evaluations of the performance in use of selected buildings were made using the Building Use Studies methodology discussed in panel 3.5. By then, the estate consisted of 1.5 m m^2 of accommodation on a campus of 9.2 ha.

The development of the mission for the University of York projects over the period can be illustrated by the application of the product integrity model from Fig. 3.3. While the mission for the halls of residence has stayed largely the same through the development programme, there has been an evolution in those for the departmental buildings, with greater emphasis being placed on quality of conception issues in later developments as client representatives sought to distinguish their new buildings from the existing ones on the campus. The missions for the symbolic buildings differed in that they included refurbishments of listed buildings from different eras of architectural history, a spaceship style Central Hall and a library deriving much of its impact from its high location. However, they can be broadly grouped together in the manner shown.

A number of important lessons can be identified from the experience of the university over more than 30 years (all citations are from contemporary documentation):

- Project development needs to take place within a clear corporate strategy for the client, and the role of constructed assets within that strategy. Thus, 'a university should be a society of people living and working together to advance learning and spread knowledge at the highest level, and further that it must embrace those who teach as much as those who learn and those who work to extend the limits of knowledge. It is essentially a meeting place . . . and while the social and academic life it fosters must be coherent it must not be self-contained'.
- A transparent framework for project development is required, which sets out expectations, procedures and performance measures to guide decision-making and allow evaluation and improvements. For the University of York, the

university funding bodies laid down clear yardsticks of acceptable expenditure levels, and the Development Plan of 1962 and Development Plan Review of 1995, both established appropriate frameworks.

- A clear project sponsor within the client side who can authoritatively articulate the essence of the brief is vital. During phase 1, this was the vice-chancellor himself who provided a vision of what the new university stood for through the university development plan, and a sounding board for decision-makers of when he 'would not like it'. Projects during phase 3 had clear departmental level sponsors. Thus, the Psychology Department's sponsor summed up the brief for the new psychology building as a 'distinctive, stand-alone architectural building with its own front door entrance and a building block which was not to look like the [adjoining] James College Block'. The brief for the Computer Science Building was to 'stimulate the imagination of the Department's researchers, respond to the enthusiasm of the Department's students, signal to the visitor that the Department has a sense of purpose and intellectual distinction, create a sense of belonging in those who work in the building, and reflect the best of British design and craftsmanship'.
- Facilities need to be capable of adaptation and change, as future use is impossible to predict, even in one of the most stable organisational types of the last millennium – with its roots deep in The Enlightenment – the research university. Thus, the 'future of an organisation so complex and liable to change as a university can never be predicted except within broad limits'. Even facilities as eternal as residential accommodation had to adapt to two distinct changes during my time at York – the abolition of single sex accommodation (with important implications for services such as showers) and the conversion of all shared study bedrooms into single ones. These changes were driven both by student agitation and increasing attention to the conference market during vacations.
- Feedback loops from the facilities in use are essential on a routine basis, so that lessons can be learned for future projects. Even for a performance criterion as central as energy usage, data are not available at York broken down by building, only for the university as a whole. The data from these feedback loops should then be subject to continuous review.
- Consistency of medium-term funding is essential. Development projects take up to 5 years to reach fruition, and funding horizons shorter than this lead to the abandonment of projects and associated abortive work, and expensive compromises and changes in brief are made during project execution.

Source: developed from Phiri (1999).

Notes

1 Interview, Financial Times 19/09/95
2 Ramirez (1999).
3 Smith (1970, p. 131).

4 See the useful review by Thompson *et al.* (2003), and Rouse (2004) and Macmillan (2006) in particular.
5 See Barber (1967) for an account of these developments.
6 See also Evans *et al.* (1998) who developed the widely cited 1:5:200 ratio, Spencer and Winch (2002) for a review of this literature and Ive (2006) for a critique of these types of calculation.
7 Kaplan and Norton (1996, p. 22); this line of argument has long been in gestation at Harvard Business School.
8 *Financial Times*, 09/11/01.
9 See the comprehensive review by Ulrich *et al.* (2004).
10 Johnson and Scholes (1999, exhibit 1.1)
11 The citations are all available at http://www.architecture.com (accessed 10/04/07).
12 Giles Worsley, 'Give it to Us Sexy, Shiny and in Public'. *New Statesman*, 26/09/05.
13 http://www.channel4.com (accessed 11/04/07); *BBC News*, 15/10/05. This was before the news was announced that extensive refurbishment of some of the artistic features would be required (*Building*, 16/05/08).
14 These were first introduced in Spencer and Winch (2002).
15 See Stanwick and Fowlow (2006) for a thorough review of other winery projects by signature architects.
16 http://www.mclaren.com/technologycentre/ (accessed 28/07/08).
17 The promotion of *Better Public Building* (HM Government 2000, 2006) in the UK is one example of the renewed emphasis on design by public sector clients – the remarkable Civil Justice Centre in Manchester (Spring, 2007) is an excellent example of this.
18 See Carmona (2004) for one analysis.
19 The French is 'fait du Prince'. Chaslin argues that architecture is 'surtout l'enjeu le plus visible, le plus spectaculaire, des ambitions politiques, qu'elles soient positive et témoignent un désir du porter une empreinte de laisser un trace d'un passage au pouvoir' (Chaslin, 1985, p. 13) [the stake the most visible, the most spectacular, of political ambitions, even if they be positive and witness a desire to make a mark, to leave a trace of a period in power].
20 Henderson (1987) most clearly articulates this view.
21 Hillier (1996, Chapter 6).
22 See, for instance, HM Treasury Green Book.
23 Cherns and Bryant (1984) were the first to comment on this organisational dynamic, while the term strategic misrepresentation is from Flyvbjerg *et al.* (2002).
24 See, for instance, Magnussen and Olsson (2006).
25 Department for Transport (2004).
26 Flyvbjerg (2006); see also Kahneman and Lovallo (1993) on the 'outside view'.
27 Flyvbjerg *et al.* (2003).
28 Clark and Fujimoto (1991, p. 340).
29 Interview in the *Financial Times,* 19/09/95.

Further reading

Richard A. Brealey and Stewart C. Myers (2006) *Principles of Corporate Finance* (8th ed.). New York, McGraw-Hill.
A standard reference on corporate finance that covers well the issues in investment appraisal.

Derek Clements-Croome (ed.) (2000) *Creating the Productive Workplace*. London, Spon.
A stimulating collection of research papers on various aspects of IEQ and their impact on the efficiency and effectiveness of building occupants.

Bill Hillier (1996) *Space is the Machine*. Cambridge, Cambridge University Press.
Redefines the architectural problem – and hence the problem of product integrity – in terms of the spatial configuration of domestic, commercial and urban form.

Chapter 4
Managing Stakeholders

4.1 Introduction

'delivering some chunk of mastodon meat back to the tribe'

The previous chapter treated the process of defining the project mission as if it were a problem that one decision-maker – the client – could solve in isolation from the other stakeholders on the project. This happy situation is rarely the case, and so the client will need to manage the other project stakeholders in order to see its project through to successful completion – a problem of some considerable delicacy at times. The Boston Central Artery/Tunnel project presented in Case 13 faced enormous challenges of stakeholder, and Fred Salvucci[1], the Boston Secretary of Transportation who championed the project, allowed himself to express a fairly cynical view of the gaming by external stakeholders around the project to bury the elevated highway through the centre of Boston in a tunnel.

This chapter will address the management of stakeholders on construction projects from the client point of view. First, the stakeholders will be defined, before techniques for mapping and analysing the differing positions of the stakeholders are presented. The role of the regulatory context in institutionalising the voice of some of the weaker stakeholders will follow, before a brief closing discussion of the ethical issues in construction project management.

4.2 Which are the project stakeholders?

The project stakeholders are those actors which will incur – or perceive they will incur – a direct benefit or loss as a result of the project[2]. Construction projects create new value, but they can also destroy value – noise and dust disturb local residents during construction; amenity (here defined as social utility) is permanently lost as a result of construction as symbolically important buildings are demolished and the landscape is changed. Moreover, where existing facilities are replaced, stakeholders in those facilities may not share in the value generated by the new one – jobs may be lost and profit opportunities move out of reach. This problem is common, for instance, when fixed links replace ferries across rivers and sounds.

It is useful to categorise the different types of stakeholder in order to aid the analysis, and hence management, of the problem. A first-order classification places them in two categories – *internal* stakeholders which are in legal contract with the client, and *external* stakeholders which also have a direct interest in the project. Internal stakeholders can be broken down to those clustered around the client on the demand side and those on the supply side. External stakeholders can be broken down into private and public actors. This categorisation, with some examples, is shown in Table 4.1.

Table 4.1 Some project stakeholders.

Internal stakeholders		External stakeholders	
Demand side	*Supply side*	*Private*	*Public*
Client	Architects	Local residents	Regulatory agencies
Financiers	Engineers	Local landowners	Local government
Client's employees	Principal contractors	Environmentalists	National
Client's customers	Trade contractors	Conservationists	Government
Client's tenants	Materials suppliers	Archaeologists	
Client's suppliers		Non-governmental organisations (NGO)	

A fundamental premise of construction project management is that the client is capable of fully articulating all the stakeholder interests on the demand side – in other words, the client has the capability to authoritatively brief the team. Yet, as the demand side list shows, the client is a complex organisation and may not be fully aware of the range of demand side stakeholder interests, particularly from its employees and tenants. As the discussion of IEQ in section 3.4 indicated, the needs of building users are often not fully understood or articulated by clients. Different interest groups within the client organisation may also have different functional requirements, meaning that any project definition is a compromise which may unravel as more information becomes available to those groups through the project life cycle regarding what the facility will be like. Similarly, financiers may have a different view from the client's employees of what is important in an NPV calculation.

On the supply side, a whole coalition of interests is arrayed. As illustrated in Fig. 1.1, the supply side receives its benefit through the income stream generated by working on the project, and the learning acquired through solving project problems. For those organisations whose marketing is reputation rather than bidding based – typically those working on a fee basis – the reputation generated through working on the project is one of the biggest benefits gained. So, it is no accident that those suppliers which specialise in meeting the requirements of the symbolic aspects of the building – particularly architects – sometimes appear to care more about their own reputation than meeting client needs, for it is embedded in the structure of the market for architectural services. These differing stakeholder perceptions are well illustrated by the Worldwide Plaza project, in panel 4.1. It is immediately clear that there is an inherent conflict of interest between the stakeholders on the demand side and on the supply side as they compete to appropriate the income stream from the project – what Michael Porter[3] calls margin in the value system. Much of this book is about how this conflict is mitigated to allow both the demand and supply sides to meet their joint and several objectives through a win–win rather than zero-sum game.

Among the external stakeholders, there is more diversity. By and large, the internal stakeholders will largely be in support of the project, although there may be factions within the client which are backing alternative investments. External stakeholders may be in favour, against or indifferent. Those in favour may be local landowners who expect a rise in the value of their holding, and local residents supporting a rise in the general level of amenity. Those against may also be local residents and landowners who fear a fall in amenity and hence

Panel 4.1 Supply side stakeholder views on the Worldwide Plaza project

On the Worldwide Plaza project the principal participants had very different views about the key to project success. For the architect, it was whether his vision had been realised and his reputation thereby enhanced:

> 'Architects deal in all sorts of magical materials and models and mirrors and two-dimensional drawings that are made to look three dimensional. We can never work in the final medium of our art, as painters and sculptors do, so it's frightening to see the final thing come together being crafted by other hands than your own. And not everything is right, things change constantly and so we're constantly losing pieces of the project. There's a constant fight to keep what you can . . . It's a frustrating and slightly scary process to go through'.
>
> For the client it was about whether the investment makes a return:
>
> 'What it comes down to is pieces of paper, numbers, internal rate of return, the net present value, discounted cash flows, that's what it's all about. It's not about whether or not the construction manager has gone from here to there or vice versa. It's not about the architect becoming the rock star of architecture . . . What it's about is dollars and cents. Sure, we want to build quality and we want to build something that is going to be a statement . . . but what it boils down to is whether it is financeable and whether there is a return to the [client] partnership'.

For the construction manager it was about getting the job done:

> 'Our major responsibilities are not really to design the building or to critique the building. We are hired to build the building. Obviously we all form opinions of the building [but] . . . there isn't much that we could do once the building is set.
>
> You can't tell an architect not to draw something. If he wants architecture and the owner's willing to pay for it, hey, it's our job to execute it'.

Source: Sabbagh (1989, pp. 230; 291; 65; 247).

the value of holdings. Splits may well occur among these groups – if someone lives 1 km from a proposed motorway junction, they may have a very different view than if they live 100 m away. Such objectors are known as NIMBYs (not in my back yard). Environmentalists and conservationists may take a more principled view than local losers, while archaeologists are concerned about the loss of important historical artefacts.

The public external stakeholders – in those situations where the public sector is not also the client – will tend to be indifferent. The agencies which enforce regulatory arrangements such as those for urban zoning, quality of specification (the construction codes) and heritage assets will tend to be indifferent to any particular project definition, so long as it complies with the codes. National and local government may, however, wish to encourage development, particularly in regeneration areas. At times, there may be conflicts of interest within the public sector between its promoter and regulatory roles, as will be explored in Case 9.

4.3 Mapping stakeholders

The first step in managing the stakeholders is to map their interest in the project – the essence of the technique is to identify the perceptions stakeholders themselves hold in order to identify potential levers for action. This can be done using the framework illustrated in Fig. 4.1[4]. The focus of the approach is the project mission as represented by the asset to be created – it is the asset rather than the mission itself which tends to be the source of contention between stakeholders. Stakeholders can be considered as having a problem or issue with the project mission, and as having a solution (tacit or explicit) that will resolve that problem. Where such solution proposals are inconsistent with the client's proposals, they can be defined as being in opposition to the project. An important part of stakeholder management is to find ways of changing opponents to supporters by offering appropriate changes to the project mission, and preventing possible supporters defecting to the opponent camp by offering to accommodate more explicitly their proposed problem solutions. The role of such mitigations is explored in Case 13.

Once the stakeholder map has been drawn up, the power/interest matrix can be used to develop a strategy towards managing the different stakeholders. It consists of two dimensions – the power of the stakeholder to influence the definition of the project, and the level of interest that the stakeholder has in that definition[5]. Both dimensions are better perceived as continua between poles rather than binary options. The level of interest is conceptually simple – it is a function of the size of the expected benefit or loss from the project. As discussed in section 16.6, power is a more slippery concept but for our purposes here it can be considered as the ability to influence the project definition process for a given project – the overt face of power. The ability to set the agenda for the project definition process or whether certain types of project are simply not proposed, are broader socio-economic and political questions beyond the scope of this book.

This matrix categorises the stakeholders into one of four types, but the discussion here can only be indicative – where a particular stakeholder sits in

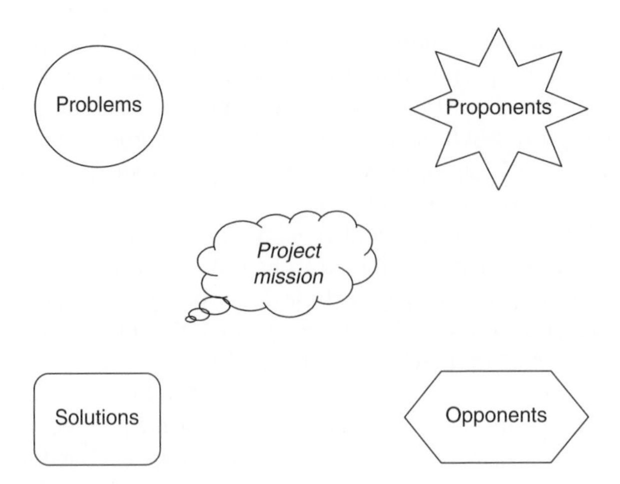

Fig. 4.1 Mapping stakeholders (source: Winch and Bonke, 2002).

relation to the project depends entirely on the specific context of that project. The first group is those which require *minimal effort*, such as the client's customers, or local and national government. A public relations approach to this group will often suffice, aimed at ensuring that those which might be opposed to the project stay in the low-interest category, while those which are likely supporters are tempted to move to the high-interest category.

The second group is that which needs to be *kept informed*. Groups which may be opposed to the project, such as local residents, conservationists or environmentalists need to be carefully managed. If such groups coalesce into well-organised movements, and are able to mobilise the press behind them, then they may well be able to move into the key player category causing the project severe disruption, or even cancellation. To a certain extent, such groups can be bought off to prevent this happening, with inevitable consequences for the NPV calculation. For instance, it is now standard practice for clients building in the City of London – which contains important Roman and medieval remains – to finance an archaeological dig prior to works commencing on site. Similarly, the concept of planning gain within the UK regulatory system is common, where a project promoter provides additional utility for the benefit of the local community to defuse potential opposition. Some groups – typically environmentalists – cannot, however, be bought off and can go on to disrupt physically the project during execution on site. The impact of these groups is two-fold:

- they upset the NPV calculation as a result of delays in the schedule and additional costs of security;
- they dissuade future clients from coming forward with similar projects.

Those which need to be *kept satisfied* usually fall into two main groups – regulatory bodies and the supply side stakeholders – which require very different management approaches. Regulatory bodies are, in essence, the institutionalised interests of the low power stakeholders. They provide forums in which local residents, landowners and government can have their voice (planning enquiries), in which the safety of clients, employees, tenants and customers is ensured (construction codes), and in which environmental and conservationist interests are heard (environmental impact assessments, EIAs). The latter even allows the purported claims of stakeholders which do not yet exist – future generations – to be heard. The first task of the client is to ensure compliance with the regulatory requirements, supported by lobbying tactics where the requirements are open to interpretation. The project management of supply side stakeholders is the subject of Part III. They are placed in this category, rather than the key players category, for two reasons. Firstly, most of them are mobilised after the project mission is defined; secondly, as shown in Fig. 1.4, they will typically have a portfolio of projects at any one time – while their power to influence the outcome of any one project is very high, their interest in the definition of any one project mission is typically limited.

The final category is that of the *key player*. Here the client is central – the analytic questions revolve around which of the other demand side stakeholders are also in this category. Where finance is raised from the traditional sources – equity and debt secured through a floating charge for the private sector, and the taxpayers for the public sector – then financiers are typically in the keep satisfied category. However, where project finance techniques are used as the source of capital – as is increasingly common in both the public and private sectors – then such financiers move into the key player category. In commercial property development where the asset is pre-let, the client's tenant can become a key player in definition, while in the provision of social housing this is rarely the case. The client's customers are also usually in the key player category, but through the proxy voice of the corporate marketing department – if it misunderstands the market for the facility, then it is unlikely to be successful because those customers will simply use other competing facilities. Whether the client's employees are in the low or high power categories depends on the internal organisation of the client, and its understanding of its business processes.

The degree of *integration* of the stakeholder map will make a large difference to its manageability. If the stakeholders are at the far corners of Fig. 4.4 (in the case study at the end of the chapter), then the definition process is likely to be turbulent and the stakeholder map unstable. If the stakeholders are clustered near the centre of Fig. 4.4, then the map will appear as relatively stable. The dispersion of the different stakeholders in the power/interest matrix will indicate the options of manoeuvrability in the project manager's decision and planning processes, and the ability to broker compromise by renegotiating the project mission. The application of these two mapping techniques is illustrated in Case 4.

4.4 The regulatory context

As discussed in section 2.4, the regulatory context of the nation state in which the constructed asset is to be located sets important parameters on both the definition and execution of the project. Our focus here is on its impact on definition, which broadly covers four areas:

- implications for the natural environment;
- implications for the rural or urban context;
- implications for the safety of users;
- the broad set of issues under the rubric of sustainability.

A fifth important area of regulations concerns those regulating the labour market and working conditions of staff; these are discussed in section 7.3 and Chapter 12.

4.4.1 *Environmental impact*

Environmental impact assessments (EIA) are a developing tool in understanding the implications for the natural environment of realising the project. A new dam can have a profound impact on the ecosystem of a region[6] and displace farmers and villagers; a new road may pass through woodlands valued by local walkers, and generate noise and pollution. Within the European Union (EU) EIAs are now obligatory on the grounds that:

> 'the effects of a project on the environment must be assessed in order to take account of concerns to protect human health, to contribute by means of a better environment to the quality of life, to ensure maintenance of the diversity of the species and to maintain the reproductive capacity of the ecosystem as a source for life'[7].

A central element of the EIA process is the assessment of alternative definitions of the project – whether a tunnel would have less impact than a bridge, whether the road should go round or through a particularly sensitive area or whether the whole thing is a bad idea in the first place. Indeed, as Case 13 shows much bargaining with external stakeholders revolves around such issues.

The acceptance of the legitimacy of such an approach within the advanced countries has led to attempts to extend their principles to developing countries through the financiers, rather than clients of projects. The major international project funding organisations such as the World Bank also have requirements for EIAs on the projects they finance. This has the effect of giving stakeholders which are unable politically to express their opposition to the project within their country the opportunity to influence it by informing and lobbying stakeholders which would otherwise have no interest in the project, as indicated by the case of the Ilisu Dam in Turkey, presented in panel 4.2.

4.4.2 *Urban and rural context*

While EIA issues tend to have implications mainly for civil engineering, a longer tradition of the regulation of projects is the attempt to conserve and control the evolution of the urban and rural environment through the control of building.

Panel 4.2 Stakeholder power on the Ilisu Dam

The Ilisu Dam in the troubled Kurdish region of Turkey was first mooted in 1954. In the 1980s, a consortium led by the Swiss firm Sulzer Hydro, including Balfour Beatty (UK), Impregilo (Italy), Skanska (Sweden) and three Turkish companies contracted for the project but it failed to meet the recommended standards of the World Commission on Dams (WCD). Launched by Nelson Mandela, the WCD is backed by the World Bank and the International Union for the Conservation of Nature (IUCN). The proposed £2bn hydro-electric dam is to be built on the Tigris River; it is claimed that it will flood 15 towns and 52 villages, while displacing approximately 78 000 Kurdish people. Export Credit Guarantees were sought from the governments of Germany, Switzerland, Italy and the UK – £200m from the latter. The Friends of the Earth (FoE) decided to show how serious they were about challenging corporate attitudes to the project by buying £30 000 worth of shares in Balfour Beatty, hence shifting the Kurdish interests as private external stakeholders down the power/interest matrix from 'Keep informed' to 'Key players' by demonstrating at Balfour Beatty shareholder meetings. Thus they provided a public lobbying platform, which the supply side stakeholders could not ignore, while raising opposition from other 'Proponent' stakeholders, many of which would rather opt out of funding the project than receive a bad press. Many calls were made to the UK Government to put a stop to this 'dam disgrace'. In November 2001, Balfour Beatty announced that it was pulling out of the project after conducting its own internal EIA. Skanska had already pulled out, and Impregilo co-ordinated its withdrawal with Balfour Beatty. However, in 2005 the project was revived by a consortium led by Siemens and is now the subject of a renewed campaign of opposition.

Sources: http://www.foe.co.uk; http://www.ft.com; http://www.dams.org; http://www.constructionplus.co.uk (accessed 23/05/01); *Financial Times*, 14/10/01; http://www.ilisu.org.uk/ (accessed 30/07/08).

Developments that might not have any environmental impact under the definition above may still threaten the rural or urban environment. Most advanced countries have zoning procedures which specify what sort of facilities may be sited in a particular area; typical categories which are placed in separate zones are residential, commercial and industrial facilities. Such zoning regulations may be either 'plan led' and provide absolute constraints on project definition, thereby making it a futile exercise to propose to build an office block in a zone reserved for housing; or, they may be more flexible and allow clients to negotiate within constraints the types of projects that might be allowed.

Linked to the overall rural and urban planning regime may be specific protections for particular quarters of a city (conservation areas in UK parlance), particular parts of the landscape (sites of outstanding natural beauty in UK parlance) and particular existing buildings (listed buildings in UK parlance). Levels of protection may also vary from a complete interdiction on change to the requirement that any development should be 'sensitive'. Such specific protections can place considerable constraints on the quality of conception, restricting innovations in design and specifying the use of vernacular materials. Most advanced countries have very active non-governmental organisations (NGOs) which are vocal and active in protecting such buildings, and which can become key players in the project definition process.

4.4.3 Safety and amenity of users

The third category of regulations has a direct impact on the quality of specification. Most authorities specify minimum standards that facilities should meet covering a number of policy areas such as:

- structural integrity, where the authorities are concerned to protect the safety of the users of the building and passers-by;
- access standards for less able users;
- minimum space standards for particular activities;
- ensuring energy efficiency and the use of sustainable materials.

Figure 2.3 provides a comparison of some of the different European regulatory frameworks for the safety and amenity of users.

4.4.4 Sustainability

The issue of sustainability has been rising rapidly up the agenda of policy-makers over the last decade. Spurred by international agreements such as the Kyoto Protocol of 1997, and supported by growing scientific evidence on climate change, governments are increasingly taking action to improve the sustainability of facilities – one national response is presented in panel 4.3[8]. The introduction of a virtual stakeholder – future generations – adds another dimension to the definition of the project mission. There are few incentives for decision-makers to take this stakeholder into account, and so, like most low power stakeholders, future generations rely on regulators to express their interests. Thus the main impact of the sustainability agenda on the management of construction projects will be through the regulatory context discussed in section 2.4.

The issues involve both how the product and process should change to enable greater sustainability so that climate change is minimised, and how specification practice should change to meet the expected changes in climate. For instance, by 2050, the UK is expected to be significantly warmer, with wetter and windier weather in autumn and winter. This has important implications for existing as well as new facilities. Most constructed assets which will exist in 2050 already exist and will need to be adapted to respond to problems as varied as a greater tendency for mould growth in winter, higher wind loadings and faster solar degradation of materials.

4.5 Managing consent

The management of consent within the regulatory framework is a strategic matter within project definition, particularly with respect to the first two categories of regulations; for projects such as the fifth terminal at London's Heathrow Airport,

Panel 4.3 Construction sustainability targets

Announced in June 2008, the Strategy for Sustainable Construction England (this is an area of devolved responsibility) set the following targets with dates of intended achievement:

Product

- All new homes to be zero carbon (2016)
- All new buildings to be zero carbon (2019)
- Water consumption in homes to an average of 130 L per person per day (2030)
- Central government office estate to be carbon neutral (2012)
- Reduce water consumption in central government estate by 25% on 2008 (2012)

Process

- Reduction in water consumption during construction by 25% on 2008 usage (2012)
- All construction projects over £1m to have a biodiversity survey (2012)
- Reduction in construction packaging waste by 20% (2012)
- Reduction in construction site waste to landfill by 50% (2012)
- Reduction in carbon emissions from construction processes and transport by 15% on 2008 (2012)

Source: HM Government (2008).

the management of consent amounts to a major project in its own right, as shown in panel 4.4. Three basic approaches are possible[9]:

- define and enquire;
- consult and refine;
- bribe and ignore.

Where the regulations are unambiguous and prescriptive, then the *define and enquire* approach is appropriate – the codes are published, and simply require to be interpreted. The *consult and refine* approach is more appropriate where codes are not prescriptive or where there are significant uncertainties. As discussed in section 2.5, a notable feature of the ways in which regulatory systems vary between countries is the extent to which they are prescriptive. For instance, the French building codes are very performance orientated compared to other countries, and hence open to interpretation. Similarly, the British zoning regulations are more open to interpretation and negotiation than those of many continental countries. In both countries, clients appoint specialist advisors to negotiate with the regulatory system. In France, *bureaux de contrôle* are appointed by the client to ensure that the designs prepared by both consultants and contractors comply with the regulations, as described in panel 2.1. In the UK, an important reason for the relatively late appointment of the contractor in the project process is the necessity for a relatively large amount of design work to be completed to facilitate negotiations with those responsible for safeguarding the zoning plan. Until such negotiations have been completed, the uncertainty around the definition of the project makes it appropriate to use fee-based remuneration of independent designers.

Panel 4.4 The fifth terminal at London's Heathrow Airport

BAA's proposal to build a fifth terminal (T5) – presented in Case 12 – at London's Heathrow Airport, on a 121-ha Green Belt site inevitably came under much opposition. This came from a variety of groups, including local inhabitants, community groups and local councils as well as action groups such as West London Friends of the Earth (WLFoE) and Heathrow Association of Control of Aircraft Noise (HACAN), with issues ranging from noise pollution to increased levels of traffic. As a result, the longest planning inquiry ever held in the UK started in May 1995 and finished in March 1999. Its findings were reviewed by the Inspector, who has in turn passed on his report to the Government which approved the development in November 2001 while setting a cap on the maximum number of flights operating from the airport. BAA justified the necessity for T5 by claiming that the number of passengers wanting to fly would double over the next 15 years. It predicted that the passenger flow would increase from 60m per year to 80m, requiring appropriate facilities to meet the demand. In order to manage this consent process, BAA continuously redefined the project in response to opposition from stakeholders, by addressing the key issues and taking action to incorporate them into the project mission. Such issues were dealt with by an array of teams set up by BAA – public opinion communicators, public inquiry representatives, environmentalists, anti-airport mediators and so on. T5 opened in March 2008, nearly 13 years after the start of the regulatory process.

Sources: http://www.heathrow.co.uk/main/corporate; http://www.money.telegraph.co.uk; http://www.wlfoe5.demon.co.uk; http://www.thisislocallondon.co.uk; http://www.news.ft.com; http://www.business.com (accessed 05/06/01); *Financial Times* 21/11/01.

The *bribe and ignore* strategy is unfortunately widespread – zoning codes are routinely ignored in many countries as recent tragedies where shanty towns have been engulfed by mudslides show. Recent earthquakes in a number of countries have indicated widespread ignorance of – or at least failure to implement – the codes relating to structural integrity. Bribery is also widespread in many countries to obtain zoning consents. These problems are symptomatic of a more general malaise in the political system of the country where the asset is to be constructed, and raise the problem of business ethics which are discussed in section 5.8.

4.6 Ethics in project mission definition

Corporate social responsibility can be defined as the extent to which 'an organisation exceeds its minimum required obligations to stakeholders'[10]. One of the strongest arguments for an ethical approach to business is the damage to the brand that can occur when corporations fail to act responsibly, so much of the debate about business ethics has concerned the consumer goods sector. The immediate self-interest in a business-to-business sector such as construction is less clear, but nonetheless persuasive. Perhaps project social responsibility can be defined as:

> 'the extent to which the project definition exceeds the minima established in the NPV calculation and those required to obtain regulatory consents'.

Such a project ethic has a number of possible dimensions, including:

- refusing to fund or participate in a project that has not been through a full EIA, as the World Bank now does;
- refusing to accept or give bribes;
- experimenting with new or more sustainable designs and materials, such as Essex County Council did with its new primary school at Notley Green;
- providing superior facilities for the construction workforce, as Marks and Spencer committed to doing after a main board director was caught short during a site visit and was horrified by what the operatives had to use – see Case 5;
- providing public art as part of a new development, as Rosehaugh Stanhope did on the Broadgate development in the City of London;
- preserving the character of the existing building during refurbishment;
- commissioning exciting and original architecture;
- participating in performance improvement programmes such as the UK's Constructing Excellence networks.

4.7 The role of visualisation

The diversity of the stakeholders in project definition and the necessity to keep stakeholders informed and satisfied have encouraged the development of visualisation techniques for communicating. Of course, the classic visualisation for the professional is the sketch or scale drawing – see panel 14.1 – but many stakeholder representatives are not trained to read architectural and engineering drawings. The 'artist's impression' is a classic visualisation technique, placing a proposed facility in its context and populating it with occupants and passers-by. At least since the Renaissance, wooden scale models have been used of buildings both as ways of explaining to clients what their new building will look like inside and out, and also as ways of communicating design information to those who will have to construct it. In naval architecture – at least for nuclear submarines – it is a 1:5 plastic scale model of the vessel that is signed off for construction, not scale drawings[11].

A more recent development is the photomontage, where a photograph of the site is taken from a known grid point, and a perspective drawing of the proposed facility is then superimposed on the photograph. This technique is much favoured by those responsible for safeguarding the urban and rural environment, because it allows non-specialists to quickly grasp the impact of the proposals on the existing scene. The development of computer-aided design (CAD) techniques and digital grid referencing has made this technique all the more effective. The capabilities of the latest generation of building information models described in panel 14.4 are taking these capabilities to a whole new level.

The development of virtual reality (VR) interfaces to CAD has pushed the potential of the visualisation of the proposed facility for the benefit of clients and other stakeholders much further since the early 1990s. The development and diffusion of CAD systems capable of modelling the proposed facility in three

Fig. 4.2 Centuria building visualisation (source: University of Teesside).

dimensions have meant that a visualisation of the facility can be projected on to a large screen. This, potentially, can alter the dynamic of the design review completely. Design reviews typically consist of a large table strewn with drawings or a small-scale model, supported by pinned-up drawings and sketches. The much greater sense of scale available from large-screen projection allows a much more intense experience for the client and others as they explore what the designers are proposing[12]. Taking the audience through the 3D model in a 'fly-thru' can give an additional sense of reality, as illustrated in Fig. 4.2. This approach has been taken furthest in the CAVE which typically has projection on five walls of a cube, further enhancing the feeling of actually being in or outside the proposed facility. Panel 4.5 shows how these advanced computer-aided visualisation techniques can be used by clients to convince stakeholders of the merits of proposed projects.

4.8 Summary

Construction projects – particularly major ones – are highly contentious. Few people welcome a housing estate where they used to walk their dog, or a motorway flyover past their bedroom, although even here there are those who find opportunity in apparent adversity[13]. The days are long past when project promoters – particularly if they were state agencies – could ignore the interests of the various stakeholders on the project. Managing project stakeholders has become much more challenging over the last 30 years or so for two very different sets of reasons:

- External stakeholders now have much more power in the process. This is manifest in both the growing institutionalisation of external stakeholder rights through an ever tightening regulatory context, and, following the collapse of socialist mass movements, the rise of environmental activism.
- The shift to concession contracting with finance secured on the asset being created by the project. While financiers used to fund projects through loans to corporate bodies – be they public or private – they now fund projects directly. As a result, they are now paying much more attention to the definition of the project mission to ensure that their investment will actually yield the promised returns.

Panel 4.5 The Centuria building VR model

The award-winning Centuria building opened in 2000 to house the University of Teesside's School of Health. Details of the project and its management are presented in Cases 10 and 11. Here, our focus is on the way VR techniques were used to convince stakeholders to fund the project.

A major stakeholder in the School of Health is the UK's National Health Service (NHS), which commissions the supply of skilled health staff from the university, and provided the funding for the project. The university invested in a VR model of the proposed new health education facility in early 1999 to show the NHS their vision of health education in the future, part of which is shown in Fig. 4.2. The VR model helped the successful funding of the project. The same VR model was then developed to allow the users of the building – represented by the different subject leaders – to participate in the fit-out of their teaching laboratories. So, for instance, the radiography subject leader participated in the design of the fit out of the radiography laboratories by interacting with the VR model.

Source: interview 29/11/01.

There are many stakeholders interested in the definition of the project – internal and external. Some will be opposed, but all will have their own view of what the final facility should look like and how it should operate. The role of the project management team is to champion the definition of the project mission among the stakeholders and to facilitate the negotiation of a compromise between those whose interests can be accommodated, and to outmanoeuvre or nullify those whose interests cannot. Managing stakeholders is one of the principal challenges in the management of projects as the case of the Boston Central Artery/Tunnel shows in Case 13. Once the project mission has been defined, the project manager can move on to procure the resources required to deliver against that mission. Part III addresses that process next.

Case 4
The Rebuilding of Beirut Central District

By 1991, over 15 years of civil war had left the central district of Beirut in ruins. Formerly the most cosmopolitan commercial and cultural district in the Middle East, Beirut Central District (BCD) was to be redeveloped with the aim of regaining its former role in the regional economic and social life. The master plan was developed by the Egyptian firm of consulting engineers, Dar Al-Handasah, with a scope including:

- responsibility for a total area of 1.8m m^2 of prime real estate;
- reclamation of 608k m^2 of unofficial landfill on the coast – the reclaimed land;
- provision of a modern urban infrastructure of roads, parks and telecommunications;
- restoration of those existing buildings that were not beyond saving which were of historical value – the retained buildings;
- extension of the *corniche* and the provision of two marinas;
- reconstruction and expansion of the traditional *souks*.

There were two main problems that influenced the choice of organisation for the project. Firstly, the financial and managerial resources of the Lebanese state at the end of the war were completely inadequate for the challenges of delivering this master plan. Secondly, Lebanese property rights meant that the former tenants of the ruined buildings had the right to take up their tenancies again at the pre-war levels of rent. A further, but less intractable, problem was that many of the surviving buildings were squatted by those dislocated by the war, and to whom the state had a moral obligation. For these reasons, a private company was incorporated in 1994 – the Société Libanaise pour le Développement et la Reconstruction du Centre Ville de Beyrouth (SOLIDERE) – as concessionaire for the redevelopment works. Essentially a form of public/private partnership, SOLIDERE had the right to:

- expropriate land and buildings in return for A-class shares in its equity;
- raise equity capital through the sale of B-class shares to Lebanese nationals and firms and some other categories of Middle Eastern investor – the eligible persons;
- raise loan capital on the international markets;
- make profits on the sale and rental of its assets;
- be exempt from taxes on its profits for 10 years, while its shareholders are exempt from taxes on their dividends and capital gains for the same period.

The scale of the project is indicated by the fact that the final capitalisation of SOLIDERE is equivalent to roughly one-third of the total annual gross domestic product (GDP) of Lebanon. It is broken down into three main phases:

(1) 1994–1999 – stabilisation of the landfill, completion of infrastructure works in the traditional central district and restoration of the retained buildings;

(2) 2000–2009 – infrastructure works on the reclaimed land and further development of the traditional central district;
(3) 2010–2019 – development of the reclaimed land.

The first phase of the project ran over both schedule and budget, but within the bounds of available finance. The main schedule slippages were a result of the extent of the archaeological schedule and problems with the stabilisation of the landfill, while the principal sources of budget variances are the squatter relocation programme and the restoration of the retained buildings.

The stakeholder map for the project is shown in Fig. 4.3. The principal *internal* stakeholders are:

- Class A shareholders, who are angry at the expropriation of their property and its alleged undervaluation by the appraisal committee;
- Class B shareholders, whose principal aim is a return on their capital;
- The project sponsor – the Lebanese government. Allied to this group were President Hariri, who was himself the largest B-class shareholder and an owner of construction firms, and pro-government political groups. The concern of

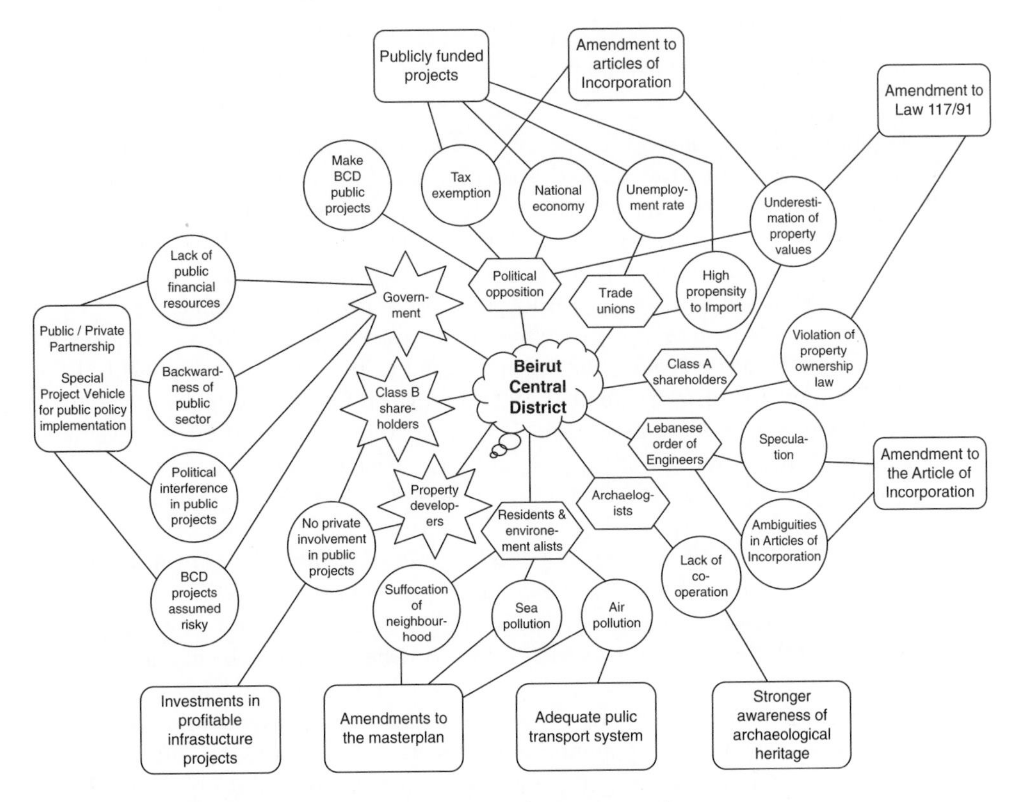

Fig. 4.3 Stakeholder map for Beirut Central District project.

this stakeholder is the revival of Beirut as an international commercial centre and the economic benefits that entail. Hariri played a crucial champion role in the incorporation of SOLIDERE.

The principal *external* stakeholders are:

- the international community, in the shape of many western leaders and the United Nations, who are openly backing the project as a major contribution to political stability in the region;
- international merchant banks providing loan capital;
- the Lebanese Order of Engineers (LoE), voicing concerns over the lack of accountability of SOLIDERE;
- local banks, which are financing individual property developments;
- the national and international archaeological community, concerned to capture the heritage of the area;
- residents and environmentalists, particularly concerned with the land reclamation aspects of the project, and the lack of mass transport in the scheme;
- property developers;
- the political opposition, consisting of various left-wing, religious and nationalist parties voicing concerns about the use of a public–private partnership, tax holidays and the up-market character of the developments;
- trade unions, concerned about the use of cheap foreign workers in the construction.

Figure 4.4 shows the power/interest matrix for the BCD project. It shows how the key players are the class B shareholders, the Lebanese government itself and the property developers which will carry out the individual developments once the infrastructure is provided. President Hariri lost a vote of confidence in December 1998, and since then there have been worries about government support for SOLIDERE. The return to power of Hariri in November 2000 helped to restore confidence. The dissatisfaction of class B shareholders and property developers with the progress of the project led to the share price of SOLIDERE dropping to a nadir by September 1999 on low profits as a result of an economic slowdown and continuing tension with Israel. International property developers such as Prince Alwaleed of Saudi Arabia have engaged in protracted negotiations before agreeing to invest. Because of continuing tension in the Middle East, the company made its first ever loss of $31.8m for the fiscal year 2000.

The international banks are happy so long as their loans are secure. A number of groups have a high interest, but the political balance of power nullifies the ability of the trade unions, the political opposition and the LoE to lobby for stronger government control over SOLIDERE, and the class A shareholders to obtain relief for their grievances. The government of Salim el-Hoss did not attempt to alter the structure of SOLIDERE, but did hold up the issue of construction permits to property developers. The archaeological community has received concessions, but little has been offered to environmentalists and residents. However, the failure by SOLIDERE to protect some of the retained buildings is the subject of litigation. While the

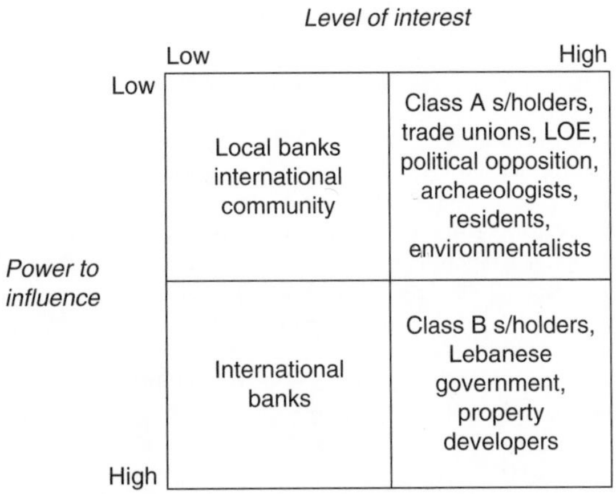

Level of interest

Power to influence	Low	High
Low	Local banks international community	Class A s/holders, trade unions, LOE, political opposition, archaeologists, residents, environmentalists
High	International banks	Class B s/holders, Lebanese government, property developers

Fig. 4.4 The Beirut Central District power/interest matrix.

international community wishes the project well, supported by high-profile visits from the likes of Jacques Chirac and Kofi Annan, they have relatively little interest in the project as such. The local banks would in any case find investments for their funds without the intervention of SOLIDERE. The main criticisms of SOLIDERE come from local banks and property developers who continue to be concerned about the high prices being charged by SOLIDERE for development land.

Despite the assassination of Hariri in 2005, by 2006 SOLIDERE was booming following the withdrawal of Syrian forces from the country around a strategy based on making Beirut the second home for oil-rich Gulf Arabs attracted to a comfortable blend of East and West. However, the return of war later that year undermined this strategy because sales of property stalled, and the lack of visitors to Beirut meant that shops could not cover their rents. This prompted SOLIDERE to look abroad for opportunities, particularly in Dubai, amid much complaint from those who believed that they were expropriated, but at least believed in the national regeneration projects. With the return of peace in mid-2008, SOLIDERE's shares soared on the Beirut stock exchange. It remains to be seen whether it can return to the pre-2006 strategy for development.

Source: developed from Hoballah (1998) and updated from *ft.com* (accessed various dates).

Notes

1 Cited in Hughes (1998, p. 221).
2 See Cleland (1998) for a fuller discussion.
3 In his 1985 book, *Competitive Advantage*.
4 The technique was originally developed from work in the social construction of technology (Pinch and Bijker, 1987) by Bonke (1996), and further developed in Winch and Bonke (2002) and Winch (2004).

5 Johnson and Scholes (2002, exhibit 5.5).

6 The Chinese authorities finally admitted in September 2007 that the Three Gorges Dam – the brainchild of Mao Zedong – threatens to create an environmental catastrophe (*Financial Times* 27/09/07).

7 Cited in Burningham (1995).

8 Residential and domestic buildings in use generate around 8% of greenhouse gases annually or about 20% if 'upstream' electricity generation is taken into account. These proportions vary greatly between countries depending on the proportion of renewable and nuclear fuel in the electricity generation mix, and the extent of the use of gas in contrast to biomass for domestic heating (Stern, 2007, Appendix 7e). It should also be noted that while the construction process itself is not particularly carbon-intensive, the manufacture of some key inputs such as concrete and steel is very intensive.

9 The first two of these are taken from Stringer (1995).

10 Johnson and Scholes (2002, p. 208).

11 Interview VSEL, 15/06/89.

12 Fischer (2008).

13 The Hôtel Campanile Bagnolet was built in the middle of the intersection of the Parisian *Périphérique* with the A3 *autoroute*, *after* its construction. The approach by foot from the Métro Bagnolet is like a scene from the dystopian film *Bladerunner*, and hotel guests wake up to the sight of cars whizzing past their bedroom windows on both sides of the building – silently thanks to the investment in triple glazing. So impressed were a team of European researchers with this robust approach to the urban environment that they named themselves Le Groupe Bagnolet. Their research provides many of the vignettes and one of the cases in this book; the research reports are available for download at http://www.chantier.net/europe.html.

Further reading

Layard, R. and Glaister, S. (eds.) (1994) *Cost-Benefit Analysis* (2nd ed.). Cambridge, Cambridge University Press.

The standard reference in cost-benefit analysis, containing a variety of reviews of the topic from a number of viewpoints.

Stern, N. (2007) *The Economics of Climate Change: The Stern Review*. Cambridge, Cambridge University Press.

The seminal analysis of the costs and benefits of responding proactively to the challenge of climate change.

World Commission on Dams (2000) *Dams and Development: A New Framework for Decision Making*. London, Earthscan.

An authoritative analysis of the costs and benefits of constructing large dams over the last century, and ways of addressing the issue of water and its supply in the new one.

Part III

Mobilising the Resource Base

In Part II, we tackled the problem of defining the project mission so as to meet the client's requirements. We now turn to the problem of how clients mobilise the resources required to deliver on that mission, and how, in turn, first-tier suppliers then mobilise the resources they require in order to meet their commitments to the client.

The conceptual framework for the analysis in this section is derived from institutional economics, particularly the work of Oliver Williamson[1]. Williamson's basic proposition is that total costs of supply are derived from two main components – production costs and transaction costs. Production costs are well understood and, in essence, involve the efficient transformation of inputs into outputs, where prices are used to signal the most efficient choice of technology. Transaction costs are the costs of co-ordinating any complex production process, and occur when a good or service crosses a 'technologically separable interface'. This point is easily illustrated. In the supply of concrete structures, the placing of formwork and the manufacture of concrete are technologically separable – the formwork merely has to be ready to receive the concrete. However, the first half of a concrete pour is not technologically separable from the second half – the technology of the production process demands that the pour be completed as one operation. Thus the choice of the most efficient method of pouring concrete is a production cost problem, while the choice of the most efficient method of co-ordinating the placing of formwork to ensure that it is ready to receive the separately manufactured concrete is a transaction cost problem.

Williamson argues that there are two basic options for co-ordinating – or governing – transactions. A *market transaction* is where independent buyers and sellers meet in the market to negotiate the price for supply of a good or service in a *spot contract* – prices are set by what Adam Smith called the 'invisible hand' of the market. An *hierarchical transaction* is where the transaction is governed internally by administrative means – prices are determined by what Alfred Chandler[2] called the 'visible hand' of management through an *authority relation*. In between these two polar forms of transaction governance lies a wide variety of mixed forms of *relational contracts*. It is these relational contracts which will form the focus of the discussion in the following three chapters, as the pure market and pure hierarchy have relatively limited application in construction projects.

What determines the most efficient governance mode on a project? Williamson argued that there were three main characteristics of transactions which influenced the choice of how they are governed: uncertainty, asset specificity and frequency. Uncertainty – whether mission or dynamic – affects transactions because it creates *bounded rationality* for decision-makers. This bounded rationality makes writing a complete and unambiguous contract between the parties impossible because of uncertainty regarding the precise conditions under which the contract will be executed, and also makes it impossible to measure fully the performance of the contract. Asset specificity is the condition where either the buyer or the supplier is limited in their choice of transaction partner because of the specific nature of the resources to be supplied. This asset specificity may be pre-contract, in which case the problem is one of monopoly or monopsony in the market, or it may be generated post-contract because contract-specific investments are made by one or both of the parties. This generates the possibility of *opportunism* on the part of one of the parties as they exploit the other's disadvantage. This opportunism often takes the form of withholding information from the other party. Frequency affects transaction governance because one-off transactions provide no opportunity to learn about the other party, while repeated transactions allow *learning* about the behaviour of the other party and hence the generation of trust. Thus the most appropriate choice of transaction governance mode can be thought of as occupying a three-dimensional space in the manner indicated in Fig. III.1, which reprises the governance level of Fig. 1.5.

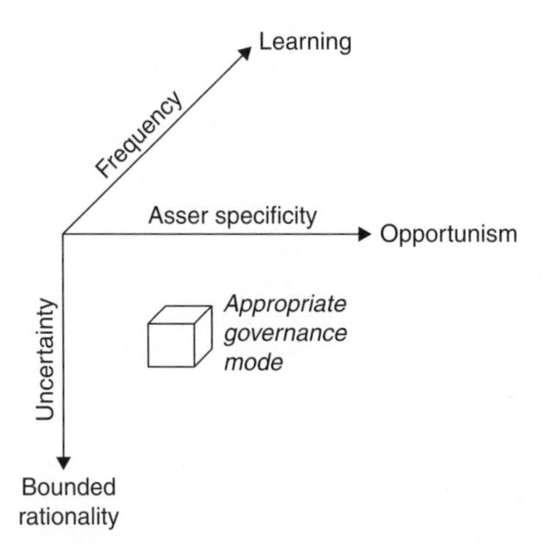

Fig. III.1 The governance level (source: Winch, 2001, Fig. 1).

Governance through relational contracts has, as its designation implies, two distinct aspects.

- The *contractual* which captures the underlying legal basis of the relationship. Although the precise formulation of these legal relationships varies significantly between countries, there is a large degree of functional equivalence in all developed economies between these formulations.
- The *relational* which captures the interpersonal and interorganisational aspects of the governance arrangements around issues such as trust and perceived equity in governance.

Within this perspective the extremes of the governance continuum can be considered to be tending to zero on relationship aspects at the market end (pure spot contracting) and tending to zero on the contractual aspects at the hierarchy end (pure autocracy). Although some have argued that the contractual aspects can undermine the development of the relationship aspects, recent research has shown that they are more complementary than antagonistic dimensions of transaction governance[3].

The next three chapters will use this simple, but profound, framework to explore how clients select and motivate their suppliers of construction services. Chapter 5 will explore the client's selection problem, while Chapter 6 will move on to how clients can motivate their suppliers once selected. Chapter 7 moves beyond the first tier of suppliers to the client and applies the same principles to the resource supply chain mobilised by first-tier suppliers to meet their commitments to the client.

III.1 A note on trust

An important aspect of transaction governance choice is the level of trust between the parties. In the context of transaction governance, trust is the confidence that the parties to the transaction will not take advantage of asset specificities to behave opportunistically[4] – either by withholding information or by seeking monopoly rents. Where transactions are made under high uncertainty, trust is essential for their effective governance. Broadly, two types of trust can be distinguished.

- *Transactional trust* is essentially future orientated in terms of the expectation that one's transaction partner is trustworthy and will not behave opportunistically in future transactions. Such trust is generated through learning in interaction with the transaction party over time; panel 5.4 explores the evolution of this type of trust.
- *Contextual trust* is more pervasive in that it is embedded in the dynamics of the business system as defined in section 2.3, and is created through obligations generated within social and family networks, reinforced by the placing of a high value on reputation. Although all transactions are embedded in a business system, the ways in which particular business systems favour high or low trust transactional relationships remain a matter of considerable debate[5].

Transactional trust predominates in business transactions, but can be supported at crucial points by contextual trust[6]. The generation of transactional trust is largely a function of frequency, because only through repeat transactions can parties come to know each other, and only when there is the prospect of further transactions does enlightened self-interest preclude opportunism. In the governance framework presented in Fig. 1.5, transactional trust is a feature of the governance level, while contextual trust is more a feature of the institutional level.

Contextual trust tends to play a larger role in countries with corporatist business systems, as defined in section 2.3, than in countries with Anglo-Saxon systems – this is the essence of what is known as the polder model in The Netherlands, for instance. Clients in countries or regions where contextual trust is important tend to favour local contractors whom they know well, or to favour the allocation of work through price rings rather than competitive tendering[7].

Notes

1. See, especially, Williamson (1975, 1985).
2. Chandler (1977).
3. See Poppo and Zenger (2002) for discussion and evidence, and Husted and Folger (2004) on the importance of perceived equity in transaction governance.
4. Lyons and Mehta (1997).
5. See North (1990) for the importance of political institutions and the rule of law; Fukuyama (1995), for a much more culturally orientated analysis.
6. The classic problem is how to govern the last transaction in a series. If transactional trust relies upon the expectation of future relationships, yet it is known that this is the last time the parties will do business, then the whole development of transactional trust can unravel.
7. See Syben (2000) for the former, and Bremer and Kok (2000) for the latter. An interesting question, which cannot be answered with the data available, is whether the savings on transaction costs available in a contextual trust environment outweigh the higher production costs because of lack of competition.

Chapter 5
Forming the Project Coalition

5.1 Introduction

> 'When people seldom deal with each other, we find that they are somewhat disposed to cheat, because they can gain more by a smart trick than they can lose by the injury which it does to their character'.

Having determined the project mission, the client's next task is to select the firms that will provide the resources required. Clients are typically faced with an overwhelming variety of firms claiming that they have the competence to meet their needs, some offering just one resource, others bundling the required resources in different ways. How is the client to choose those that are appropriate, competent and trustworthy? This is what is known as the principal/agent problem, so the chapter will start by defining the problem more precisely. In construction, the process of selecting competent suppliers is known as the process of *procurement*, and it is to a review of the different ways of procuring construction services that the chapter will then turn. The solutions to the procurement problem have evolved over time, and differ throughout the world, so it is appropriate to review these different solutions. From this world-wide experience, some principles of resource-base selection can be proposed, defined in terms of the risks and rewards that each solution offers for the client. The chapter can then focus on the details of how to select resource bases, focusing on criteria such as price, value and reputation. Finally, a recent development in resource-base selection – partnering – which attempts to address the problem identified over 200 years ago by Adam Smith (in the epigraph above) will be explored[1].

5.2 The principal/agent problem in construction

The problem is simply stated; its solution is profoundly difficult. The client (the principal) wants to hire the most competent (efficient and effective) suppliers (the agents) of the required resources, yet the agent knows more about its real competence than the principal. The problem is one of *asymmetry of information*[2]. This asymmetry generates two problems, well known in the world of insurance, which can be posed in construction terms in the following way.

- *Moral hazard*. How can the client be sure that the firm, once hired, will fully mobilise its capabilities on the client's behalf, rather than on behalf of the firm itself or some other client?
- *Adverse selection*. How can the client be sure that the most enthusiastic offer of the required resources is not also the most desperate; that the lowest price is offered because nobody else will contract with the supplier because the other clients know more about its real capabilities?

This chapter focuses on the problem of adverse selection, or how to ensure that the client hires competent suppliers of construction services, while Chapter 6 will focus on the problem of moral hazard. Adverse selection is known by economists as 'the lemon problem', described in panel 5.1[3]. Suppliers of poorer-quality goods have the greatest incentive to charge the keenest prices in order to offload goods that the supplier knows to be inferior but which the buyer cannot know to be inferior, because the inferior characteristics are hidden. Thus price signals in the market do not work and, in a version of Gresham's Law (bad money drives out good money), poor-quality goods will drive out better-quality goods in markets where there are asymmetries of information. The lemon problem is particularly severe in construction because of the very high levels of uncertainty inherent in the construction project process, and it is highest in the earlier stages of the project.

The first step in addressing the requirement for a new facility in the context of the principal/agent problem is whether to buy or procure. Buying involves going out into the market for existing factilities – either new or used – and buying it either outright or through a lease. This is the lowest-risk option and places the construction client closer to the car-buyer, but does depend on the existing market for suitable facilities in the right place and the role of property in the financial structure of the client. A slightly more bespoke solution is to enter into a pre-let arrangement with a property developer which can then adapt the fit-out of the facility to the client's particular needs. One of the most important contributions

Panel 5.1 The lemon problem

The seller of a used car knows more about its performance and condition than the buyer can find out in a test drive; indeed the seller may be able to hide poor or even dangerous features of the car on offer. This explains the large price premium of new cars over second-hand ones, because new ones come with a manufacturer's warranty. Solutions to this problem include trust ('would you buy a used car from this man?') which may be generated by: branding the dealer or offering additional warranties; buying a car with a known history; use of expert third parties to inspect the car and sticking to bicycles.

Source: Akerlof (1970).

of value management – see section 9.7.2 – in the very early stages of project decision-making is to identify whether there really is a requirement to initiate a construction project at all, or whether lower risk options should be preferred.

5.3 Procuring construction services

As shown in Chapter 1, the project process is one of progressive reduction of uncertainty through time. Therefore, contracts for those services procured in the early phases of the project – principally associated with design – will face higher uncertainty than those for services procured later in the project which are principally associated with execution on site. Thus we would expect that design services will tend to be procured in a different way from site execution services, and indeed this is typically the case. There are four main ways that clients have found to procure construction services:

- maintain an in-house capability;
- appoint a supplier;
- launch a *concours*;
- issue an invitation to competitive tender.

5.3.1 *In-house capability*

This option is widely used in many countries by clients that undertake large volumes of construction work, very often the state and its agencies. It is a classic example of the use of hierarchy as a result of uncertainty in transaction governance. Two of the most famous examples are the US Army Corps of Engineers, presented in panel 5.2, and the Directions d'Equipement that cover France at the national, regional and departmental levels. Such agencies exist in many countries – the UK is an exception – handling differing proportions of the overall construction workload, particularly in high-uncertainty tasks such as concept design and project management. Another widespread application of the in-house capability option is at the other end of the project life cycle in the facility management group within firms. Although there has

Panel 5.2 US Army Corps of Engineers

The US Army Corps of Engineers was founded on the French model of the Corps des Ponts under Jefferson, and is largely staffed by civilians. As well as designing and managing projects associated with military activities, it is also responsible at the federal level for projects associated with waterways, power generation, flood control and coastal protection, and it maintains an extensive construction engineering research capability. Further, it provides consultancy services to a number of other US state agencies and to governments abroad. Its demonstration and advocacy of value engineering, partnering and e-procurement has been influential worldwide.

been a shift towards outsourcing of this function, many firms which own their own property employ a group responsible for the repair and maintenance of the property portfolio; this group may also undertake smaller renovation projects.

There are a number of important advantages offered by the in-house capability option:

- It is not necessary to write a complete contract prior to starting the project because the authority relation between client and supplier allows continual adjustment.
- The client has the administrative capability to audit the detailed operations of the supplier of construction services, and to control its expenditures.
- The risk of opportunistic profiteering at the expense of the client is minimised as the in-house operation is only a cost-centre (if it were a profit-centre it would not be a true in-house service, but merely a commonly owned entity).

However, there are also major disadvantages:

- If construction is not the core business of the client, then managing construction activities directly may be a diversion from that core business.
- Low transaction frequency may make investment in an in-house capability unviable or inefficient.
- Lack of competition may lead to production inefficiencies within the in-house supplier, thereby raising production costs.

The Network Rail case shown in panel 5.3 shows how the advantages eventually outweighed the disadvantages for one client.

Panel 5.3 In-sourcing maintenance projects at Network Rail

The privatisation of British Rail under the Railways Act of 1993 is widely seen as a failure of either principle, or execution, or both, depending on political viewpoint. Most commentators agree that the fragmented organisation structure of the industry fits poorly with the inherently integrated nature of the rail network as a complex system. A governance analysis would doubtless yield many insights into the apparent failings of this organisation structure. However, this case will focus on one area where the failings have been recognised, and a marked shift from market to hierarchy has taken place – contracts for the maintenance of the permanent way.

British Rail had carried out its own rail maintenance as it presided over a slowly declining system. The policy of introducing 'market disciplines' into the operation of the network when the private entity Railtrack took over as network operator in 1994 led to the outsourcing of maintenance to a number of construction companies on the basis of 20 framework contracts dispersed geographically around the network. Contractors were contracted – in essence – to maintain the permanent way to the standards existing at the time of privatisation by providing track fit for purpose against predetermined standards. However, those standards were already in decline by the time the new companies took over during the mid-1990s – for instance rail breaks per million train-miles had been rising since 1989 and reached crisis levels in late 1998. By 1995, the track replacement rate was the lowest than it had been since the 1940s, and had not surpassed the steady-state rate of renewal of 800 km per year since 1983.

Although performance did improve once Railtrack entered the private sector and additional investment funds became available, concern was expressed that it was not spending enough, and was focusing its efforts on performance criteria that led to bonuses for reducing delays to trains, rather than addressing the underlying quality of its assets which continued to deteriorate. Moreover, Railtrack – like British Rail before it – had a poor idea of the overall condition of the network, which made both its own setting of priorities difficult and the job of the Office of Rail Regulation (ORR) charged with overseeing its performance near impossible in this area. The most detailed knowledge lay with the contractors rather than Railtrack itself, yet Railtrack also moved to an asset renewal policy based on actual condition rather than age. At the same time, utilisation of the network increased dramatically, with passenger traffic increasing by 27% and freight by 35% between 1995 and 1999. This poor asset management interacted with rising utilisation of the network, leading to the crisis which came to a head because of the fatal Hatfield derailment in October 2000.

Railtrack was effectively forced into administration by the government in October 2001 because it proved incapable of meeting the challenge of the crisis in maintenance. Its successor – a not-for-profit company called Network Rail – started to address the problem by reviewing the governance of rail maintenance contracts almost as soon as it took over in October 2002. In little over a year it entirely reversed the outsourcing policy in a shift that can be usefully interpreted through the governance framework. In order to understand this, we first need to identify the principal contingencies of the transaction to be governed. Rail maintenance – as compared to renewal and upgrade – combines high frequency and asset specificity with medium levels of uncertainty because:

- it is repetitive;
- it is small scale;
- it requires well-located depots with few alternative uses;
- it is geographically dispersed around the network;
- it is typically conducted at night against tight deadlines to ensure network availability the following day.

Although task uncertainty itself is relatively low, the physical dispersion and constrained 'possession' periods combine with the more familiar contingencies to make transaction governance in the absence of high trust very costly. Direct supervision of contractors by the client – a classic transaction cost – would be prohibitively expensive, so consummate performance has to be left to the contractor. The need for trust is reinforced by the performance nature of the contract – it is up to the contractor to determine fitness for purpose. However, the repetitive nature of transactions also presents the possibility for learning.

The initial rationale offered by Network Rail for in-sourcing maintenance contracts was a learning one. It argued that it did not know enough about the cost drivers of rail maintenance to act as an informed client, and so decided to take in-house one contract that was a 'microcosm of the network' when it expired in March 2003. This was followed by the announcement some months later that two further contracts would be taken in-house. A senior director of Network Rail stated: 'Our objective is clear – to drive down maintenance costs and become a more informed and intelligent buyer. The maintenance contracts we have inherited from Railtrack do not give us a clear understanding of cost and efficiency issues. That is why we are changing them'. By October 2003, the argument had moved on – Network Rail had undergone its learning and concluded that in-house maintenance was cheaper. It was announced that all rail maintenance contracts would be taken in-house as part of the New Maintenance Programme, with implementation over the following 12 months. It was claimed that this could save £0.3bn on the £1.3bn annual maintenance budget. It is, perhaps, also worth noting the argument upon which Network Rail did not rely – that the changes to contractual relationships offered advantages in safety in the febrile atmosphere of the aftermath of the Hatfield accident and a later fatal one at Potters Bar in May 2002 which was also related to the rail maintenance regime.

What are the sources of this potential saving? Clearly, Network Rail had learned what the contractors already knew – the production cost structure of maintenance contracts. That these were lucrative is indicated by the large falls in the share prices of the contracting companies which typically derived between 10 and 30% of their turnover from this source when the decision was announced. However, it is unlikely that profits were at the level of over 20% on turnover under the watchful eye of the ORR, and the potential for efficiency savings is limited because of the fragmented nature of the work. It can, therefore, be suggested that many of the savings were in transaction costs:

- multiple layers of inspection in a context where there was very low trust of the contractor's ability to perform work to the expected standards and where Network Rail was already committed to raising levels of inspection and planning;
- simplified administration of contracts in a context of high transaction frequency;
- the efficiency benefits of administrative fiat in a context of many low-level decisions needing to be made;
- the opportunity to generate trust through an employment contract, rather than a commercial contract.

It is also worth noting that Network Rail was very clear that it had no plans to take in-house renewal and upgrade work for two reasons:

- Frequency levels are much lower in renewal and upgrade transactions, and projects are significantly larger.
- Transaction costs as a proportion of total costs are lower, because 'you can measure performance very clearly in renewals. You can structure renewals contracts so that your contractor has a real incentive to show he's improving efficiency'.

The implementation of the New Maintenance Programme was completed in July 2004 with a projection of savings of a more modest £700m over the following 5 years. This announcement was accompanied by reports that train delays had fallen by 21% in those areas that had been taken in-house earlier.

Source: Winch (2006a, pp. 336–338)

5.3.2 Appointment

Appointment is used throughout the industry worldwide. In conditions where there is not enough information to allow the preparation of tender documents, the appointment of the supplier on the basis of reputation for having previously completed similar projects is often the only option if in-house capabilities are inadequate. This approach is most often used in the appointment of suppliers of design and project management services; indeed it is the predominant approach to the appointment of such suppliers in most countries. Appointment may also be used when the requirement to mobilise the resources for execution on site is so urgent that time cannot be devoted to preparing tender documents and waiting for the responses to come in. Appointment does not mean that hard-headed negotiations regarding the precise terms upon which resources are supplied cannot take place, and that a number of possible suppliers are not included in the discussions, but it does mean that the process of selection is not always very transparent.

The advantages of appointment are as follows:

- The search for suppliers can be easily restricted to those with a proven capability – reputation for delivering similar projects in the past.
- The search and selection costs can be minimised, and the risks of making a major mistake reduced, by restricting the choice to suppliers with known capabilities on the basis of previous track record.
- The repeat transactions between clients and suppliers can be used to enable high-trust relationships to be built up.

The disadvantages are the following:

- Lack of competition can lead to lower levels of production efficiency and effectiveness among suppliers.
- Appointment criteria are often not very transparent, leading to difficulties in auditing the rationale for the choice. This is particularly a problem for public sector and regulated private sector organisations.
- Relationships can become too cosy, leading to the use of inadequately rigorous appointment criteria.
- Cosiness can degenerate into corruption, which became systemic in the Japanese, Dutch and Italian construction industries – see section 5.8.

5.3.3 Concours

This is another widely used way of selecting suppliers, particularly of design services. The essence of a *concours* is that competition is based around the quality of the solution offered to the client's problem, rather than its price. It tends to be used where the symbolic quality of the solution is paramount and original and exciting solutions are sought. This is mainly in situations where the client wants a 'signature' building, although it is also used for some civil engineering works with high symbolic quality, such as bridges and train stations with which the Swiss-based architect/engineer Santiago Calatrava has been particularly successful. As discussed in section 3.5, this desire may be commercially driven as the statement of corporate principles often seen in corporate headquarters buildings, or socio-politically driven as a statement of the wider responsibilities of clients to commission buildings that make a broader contribution to the urban fabric or national culture. In some countries, such as France, the *concours* has become embedded as one of the mainstream ways of selecting suppliers of architectural services, while in others, such as the UK, its use is much less widespread[4]. The *concours* is also notable for the way in which it shifts responsibility for the selection of the supplier away from that client alone, towards a jury representing a broader body of opinion, albeit one often dominated by other design professionals. Panel 5.4 describes the organisation of the *concours* for the Tate Modern project.

The advantages of the *concours* are that they:

- broaden search beyond immediate networks;
- allow talented young professionals their first major project;
- stimulate public debate about the sort of buildings that are appropriate contributions to the urban fabric.

The disadvantages of *concours* are that they:

- are expensive to run, particularly if competitors are reimbursed for their efforts;
- can lead to a formalism in design where aesthetically striking solutions which emphasise symbolic quality are preferred at the expense of other aspects of product integrity which are less easily judged from concept models and drawings, such as spatial and indoor environmental quality;
- rupture the briefing process which means that the mutual adaptation of client desires and constructive possibilities is less thoroughly worked through than when the design services supplier is appointed.

Panel 5.4 The concours for the Tate Modern

Once the Bankside Power Station had been identified as the new location, the Tate ran a concours to choose the architect for the Tate Modern. The Tate emphasised that the concours was to choose an architect, and not a design. Some 149 architectural practices entered the open competition, and after a sifting of the proposals, a shortlist of 13 was drawn up for a more detailed examination. These 13 were given a month and a small fee to develop ideas for presentation to the Tate team. The more experienced architects such as Piano and Koolhaas were at ease with this approach, while some of the younger ones were more cynical, believing that their ideas could be pinched and that the Tate was not playing by the rules they had set for the competition. Following the presentations, the final shortlist of six was drawn up, and their proposals went through further workshops. The perception that Herzog and de Meuron were always moving forward through these workshops, and their empathy with the industrial character of Bankside, won them the commission. Pierre de Meuron was told the news by his mother, who happened to be listening to the local radio in Basel and caught the news item.

Source: Sabbagh (2000).

5.3.4 *Competitive tendering*

This is the most commonly used means of selecting suppliers of construction services. Competitive tendering is distinguished by the formalisation of both the selection process and the criteria upon which the final decision is made. Under competitive tendering, the client, or its advisors, issues a codified set of documents which provide a detailed description of the construction service to be rendered. These documents allow prospective suppliers to calculate their price for supplying the services. It is on the basis of this price that the supplier is selected. The lowest price is typically the key selection criterion, but it is usually necessary to ensure that the offer is compliant with the tender documents and errors have not been made in the calculation of the price. Competitive tenders may be open to all comers, or selective on the basis of a pre-established tender list. Where competitions are open, supplier selection tends to include some sort of evaluation of supplier competence. In selective tendering such matters are usually handled in the pre-qualification process for being invited to tender. Competitive tendering is the most widely used method of selecting suppliers of on-site realisation resources, and actively promoted by many governments in pursuit of transparency – notably under the public procurement directives of the European Commission.

The advantages of competitive tendering are as follows:

- Keen price competition among suppliers encourages production efficiency.
- The transparency of selection criteria facilitates audit of supplier selection decisions, particularly important for public sector clients.
- The low barriers to entry minimise the risks of supplier cartels forming.

However, these highly persuasive advantages are counterbalanced by some less widely understood disadvantages:

- The risks of encountering the lemon problem are high, because only limited information is available on the competence of suppliers, particularly with open tendering.
- The costs of search and selection are high – one estimate in the UK by KPMG put tendering costs at 10% of construction firm turnover – costs which are paid in the end by the client in higher prices to cover their suppliers' overheads. In addition, the client has to cover its own search and selection costs. An overall figure of 15–20% of total project value may not be excessive for the transaction costs of competitive tendering.
- The requirement to prepare complete and unambiguous tender documents limits the use of competitive tendering for high-uncertainty transactions.
- The 'winner's curse' – that errors of omission will win tenders, while errors of inclusion will lose them – means that there is a systematic bias towards underestimated tenders leading to later motivational problems in contract delivery.

In theoretical terms[5], competitive tendering is an (reverse) auction in which there are one buyer and many sellers where the buyer induces sellers to reveal their valuations of the contract so as to eliminate information asymmetries between buyers and sellers, so the the buyer pays the lowest price to the most efficient seller. Through this process the criterion of market efficiency in asset acquisition is met. In the sealed bid auction typical in competitive bidding, price information is revealed to the buyer but not to other sellers. Sellers are induced to enter the auction through binding commitments from the buyer to obey the pre-announced decision rules on how the acquired information is to be used to select the successful bidder without exploiting that information to sellers' disadvantage[6]. Neoclassical auction theory shares a fundamental assumption with principal/agent theory – that information is asymmetrical but complete, and that the game is about revealing the preferences of the players. However, the situation on projects during competitive tendering is that information is not complete, that is there is uncertainty. Typically clients are not fully aware of their requirements for the facility at the termination of the project and contractors are not fully aware of the demands of riding the project life cycle. In competitive bidding in construction, information is both asymmetrical and incomplete and no scheme of inducing suppliers to reveal their preferences will reduce that underlying uncertainty although as we shall see, some arrangements are more effective than others in coping with that uncertainty. For these reasons the claim that the competitive tendering is always optimal in minimising adverse selection is not tenable.

5.3.5 Appropriate procurement

In broad terms, the appropriate form of resource – base selection is a function of the level of uncertainty in the specification of the resources required at the time of selection. There are two main interacting dimensions to be taken into account:

- *The level of mission uncertainty*. Large one-off projects will face much higher levels of mission uncertainty than small repeat projects; new-build will face lower levels of mission uncertainty than refurbishment and so on. For instance, the level of uncertainty in providing the design for a portal frame shed for a factory unit is much lower at project inception than a tunnelling project is during on-site execution.
- *The phase in the project life cycle* – and hence the extent to which dynamic uncertainty has been reduced from the initial mission level – at the time of selection. Thus the selection of designers will usually take place under higher levels of uncertainty than the selection of those responsible for on-site resources.

These relationships are illustrated in Fig. 5.1, with level of mission uncertainty at selection on the vertical dimension, and the phase of the project life cycle (i.e. dynamic uncertainty) on the horizontal. The lower the levels of mission and dynamic uncertainty, the more appropriate is the use of competitive tender; the higher the levels of mission and dynamic uncertainty, the more appropriate is appointment or the use of in-house capabilities[7]. The most frequently adopted method at each phase is shown in bold. However, clients will only establish in-house capabilities if transaction frequency based on an adequate supply of similar enough projects is high enough to justify the investment in an in-house capability. The most uncertain projects are likely to require the appointment of highly specialised resources with experience of the particular problems to be solved. On the other hand, where projects are very repetitive and hence mission uncertainty is low, particularly where multiple clients share the requirements for the particular building

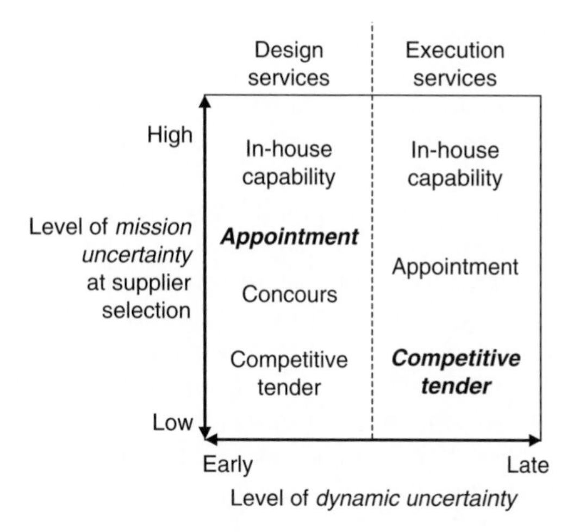

Fig. 5.1 Supplier selection methods (methods in bold are the most common at each phase).

type, the selection of external suppliers through competitive tenders may well be more efficient than using in-house facilities. Where uncertainty remains high right through execution – either in terms of the task to be performed or when it will be required – as in repair and maintenance work, then in-house capabilities may be more effective and efficient than hiring external resources as shown in panel 5.3.

5.4 The formation of project coalitions

In the half millennium since the decline of the craft system and the emergence of modern construction project organisation, a number of different combinations of selection criteria have been used to form project coalitions. This evolution in the UK can be traced in Case 2 and four basic types of project coalition structure have emerged[8].

- *Separated project coalitions*, characterised by the appointment or use of *concours* for the selection of suppliers of design resources, and competitive tendering for on-site execution resources.
- *Integrated project coalitions*, characterised by the letting of a single contract for both the design and execution of the project on a competitive tender basis.
- *Mediated project coalitions*, characterised by the appointment of design and project management resources, coupled with competitive tendering by trade package for execution.
- *Unmediated project coalition*, which are typically used by private sector clients with in-house project management skills to gain much greater leverage over their suppliers moving towards explicity supply-chain management approaches.

5.4.1 Separated coalitions

The traditional *separated* form of project coalition is trades contracting, illustrated in Fig. 5.2 and panel 2.5. In this form of project coalition, the architect and any other designers are appointed, with the architect very much leading the design team. The architect is then responsible for selecting the trade contractors who will execute the site works on the basis of either competitive tenders or appointment. The architect remains responsible for the overall co-ordination of the activities of the trade contractors, but is not usually liable for any failings on their part. This form of project coalition is having something of a revival in France, where the 1985 law on the public sector client (loi sur la Maîtrise d'Ouvrage Publique) is reinforcing the use of this traditional form in preference to general contracting.

General contracting, as shown in Fig. 5.3 and panel 2.6, is distinguished from trades contracting by the main or general contractor who enters into contract with

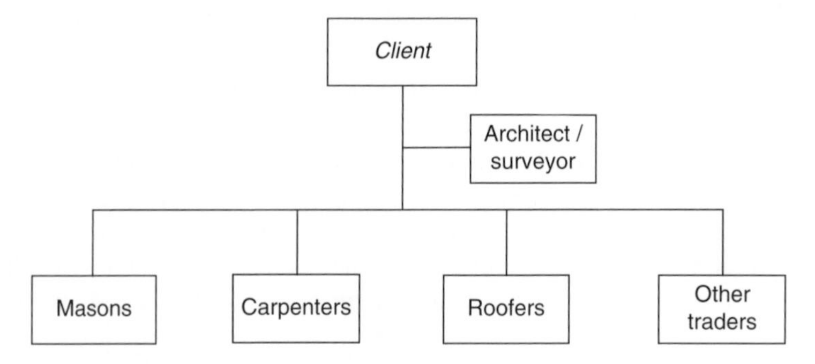

Fig. 5.2 Separated project coalition: trades contracting.

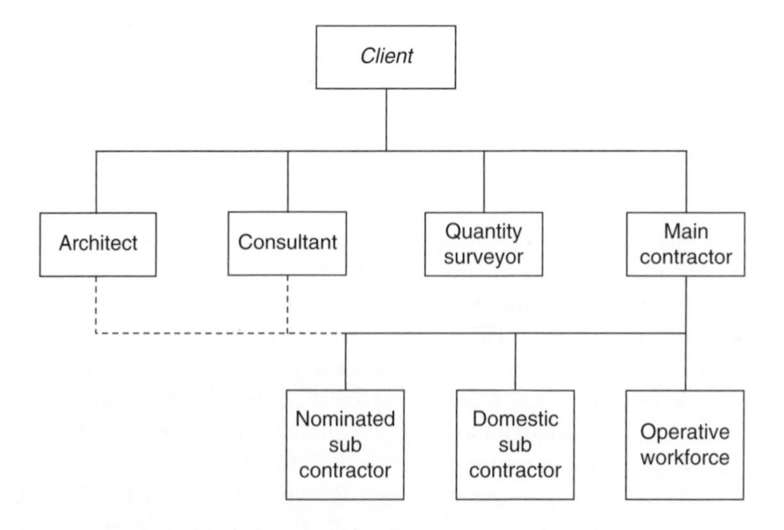

Fig. 5.3 Separated coalition: general contracting.

the client for the whole of the execution of the project on site, typically for a fixed price. As described in Case 2, this form of project coalition rapidly became the norm in the UK during the nineteenth century, although it did not become widely established in the USA and on the continent of Europe until after 1945. Today, it remains the principal form of procurement in the USA, the UK and internationally, and its use in continental Europe is favoured by EU procurement directives.

5.4.2 Integrated coalitions

Integrated coalitions, known variously as turnkey contracting, design and build, and single-point responsibility, are a long-established form of project coalition structure, favoured by clients who wish to transfer the maximum of risk to the supplier. The structure is illustrated in Fig. 5.4. Because of its integrated nature, supplier selection must take place relatively early in the project life cycle. It is not, therefore, appropriate for projects where there is high mission uncertainty, and tends to be used for projects where building types are largely repeated, thereby reducing mission uncertainty, such as industrial and commercial facilities. However, many commentators, such as Bowley, have also argued that the organisational integration of design and execution that such an organisational structure allows ought to yield efficiency savings in production costs.

It should be noted that, at least in the UK, there is a tendency to describe as integrated any procurement route that shifts from the traditional route and gives the contractor responsible for site execution any responsibility for the design. Typically, this responsibility is limited to detail design, rather than concept or scheme design. It is also common for the design team procured by the client to work on the early phases of the design to be 'novated' to the contractor in order to complete detailed design. In essence, this is a process of risk transfer for site-related risks, and is unlikely to stimulate innovation unless the supply-side team integrates, yet this opportunity is undermined by the practice of novation.

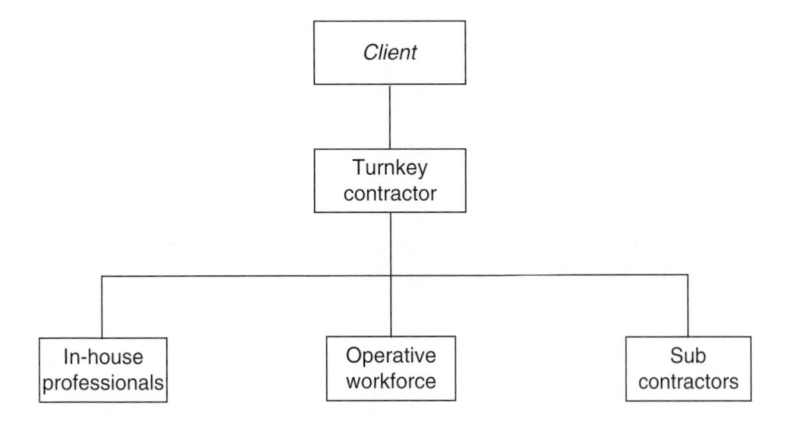

Fig. 5.4 Integrated coalition: turnkey.

The loss of the trilateral governance by the design team described in section 6.6 that novation implies has also generated a demand for a new actor in the system – the Employer's Agent – to act as a third party in issues such as whether a change in the specification is a change in requirements by a client or 'design development' in which case it is the responsibility of the design and build contractor.

5.4.3 Mediated coalitions

These are of growing importance, particularly where mission uncertainty is high because of either the technical challenge of the project or the necessity to deliver it quickly. Mediated coalition structures are characterised by the appointment of not only the designers but also the construction manager who will be responsible for managing the trade contractors mobilised for execution on site. These trade contractors are usually selected on the basis of a competitive tender organised by the construction manager. Precise arrangements and terminologies vary considerably – where the trade contracts are placed with the construction manager, this is often known as *management contracting*; when they are placed with the client directly, this is usually known as *construction management*. Where the trade packages include significant design resource bases managed by the construction manager, then the structure may be known as *design and manage*. Figure 5.5 shows the basic construction management coalition structure, and section IV.ii describes the construction management process for the Tate Modern project.

Within the UK public sector, mediated routes have become known as prime contracting, sometimes in the context of framework agreements for programmes of work defined, for instance, regionally. An important difference between prime contracting and some 'pure' mediated routes is that the prime contractor is expected to take some risk associated with budget and schedule through appropriately structured incentive contracts – see sections 5.6.3 and 6.5.3. This risk may also extend to the satisfactory operation of the facility in its early years – see Case 7. Clients may also require the construction manager to offer a guaranteed maximum price for the facility.

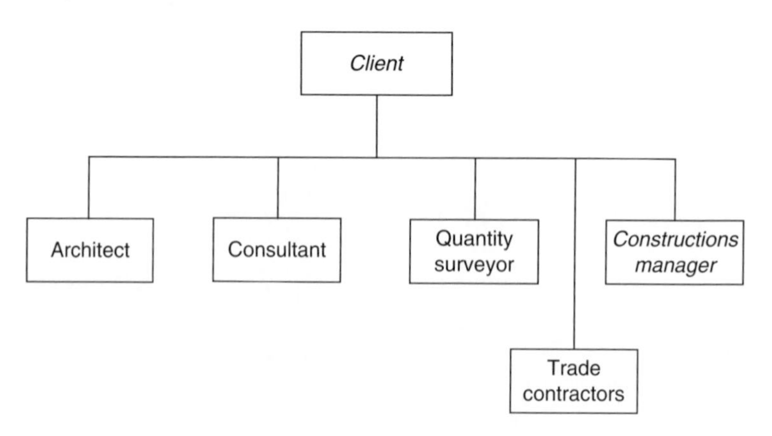

Fig. 5.5 Mediated coalition: construction management.

5.4.4 Unmediated coalitions

For some clients, the distance between themselves and the project coalition implied in the separated, integrated or mediated project coalitions, where systems integration is handed over to one or two principal suppliers, does not meet their needs. Such clients prefer to contract directly with a number of suppliers and co-ordinate these suppliers themselves which is why we have dubbed this arrangement the unmediated coalition. Unmediated coalitions require high levels of in-house project management capability – see section 15.4 for a discussion of this – and a continuing programme of projects to be effective. For these reasons, they tend to be found in the private sector, particularly amongst property developers such as Slough Estates and Stanhope[9]. BAA also uses this approach as illustrated in T5 in Case 12.

Figure 5.6 presents a social network analysis diagram of the information flows within a Slough Estates (SE) unmediated project coalition. It shows how central the client – the node at the centre of the network – is within the network of project coalition information flows both visually and by calculation of the centrality value for that node. Social network analysis offers a significant potential for new insights into the dynamics of project coalitions, and for comparing procurement routes on their information processing capabilities, and shows how separated project coalitions place the client much less centrally in the project coalition network[10].

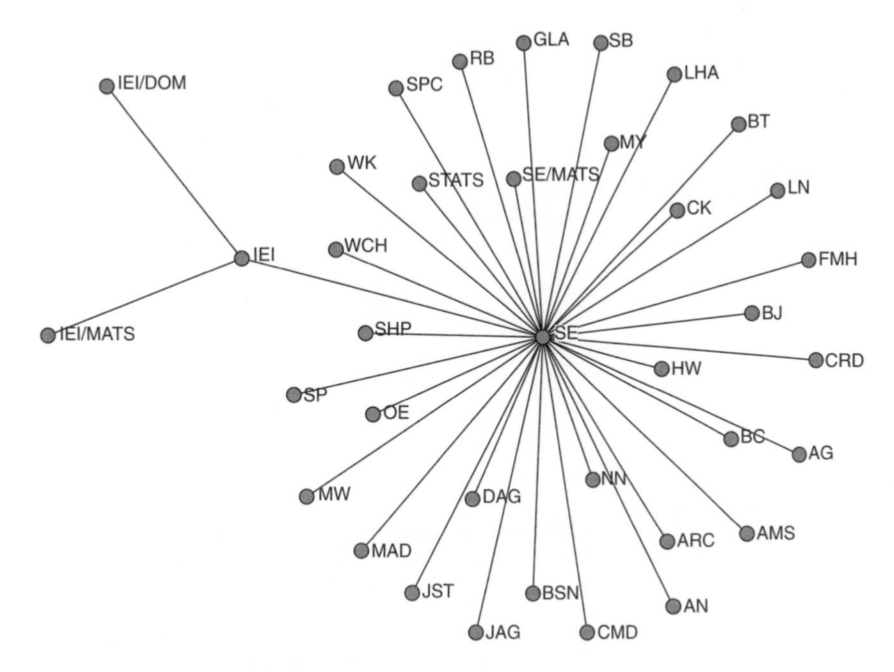

Fig. 5.6 Unmediated coalition: social network analysis (source: prepared by John Kelly from data kindly supplied by Stephen Pryke).

5.4.5 Appropriate project coalition structures

These three generic types of project coalition structure all combine different ways of selecting the different actors. Figure 5.7 compares them in terms of the risks the client faces in using each of them. The most important criterion for choice is the level of mission uncertainty. If large amounts of information are available very early – perhaps the building is a very simple type, or it is a repeat of one already completed, or it can be designed from a modular kit of parts – then the integrated route is the most appropriate. As will be discussed in Part IV, this yields low risks from the client point of view of schedule and budget overruns and lack of conformance to specification (the quality of realisation issues), yet poses specification and conception quality risks as it is difficult to change the design if new information becomes available, or the client is not fully able to articulate its needs through the contract. If there are larger mission uncertainties because of regulatory requirements such as planning permission, short schedule, novel technological challenges or the requirement for an exciting signature building, then separated or mediated coalitions are more appropriate. This is because most uncertainty reduction is achieved through design, and the separation of the contracts for design and execution allows iterative design in the search for the most effective solutions to the problems posed and hence high quality of specification and conception. Where mission uncertainties are highest, then the mediated coalition is most appropriate, but this poses realisation risks for the client, particularly with respect to budget.

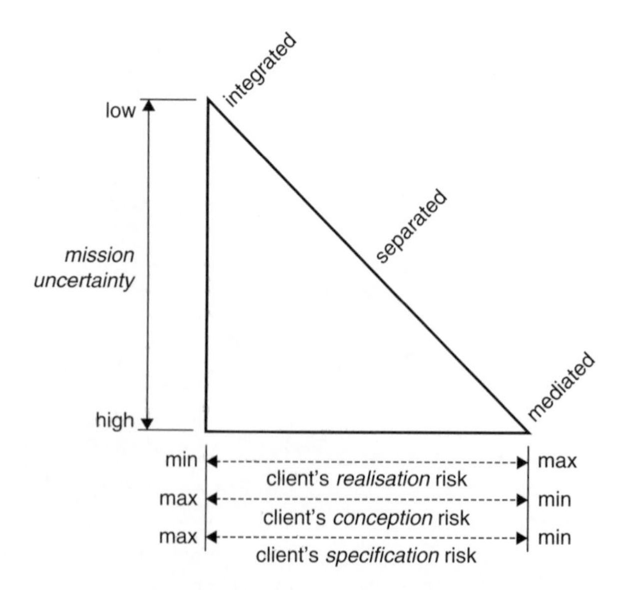

Fig. 5.7 Client's risks associated with different project coalitions.

5.5 Selecting resource bases

Selection procedures for suppliers vary considerably depending on the method selected. Such procedures are not required for in-house capabilities, and they tend to be relatively informal for appointment. Those for the more open methods of *concours* and competitive tender are much more formalised; indeed, failure to follow the prescribed procedures can lead to the tendering process being invalidated. The appointment process typically rests upon two principles – certification and reputation. In most countries, the suppliers of design services – mainly architects and engineers – are certified. This can be achieved through self-regulation such as through the British-type professional institution which confers the status of 'chartered' practitioner upon completion of an accredited programme of education and supervised practice. More commonly, it is achieved through the state registration of practitioners such as in the German *Kammer*, where satisfactory performance in official qualifying examinations is the criterion for certification. Some countries, such as the USA, combine the professional and registration approaches to certification. Firms offering design services then commit to ensuring that the design work is at least supervised by certified practitioners, and usually to its performance by such practitioners. Certification arrangements are much less developed among suppliers of on-site execution services, although whether this is a cause or result of the reliance upon competitive tendering for such suppliers is moot. An important exception is the German *Meister* qualification, which must be held by the principals of all trade contracting firms[11].

Although certification can assure a minimum level of competence and goes some way to avoid the lemon problem, the level of expertise required to meet the challenges of the needs of most clients is beyond that which can be measured through certification. The reputation of suppliers – or more precisely, the reputation of their principals (partners or directors) – then becomes critical. Reputations can be based on competence in specific building types (such as HOK and mixed-use stadia); fame as a star designer (such as Frank Gehry or Richard Rogers) or competence in solving specific technical problems (such as Ove Arup for complex structures). Reputations are hard gained and easily lost. One of the major barriers to entry in markets where appointment on the basis of reputation predominates is establishing a reputation. For start-up suppliers this can seem like a Catch-22, and good contacts with clients and other suppliers are essential for getting a break. Once gained, a reputation can be easily lost through a project that goes wrong. Reputation is, in its nature, ephemeral; a supplier is only as good as its completed projects, and clients need to work hard to ensure that their sources of information are up to date. By far the best way of minimising surprise in the appointment of suppliers is for clients to use suppliers that have successfully completed for them earlier projects of a type similar to the one under consideration.

Concours are subject to more codified procedures – based usually on long-evolved custom and practice. They are normally used for selecting architectural resources, which may also team up with engineering resources as required by the *concours* brief. In countries such as France, these procedures are highly formalised, and deviation from the procedure is grounds for invalidating the selection made.

Concours can be either open or invited. The aim of the former is often explicitly to help new start-up suppliers break the reputation Catch-22, and many star architects have started their careers by winning such a *concours*. Invitation to participate in selective *concours*, however, is almost entirely based on reputation and influence.

Perhaps the most highly codified approach is used in competitive tendering; Table 5.1 provides a summary of the national code of practice in the UK. Formalised procedures facilitate transparency and most countries have such procedures, but they are not, of course, necessarily followed. The process of competitive tendering is essentially one of price formation under information asymmetry and uncertainty: the client does not know which is the most efficient supplier, and the suppliers do not know how efficient they are in relation to their competitors; and neither party is aware of all the risks associated with the project. Bidders do not know the budget ceiling of the client. They are constrained to bid as high as possible in relation to the client's budget ceiling – above which the client will cancel the project because the net present value (NPV) becomes negative – and as low as possible to beat their competitors. Problems of information asymmetry also exist in relation to the specification of resources as the client has had much more time to explore these issues during design development than the bidders have during a tender period usually measured in weeks rather than months, but most bidding procedures provide for full disclosure here, and for emendations to the contract if new information becomes available later.

In deciding on the appropriate mix of selection procedures for any procurement route, clients are able to take advantage of an important property of the project life cycle – that the early stages of the project involve the strategic decisions taken under relatively high dynamic uncertainty, but the cost of the resources required to make these decisions is a relatively small proportion of the total budget of the project. It is the decisions taken later in the project under lower dynamic uncertainty that account for the largest proportion of budgetary spending. The implication of this is that the returns to minimising production costs through using competitive tender as a selection method are higher for the later decisions in the project life cycle. Competitive tendering for the supply

Table 5.1 Code of practice for competitive tendering (source: summarised from CIB Working Group 3, 1997).

- in a single round of tendering follow clear procedures that avoid collusion
- conditions for all tenderers should be the same
- confidentiality should be respected by all parties
- standard forms of contract from recognised bodies should be used
- maximum number of invitations to tender should be six – fewer for design services
- minimum tender period should be 6 weeks – more for design and build
- minimum number of compliant tenders received should be four – fewer for design and build
- selection should be made on quality as well as price
- tender prices should be fixed on an unaltered scope of work
- there should be a commitment to teamwork from all the parties

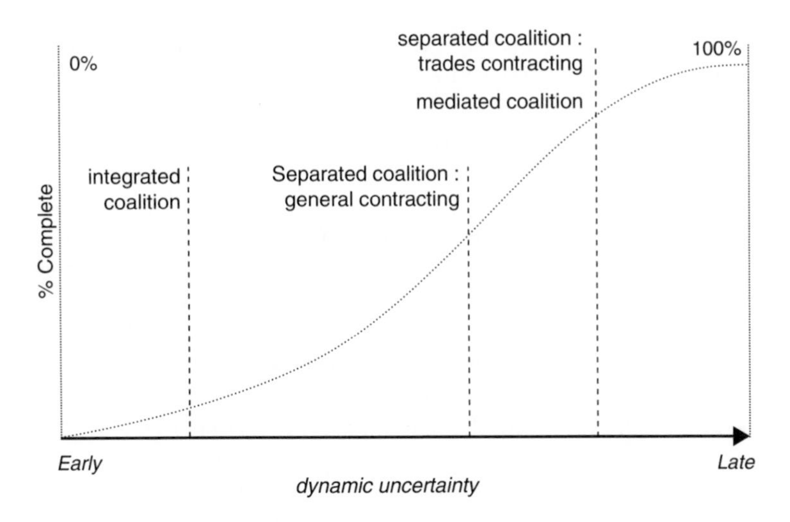

Fig. 5.8 The formation of firm contract.

of architectural resources to complete scheme design may reduce the production cost of those resources by 5%, but as they only account for 5% of the total budget, this will achieve a saving of only 0.25% on the total budget. Against this must be weighed the transaction costs of competitive tendering, and the greater risk of the lemon problem at the higher level of transaction uncertainty. On the other hand, the supply of mechanical and electrical equipment may account for 30% of total project budget, so a 5% saving on production costs through competitive tendering will save 1.5% of the project budget. Project coalitions differ in the point in the project life cycle at which they switch from relying largely on appointment, to relying on competitive tendering to produce firm-price contracts, and hence the benefits of competition in minimising production costs. This is illustrated in Fig. 5.8, which shows the point in the project life cycle at which the switch to a firm-price contract typically takes place within the different project coalitions.

5.6 Forming more effective project coalitions[12]

As illustrated in Fig. 5.1, the use of appointment for the procurement of design services, and competitive tendering for the procurement of execution services, became the predominant means of selecting the project coalition in most countries. National variations, such as the French use of *concours*, the Dutch use of appointment through the honour code and German preferential treatment of local suppliers in competitive tendering do not alter this broad conclusion. More recently, the role of competitive tendering has been reinforced as the European Union and World Trade Organisation have tried to open up construction markets and have extended competitive tendering to the supply of design services as well.

However, as discussed in section 5.3.4, there are serious limitations to the use of competitive tendering as a means of selecting resource bases.

- It tends to release the dynamics of adversarial behaviour, which will be discussed in section 6.7.
- It is very expensive in transaction cost terms, and there is little evidence that it is an effective means of identifying the most efficient supplier, as opposed to the supplier which is most willing to trade off all other project performance criteria against the order-winning one of lowest tender price.
- The client's definition of the project mission, as discussed in section 3.8, includes multiple performance criteria, yet competitive tendering largely motivates against one – price.
- It does little to solve the lemon problem, defined in panel 5.1.

These limitations of the established methods of procurement have led to a search for better ways of selecting resource bases, thereby reducing the lemon problem. All have the principal property that they enable the client to learn more about the capabilities of their suppliers, but some also facilitate joint problem-solving which can tackle the underlying uncertainties inherent in the project. They also tend to place significant additional demand on the client's capability to manage its projects, particularly in the clarity of the project mission and the ability to make appropriate decisions while riding the project life cycle. We can identify at least four approaches here. One is to enable clients to obtain more information about their suppliers through best value criteria for selecting suppliers in competitive tendering, rather than lowest price. A second is to engage in a competitive dialogue with suppliers to facilitate co-learning. A third is to invest in uncertainty reduction with the selected supplier prior to forming a firm contract in two-stage tendering and a fourth is to raise transaction frequency and thereby enable learning between the parties through framework agreements.

5.6.1 Best value procurement

Combining quality and price criteria in competitive tendering has been of increasing importance for two linked reasons. The first is the spread of competitive tendering to the supply of design and other professionally certified resources; the second is the growing use of integrated coalitions. Both imply that competitive tendering is used under higher levels of dynamic uncertainty rather than in the case of the supply of on-site execution services alone. Quality is typically assessed on a scoring principle covering those areas that the client believes to be important to the success of the project, such as supplier track record, qualifications of key project personnel, references and proposed approach to the problem. Scores on these criteria are then weighted, together with the price offered, to provide an overall score to identify the best overall value. The ratio of quality to price is important, and varies with the level of uncertainty prevailing

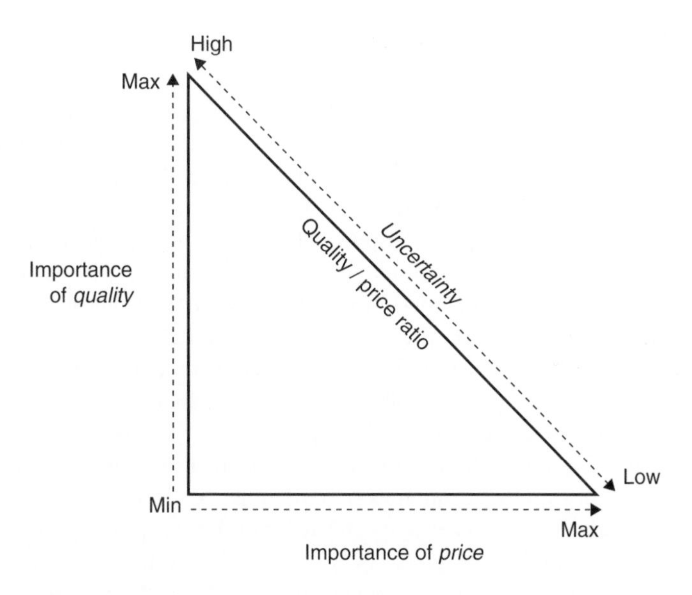

Fig. 5.9 Appropriate price/quality ratios in competitive tendering (source: adapted from CIB Working Group 12, 1997, Fig. 4).

at the time of selection, as illustrated in Fig. 5.9. Such quality/price ratios might be expected to vary from 85/15 for feasibility studies to 20/80 for design work on repeat projects.

5.6.2 *Competitive dialogue*

Although various forms of negotiation have been allowed under EU procurement regulations, the recent introduction of competitive dialogue in January 2006 formalises these processes and incentivises joint problem-solving prior to award of the contract for the execution works. It has been developed explicitly to try to formalise processes where high levels of mission uncertainty mean that some sort of dialogue in order to clarify requirements is inescapable, yet the requirements for accountability and transparency remain. Where mission uncertainty is high, multiple options for the definition and delivery of the facility will, almost by definition, exist. Competitive dialogue encourages the supplier to reveal these options to the client in return for safeguards on due process and, in particular, confidentiality around any proprietary ideas revealed. Competitive dialogue is particularly suitable for the selection of concessionaires on privately financed projects where suppliers compete on the basis of both their financial and technical solutions to the client's needs, but they are appropriate wherever mission uncertainty is high. The selection of the Olympic Delivery Partner for the London 2012 Olympics was the first time this process was formally used in the UK.

5.6.3 *Two-stage tendering*

This option can be used when the client is pressed on schedule, or the project requires early contractor involvement. It was used on the successful Emirates Stadium presented in panel 6.5, and, in effect, this was the arrangement on the Eden Project presented in Case 17. Two-stage tendering involves organising a competitive bid for the supply of construction project management services typically covering detail design and execution on site. This typically fixes the costs of pre-contract services, design fees, risk premia and the like. A preferred contractor is then selected who then organises competitive tendering. Keen pricing at this stage is dependent on the effectiveness of this tendering process at the second tier of suppliers. The second stage significantly reduces uncertainty which allows the negotiation of either a lump sum or guaranteed maximum price contract – see section 6.5 – for execution on site, which can also include detail design.

There are many advantages to two-stage tendering such as speed of procurement and early involvement of the contractor in the project. The problem is that the client is effectively locked in to the contractor after stage one yet the contractor is free to walk away from the project should the client be too tough in negotiating the stage two contract – in other words, it puts the client at risk of the hold-up problem discussed in section 6.3. It is also not appropriate where there is expected to be continuing high levels of uncertainty through execution, and demands discipline from the client, particularly between stages one and two. Arguably, two-stage tendering is most likely to be effective within some kind of alliancing relationship as discussed in section 6.8.

5.6.4 *Framework agreements*

Over the past 10 years the concept of partnering has diffused widely since its initial development by the US Army Corps of Engineers, and is now widely seen as a way of avoiding many of the negative consequences of competitive tendering[13]. However, diffusion has also meant confusion, with the term *partnering* meaning many different things to different people, and it is seen by some as a panacea for all the industries ills, rather than a procurement option[14]. For our purposes here, partnering will be defined as formal arrangements between at least two members of the project coalition to work together on a programme of projects for a defined period, usually called a *framework agreement*. Thus, the essence of partnering is to raise transaction frequency so that clients and suppliers can learn more about each other's requirements, minimise search and selection costs, make transaction-specific investments that can lead to reduced production costs, and generate trust, and requires that the client develop a programme management approach as discussed in section 15.3. It is based on the basic social process of cooperation, presented in panel 5.5. Partnering can also be distinguished from *alliancing* – discussed in section 6.8 – which provides non-adversarial incentive arrangements for the coalition members on a single project.

Panel 5.5 The evolution of cooperation

Experimental studies based on the famous Prisoners' Dilemma game have shown that the essential prerequisite for the emergence of cooperation rather than defection between two parties is repeated games. In Prisoners' Dilemma the highest rewards for a single party come from defection when the other party cooperates; the cooperating party then receives the lowest reward. Both parties share low rewards when both parties defect, and medium rewards when both parties cooperate. In a single game, the strategy bringing the highest reward is defection because if you cooperate, the other party may defect, delivering the lowest level of reward. However, the mirror image situation also applies to the other party, so both defect and jointly receive very low, but not the lowest, rewards. Where games are repeated – and are known to be repeated into the future – then it is possible, and indeed very common, for mutual cooperation strategies to emerge, even without communication between the parties. Under repeated games, the strategy with the highest returns in the long run is tit for tat – never to defect first, but always to respond to a defection by the other party with a defection. Interestingly, better long-run returns are also experienced by a party which is prepared to forgive a single defection by the other, but not more than one. Although more aggressive strategies can yield higher returns in the short run, they all hit problems of repeated defection as other parties learn that they will defect – in the long run, cooperation is the better strategy.

Source: Axelrod (1990).

Where appointment is used as the method of selection, then partnering, in deed if not in name, is frequently the norm. Repeat clients are the most important source of new contracts for most suppliers of design services. The selection of consultants which have successfully completed earlier contracts greatly reduces adverse selection through minimising information asymmetries in favour of the client – lemons can more easily be spotted because they leave an acid taste once chewed. As Adam Smith implies in the epigraph to this chapter, moral hazard is also reduced because the failure to deliver on the part of a supplier will entail the loss of goodwill, reputation and future contracts. However, these arrangements are frequently not formalised – see Case 6 for an exception – and would not be considered true partnering by many observers.

The importance of the growth of formal partnering is in the context of the use of competitive tendering as the predominant means of supplier selection. Many clients are realising that the adverse selection problems inherent in competitive tendering outweigh the benefits of tough price competition, and they are moving towards selecting suppliers for programmes of work rather than issuing tender invitations for each project. In the UK, it was the privatised utilities and retail clients which were at the forefront of this movement during the 1990s. The privatised utilities moved to correct the years of underinvestment in constructed assets, which was the heritage of their period as nationalised industries – see panel 6.3 for the case of BAA. The retail clients mobilised to take advantage of new market conditions – see panel 5.6 for the case of Esso Europe. Case 5 illustrates these principles in the evolution of the relationship between Marks & Spencer and Bovis over half a century.

Panel 5.6 Esso's Blue Ribbon Task Force

In 1994, Exxon set up the Blue Ribbon Task Force to reduce the schedule and budget for the construction of its retail service stations. They quickly realised that prefabrication was the key to meeting their targets, and invited a German company to adapt an existing UK design to meet the tough German regulations. This design was then checked against national building regulations across 14 European countries. Pan-European standardisation was achievable, so in 1996 Exxon established a European Retail Engineering Skill Centre (ERESC) in Brussels, and partnered with a number of suppliers across Europe to roll out the programme. Suppliers are continually benchmarked against each other across Europe, and weak suppliers were supported to bring them up to the mark.

For the shops at service stations, trials were conducted by 8 suppliers building 25 different shops using a variety of volumetric and flat-pack designs. As a result of these trials, potential suppliers were asked to price two different designs to be constructed across whatever geographical area the supplier believed it could serve. From 17 bids, 2 were selected – Rousseau Stewing and General Electric Capital Modular Space (GECMS). These two companies then bid country by country to deliver actual shops, with the allocation of work carefully balanced between them by the ERESC. Savings of up to 60% were achieved in the very high cost countries such as Germany. GECMS builds its modules in a factory in Belgium with 50 staff. It partners with its own supply chain clustered around the factory, and employs directly most of its on-site erection workforce.

Source: Bennett and Jayes (1998).

The benefits of framework agreements include:

- the development of trust between the parties as they learn about each other through repeated transactions;
- the opportunity to develop standardised component systems to be rolled out on a programme of projects, and to progressively reduce the production costs of these systems down the learning curve;
- the elimination of large areas of transaction costs associated with supplier selection and dispute resolution;
- more predictable workloads for suppliers, allowing them to make investments in processes, equipment and people;
- greater opportunity for the alignment of objectives within the project coalition, enabling a more collaborative approach to problem-solving;
- the ability to implement dedicated IT systems for process and project, thereby reducing the problems of interoperability discussed in section 14.3.

However, successful framework agreements rely on a number of specific factors:

- Clients with investment programmes requiring a number of closely related facilities in technological terms. It is notable that many of the widely publicised partnering relationships have been for relatively simple building types such as out-of-town stores and McDonald's restaurants.

- A willingness to change and shift from a win–lose to a win–win relationship. Managers who have honed their skills and made their careers winning adversarial battles may have great difficulty in switching to new ways of working, and accountability requirements may also have to change. This last requirement is a particular difficulty in the public sector. Significant investments in training and relationship building may be required before benefits are realised.
- A willingness to swallow losses on a particular project and evaluate the benefits of the partnership over the programme as a whole.
- Clients with strong project and programme management capabilities, who are capable of understanding the risks they are sharing with the resource bases. As BAA's Construction Director put it, 'partnering is tough; it requires a huge investment in time, training and coaching'[15]. We will revisit this issue of client capabilities in Chapter 15.

5.7 The development of e-procurement

E-procurement is the latest development with potential for changing the procurement process; further details are provided in section 14.6. The selection and motivation of contractors involves the embodiment of information in an enormous amount of paper; if this information can be captured and transmitted in digital form, then there is the potential for significant savings in transaction costs. In order to recoup the investment costs in such systems, much greater standardisation of the procurement process is required than has been the practice to date. The two main aspects of e-procurement are *e-sourcing*, for the identification and selection of potential suppliers and purchase-to-pay (P2P) systems for the administration of the contract once let. One e-sourcing initiative in the UK is the government-sponsored Constructionline which provides a database of 14 000 suppliers of construction services and 1700 clients supported by a standardised pre-qualification service for suppliers[16]. P2P systems need to be integrated into the enterprise systems of the client organisation such as to ensure that all transactions are properly accounted. Figure 5.10 shows CGI's integrated e-procurement system, but it is only implemented at present for 'commodity' purchases. Initiatives that include complex asset procurement are presently taking place on both sides of the Atlantic, such as Construction Industry Trading Electronically (CITE)[17] in the UK and the PD^2 in the USA, described in panel 5.7, but they remain limited in their functionality beyond relatively low uncertainty transactions.

Panel 5.7 PD^2 e-procurement system

Procurement Desktop Defense (PD^2) was implemented to meet the US Department of Defense's (DoD) procurement requirements as part of its Standard Procurement System by providing a single application for the whole of the DoD. After much criticism of its cost-effectiveness, the system was relaunched in 2003 and moved to a web-based format in 2006. In essence, PD^2 is an interface between the internal applications of the client and the systems operated by suppliers. In combination with AcquiLine, it has the potential to provide seamless, web-based transaction governance. PD^2 enables the paperless selection and management of contractors, also making it much easier to maintain the integrity of databases on contractor performance. It 'provides graphical document management, electronic routing and approval, online acquisition regulations, workload management and powerful ad-hoc reporting' and is currently deployed to over 20 000 users at approximately 800 sites around the world. It remains unclear whether the investment has been worthwhile.

Sources: http://pd2.amsinc.com/pd2web.nsf (accessed 10/11/01); Wikipedia (accessed 07/08/08); http://www.caci.com/business/systems/ (accessed 07/08/08).

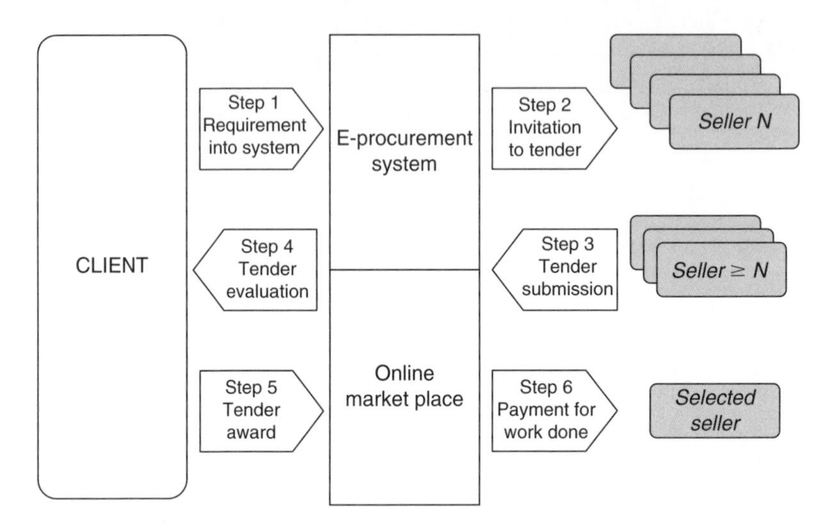

Fig. 5.10 Momentum reverse auction system (source: adapted from http://www.cgi.com/web/en/library/solutions/governments/74095.htm(accessed 07/08/08)).

5.8 Probity in procurement

One of the major issues in the selection of resource bases is probity in the selection process. Bribery and corruption have long been associated with construction procurement, particularly when the client is in the public sector[18]. Such corruption is, broadly, of two kinds:

- *Personalised* – where one or more public officials extract bribes from suppliers in return for favouritism in the award of contracts. It can usually be controlled, even if not eliminated, through the maintenance of effective independent audit procedures.
- *Institutionalised* – where work is shared out between suppliers and in return these suppliers make donations to political parties. This is inherently corrosive and can only be resolved through reform of political institutions.

Personalised corruption is widespread, particularly in developing countries where it can become so pervasive that the effective management of projects becomes impossible. Table 5.2 shows selectively the rankings of perceived levels

Table 5.2 Scores for levels of perceived corruption in 2007 (10.0 = lowest) (source: Transparency International: http://www.transparency.org/).

Ranking	Country	Corruption score
1	Finland	9.4
1	Denmark	9.4
1	New Zealand	9.4
7	The Netherlands	9.0
12	United Kingdom	8.4
20	United States	7.6
25	Uruguay	6.7
34	United Arab Emirates	5.7
38	Botswana	5.4
41	Italy	5.2
72	China	3.5
143	Russia	2.3
179	Somalia	1.4

of corruption of public officials across all sectors for 179 countries for which data are available. The rankings can be explained by factors deep in the national business systems of the countries concerned, and broadly suggest that the Nordic and Anglo-Saxon countries tend to be less corrupt on this measure, and some of the most impoverished countries in Asia and Africa more corrupt.

The data on perceived corruption do not measure levels of institutional corruption, yet this is a major problem in construction procurement. In the early 1990s, both Italy and Japan were rocked by public procurement scandals. In both cases the basic mechanism was the same. The political parties which had dominated national politics since 1945 – the Christian Democrats in Italy and the Liberal Democrats in Japan – were systematically extracting donations to party funds from a selected group of construction contractors in return for the available contracts being shared between them. The process – known as *dango* in Japan and *tangentopoli* in Italy – completely undermined the selection of resource bases on the basis of efficiency or effectiveness, and created very cozy conditions for the leading construction corporations in both countries. Other countries have also suffered in similar ways, as panels 5.8 and 5.9 show.

Panel 5.8 Jacques Chirac and procurement in the Ile de France

Between 1989 and 1995, the procurement of contractors for the construction, refurbishment and maintenance of schools in the Ile de France – the region that includes Paris – was institutionally corrupt. At least nine French construction firms – including what were then the local subsidiaries of some of the world's largest construction corporations – are implicated, including what are now Bouygues and Vinci. Contracts to a total of 23.4bn francs (£2.34bn) were shared out between these firms, which were then obliged to pay a sum equalling 2% of the contract value to a fund. From this fund, the three main political parties shared pay-outs – 1.2% to the ruling coalition of the Rassemblement pour la République (RPR) and the Republican Party, and 0.8% to the opposition Socialist Party. Unsurprisingly, it was the

Greens who exposed the scandal in 1996. Payments were made both directly in cash and through the creation of fictional employees, and the process was coordinated by the office of the Mayor of Paris. At the time, Jacques Chirac was both Mayor of Paris and President of the RPR.

The trial finally started in 2005 and some 43 people were convicted following much comment about the 'empty chair' at the hearings because of the absence of Jacques Chirac – then president of France and hence immune to prosecution or being asked to testify. Appeal hearings in 11 cases confirmed the convictions, including prison sentences, in early 2007 and the construction firms were fined a total of €47m later that year for their role in the scandal. Jacques Chirac was finally indicted for this and other alleged corruption offences in November 2007 after he had left office.

Source: *Le Monde* (25/07/01; 23/03/05; 07/07/05; 11/05/07; 21/11/07); Wikipédia (accessed 07/08/08).

Panel 5.9 Zembla: The collapse of bid ringing in Dutch construction

During a TV programme transmitted in November 2001, a former director of a Dutch contractor sensationally revealed the extent of bid rigging in the Dutch construction sector. A parliamentary enquiry reported in December 2002 that forgery, collusion, tax evasion, cover-pricing, predatory pricing and bid rigging were pervasive in the Dutch construction sector with the tacit approval of the public authorities. Rooted deep in the corporatist culture of the Dutch society and economy – the polder model – the revelations came concurrently with the broader political breakdown of that system. The UPR (Uniform prijsregelend reglement) and Erecode (honour code) had created a price-ringing cartel which was openly defended on the grounds that it was more efficient than competitive tendering because of the savings on the transaction costs – identified in sections 5.3 and 6.6 – that it allowed.

The practice grew until it became the 'industry recipe' for Dutch construction in which risks were systematically transferred from the supply side to clients at an estimated average cost increase of 8.8% of construction turnover. The policy response was the launch of a reform programme for the Dutch construction sector explicitly modelled on the UK reform programme described in Case 2, and to reassert the importance of competitive bidding in the context of EU procurement regulations.

Sources: Bremer and Kok (2000); Dorée (2004); Priemus (2004).

It is for these reasons that most countries have public sector procurement policies which stress the importance of competitive tendering on the basis of lowest cost, reinforced by international bodies such as the European Commission and the World Trade Organisation. Those countries that have tried to move away from competitive tendering, and allow tenderers to collaborate in setting prices, can fall foul of international regulatory bodies.

The issues of transparency and probity in procurement make the development difficult of new procurement routes such as partnering and concession contracting where price is not the sole selection criterion. For instance, the contracts which caused the trouble in the Ile de France were METP (Marché d'Entreprises Travaux Publics) contracts through which the supplier undertook construction and then managed the facility for a period before it reverted to the public authority, reimbursed by regular payments. Great attention has to be paid – and be seen to be paid – to objective selection criteria, and partnering relationships have to come up for regular and public review and audit.

5.9 Summary

The aim of any procurement route is to motivate the best in efficiency and effectiveness from suppliers while mitigating the worst in opportunism. Under uncertainty, it is difficult to come to the clarity of agreement in advance that would render opportunism nugatory, yet a reliance on trust alone to mitigate opportunism can only be described as naïve. Even the Dutch system, which apparently started with good intentions, degraded as firms took advantage of the system for their own advantage. No procurement route is perfect and always depends fundamentally on the interpersonal relations of those responsible for interfaces within the project coalition, but some procurement routes are better than others for particular situation in terms of the uncertainty inherent in the project mission and the level of dynamic uncertainty at the formation of firm contracts. As Case 5 shows, even a client such as Marks and Spencer historically committed to collaboration commissioning a relatively stable building type strategically varies its procurement route depending on the actual requirements of the particular project being commissioned.

This chapter has explored the ways in which clients select their suppliers of design and execution resources in terms of the principal/agent problem using concepts from transaction cost economics. In particular, it focused on the problem of adverse selection, or how clients can ensure that their suppliers are appropriate, competent and trustworthy. Four main methods of selection were identified – in-house capabilities, appointment, *concours* and competitive tendering – and the strengths and weaknesses of each in relation to the levels of mission and dynamic uncertainty on the project were explored. The configurations of project coalitions that arise from these selection decisions were then analysed and grouped into four main types – the separated, integrated, mediated and unmediated coalitions. These were again analysed in terms of the level of uncertainty inherent in the project mission. Finally, some more recent developments – selection on the basis of best value, competitive dialogue, two-stage tendering and framework agreements as well as e-procurement – were explored. We concluded with some cautionary tales on the frailities of procurement systems in the face of corrupt and opportunistic forces. Once selected, the supplier must be motivated to provide consummate performance in the delivery of the resources promised in the contract, and it is to this problem that we will now turn in Chapter 6.

Case 5
Partnering Between Marks & Spencer and Bovis

Marks & Spencer plc (M&S) is a leading UK retailer of clothing and household goods which has added financial services to its portfolio. It is famous worldwide for its skills in supply-chain management and the way in which it has developed close relationships with its suppliers of retail goods. It has applied similar principles to its suppliers of the stores in which these goods are sold. M&S is first and foremost a retail business. Store development is no more than a vehicle for the company's core business. The building is not seen as an end in itself; every element in that store is there to enhance the business by attracting customers to buy.

The fundamental principles of M&S procurement policy are flexibility and integrity. The traditional M&S policy was based on the scheduled refurbishment of the existing stock; even where complete demolition was involved, this was typically based on a phased schedule to allow the store to continue trading. These demands required considerable flexibility in the procurement process, which was achieved by establishing continuing relationships with suppliers, and a mediated procurement route – the estimated prime cost (EPC) contract, more commonly known as *management fee*. In 1970, many of M&S's principal suppliers had worked with the company for up to 40 years. Most famously this included Bovis as the EPC contractor, but also applied to the small number of architectural practices, principal quantity surveyors and key trade contractors employed on a continuing basis. Integrity is achieved by building up high levels of trust through personal relationships, and ensuring that the available work is shared fairly among the regular suppliers. It is also achieved by ensuring a high level of site facilities for the operative workforce – a policy instigated by M&S board directors themselves who were shocked at the low quality of facilities available following a site visit.

As the store development policy shifted towards new build in the mid-1980s, with the advent of the out-of-town shopping centre, M&S began to shift to a broader base of suppliers and to a more conventional and less flexible integrated coalition. In particular, two-stage tender design and build, with and without novation of the designers, based on the UK's JCT 81 standard form of design and build contract, was preferred. Bovis now shared the workload with three other contractors, while obtaining around 50% of the overall M&S workload. The effect of this shift away from EPC was that contractors were less empowered and design was more closely controlled within the client organisation, resulting in the development of a fully specified store 'template'.

The first M&S store on the European continent was opened in Pans at the Boulevard Haussmann in 1975. The approach to procurement in France was to apply as far as possible the same arrangements in France as in the UK – as one (French) informant put it, M&S comes with 'the Union Jack stamp on it'. This was in line with the overall M&S strategy of applying the distinctive M&S business model to continental markets, rather than extensively adapting that model to the different markets. In 1988, as part of its new European expansion policy, the M&S French development office began looking for French consultants with whom to work in France and Belgium. One of the selection criteria was that the consultants had either worked for other large British clients or had links with an office in the UK. This was to ensure that the consultants would be able to follow the methods used by M&S in its development process – it tended to choose architectural practices associated with its British suppliers and retained the Paris offices of British quantity surveying firms. M&S 'took Bovis in its suitcase' when it entered the continental European market. In an attempt to replicate the EPC type of contract, the role of the *assistance au maître d'ouvrage* (AMO) was developed by M&S.

In 1968, M&S operated an estate of 241 retail stores with a sales surface of 3.9m ft^2; by 1997, the total number of M&S branded stores was 373 – the company operated as Brookes Brothers in the USA and Japan – providing 13.3m ft^2, 286 of the stores being in the UK and providing 10.6m ft^2. A 3-year capital programme launched in 1997 aimed to add a further 1.9m ft^2 in the UK and 1.1m ft^2 in the rest of the world – including a further 0.18m ft^2 in France – an increase of 23% on the 1997 figure.

A major part of this expansion in the UK was the acquisition of 19 stores from Littlewoods, all of which were to be refurbished at once and re-opened by the end of 1998 – providing an additional 0.6m ft^2 of retail space. This fast-track programme was dubbed Project Robin. The decision to go ahead was based on 3 hour 'non-finessed' surveys of each store, so design development inevitably contained surprises. To deal with the risks and tight schedule of Project Robin, it was clear that a new management approach was required. As a result, a dedicated team was set up to manage the programme in a tightly coordinated manner. The rationale behind Project Robin was to improve communications between in-house project team members from different departments as well as partnering with, and increased empowerment of, contractors.

The Project Robin team was set up in September 1997 and remained in place for the duration of the refurbishment programme for the 19 stores. This team included members from the M&S store development function, as well as store operations and people seconded from leading construction firms. This organisational structure was a totally new initiative on the part of M&S, and even more radically, it was led by a manager from store operations, and not store development. This cross-functional approach provided a forum for project innovation and new forms of project management. Decisions concerning all Project Robin stores could now be made centrally by one project team, rather than projects being managed by regional core teams. This enabled a fast-track approach whereby main contractors were appointed by negotiation as opposed to competitive tender.

The dedicated team structure represented a shift in M&S development management policy. In the past, the single partner arrangement with Bovis meant that M&S could make changes in the knowledge that these would be accommodated by the building team. During the 1980s, with the shift to design and build, this flexibility was reduced in the search for efficiency, and contracting relationships became more traditional. As the Project Robin manager remarked, 'the pendulum is now returning to the centre' as strong partnering agreements are established with a number of preferred firms. He feels that empowerment should be encouraged, that the store development policy should be 'brought back to type' and that there is a need for a flexible approach to store development where the triad of time, cost and quality can be evaluated for each project.

Procurement innovations in France – M&S's largest overseas market – went in a different direction. Bovis for many years retained the role of AMO, but as one (consultant) informant commented, '(Bovis has) now been put into competition with other firms as on some jobs it was felt they were flooding the project with people'. Additionally, the use of the AMO by M&S has come up against problems as some French contractors had refused to tender, leaving M&S more reliant on Bovis than it would like. For this reason, M&S widened their list of preferred contractors and experimented with *entreprise générale* (general contracting). This was first tried out in a new-build fit-out in 1998, where as-built drawings of the commercial centre were available.

Thus a clear pattern emerges in M&S procurement policies for the selection of principal contractors. Where the programme of store development was dominated by the refurbishment of existing buildings, as was the case from the 1950s to the 1980s with town-centre stores, a mediated coalition was preferred, with partnering in effect if not in name between M&S and Bovis, and the principal consultants. When store development strategy switched to new build in out-of-town stores, risks reduced and better value for money could be obtained by using a more integrated coalition, although this was always on a relatively open alliance-type basis. Entering foreign markets raises uncertainty, and so M&S preferred to rely on its UK supply base's French offices and associates, and to adapt its favoured UK procurement route to French practice. Having gained confidence in the French market, M&S chose to switch to a more typically French integrated coalition on a new-build store. When a major part of its store development programme included 19 fast-track refurbishments, with information based on brief surveys, it chose to switch back to a much more mediated coalition, with formal partnering with a small group of suppliers.

In 2001, M&S announced the closure of most of its overseas operations, the sale of property and a move from its flagship headquarters in Baker Street to new premises in Paddington, West London. M&S requirements as a client for construction were now entering another phase. Experimentation in store formats to try to regain market share was reflected in the approach to procurement. For the short-lived Lifestore concept, they selected a 'star' architect John Pawson in an attempt at 'designer' credibility. A more sustained – and sustainable – strategy has been Plan A, launched in January 2007 in the context of a reassertion of the role of property in the financial structure of the business in its 2006 Annual Review:

> 'Our biggest tangible asset is our store portfolio. It is also the way that our customers experience our brand. Our stores are in great locations, but they have been subject to underinvestment in the past years. We have embarked upon a major refurbishment programme to rectify this and bring our stores up to the standards out customers expect'.

Plan A intends to make M&S carbon neutral by 2012, and it became a founder member of the UK Green Building Council and opened the first of four eco-stores in Bournmouth at the end of 2007 by refurbishing an existing high street property. In view of the uncertainties involved in sustainable specification and its pilot nature, the project apparently used a traditional procurement route of architect design followed by the selection of a general contractor. A new factory in Sri Lanka was also claimed to be carbon neutral in mid-2008. Learning from these projects prompted M&S to appoint the Building Research Establishment to a 5-year framework contract to develop its green building strategy.

Sources: MPBW (1970); Sieff (1990); Carr and Winch (1999); http://www.mark-sandspencer.com (accessed 24/04/07; 10/08/08); Building (accessed 10/08/08); *Contract Journal* (accessed 10/08/08).

Notes

1. Cited in Milgrom and Roberts (1992, p. 257).
2. See also McAfee and McMillan (1986).
3. The research programme stimulated by 'the lemon problem' resulted in the award to Akerlof and two colleagues of the Nobel Prize for Economics in 2001.
4. See the survey by Biau (1998) of the organisation of *concours* in nine European countries.
5. This argument is largely derived from McAfee and McMillan (1987) and Samuelson (1986). Lowe and Skitmore (2006) review the construction-specific literature.
6. Where such information is used in negotiations with sellers, it is pejoratively known as a 'Dutch auction' – a form of open bid auction.
7. Sadeh *et al.* (2000) make a similar argument for defence acquisition projects.
8. See also Masterman (2001).
9. See the case study of Stanhope provided by the NAO (2005b).
10. See Pryke (2004, 2005); Pryke and Pearson (2006).
11. In effect, the Chartered Institute of Building's *Chartered Building Company* scheme implements a similar principle.
12. Many of the insights in this section are taken from Simon Rawlinson's excellent series on procurement which started in *Building* in 2006. See also http://www.davislangdon.com/EME/Research/
13. Research projects by a University of Westminster team – Barlow (2000), Barlow *et al.* (1997), and led by Mike Bresnen – Bresnen and Marshall (2000a, b) – review the evidence.
14. Bresnen (2007) wittily punctures some of these pretensions.
15. Andrew Wolstenholme, presentation *Construction Innovation* Conference, Ottawa, June 2001.
16. http://www.constructionline.co.uk (accessed 07/08/08).
17. http://www.cite.org.uk/(accessed 30/08/09)
18. Much of the national information in this section comes from two special issues of *Building Research and Innovation*: Construction Business Systems in the European Union 2000, **28**: 2 and Global Construction Business Systems 2002, **30**: 6. See also Stanghellini (1996) on Italy and Sha (2004) on China.

Further reading

Gruneberg, S. and Ive, G. (2000) *The Economics of the Modern Construction Firm*. London, Macmillan.
A detailed analysis of the behaviour of construction firms and their markets.

Milgrom, P. and Roberts, J. (1992) *Economics, Organization, and Management*. Upper Saddle River, Prentice-Hall.
The standard text on institutional approaches to the economics of firms and sectors.

Walker, D. and Hampson, K. (2003) *Procurement Strategies: A Relationship-Based Approach*. Oxford, Blackwell.
An extended reflection on the learning from the successful National Museum of Australia project in Canberra showing how innovative partnering relationships delivered a highly successful facility.

Chapter 6

Motivating the Project Coalition

6.1 Introduction

> 'As there is a certain degree of depravity in mankind which requires a certain degree of circumspection and distrust, so there are other qualities in human nature which justify a certain portion of esteem and confidence'.

The words of James Madison[1], commenting in 1788 on the struggle to ratify the new constitution of the United States, sum up well the dilemmas generated by adverse selection and moral hazard in construction procurement. Once a supplier firm has been selected in the manner discussed in section 5.3, it needs to be motivated to give its best on the project so that it acts in the client's interests as well as its own best interests. Higher qualities in suppliers need to be encouraged and tendencies to depravity discouraged. Although there would be no contract if there were no coalition of interest between the two parties, there remains plenty of room for divergence of interest unless appropriate incentive arrangements are in place. This is the problem of *moral hazard* – how can the client be sure that the firm, once hired, will fully mobilise its capabilities on the client's behalf, rather than on behalf of the firm itself or of some other client? In other words, how does the client encourage consummate rather than perfunctory performance of the contract? Perfunctory performance is that which is not in unambiguous breach of the contract, but is less than satisfactory in execution; consummate performance is the best available under the circumstances.

The traditional answer to these questions is to use complex contracts. However, such contracts frequently set up perverse incentives, generate high transaction costs in attempting to manage the contract, and lead to a climate of low trust and adversarial relations. This chapter will first define the problem of moral hazard in construction contracting, and then identify why its solution is made more difficult because of post-contract asset specificities. It will then go on to investigate how these problems have been solved through the development of complex forms of contract which have important properties of turning markets into hierarchies. Finally, new ways of managing the relationships between clients and their suppliers, which are aimed at aligning incentives within the project coalition, will be discussed.

6.2 The problem of moral hazard in construction projects

The incentive framework in a contract for the supply of goods or services is best defined in terms of who takes the risk if things do not go according to plan. In a world of perfect information, *simple contracts* are all that are needed between buyers and sellers. So long as the buyer can fully measure the quality of what is offered on the market, he or she can buy on price with confidence, and the adage that 'you get what you pay for' is true. The wise buyer smells and pressure-tests the melon on the stall, and compares melon prices on adjacent stalls, before parting with money and completing the transaction. Prior to purchase, all the risks associated with the melon lie with the supplier; after purchase they all lie with the buyer – risk is transferred cleanly and unambiguously, and all the information about the quality of the melon is in the price negotiated. In such cases, there is no problem of post-contract incentives; the transaction is timeless and only needs the lightest governance to be viable, such as redress if the seller refuses to hand over the melon once the buyer pays the agreed price for it (or vice versa). Such information-perfect markets are rare, however, and most need some sort of post-contract incentive framework to ensure consummate delivery – even the advocates of neoclassical auction theory discussed in section 5.3.4 see post-contract incentives as vital to efficient contracting.

The central tenet of this book is that construction project management is a problem of the management of uncertainty through time, and so it will be no surprise to find that melon-type contracts are rare in construction. Only contracts for the most straightforward 'commodity' materials – such as cement, common bricks, and standard timber and steel sizes – are close to the melon-type, and even here delivery reliability and quality assurance can involve post-contract incentive problems. Most construction contracts are incomplete when agreed, and so motivating consummate rather than perfunctory performance of the contract is a major management problem to which a wide variety of governance arrangements has evolved to address.

The problem of moral hazard arises post contract for the following reasons:

- Suppliers have information that is critical for effective client decision-making, but are not motivated to fully share that information.
- Buyers cannot easily monitor the quality of the goods or services received, and so suppliers are tempted to substitute lower quality goods or be less than diligent in the supply of services.
- Clients find it difficult to clearly measure the relative performance of the contractually separate members of the project coalition.
- Uncertainties regarding the utility of the facility mean that the client may wish to make changes in its functionality as new information becomes available through the project life cycle.

For these reasons, that will be discussed in this chapter, the problems of moral hazard are rife in construction procurement, and elaborate governance mechanisms have been developed in the attempt to mitigate them; these are typically embodied in *complex contracts*. They also exist in the relationship between employers and employees, an issue which will be tackled in Chapter 7. Negotiating complex contracts from scratch would be a very time-consuming process, and so in most countries standard forms of contract exist. While these are necessarily adapted to the requirements of the national construction business system, they all share a number of common features, which will be discussed in section 6.4. For the purposes of illustration, the standard form of contract used in this chapter will be the 2005 NEC 3rd Edition, *Engineering and Construction Contract* (ECC), which is becoming widely used in the UK for both building and civil engineering, is recommended as the best practice choice by the Office of Government Commerce and is also used in other countries which have contracting systems based on UK practice. Software applications are available to support its administration – see panel 14.6.

6.3 The problem of switching costs

The achievement of consummate performance is made more difficult to manage as a result of the *fundamental transformation*[2] between pre-contract and post-contract relations between the parties. During the selection of the supplier, opportunistic behaviour is constrained through the market. In the absence of monopolies or cartels in supply (pre-contract asset specificities), signs that a supplier is behaving opportunistically during contract negotiation are likely to mean that the contract is awarded to a competing firm. However, once the contract is signed, the parties make transaction-specific investments – the classic case is the new factory with specialist machinery dedicated to the needs of a particular customer. These may, at first sight, appear to be trivial in construction compared to other industries, but they are not inconsequential in the governance of construction transactions – indeed they are surprisingly important. The problem is that the fundamental transformation generates asset specificities which increase the possibility of opportunistic behaviour by the parties. Capital sunk in the project cannot always be easily switched to other uses, and the ability to switch is unevenly distributed between the parties – *temporal specificities* post contract generate the hold-up problem[3].

From the supplier side, one of the most important reasons for making minimal client-specific investments is to retain the possibility of switching. Construction resource bases typically make general-purpose investments – construction plant is multi-purpose, and human resources are formed around broadly based crafts rather than specific skill sets. This is so that they can take advantage of the portfolio effects identified in section 1.5 and quickly redeploy resources in the event of a problem on a particular project. For instance, if earlier trades do not finish their work on time, resources can be switched at short notice to a project that is ready to go ahead. Similarly, if the client is slow in making a key decision, or there are delays in obtaining regulatory approvals, architects and engineers can be allocated overnight to the work of another client, which can be progressed. The system of interim payments while the work is in progress also has the effect of reducing post-contract asset specificities for the supplier – the less money the supplier is owed by the client, the easier it is to switch resources away from that client should problems in the relationship arise. Where a client has a reputation for poor performance on interim payments, suppliers typically charge a risk premium.

From the client side, the situation looks very different. Switching is not an easy option because capital is sunk in the existing project through the interim payments to suppliers for work already completed. The costs of replacing a supplier can be quite large, and this gives the supplier a significant margin for opportunistic behaviour. This is known as the hold-up problem because bargaining within the opportunistic margin typically takes the form of the threat of delayed payment or delays to the progress of the works. The costs of switching are high and include:

- transaction costs of retendering;
- inability to recover the additional costs generated by the original supplier's failure from the new supplier;
- associated litigation;
- the premium likely to be charged by the new supplier to complete the works because of the uncertainties around what work has actually been successfully completed.

Thus the size of the margin for opportunistic behaviour by suppliers illustrated in Fig. 6.1 is, in formal terms, equal to the difference between the contract price and the value of the facility to the client less the client's switching costs. Of course, on actual projects the uncertainties in investment appraisal calculations discussed in section 3.6 make the precise parameters of this margin very difficult to determine and bias clients towards negotiation rather than switching. These problems exist where there is no change in the contracted scope of works, but where it is necessary to renegotiate the contract this opportunism is given full play up to the cap of the client's switching costs, and suppliers are well practised in exploiting this margin. The evidence is that switching costs are much greater for suppliers of on-site services than design services. Clients can, and regularly do, sack architects, but this is relatively rare for suppliers of site-related services.

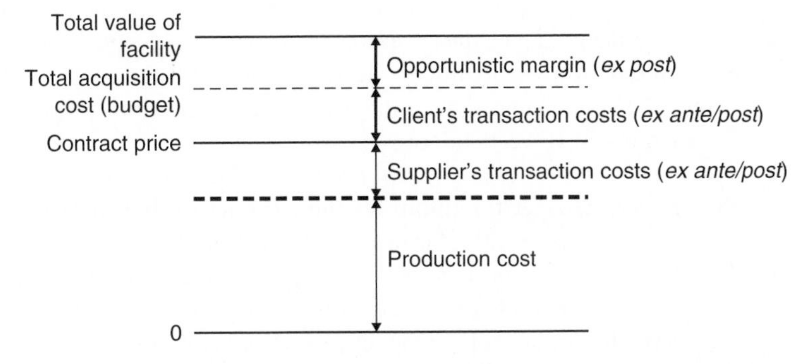

Fig. 6.1 Opportunistic margin in transaction governance (source: adapted from Winch, 2006a, Fig. 14.5).

The existence of these switching costs has two very important results for the management of the construction project. Firstly, in the situation where performance is perfunctory, it tends to be indulged so long as the additional costs generated by poor performance lie within the opportunistic margin – better the devil you know than the devil you do not. Secondly, where uncertainties need to be resolved in the contract as more information is acquired through the life cycle, or actual changes in the contract are required because of new information arriving, the negotiations around the adjustment of the contract allow ample scope for opportunism up to the switching cost limit. While clients are not entirely helpless in this situation – and indeed may withhold information from suppliers – the information asymmetries in this situation typically favour the supplier. As a result, prices charged for additional and changed works tend to carry a higher profit margin than those for the works as originally specified. Much of the governance of transactions in construction is aimed at minimising opportunistic behaviour by suppliers.

6.4 Managing the problem of moral hazard

Under the ECC, the suppliers responsible for execution are obliged to provide 'the Works in accordance with the Works Information', while the designers who generate the works information are obliged to provide 'the Services in accordance with the Brief'. Clear as these obligations are, they require much more complex contracting than the obligation to 'hand over a melon'. Both the brief and the works information are, inevitably, incomplete documents – the brief more so than the works information. Hence, the contracts between the client (employer under the ECC) and its suppliers (contractor or consultant under the ECC) need to be continually adjusted as information is acquired through the life cycle. Complex contracts, therefore, provide a number of mechanisms by which principals and agents can continually adjust the detailed terms of the contract, which has the effect of turning market into hierarchy[4]. Complex contracts typically enable this adjustment process by:

- *Specifying the conditions under which adjustments can be made.* These are known as compensation events under the ECC and are listed in some detail, ranging from the client changing its mind, to weather conditions, which have occurred less than every 10 years on average.
- *Specifying authority systems to facilitate change.* The ECC provides for the role of the project manager to administer the contract on behalf of the client, and to interface with the contractor or consultant should compensation events arise.
- *Providing incentive systems to motivate suppliers.* These can be either negative in the form of penalties for failure to meet commitments, or positive providing incentives for good performance. The ECC contains optional provisions for damages, such as penalties for the late completion or poor performance of the facility; positive incentives are discussed in section 6.5.

- *Using administered pricing systems to handle uncertainties.* The ECC provides for the option of bills of quantities to be used to specify the unit rates for the work to be done without contractually fixing in advance the amount of work. Similarly, arrangements for adjusting agreed prices for inflation are included.
- *Providing conflict-resolution procedures.* Although standard forms do not replace the option of using the national legal system for redress, they often provide for less expensive and time-consuming conflict-resolution procedures within the contract. The ECC provides for the appointment of an adjudicator jointly by the employer and the contractor or consultant.
- *Specifying standardised operating procedures.* The ECC provides a whole book of flow charts showing how each decision to adjust the contract in detail should be handled.
- *Posting credible commitments.* One of the parties offers another a hostage which is forfeited if relations break down. This is normally done through bonds in construction projects, and these are an option within the ECC.
- *Providing arrangements for the measurement of supplier performance.* The supervisor is allocated the responsibility of providing quality control of suppliers' work in the ECC.

The ECC recognises that uncertainty is reduced through the project life cycle, and contracts formed later in the project life cycle have more fully specified conditions than those formed earlier. For instance, the provisions relating to the achievement of the desired quality of conception and specification by the consultant are less onerous than those relating to the quality of realisation on site by the contractor. Suppliers of the former are only obliged to perform their obligations with 'reasonable skill and care', while the obligations of those for the latter are strict and must be in accordance with the works information. Similarly, detailed requirements for the management of schedule, budget and quality of conformance by the contractor are provided, while those for the consultants are much looser.

Experience[5] with the second edition has shown that the contract cannot just be left in the drawer – it is designed to be a proactive tool for governing the project coalition. In particular its time-paced processes are relatively demanding in terms of managerial capacity and some under-staffed project teams have inadvertently generated liabilities for themselves through not meeting the deadlines they contain. The collaborative basis of the contract also caused problems for project managers used to working with forms of contract which require different behaviour for their effective operation. Some contractors lost money on ECC contracts, and adjudication levels associated with the contract were relatively high. The third edition attempts to clarify some of the issues around 'compensation events', and clarify wording of clauses more generally.

6.5 Contractual uncertainty and risk allocation

Where the client decides that its workload does not justify establishing an in-house capability, or where internal governance costs outweigh the gains from reducing information asymmetry, the client as principal needs to hire one or more agents to provide the resources required for the project. If the level of information available at contract formation means that simple contracting is not viable, then a choice of three basic types of complex contract is available: fee-based, fixed-price and incentive contracts.

6.5.1 Fee-based contracts

Fee-based contracts are those where goods and services are provided at an agreed rate as a function of an agreed parameter. Fee-based contracts are used where it is possible to identify broadly the type of resources required but not the amount required. Such contracts are typically used in high-uncertainty situations, such as in the early stages of design[6]. Indeed, they are the predominant way of procuring architectural and engineering design services. In construction, there are two main ways of letting fee-based contracts. The first is what is known under the ECC as a *cost-reimbursable* contract, where the parameter is the costs incurred by the supplier itself on the basis of an agreed rate (frequently time-based) for the provision of the required resources (typically skilled people). The second is where the parameter is the price of the contract let for the execution of the works on site. This *percentage fee* approach is more common and many countries have national systems for establishing these rates – the German federal Honorarordnung für Architekten und Ingenieure (HOAI) is one of the most comprehensive (see panel 6.1). Such an approach does not exist under the ECC. Fee-based contracts are strongly associated with supplier selection through appointment and *concours* in the context of a separated procurement route.

Panel 6.1 The HOAI

The HOAI has the force of law for all publicly funded projects, and is widely followed by the private sector. The version of January 1996 provides for the calculation of the remuneration of suppliers of 12 basic types of architectural and engineering design service (interiors, acoustics, geotechnics, etc.) at five levels of project complexity for both new build and refurbishment. Remuneration is based on a percentage of the value of the works on site, broken down by nine project phases. Where additional services are required by the client, these are to be supplied on the basis of an hourly rate. Negotiations between principal and agent typically revolve around the complexity of the project proposed.

6.5.2 Fixed-price contracts

Fixed-price contracts are those where the price is fixed for the supply of an agreed amount of work. They can either be a true lump sum, where the contract price is fixed, or be subject to after-measurement when the precise quantity of work to be done is not known in advance. The latter situation is common in groundworks. Fixed-price contracts are used in situations where a large amount of information is available, and so the contract is relatively complete at the time of agreement. Such contracts frequently contain provisions for minor adjustments to the price to take account of inflation or variations in the quantity of work to be done, through the use of bills of quantity. This type of contract is the most widespread for the procurement of execution services on site, and is strongly associated with competitive tendering as a selection method, discussed in section 5.3.4.

6.5.3 Incentive contracts

Incentive contracts mix features of both fee-based and lump-sum contracts, and are considered optimal under neoclassical auction theory[7]. There is a wide variety of such contracts, but what unites them is the attempt to provide positive incentives within the contract to motivate consummate supplier performance through *gainsharing* between the parties. Incentive contracts usually consist – in the terminology of the ECC – of a target price (TP) for the facility consisting of an estimated actual cost (AC_e) for inputs required to construct the facility, plus a percentage fee (F) to cover suppliers' overheads and profit. If outturn AC_o is greater than the estimated AC_e, then the contractor pays (i.e. not reimbursed for) an agreed share of the excess; if AC_o is less than AC_e, then the contractor is paid an agreed share of the saving. These relationships need not be linear and can be capped to limit the risk of one of the parties relative to the other. For instance, a guaranteed maximum price (GMP) for the facility caps the risk of the client, as illustrated in Fig. 6.2, while a guaranteed maximum liability (GML) caps that of the supplier, as illustrated in Fig. 6.3. To encourage cost-saving proposals from the

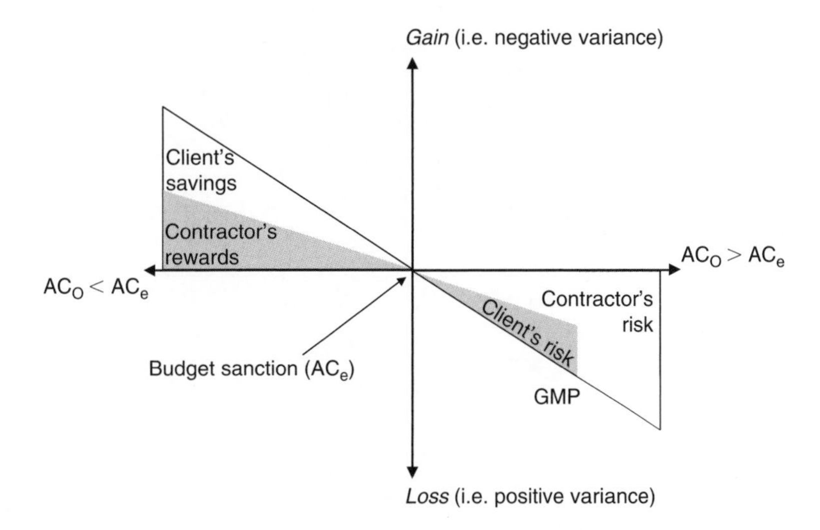

Fig. 6.2 Incentive contract with guaranteed maximum price (GMP).

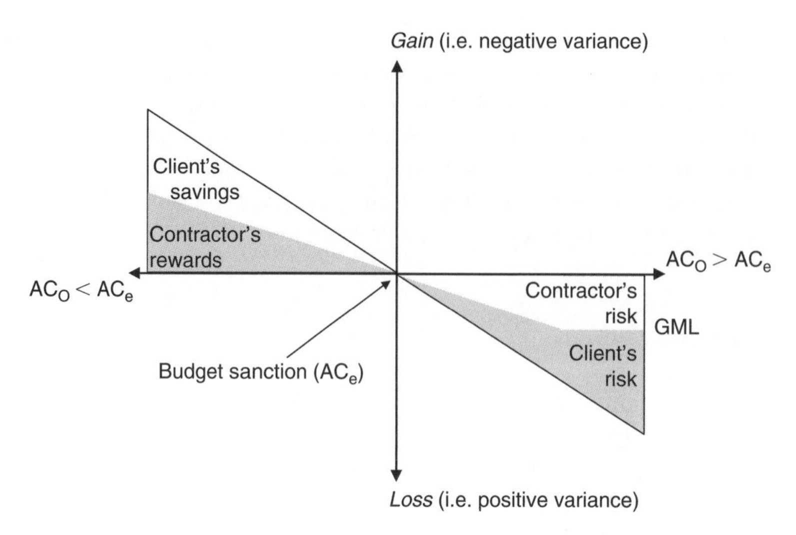

Fig. 6.3 Incentive contract with guaranteed maximum liability (GML) (source: adapted from Knott, 1996).

contractor, the ECC provides that the percentage fee on the AC_e is ring-fenced on any item where proposals from the contractor result in savings on AC_e, and is paid as if the saving had not been achieved. Other performance criteria, such as schedule, may also be gainshared in the manner illustrated in Fig. 6.4, and the incentives may also include bonuses on criteria such as number of defects or lost hours because of accidents. Case 6 illustrates in more detail some of the tools of incentive contracting.

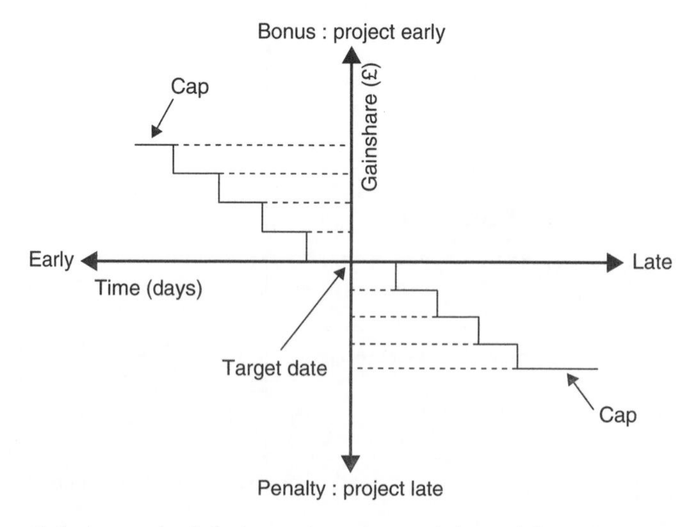

Fig. 6.4 Gainshare schedule incentive scheme (adapted from Scott, 2001).

6.5.4 Appropriate contract type

Different contract options offer different risk profiles in terms of allocation of responsibility for the costs of changes in the specification, illustrated in Fig. 6.5. The full cost-reimbursable contract effectively writes an open cheque to the supplier, and so detailed auditing systems are typically in place to justify invoices and minimise opportunism – time-sheet management becomes a major task. However, such contracts tend to be of relatively low value and are used under conditions of high uncertainty.

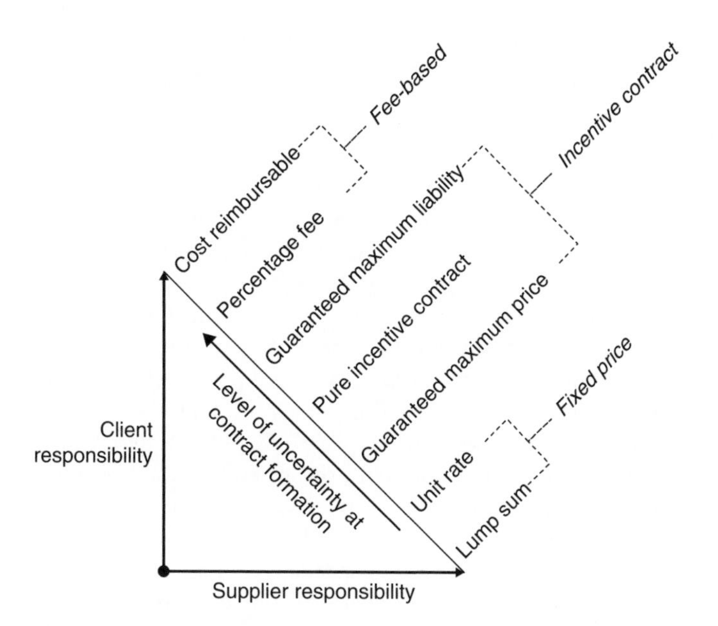

Fig. 6.5 Responsibility for changes in the specification.

In effect the client is investing in a process of uncertainty reduction so that in later contracts, typically let for much larger amounts, greater risk transfer to the supplier can take place. Percentage fee contracts tend to offer more protection for the client because they are much less open to opportunistic behaviour on the part of the supplier, which cannot directly influence the parameter used to calculate the fee. Fixed-price contracts transfer most risk to the supplier, although even here the supplier is usually protected against changes generated from sources beyond its control. Where the fixed price is based on unit rates, such as a bill of quantities, the uncertainties in the precise amount of work to be done are shifted to the client, which opens up a margin for opportunism in the pricing of the unit rates. Incentive contracts allocate responsibility more equally between the parties, with a GML favouring the supplier and a GMP favouring the client, as shown in panel 6.2.

Panel 6.2 Laing in Cardiff

If Laing's managers could have better judged and controlled the risks involved during the construction of the Cardiff Millennium Stadium, the loss of almost £30m on a £100m project could have been prevented, thereby helping to avoid the sale of Laing's 150-year-old construction company for £1.00 in September 2001. Apparently, the initial mistake was to greatly under-price the job by tendering on the basis of 'sketchy' designs, coupled with the use of the GMP contract. It is claimed that the numerous project risks included problems such as managing a 'tricky site'; the long-standing conflict between the owner of a neighbouring site and the client, which led to a refusal to allow crane overswings; and problems in ensuring safe egress from the stadium in the case of an emergency. Laing, under the terms of their contract, could neither speed up the delayed works at a later stage, nor buy time, without incurring large penalties which they could not recover.

Sources: http://www.barbourexpert.com; http://www.ft.com (accessed 07/06/01); *Financial Times* 28/09/01.

One of the main problems with both fee-based and fixed-price contracts is that they create perverse incentives for the agent. Under fee-based contracts, it is in the interests of the agent to generate work to be done, for themselves in cost-reimbursable contracts and for the contractor in the cases of percentage fee contracts. There is certainly no direct incentive to minimise the costs of the project, subject to the constraint that the client will cancel the project if the price rises to the point where the NPV becomes negative. Third-party value engineering exercises, which are discussed in section 10.5, are an attempt to manage this type of opportunistic behaviour. Similarly, those tendering for a lump-sum contract have an interest in maximising the contract value, subject to the constraint that they are competing with others for the contract. For this reason, different clients pay different prices for technically similar facilities in accordance with tenderers' perceptions of what the market will bear[8]. Incentive contracts are an attempt to remove these perverse incentives by creating *efficient contracts* which balance the costs of risk bearing against the incentive gains that result.

For incentive contracts to be efficient, the following conditions need to be met:

- Improved performance needs to provide a positive benefit to the client greater than the cost of the incentive – there is no point in motivating the early completion of part of a building if the building is not usable until it is entirely complete.
- The drivers of performance need to be within the control of the party motivated – it is not appropriate, therefore, to include additions to budget and schedule generated by problems with regulatory approval within the gainshare calculation. Moreover, as most savings come through changes in specification, incentive contracts are more appropriate for motivating the members of integrated and mediated coalitions, rather than separated project coalitions.
- The rewards from consummate performance need to be greater than the penalties associated with perfunctory performance, otherwise effort will focus on minimising loss rather than seeking gain.

- Gains and losses need to be accurately measurable – this is a major reason why they are largely inappropriate for contracts for the supply of design services.
- The benefits of improved performance need to be greater than the costs of measuring performance. This is not always the case – see panel 6.3.

Economic theory generally assumes that principals are risk neutral and agents are risk averse[9]. As defined in section 13.4, risk neutrality is the condition where the decision-maker is not prepared to pay a premium to reduce risk; risk aversion is the condition where the decision-maker is prepared to pay a premium to reduce risk, or requires a premium to accept risk. As a result, the gainshare in incentive contracts tends to be skewed towards the supplier, and GMLs tend to be lower than the potential rewards as shown in Fig. 6.3 where the supplier is being induced to accept more risk. However, public sector clients tend to be more risk averse than theory presumes, and so favour GMPs as discussed in Case 6.

Panel 6.3 BAA pavement team

All runway and apron work for BAA – the leading UK airport operator – was undertaken through a 5-year partnering agreement with AMEC, signed in 1995. The partnership was run by an integrated management team, allowing AMEC to spend much more time on planning individual projects. By 1998, costs had been reduced by 24%, and no extensions of time were required. Initially an incentive contract was in place, but after 2 years it was found that the quantity surveyors were spending 30% of their time calculating the gainshare. However, enough trust had been built up that the partnership switched to a cost-reimbursable contract, thereby reducing transaction costs in the shape of four quantity surveyors.

Sources: Bennett and Jayes (1998); *Construction Productivity Network Report*, Technical Day 2 (1999).

6.6 Governing the contract and the role of third parties[10]

The additional arrangements required to govern complex rather than simple contracts are transaction costs borne by the parties. On top of the transaction costs identified in section 5.3 incurred in minimising adverse selection problems in supplier procurement, those incurred in minimising moral hazard are as follows:

- Costs of preparing and agreeing the contract with the selected supplier. These are much reduced if standard forms of contract are used.
- Costs of dispute resolution, including adjudication, arbitration and litigation. These costs are not only the fees paid to adjudicators, arbitrators, courts and legal advisors, but also the opportunity costs of the staff time involved in participating in such actions.
- The costs of hiring third parties to undertake measurement of contractual performance and enable adjustment of the contract.

It is this last group of costs which is, arguably, most substantial, perhaps adding a further 10% to total contract price in addition to the transaction costs associated with the selection of suppliers from the relevant market. Thus making a total of 25–30%. So it is to this group of costs that we turn our attention.

Third parties are used extensively in the administration of construction project transactions in three main ways:

- measurement of performance by the supplier against contractual commitments;
- speedy adjustment of minor details of the contract;
- a first line of dispute resolution between the client and supplier.

The provenance of these third parties is important. Unlike the situation in France and Belgium, under the professional system in the UK – see Case 2 – there is no third party during design for product integrity issues. For execution, the third party responsible for quality under the separated procurement route is the agent from the design contract, typically called the 'engineer' or 'supervising officer'. This is illustrated in Fig. 6.6.

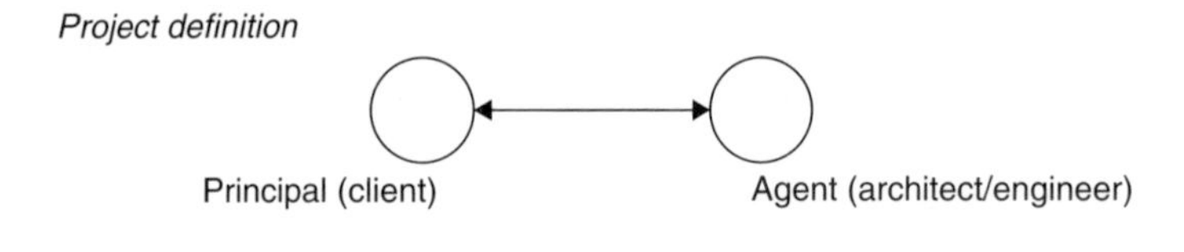

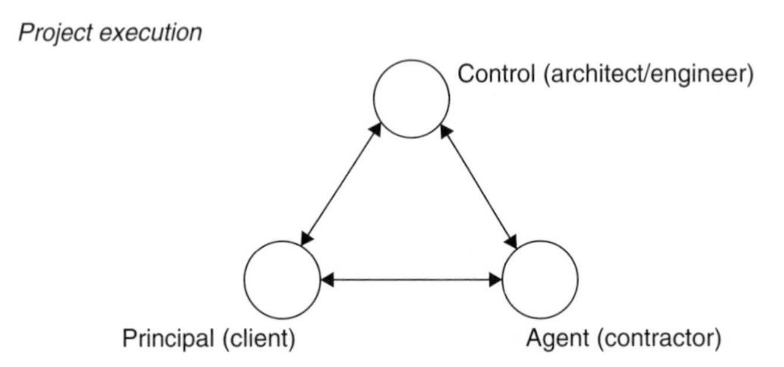

Fig. 6.6 Principal, agent and third party in the professional system.

The fundamental problem with this arrangement is that should problems arise because of the inadequacies of the design – and this is a frequent problem – the third party is naturally tempted to resort to its agent role and protect its position in relation to the client, rather than to act impartially to resolve the problem. The ECC has addressed this issue by separating the role of the supervisor from responsibilities for design.

Measuring the quality of realisation poses a number of difficulties in construction, particularly while the project is in process. Such interim measurement of achievement against schedule and budget is vital to allow interim payments to be made to the supplier, and is also required where the contract allows for after-measurement to determine actual quantities of work against the estimate. Measuring the quality of conformance in-process is required because many vital aspects of conformance quality are hidden once the facility is complete.

The supervisor under the ECC is responsible for measuring the quality of the work produced by the supplier. The French *bureau de contrôle*, presented in panel 2.1, verifies that the design solutions proposed by the architect are structurally sound and conform to the building regulations. In the UK and comparable systems the principal quantity surveyor is typically charged with the responsibility for measuring the work completed in a period for the valuation of interim payments; this is the responsibility of the project manager under the ECC. Such valuations are even more important in the case of incentive contracts, because AC_o also needs to be measured.

Uncertainty is never fully removed from the contract, even after works have started on site. The client requires a delegated decision-making capability to make adjustments to the contract should compensation events arise – these are the responsibilities of the project manager under the ECC. Under the UK system, the principal quantity surveyor is charged with valuing the financial implications of such adjustments. In France, the *bureau de contrôle* approves any changes in the specification arising from such adjustments. One of the reasons that the client hires a third party to carry out these measurement and adjustment functions is that the actor can then also provide a first stage of dispute resolution between the principal and the agent.

A notable feature of governance in the UK and many other countries is that while it is normal for there to be provisions for third parties in fixed price and incentive contracts, this is much less common with fee-based contracts, although France, with the *bureau de contrôle*, is an exception here. Yet, given the high levels of uncertainty such contracts are typically formed under, and the potential for opportunism that a cost-reimbursable contract allows, this is puzzling. The answer to this puzzle is that there is, in effect, a quasi third party in the shape of the national certification system for designers. Thus in the UK, the professional institution provides quality assurance of the skills of its members and a code of ethics in the execution of responsibility. Risks to clients are also mitigated through professional indemnity insurance to cover loss and the highly risk-averse behaviour generated by unlimited liability in partnerships. However, perhaps most importantly, selection by appointment means that reputation is enormously important for fee-based suppliers. Gaining a reputation among clients for opportunistic behaviour would quickly lead to commissions drying up.

However, there are systemic problems with such third-party governance of transactions, where the third party is in contract with the principal alone but is expected to act impartially on behalf of both principal and agent. The first problem is that the requirement for impartiality is heroic, particularly as many of the adjustments required will be because of a lack of consummate performance during the design phase. Secondly, a perceived lack of impartiality means that the agent also measures its own performance as a check on the calculations of the third party reporting to the client – the 'man for man marking' of contractor's and principal quantity surveyors is an example of this – thereby generating even more transaction costs.

6.7 The dynamic of adversarial relations

The attempts by clients to manage the problem of moral hazard and reduce the scope for opportunistic behaviour by the contractor have the tendency to generate vicious circles of adversarial behaviour between the parties, as illustrated in Fig. 6.7. Chris Chapman and his colleagues[11] authoritatively demonstrate how the desire by the client to get the 'best deal' creates a dynamic of adversarial relations in which transaction costs escalated as production costs appear to be pushed down. An important element in the *adversarial relations* dynamic is that of control. Clients determined to ensure that the keen prices obtained through competitive tendering are not competed away through opportunistic behaviour by contractors deploy a third party – the principal quantity surveyor. An extensive apparatus of in-project cost control has evolved which, given that the contracts are typically fixed price, can do little more than manage the shape of the project's cash-flow *S*-curve. Contractors respond to these control techniques by developing their own quantity-surveying capabilities, adding to site overheads and generating the distinctive twin-hierarchy of site management, with the contract manager focusing on time and quality and the site surveyor focusing on cost. As costs are largely determined by design decisions in competitive tendering on complete designs, and contract prices are fixed through that process, there is little real opportunity to reduce costs – this is a cost-control rather than a cost-reduction process.

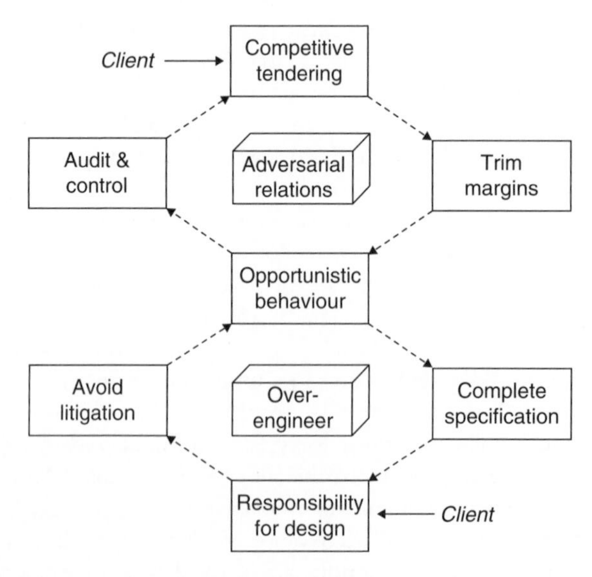

Fig. 6.7 The dynamic of adversarial relations and over-engineering (source: Winch, 2000c).

This dynamic of adversarial relations interacts with a second dynamic of *over-engineering* in the manner illustrated in Fig. 6.7. Clients are naturally concerned to pass complete responsibility for the design to their professional advisors. Designers wish to protect themselves against litigation for defects in their designs, but they face the possibility of opportunistic behaviour from contractors. The concern to minimise opportunistic behaviour by contractors obliges designers to specify the product completely, yet their (inevitable) lack of experience with site processes means that their specification decisions do not reflect site conditions or the capabilities of contractors. This creates rigidity in the design which tends not to be fully optimised in relation to the problems of construction. It also encourages designers to allow high safety margins in their designs leading to over-engineered, and hence costly, designs. In the absence of the ability to change the specification to better match their capabilities, contractors maximise their flexibility. Contractors retain, therefore, a preference for craft forms of work organisation, which delegate considerable control over the work process to the operatives, particularly in the form of the gang organisation discussed in section 7.5. This in turn freezes the construction technology, preventing contractors innovating to reduce the pressure on their margins. As a result, further opportunistic behaviour occurs.

It is through these two self-reinforcing dynamics that competitive tendering, despite its emphasis on lowest price, does not actually deliver low production costs; in addition, it generates very high transaction costs. The requirement that designs be fully specified at tender makes heroic assumptions regarding the competence of architects and other designers in the technical details of a wide range of construction technologies, and the ability of the client to keep requirements fixed over a period of years. In practice, designs are rarely fully specified, changes are inevitable, and so complex contracts develop that enable such changes to be negotiated. Principal quantity surveyors become guardians of the contract on behalf of the client, making sure that the possibility of opportunistic behaviour by contractors keen to recoup the profit margins that have been competed away during tender is minimised. The cycles of adversarial relations and over-engineering are in place, at the expense of removing any incentive for any actor to reduce budget costs as opposed to control costs against an agreed budget.

6.8 Alliancing

What, then, is the way out of this vicious circle? One way is through alliancing, which can be used even if the conditions for framework agreements identified in section 5.6.4 are not fully in place enabling the formation of an *integrated project team*. There will always be situations of relatively high or low uncertainty when fee-based or lump-sum contracts will be appropriate, but for an increasingly large middle ground of transactions, incentive contracts are appropriate because they create positive incentives for consummate performance. Effective incentives within the contract require new ways of working. In particular, they require that the supplier's accounts for the project are open to the client. Under the ECC, the contractor is obliged to keep records open to inspection by the project manager at any reasonable time, detailing the calculation of AC_o and the handling of compensation events that involve subcontractors. The scope for opportunism is much reduced because the scope for withholding information is much reduced. In return, the supplier can gain additional profit by bringing AC_o below AC_e.

In the past, clients have tried to minimise moral hazard through ever more detailed control over their suppliers, deploying more onerous contract terms and hiring third parties to watch over them. However, none of this altered the basic relationship between the parties – post contract, the supplier had more information and lower switching costs than the client. Alliancing is an attempt to redress the balance. By providing positive incentives for the sharing of information, the chance of the vicious circle being broken is increased to the benefit of both parties. If alliancing leads further to a framework agreement, then the switching costs of the supplier are also increased and power is more evenly balanced between the parties. Panel 6.4 illustrates the benefits of alliancing, while Case 6 explores alliancing in the context of the ProCure 21 framework agreement.

Panel 6.4 Alliancing for road building

The Tunstall Western Bypass Phase 2 was a relatively small project (£10m), but well illustrates the potential of alliancing, even within the context of traditional relationships. Selected by competitive tender, Birse signed a traditional ICE Fifth Edition contract with Staffordshire County Council. However, the tendering procedures were changed to provide the bidders with much more background information than usual; the bidders were encouraged to submit alternative design proposals and asked to nominate their own contract period.

Once the project was awarded, a two-day facilitation workshop was held attended by all principal parties, and a project charter was signed between the client and Birse, as presented in panel 16.6. This was followed up 3 months later by another workshop. Project staff were co-located, and also made a value engineering agreement – effectively, an incentive contract which shared cost savings of £400 000. A £6m risk associated with landfill was averted, and the contract was expected to be completed 11 weeks early on a 67 week schedule.

Source: CIB Working Group 12 (1997).

The overriding aim is to align the goals of the client and its suppliers to generate a win-win culture on the project and a high level of transactional trust. The main means of gaining alignment are as follows:

- The use of incentive contracts, linked to open-book accounting, with the gainshare calculated at the level of the project as a whole, and a clear agreement on how the gainshare is to be apportioned between the different alliance members.
- The appointment of an integrated supply team on the basis of the best person for the job; see section 7.6 for more on integrated supply teams.

- A clear statement of the project mission, so that all parties know what they are getting into.
- A clear statement of what is meant by project completion, so that the gain-share can be calculated.
- The establishment of a project supervisory board where all the alliance members are represented by senior staff not directly involved in the project. The aim is to provide a forum where differences can be aired openly but confidentially; see Case 16 for an informal version of this.
- The co-location of staff from the different resource bases to develop a common project culture; see Case 15.
- The appointment of outside consultants to facilitate team building at project inception, and also to act as coach through the project life cycle when the going gets sticky.
- The use of an underlying contract that is sympathetic to alliancing, such as the ECC.
- The agreement of a project partnering charter, which is a statement of intent to behave in a collaborative rather than competitive manner, and is not legally binding – a vital clause is 'to have fun'.

None of this can work, however, without the right quality of people on the job – fundamentally this is the difference between the success of the Emirates Stadium project and the failure of the Wembley National Stadium project as shown in panel 6.5.

Panel 6.5 A tale of two stadia

Two new stadia were completed in London during 2006 – the 60k seat Emirates Stadium for Arsenal Football Club (contract value £220m) and the 90k seat Wembley National Stadium (contract value £445m). Both used the same JCT with Contractor Design form of contract and negotiated a GMP with novation of consultants. One was finished on schedule and budget; the other over a year late with major losses for the contractor and its supply chain. Although smaller and without a signature arch, the Emirates Stadium was arguably as complex as Wembley because of the need to construct two bridges over rail lines to improve access and to work around an operating waste disposal facility in the early phases. Both suffered the same market conditions for labour and materials, in particular steel; both had risk-averse novice clients; both suffered tortuous fundraising and had to interface with multiple stakeholders; and they shared specialist stadium architects HOK. So why is there a difference? A review of the two projects suggests the following.

- *Contractor experience.* Sir Robert McAlpine is a highly experienced UK contractor with a number of similar projects to its credit, including the Eden Project in Case 17, and its project director had 37 years experience with the company. Multiplex was an Australian newcomer to the UK, and failed to recruit experienced UK people as it set up its organisation despite rumoured £1m 'golden hellos'.
- *Client capability.* Wembley National Stadium Ltd. was, in effect, a special-purpose vehicle for the project which initially lacked adequate management systems to provide oversight of the project. This was resolved by the appointment of a new chairman with a construction industry background, but only after the selection of Multiplex. Arsenal's lead on the project was its former managing director who had 56 years experience with

the club and was informally supported by a number of fans with experience at the highest level in property and construction.

- *Early contractor involvement.* McAlpine was selected early and worked on a reimbursable basis during design development until the target price was negotiated, together with the steelwork fabricator Watson. Although Multiplex was selected during design development, there is no evidence of its involvement in the process, and if the will were there, the ongoing investigation into the probity of its appointment might have prevented any actual involvement.
- *Low balling.* Multiplex is generally reckoned in the industry to have 'bought' the project. A proposal to retender the project even though the original tender process was cleared of fraudulent manipulation was rejected on the grounds that there was not a high chance of obtaining a lower one. As the Chairman of WNSL later put it 'Multiplex was the only firm that would build Wembley for a fixed-price deal available at the time. No UK contractor would touch it'.
- *Management style.* Multiplex's aggressiveness became legendary as it pincered itself between a low fixed price and rapid inflation in labour and materials. It starved subcontractors of cash, thereby provoking them to use hold-up tactics. Most notably, Cleveland Bridge, the contractor for the steel arch, walked off site in 2004 claiming systematic underpayment and bullying tactics. Multiplex found out for itself how large switching costs can be as the project was delayed for months and the replacement contractor would only work on a reimbursable basis. There have been no public complaints regarding McApline's treatment of its supply chain.
- *Trust.* During the funding crisis on the Emirates Stadium in 2004 all the consultants stood down, but McAlpine kept a small unfunded team progressing the project forward and ensured that the works completed so far did not deteriorate and no overall schedule delay resulted just as they had on the Eden Project. During this time, it also instructed Watson to buy steel in advance because price rises were expected. As a director of the structural engineers Buro Happold put it 'Buro, HOK, and McAlpine have a history of working in partnership and we know how to solve problems together. We can talk openly and frankly and create a no-blame culture. Simple stuff, really'. At Wembley, Multiplex's management style meant that trust was reduced to the point where WNSL and Multiplex would not go on site at the same time, which made the commissioning and hand-over processes very difficult to handle and engendered further delays and costs. Morale was very low amongst the workforce with festering industrial relations. As one 'site source' put it 'I've been working in this industry for 25 years and I've never seen so many people who think fuck it, who gives a shit'.

The outcome with the Emirates project is a highly satisfied client and team. The outcome with the Wembley project is indeterminate because of the amount of litigation that Multiplex has instigated after losing an estimated £100m on the project, but by the end of the project five subcontractors had already gone into administration because of their problems on the project, while a sixth had to be bailed out by Multiplex. Cleveland Bridge nearly failed in 2003, and faced massive claims from Multiplex. Multiplex was sold to a Canadian company in 2008. WNSL has done rather better – the schedule overrun will soon be forgotten and English football has a magnificent stadium at a rock bottom price.

Sources: *Building* various issues; NAO (2003).

6.9 Summary

Informed by principal/agent theory, this chapter has identified the problem of moral hazard in construction procurement and defined it in information terms. It has shown that attempts to reduce opportunism through standard forms of contract have tended to generate a vicious circle of adversarial relations. The use of incentive contracts in the context of an alliance relationship between the parties was suggested as a way of breaking that vicious circle. Having explored the relationship between the principal and the agent, our attention turns in the next chapter to relations between the agent and those it recruits to help it meet its obligations to the client.

Case 6
NHS ProCure 21

The UK's National Health Service has a massive requirement for constructed facilities of various kinds for the delivery of health care, and a considerable increase in the budget for capital investment after 1997 stimulated strategic initiatives to ensure consummate project delivery. However, individual procurement decisions are devolved to around 600 individual trusts and foundations around the country, and so clients are typically inexperienced. For large hospitals, PFI as defined in Case 2 was preferred, while for primary care facilities a form of PPP known as Local Improvement Financial Trusts (LIFT) was favoured. For the large amount of investment that was not suitable for either PFI or LIFT and therefore financed directly from taxation, the ProCure 21 framework initiative was developed in England. Piloted during 2002, the framework was rolled out nationally in 2003 for 5 years and has since been extended to 2010. It was extended to minor works (<£1m) in 2007. The framework was reviewed in 2005 as part of a more general administrative reform within the NHS which generated significant uncertainty around the framework, but it has since revived, and similar frameworks were adopted in Wales and in Scotland in 2008.

The framework is managed centrally by the Department of Health through a small dedicated office. Principal Supply Chain Partners (PSCPs) each pay £170k per year to be a member of the framework and then compete with each other for projects. Initially there were 12 PSCPs but 4 withdrew during 2006 on the grounds that they were not winning enough work. NHS trusts and foundations are not obliged to use ProCure 21, so the framework agreement needs to be sold to them on its merits as an effective way to procure facilities. The principal selling points are improved predictability in delivery, free support in dealing with VAT issues, and at least 6 months off the schedule because all PSCPs are pre-qualified at initial selection to be members of the framework and there is therefore no obligation to follow EU procedures for each project. For instance, following a fire at one hospital the PSCP was selected in 8 days. By August 2008, 179 schemes over £1m and 37 under £1m had been completed and 41 schemes were on site.

The foundation of ProCure 21 is a gateway process which provides multiple entry points for the PSCP as shown in Fig. 6.8. The selection of SOC, OBC or

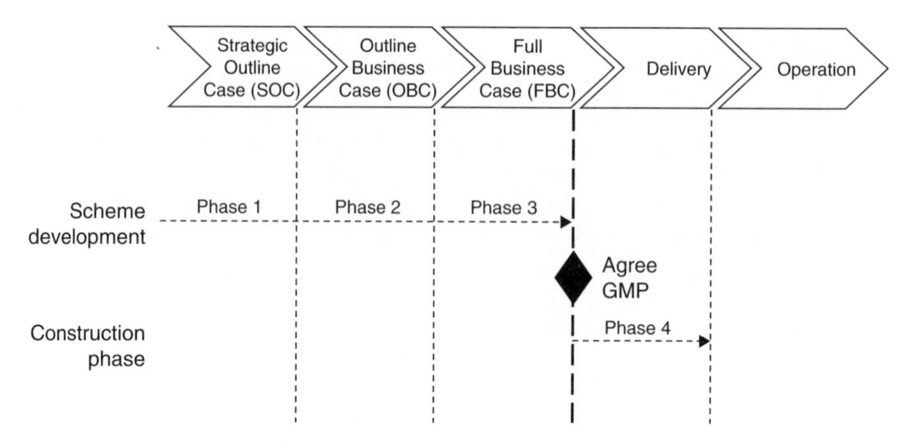

Fig. 6.8 The ProCure 21 process (source: ProCure 21 Guide v 5.0, Fig. 18).

FBC for PSCP entry depends on the ability of the PSCP to add value – the greater the mission uncertainty, the greater the benefits of early entry. PSCPs are selected by interview based on expressions of interest against a scheme information pack and shortlist selection criteria. The criteria typically include experience, proposed supply chain, capacity and proposed management team. Implicit in the ProCure 21 approach is that the PSCPs will present as integrated supply teams, but approaches vary and it is unclear how integrated the supply chains actually are.

Early formation of an Integrated Project Team is enabled through a Launch Workshop where all internal stakeholders are represented including clinicians. The project is then managed by a Delivery Team typically consisting of the Project Manager representing the client, a Site Agent representing the supply chain, and a user/clinician of the facility. The Delivery Team reports to a Principals' Group typically consisting of the Project Director, a senior director from the PSCP and a senior clinician. There may also be a Project Board within the Trust, or one Principals' Group may have oversight over a number of Delivery Teams. The Task Groups actually doing the work are managed by the Delivery Team as shown in Fig. 6.9.

One output of the launch workshop is a project charter; a second is a joint action plan that takes the project through phases 1, 2 and 3 to the signing of the construction contract mandated as an amended version of NEC 2 Option C (target contract with activity schedule) – see the first edition for a discussion of NEC 2. This occurs at the end of scheme design (co-ordinated 1:50 drawings with planning permission). Benchmarking for the GMP is provided by the departmental cost allowance guidelines from the central DoH database and the cost advisor appointed by the client. This is supported through the project life cycle by the use of the benchmarking toolkit developed in collaboration with the OGC.

It is expected that all value management proposals and associated innovations from the PSCP will have been discussed and agreed prior to confirmation of the GMP,

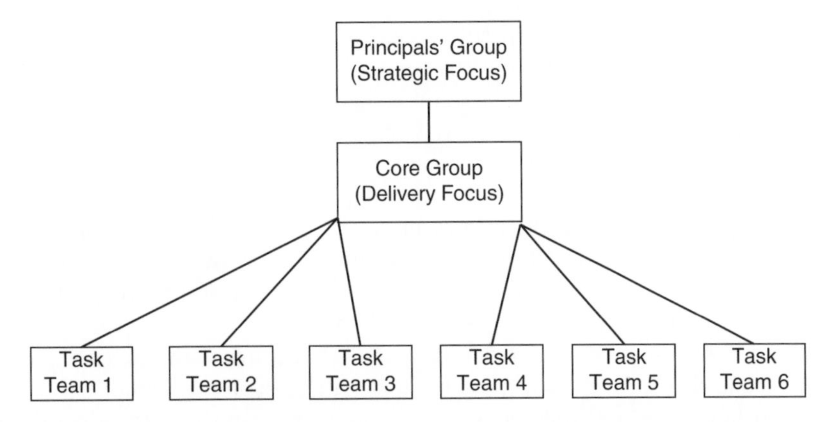

Fig. 6.9 ProCure 21 integrated project team structure (source: ProCure 21 Guide v 5.0, Fig. 14).

which allows for gainshare with the PSCP on a 50:50 basis up to a cap of 20% of the GMP. The maximum the PSCP can gain is thus 10% of the GMP. The trust can invest its share of the gain back into the project to provide additional benefits. There is no painshare with the NHS trust; the only changes to the GMP allowed are changes in scope and compensation events under the contract, which are absorbed by the client risk allowance. The structure of the GMP is shown in Fig. 6.10.

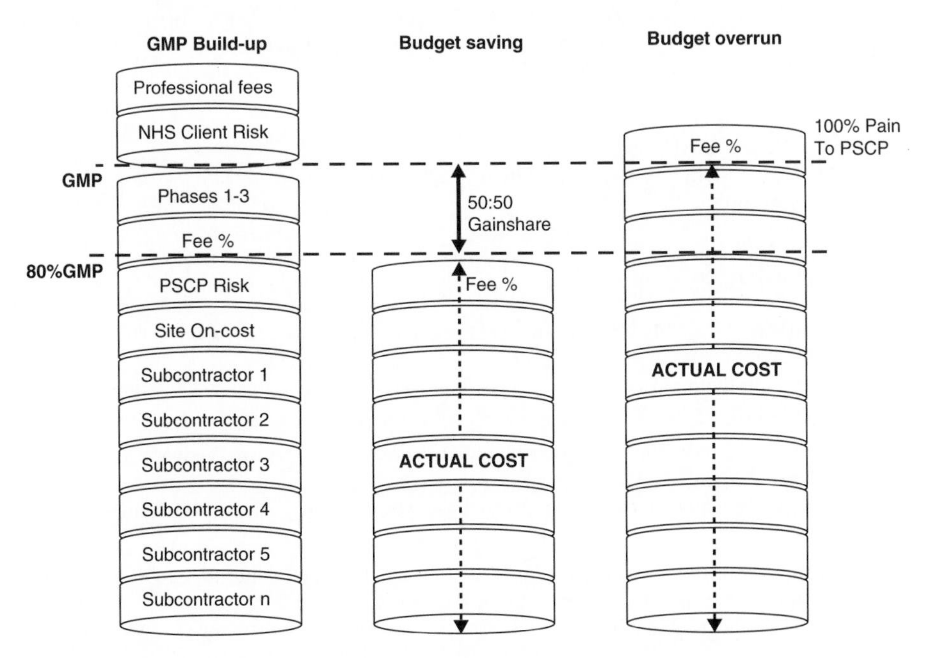

Fig. 6.10 ProCure 21 GMP structure (source: ProCure 21 Guide v 5.0, Fig. 23).

The completion of scheme design and the agreement of the GMP are the basis for signing off the full business case. That signature allows the project to proceed to the delivery phase. Realisation is managed by the project manager as defined in the contract, and the client always pays the lesser of cumulative cost or cumulative value expressed in the earned value curves (see section 10.8). Accounting is done on an open-book basis, with monthly payments authorised by the cost advisor. At project completion the outturn price and the data from the benchmarking toolkit are uploaded to the ProCure 21 team centrally and any gain or pain is apportioned. A review workshop is also recommended, and participants are eligible to join the ProCure 21 Club to share best practice.

Facilities procured under ProCure 21 are generally cheaper and quicker to build than conventional procurement with greater cost certainty (i.e. a tighter distribution around mean costs), and appear to perform better than the industry as a whole on KPI measures.

Sources: http://www.building.co.uk (accessed 13/08/08); ProCure 21 Guidelines v5.0 (2007); http://www.nhs-procure21.gov.uk (accessed 13/08/08).

Notes

1 Hamilton, A., Madison, J., and Jay, J. (1961) *The Federalist Papers*. New York, Mentor Books, p. 346.
2 The concept comes from Williamson (1985).
3 Masten *et al.* (1991) provide the classic analysis, while Chang and Ive (2007a,b) develop it in the context of construction projects.
4 The analysis here draws very much on Stinchcombe's 1985 work on North Sea oil contracts.
5 See Eggleston (2006) for a brief summary of this experience.
6 Bajari and Tadelis (2001).
7 McAfee and McMillan (1986).
8 See Fine (1975).
9 Milgrom and Roberts (1992).
10 The analysis here is developed from Winch and Campagnac (1995).
11 In Curtis *et al.* (1989).

Further reading

Gruneberg, S., and Ive, G. (2000) The Economics of the Modem Construction Firm. London, Macmillan.
The companion to Ive and Gruneberg, it provides an authoritative guide to the economics of the construction firm.

Murdoch, J., and Hughes, W. (2000) Construction Contracts: Law and Management (3rd ed.). London, Spon.
A comprehensive guide to the principles of construction contracts from a managerial point of view.

Rubin, P.H. (1993) Managing Business Transactions: Controlling the Cost of Coordinating, Communicating and Decision-Making. New York, Free Press.
Perhaps the best application of the concepts of transaction cost economics to managerial problems.

Chapter 7
Managing the Dynamics of the Supply Chain

7.1 Introduction

> 'I do as little as possible for a firm but as much as I can when working for myself'.

Max Gagg[1], a self-employed bricklayer, encapsulates one of the central issues in managing the dynamics of the supply chain – how agents can effectively motivate the delivery of the required resource inputs to the project. The previous two chapters analysed the business-to-business relationships between the client as principal and its suppliers of construction resources as agents. This chapter turns to examine how the suppliers ensure that they now mobilise the resources required to meet their commitments to the client. They may decide to use existing resources that are available in-house, or they may choose to procure them from one or more additional firms – what is known as the make/buy decision. In the first case, the process of mobilisation and motivation is essentially an operational one, with the management of human resources at its core. In the second, further firm-to-firm relationships are formed which are termed *subcontracts* to distinguish them from the direct contract between the client and its principal suppliers. In turn, these subcontractor firms then have to make the same make/buy decision. Thus, a chain of firms linked through a series of contracts can develop behind the *first-tier* suppliers in direct contract with the client. This chain of firms linked through a series of contracts has become known as the *supply chain*. On some projects, several tiers of suppliers may develop within a supply chain – in Japan in particular, four or five tiers are not unusual.

This chapter will explore how these first-tier suppliers manage the mobilisation of resources – both internally and through the supply chain. First, a more formal definition of the issues will be proposed in terms of the problem of the vertical and horizontal governance of transactions. Then the argument will move on to examine the mobilisation and motivation of internal resources – essentially focusing on the employment relation and its variants. The management of the supply chain of external resource bases will be the focus of the balance of the chapter. Most of the governance issues here are very similar to those in the relationship between the client and the principal contractor, so attention will be directed towards those features of governance specific to the subcontract relationship.

7.2 Horizontal and vertical governance

In order to make the argument here clearer, it is useful to distinguish between vertical and horizontal transaction governance. Vertical governance focuses on the shifting set of transactions between the client and its first-tier suppliers. Depending on the level of integration of the project coalition structure chosen, the number of vertical transactions can vary from one (for a completely integrated project coalition) to more than 50 (for a construction-management type mediated project coalition, ranging from the acoustics engineer to the façade package contractor). Behind each one of these vertical transactions there may be one or more tiers of horizontal transactions. Where all resources mobilised are in-house to the supplier, no further external transaction needs to be governed; where an extensive supply chain develops, there may be a large number of horizontal transactions which need to be co-ordinated. These relationships are illustrated in Fig. 7.1, which captures the different types of contracts that are deployed within the construction project value system illustrated in Fig. 1.1. The external ones in the horizontal dimension are usually called the *supply chain*; the vertical ones can usefully be defined as the *project chain*. The relative importance for the construction

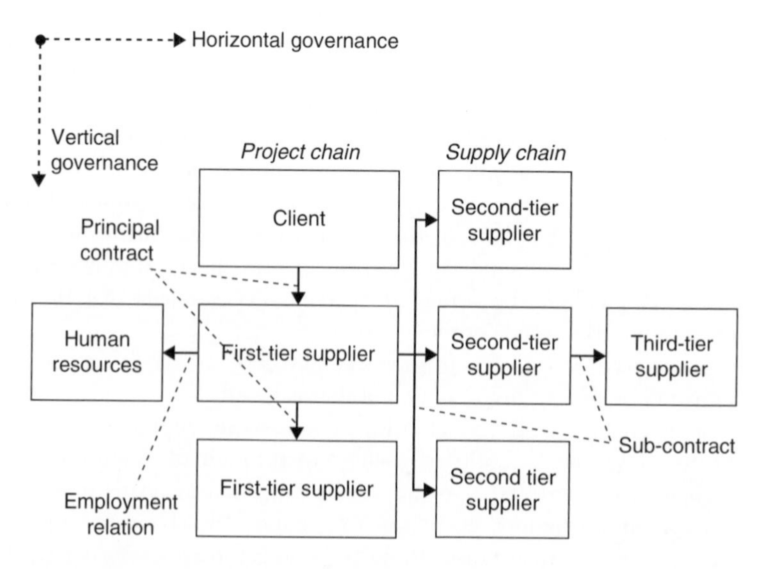

Fig. 7.1 Horizontal and vertical governance in the project value system (source: Winch, 2001).

project manager of the supply chain and the project chain will depend on the procurement route chosen by the client.

The governance of both vertical and horizontal transactions is subject to the same three contingencies that were identified in Fig. III.1 – uncertainty, frequency and asset specificity. However, choosing the most appropriate governance mode is usually easier for horizontal transactions than for principal contracts, because the scope of work for each contract is less and therefore the impact of each factor is easier to determine. While the provision of resources in-house by the client is unusual, and tends to be restricted to transactions under very high uncertainty, it is much more common in horizontal governance. Indeed, by definition, at some point in the supply chain there must be a switch to in-house resource supply.

Vertical governance becomes horizontal governance through the process of *integration*. In the classic meaning of the term, (vertical) integration means the integration of companies up and down the supply chain, such as a car assembler purchasing its component suppliers. However, in project-based industries, where projects and firms are arrayed in a matrix as shown in Fig. 1.4, integration means moving up the project chain towards a single point of responsibility in an integrated project coalition – often known as an *integrated supply team*. Its inverse is fragmentation into a large number of principal contracts, such as in mediated project coalitions. Integration may also imply integration along the supply chain – there are certainly advocates of this – but actual examples are rare in construction[2].

7.3 Internal resource mobilisation

Internal resources are essentially human resources. It is human information processing which reduces uncertainty and solves problems, thereby adding value to plant and materials. So it is on the management of human resources that our attention is focused. This is a complex and wide-ranging topic, central to the fifth first-order project process – maintaining the resource base. The discussion in this section will be limited to those aspects of the management of human resources that set the framework for the relationship between employer and employee – in particular, the employment contract.

At the centre of the employment contract is a hierarchical relationship; indeed it is this that distinguishes it from a business-to-business commercial contract and is the legal test of an employment contract – see panel 7.1. Typically, employment contracts provide for the employee to put themselves at the disposal of the employer for a period of time to undertake a broadly specified range of duties. Most commentators are impressed by how loosely defined employment contracts are compared to commercial contracts. The reason for this is that what is purchased in the (labour) market is the resource itself, not the output from mobilising that resource[3]. Once purchased, the resource has to be mobilised efficiently and effectively to meet the needs of the firm – the problem of organisation which is explored in Part V.

Panel 7.1 The employment contract

Under English law, there is a clear and important distinction between the employment contract and the commercial contract, and most jurisdictions have a similar distinction. The commercial contract is a contract *for* services. In other words, suppliers form a contract with a customer to supply a specified good or service. The customer does not desire to either know how the supply of that service is organised or be responsible for the supply. The contract is formed between two independent and equal parties which have no responsibility for each other as organisations. An individual can enter a contract for service as a self-employed person. An employment contract, on the other hand, is a contract *of* service. Here the employee puts himself or herself at the disposal of the employer, who may direct the employee to carry out any task within reason. Around the employment contract are a large number of obligations, such as the obligation to provide a safe system of work under health and safety legislation. The contract is formed between a strong and a weak party, and much of trade union action and employment legislation is designed to ensure that the employer does not exploit its relative power to the disadvantage of the employee. The test of whether a self-employed person is in a contract for service or a contract of service is made according to the facts of the case, and not the letter of the agreement, and this usually involves a test of the amount of control the alleged employer has over the daily activities of the alleged employee, or the extent to which the employee is integrated into the employer's organisation. Increasingly, the English courts have been defining self-employed construction workers as employees to ensure that legal obligations to deduct income taxes and the like are enforced.

The employment contract is subject to exactly the same set of governance factors as the commercial contract, where asset specificities take the form of job-specific skills. Broadly, skills are either general or job specific – while the former are the product of education and general training programmes, the latter are the result of learning on the job, reinforced by job-specific training. Pre-contract asset specificities are rare for human resources in construction – most employees are recruited for their general skills because job-specific skills, by definition, tend not to be transferable between organisations unless they are very similar, or the recruiting organisation wishes to copy in some way the employee's former organisation. Post-contract asset specificities arise through learning – processing information into knowledge on the job as uncertainties are reduced. Thus, for human resources, asset specificity is a function of frequency, in that asset specificities (job-specific skills) are generated when employees learn to do the job better next time. Situations of low frequency where any learning will not be required in the future will generate low asset specificity, while situations of high frequency where learning through practice rapidly increases the value of the employee to the organisation will generate high asset specificities.

Figure 7.2 illustrates the dimensions of uncertainty and asset specificity along which employment contracts are formed. Where asset specificity and uncertainty are low, *sequential spot* contracting – which usually takes the form of piecework – is viable. Task assignments are specified completely in advance and recompensed upon completion. Variations in task assignment lead to renegotiation of the contract each time. As asset specificity rises, *contingent claims* contracts become

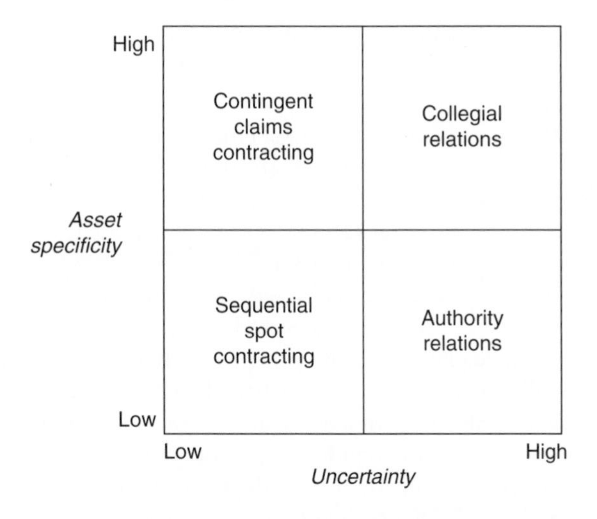

Fig. 7.2 Governing the employment relationship.

more viable, where the set of possible task assignments is specified in advance through very detailed job descriptions. Task assignments, or at least the rules for making such assignments, are frequently negotiated through collective bargaining with trade unions, rather than individual bargains struck by each employee, and such arrangements typically include formal procedures for renegotiating task assignments. The reasons that contingent claims contracts are preferred over sequential spot contracts in higher asset specificity situations are the following:

- There are economies of scale in bargaining once a year, instead of for each new assignment, particularly where the workforce is large.
- Employees tend to be risk averse and to demand a premium for accepting the risks to their future income associated with sequential spot contracting.
- Ethical considerations tend to view sequential spot contracting unfavourably – both trade union and regulatory pressures tend to disfavour it.
- Where asset specificities are generated, opportunism by employees in order to improve their bargaining position quickly undermines sequential spot contracting.

Where uncertainty is higher, but asset specificities tend to stay low, the *authority relation* is more appropriate. Here the constraints to job assignment tend to be the general skills of the employee, but within that range they can be reassigned to a number of different jobs. Hierarchical relations tend to be clearly defined, and they are often gendered in that the more junior employees are typically women. Employees in such jobs do not build up large asset specificities. The returns from learning on the job are fairly limited – an experienced employee is more valuable, but not that much more valuable, to the employer than a trained but inexperienced one. A classic example of this kind of work on construction projects is secretarial and administrative work. The greatest asset specificities can be generated when the execution of the task requires significant levels of trust, such as a nightwatchman.

Where both uncertainty and asset specificities through learning are high, a *collegial* relation is more appropriate, where the hierarchical nature of the employment relationship tends to be suppressed and a much more egalitarian culture reigns. Here, the process of tackling uncertainties stimulates considerable learning, and hence employees build up asset specificities. Under an authority relation, this learning would tend not to be mobilised – the task of the employer is to release the creative energy of the employees and this is best done through developing teamworking and leadership skills, as will be discussed in Part V. Many jobs in construction fall within this category, particularly among design and project management roles. It is also very much the culture of small building firms, where 'fraternalism' exists between the builder and his operatives, with personal relationships carrying them through uncertain periods when there is no work for a particular trade, often mediated by the builder's wife[4].

One important implication of sequential spot contracting is that, particularly in times of boom, switching costs for workers are low – they can leave as soon as they have been paid and find work on another site the next Monday. Such instability in human resource availability can pose a significant threat to the schedule, particularly in the closing phases of the project. For this reason contingent claims contracting can yield dividends, particularly as it allows the development of retention-orientated bonus schemes of the type shown in panel 7.2. A second is that different pay rates around the site can cause pay disputes on the basis of equal pay for equal work. Again, a shift to contingent claims contracting where overall site agreements are negotiated with the trade unions, as on the Heathrow Terminal 5 and London 2012 Olympics sites, can yield significant benefits[5]. A third is that workers can also generate their own version of the hold-up problem because those in key operations – typically tasks on the critical path or in logistics – can threaten to strike, thereby disrupting progress[6]. Again, reliance on sequential spot contracting rather than contingent claims contracting arrangements negotiated with trade unions makes the project much more vulnerable to such threats to budget and schedule.

Panel 7.2 Creating commitment to a (de)construction project at Rocky Flats

The decommissioning and closure of the Rocky Flats nuclear weapons manufacturing facility in Colorado was always going to be difficult. A subsidiary of CH2MHill – one of the members of the London 2012 Delivery Partner joint venture – Kaiser-Hill won the contract for the work. The facility had 800 buildings and other installations, some with high levels of radioactive pollution, and relationships with local stakeholders and regulators were very poor. Relationships with trades unions were also poor and the accident rates higher than in the construction industry generally. Yet Kaiser-Hill achieved one of the most rapid and effective nuclear clean-ups ever by 2005. There are various aspects to this story; here we will focus on how the workforce was incentivised to perform.

Kaiser-Hill negotiated a target schedule (see Fig. 6.4) contract with the US Department of Energy with very high incentive for early completion. Kaiser-Hill then needed to incentivise the workforce. It faced two different problems in doing this – workers might shirk so as to spin out their employment longer or they might quit for longer-term opportunities while their skills were still needed. The solution was an incentive scheme that paid well for early completion against targets, but not all the bonus was paid immediately. The balance – called scrip – was put into an incentive pool on a diminishing curve. In the early years up to 80% of the bonus went into the pool; in later years it dropped to 45%. If the maximum bonus in the target cost contract were earned, the scrip would be worth $1.00 per share, diminishing as earnings against the target schedule diminished. Workers who were laid off retained their scrip in the bonus pool; workers who quit lost their rights to the scrip they had earned. This arrangement also discouraged workers whose tasks were completed early in the schedule and were then laid off from completing those targets against schedule in a way that jeopardised the performance of later tasks in the schedule. The levels of bonus earned in this way were described as 'lifestyle changing'. Although there are many idiosyncratic aspects to the Rocky Flats story, the principle of the delayed bonus pool would appear to be transferable to other large projects.

Source: Cameron and Lavine (2006).

7.4 Shirking

A particular type of opportunism develops in employment contracts – *shirking*, or the Max Gagg problem. The problem rests in the combination of three factors:

- the purchase of the human resource as an employee's capacity for work, rather than work done;
- the greater information that the employee has about his or her real capacity for work and the amount of work they have completed, than the employer;
- measurement difficulties in the amount of work actually done.

When employees arrive at the workplace for the agreed number of hours for the agreed number of days per week, they have fulfilled the basic terms of their employment contract and are entitled to be paid. Whether they create any value while at work is a result of the effectiveness with which they are organised and motivated. However, they possess greater knowledge regarding both their capacity for work – particularly how fast they can work – and how much work they have actually completed. The whole discipline of industrial engineering (ergonomics, work study, etc.) was developed to try and solve this information problem for the employer. This is an old problem: a traditional interpretation of the Jewish Passover story is that the Egyptian Pharaoh held a brick-making contest among the Hebrew slaves and used the results as a standard to set a much higher daily production quota[7]. Where the output cannot be directly measured, in terms of either quantity or quality, there are even more opportunities for shirking. Perverse incentives can lead to the production of a large quantity of useless components.

Each of the different forms of employment contract is associated with different ways of solving the shirking problem. Sequential spot contracts are typically used where output is easily measurable in terms of quantity, and quality can be effectively controlled. Here piecework payments can provide efficient incentives because workers are paid directly in relation to their output, and hence shirking costs the employee, not the employer. Under contingent claims contracts, the full panoply of industrial engineering is deployed, often in conjunction with the machine pacing of the work. Under the authority relation, shirking is a much larger problem, and in a poorly managed work environment, it can become endemic. The promise of promotion is typically the main incentive to perform under the authority relation. For this reason, the authority relation is often used only where one of the other alternatives is not viable. Motivation in the collegial environment is usually through peer pressure and the fear of gaining a reputation for being a shirker. Sometimes such behaviour is reinforced through group bonus or share option schemes, and organisational culture becomes the principal source of motivation.

7.5 The role of sequential spot contracting in construction employment

A notable feature of employment in construction in many countries is the extensive reliance on sequential spot contracting, particularly for skilled operatives[8]. This can take many forms; sometimes it is temporary employment, where staff are employed for a period of days or weeks and assigned to a specific task, and then laid off when that task is completed. Rates of pay are agreed task by task, although sometimes these are regulated by trade unions as in US 'union shop' construction. The dynamics of this approach during the post-unification construction boom in Berlin are shown in panel 7.3. A second approach is the use of agency workers, but perhaps the most widespread approach is to use *gang organisation*, where workers agree to undertake an assignment for a lump sum – a pure piece-rate system. Under some jurisdictions, such as the UK, gang members have the formal status of being self-employed but, as shown in panel 7.1, this status is open to challenge because such workers remain within a hierarchical relationship with the employer and do not trade as independent businesses.

Panel 7.3 Mobilising human resources in Berlin

The construction boom in Berlin after *die Wende* drew workers from all over Europe, much to the horror of the German construction trade union, IG Bau-Steine-Erden. Pubs such as *The Oscar Wilde* at Oranienburger Tor acted as labour exchanges for Irish workers, for instance. On one project – Galeries Lafayette in Friedrichstadtpassage – almost all the workers for the structural works executed by the largest Austrian construction firm, Maculan, were recruited locally, but few were German. Self-employed gangs were formed from homogenous national gangs, led by a ganger who could speak some German. For example:

- Formwork for ceilings: Portuguese (70 workers)
- Formwork, steel placement and concreting for walls: Italian and (former) Yugoslavian (70 workers)

- Formwork, steel placement and concreting for walls: Austrian and (former) East German (25 workers each)
- Transport on site and cleaning: Hungarian, Czech and Yugoslavian (20 workers)

Only the Austrian workers had some experience of this type of construction work. They, together with the East Germans, were Maculan employees. Those from Eastern and Southern Europe were from rural areas and trained on site. To obtain a workforce of 70 workers, Maculan recruited 400 and put them to work for a week. At the end of the week, 330 were dismissed and the other 70 given further training. Maculan went into liquidation in 1996.

Sources: Syben (1996) and research by Sandra Schmidt.

These types of arrangements were common in most manufacturing and primary extraction sectors during the nineteenth century. Sequential spot contracting – known variously as internal subcontracting, the bargain system, the butty system or tasking – was the employment contract upon which the industrial revolution was built. However, construction differs from all other manufacturing industries in that it did not make the transformation which started in the late nineteenth century whereby the 'invisible hand' of the market was replaced in industry after industry by the 'visible hand' of managerial capitalism and the large-scale corporation emerged. It was these newly emerging corporations which led the shift from sequential spot contracting to contingent claims contracting. Scientific management, premium bonus schemes and collective bargaining with well-organised trade unions became the norm in industry after industry, until challenged by new managerial methods from Japan after 1970 which, in essence, took contingent claims contracting to new levels of sophistication[9].

The construction industry has not made this shift to managerial capitalism; the contractual relationships between employers and employees in the UK construction industry of the late 19th century look very similar to those that exist in the industry over a century later[10]. The reasons for this are beyond the scope of this book, but they are deeply rooted in the industry recipe analysed in section 2.2. The point to be made here is that a very important reason for the shift from sequential spot contracting to contingent claims contracting was to allow management to gain control over the work process[11]. For the more romantically inclined, this is seen as a loss of worker autonomy, but behind it lay the drive to rationalise the production process which was at the basis of Western prosperity in the second half of the 20th century. In contrast, the construction industry remained in the position where work organisation 'depended upon the skill and self-discipline of the individual navvy. The amount of real direction and supervision that came from the management and the foremen was minimal. The men had a rough idea of what was needed and got on with it'[12]. Nearly 30 years later, a site manager argued that he did not give detailed instructions on how to organise the work of bricklayers for fear that the gang would claim against the firm (i.e. opportunistic behaviour) if the situation changed[13].

There are a number of reasons why this continuing reliance on sequential spot contracting may be limiting for the development of the industry:

- One of the major technological developments behind the shift to large-scale production, in particular the assembly line, was the development of interchangeable parts[14] – components that push-fit together quickly and cleanly. Their introduction meant a shift from the general skills of the skilled machinist and fitter, to the job-specific skills of the machine-minder and assembler. The construction industry is presently trying to introduce standardisation and pre-assembly without addressing the human resource governance implications; we will return to this issue in section 12.10.
- The reliance upon the general skills of the craftsman, rather than job-specific skills, means that it is difficult to innovate in the construction process because any such innovation requires job-specific training so that the workers can learn how to use the new technique. Sequential spot contracting stifles process innovation, which is why it has became so unpopular among managers in most industries over the last century. This issue is becoming even more important as the growing emphasis upon sustainability brings unfamiliar technologies to the construction site.
- The whole range of performance improvement techniques, ranging from total quality management and *kaizen* to the theory of constraints, relies upon detailed measurement of performance and an understanding of the drivers of that performance.
- Wherever asset specificities are created through learning by employees, sequential spot contracting means that they will be used opportunistically by employees rather than captured by the firm as a whole. The use of sequential spot contracts is only efficient where skills are completely generic and freely available in an undifferentiated pool, and no learning takes place during task execution which is of value to the firm.

However, there remain important advantages of sequential spot contracting:

- Important production efficiencies arise from the motivation effects of paying purely on the basis of output – workers are encouraged to self-organise to maximise output.
- Where transaction frequencies are low, the determinate nature of contracts maximises flexibility – workers can be taken on and let go as the specific task requires.
- Many workers prefer the autonomy of gang organisation.
- In some jurisdictions, employer's social obligations are avoided – for this reason sequential spot contracting on a self-employment basis is not lawful in countries such as France and Germany.

However, perhaps the most important reason for its continuing importance is that first-line management in construction tends to lack the training to manage in any other way. It is deeply embedded in the industry recipe; arguably, this is the most significant single barrier to improved performance in the construction process.

7.6 Managing the supply chain

The *supply chain* for a given project can be defined as the set of firms engaged in external transactions commencing with a principal contractor and terminating when external transactions switch to internal ones in the employment relationship. On this definition, a project may have one or more supply chains, depending on the client's choice of number of principal contractors. Supply chains may be switched from project to project complete as in integrated supply teams (IST), or more normally they are formed anew for each project. Each firm within the supply chain has to make its own make/buy decision regarding whether to subcontract further for the resource inputs it requires, or to use its own employees to add value. In principle, the supply chain extends right back to the extraction of the raw materials – iron ore, gypsum, clay, etc. – from the earth, but in practice it is helpful to limit the analysis to the point where the production of materials and components becomes construction specific. Thus, a quarry for aggregates is part of the construction supply chain; a mine for bauxite is not.

Horizontal governance in construction has always involved market transactions for the supply of specialist skills and services; trades such as roofing, for instance, have traditionally been the province of specialist rather than general contractors. Our concern here is to understand the variety of different external governance modes for different trades. There has long been a debate about why construction firms subcontract.

- Some argue, in effect, that it is a solution to a transaction cost problem generated to maximise flexibility in coping with the uncertainties in the resources required for a particular project[15].
- Others argue that it is a solution to a production cost problem, driven by specialisation to achieve production efficiencies through economies of scale and the learning curve[16].

Recent developments in supply chain management suggest that both could be right, depending on the trade supplied. The segmentation of the supply chain into those suppliers that are relatively critical to the buying firm's operations and those which are less critical and can be treated as a commodity is central to effective supply chain management[17].

Our concern here is with *relatively* low-uncertainty transactions during realisation; high-uncertainty transactions will tend to be governed either internally or professionally, as discussed in section 6.6. Figure 7.3 identifies the external horizontal governance options under low transaction frequency. Where both of these are low, sequential spot contracts (traditional trade subcontracts) will be favoured. This is the case for many of the traditional trades where there is a ready supply of firms to do the work and large asset specificities are not generated during project execution – one bricklaying subcontractor can be relatively easily replaced with another. However, this situation should not be confused with the melon-type contract discussed in section 6.2; it remains a relational contract, even if that relationship is pretty distant.

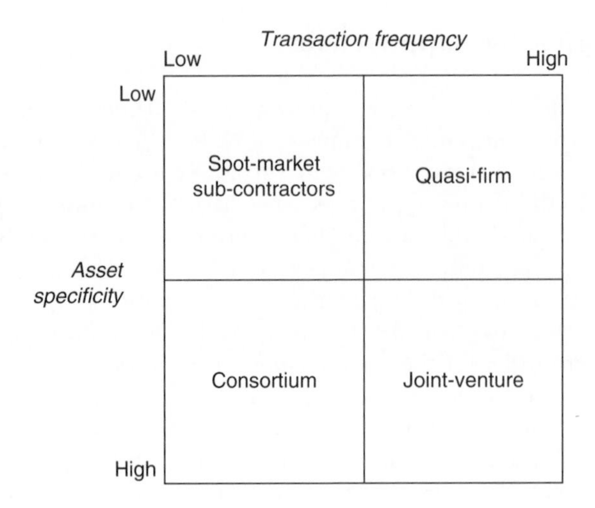

Fig. 7.3 External horizontal governance options for the supply chain (source: Winch, 2001, Fig. 4).

As transaction frequencies rise, other options become available – it becomes possible to enter into repeat relationships with suppliers. This allows the development of a *quasi-firm*, where trade subcontractors are offered repeat contracts by principal contractors conditional on satisfactory performance[18]. This is most favoured where the same resources are required project after project, such as in housebuilding. Many construction quasi-firms are fairly informal – the same firm is used simply to reduce procurement costs without any strategy for deepening the relationship over time. However, this is changing with the advocacy of partnered ISTs. The 'firm' element of the quasi-firm tells us that important hierarchical elements enter into the relationship, but without any of the obligations on the part of the employer that characterise the employment relationship. Smaller firms within the quasi-firm can find themselves treated in the most cavalier way when it suits the principal contractor. In other cases – such as Japan – the whole construction industry can form itself into a series of tiered quasi-firms around the leading principal contractors. In Shimizu, this tiered 'family' is known as the *kanekikai* which employs a total of nearly 70k people[19].

Where asset specificities are higher – either pre- or post-contract – problems emerge of potential opportunistic behaviour by the subcontractor. There are various ways in which this can happen.

- The subcontractor owns a proprietary technology required for the project.
- The client specifies the use of a particular supplier – the nominated subcontractor.
- A particular subcontractor's work forms a large percentage of the total cost of the works, or is at a key point on the critical path, and switching costs arise as a result as discussed in section 6.3
- The contract contains a local content or technology transfer clause which has the effect of generating asset specificities for the local partner

Under such circumstances, a *consortium* between the subcontractor and principal contractor will be favoured[20], thereby minimising opportunistic behaviour by the owner of the specific asset at the cost of the principal contractor. By providing for a common incentive framework between the principal and subcontractor, the consortium arrangement reduces the scope for opportunistic behaviour by the supplier of the key resource and improves the flow of information between the parties. These issues most frequently arise on building projects around the mechanical and electrical (M&E) services packages, and it is noticeable that it is common for construction firms engaged in principal contracting to own a subsidiary focused on M&E contracting. Where the same resources are required on a programme of projects, it may be appropriate to form a *joint venture*[21].

7.7 Managing consortia and joint ventures

The formation of a consortium or joint venture is more complex than the formation of a traditional subcontract or a quasi-firm. While in the latter, risks are allocated and managed through the subcontract, in the former, risks are internalised and shared between the partners without the benefit of hierarchy to resolve problems and disputes. The costs associated with managing the relationships within the alliance are transaction costs, and are typically higher than those incurred within a single organisation. Selecting and negotiating with potential partners may be called the Jane Austen problem – the problem of finding a partner of appropriate status, wealth and character, defined in panel 7.4.

Panel 7.4 In search of status, wealth and character

The six novels of Jane Austen (1775–1817) provide a witty satire on the world of the English squirarchy at the turn of the nineteenth century – a world in which a successful marriage was everything for a woman, yet one in which notions of romantic love were starting to develop. The problem of finding for their daughters husbands who combined a position in society to match their own (or better), were rich enough to support their daughter in a manner appropriate to that status, and were moral and trustworthy taxed many a parent. A potential match often failed on one or more of these criteria, and mistakes were often made. While Jane Austen's novels tend to have a happy ending, the same is not always true for strategic alliances between firms.

In terms of *status*, the partner firms have to possess complementary resources so that the whole of the alliance is greater than the sum of its parts in terms of value generated to balance the additional transaction costs incurred. This is the central rationale for the formation of a joint venture. Unless the partners to the alliance bring complementary resources, there is little reason for them to come together and every reason for them to separate should they find out that the dowry was not what was expected. For instance, Medicinq, one of the prime supply chain partners in ProCure 21 presented in Case 6, is formed from three regional builders to enable them to offer coverage over the whole ProCure 21 area; they have decided not to bid to be a member of the Scottish equivalent because they have no local partner there.

In terms of *wealth*, the firms have to be equally capable of bearing the risks shared within the alliance. *Integrated consortia* are jointly and severally liable for contract execution, they operate with a common budget and they divide profits and losses at project completion in proportion to the original investment made by each partner. The inability of one partner to meet its obligations shifts that responsibility to the others. For instance, the size of some framework agreements such as UK Ministry of Defence Regional Prime Contracts valued at up to £700m over 5 years means that suppliers prefer to share risks with others through forming joint ventures.

Not all members of alliances are able to share risks equally. For instance, principal contractors on international projects are often required to enter consortia with local suppliers as part of the market entry criteria. Sometimes these firms lack both assets to back their full participation as well as complementary skills – in effect the team is not very integrated and they are little more than subcontractors in terms of their relationships with the principal contractor. For instance, under framework contracts, the client may expect the IST to include suppliers of design and other consultancy services. Such firms are rarely able to share risks on an equal basis with the much better capitalised contractors, which is one reason why ISTs are probably closer to a quasi-firm than a joint venture.

In terms of *character*, the firms have to have working practices that are similar enough to allow them to function effectively together. Firms differ on many dimensions – organisational culture, management practices and strategic intent. Unless the partner firms have the same basic objectives for the joint venture and are able to adapt their existing management systems so that they are compatible with those of their partners, problems are likely to arise. Organisational cultures are more deep rooted and less amenable to change – they are often the biggest stumbling block to successful joint venture management. Many commentators on joint ventures focus on differences in national cultures as a major management problem in international joint ventures. However, there is no solid evidence that these problems are actually any worse than those caused by differences in organisational cultures between firms of the same nationality. International construction projects work in English and are rooted in an engineering culture that favours fact over opinion. Certainly, on the Channel Fixed Link project, organisational cultures were more of a problem than national cultures. This process of courtship played an important role in the development of the Dioguardi and Beacon relationship, presented in panel 7.5.

Panel 7.5 The Beacon Dioguardi strategic alliance

In the autumn of 1995, a 60:40 consortium of Beacon Construction of Boston, USA, and Fratelli Dioguardi of Bari, Italy, successfully bid for the construction of the Italian Chancery in Washington, DC. This success was the highest point to date of an international partnership that had been evolving for over a decade.

The seed for this alliance was meetings between the chief executives of the two companies – who also held professorial positions at local universities – at academic conferences. The motive was the deep recessions in their respective local markets which led to a formal 'International Bridgehead Agreement' being signed in early 1992. The opportunities were particularly challenging projects for which each needed additional expertise. Thus, Beacon helped Dioguardi with developing the business case for the refurbishment of the derelict Margherita Theatre in Bari, while the favour was returned through deploying Dioguardi's technical skills in constructability analyses of the New England Holocaust Memorial and the redevelopment of the Beth Israel hospital in Boston.

At the heart of the partnership are common interests and complementary skills. Both firms have experience throughout the whole facility life cycle from inception to operation, and strong operational skills. They are both medium-sized with strong regional bases and reputations. Both have reputations as good employers, and senior management interested in reflective practice as witnessed by their active involvement with local universities. However, Beacon's skills are more process orientated with its focus on feasibility studies, construction and supply chain management, and facility management. Dioguardi's skills are more product orientated with its willingness to put its own capital at risk to finance projects, strong design capabilities, patented and proprietary structural systems, and skills in refurbishing historic buildings. The development of these different capabilities reflects both the strategic decisions by senior management and the character of domestic construction markets. In appropriate combination, they strongly reinforce each other.

Source: Pietroforte (1997).

7.8 The dynamics of supply chains

The externalisation of resource – base mobilisation of resources through a supply chain generates a specific dynamic which causes a number of managerial problems where tasks are dependent on each other in a *parade of trades*[22]. The problems are generated by the interaction between two phenomena:

- Task execution is subject to random variations in duration, particularly in construction where process capability – see panel 12.3 for a definition – tends to be relatively underdeveloped.
- In facing the uncertainty generated by such variations, independent decision-makers at the resource-base level tend to act in their own interests rather than in the interests of the parade as a whole.

These two phenomena can reinforce each other so that the random variations increase in amplitude simply because of system effects, without there being external random events to affect the system (i.e. the parade) as a whole. These phenomena are the subject of well-known management games – MIT's beer game and Goldratt's matchstick game – as shown in panel 7.6.

Behind both of these games is a very serious point about the system-level effects on processes made up of discrete tasks or decision points. While each individual task within the process is executed as efficiently as possible and every decision is taken rationally, the effects at the level of the process as a whole are

Panel 7.6 Simulating supply chain dynamics

In Goldratt's *The Goal*, our hero takes a troop of scouts for a hike. Frustrated by the inability of the single-file troop to keep together, he plays a game with the scouts involving the passing of matchsticks in a sequential process made up of mess bowls ranged along a picnic table, each staffed by a scout. The transfer of matches from one stage of the process (mess bowl) to another is determined by the roll of a die by the scout responsible, unless constrained by lack of matches in the bowl. Despite stable demand for output of matches from the process, and an input of matches to the process matched exactly to demand, the process as a whole fails to meet demand, and in-process stocks rapidly build up in an unpredictable manner. The problem is that the process as a whole is not managed, and that there are no buffers in the process to absorb random shocks.

MIT's beer game has been played thousands of times since the 1960s, and shows how a single one-off doubling of demand can create wild fluctuations in order quantities. The game consists of three teams – retailer, wholesaler and brewer – which are asked to respond to an unexpected increase in demand for a particular beer. Because of time lags between order and delivery of 4 weeks, by the time the brewery hears of the increase in demand through the supply chain, the retailers are starved of product and typically respond by over-ordering. When the brewery responds to these inflated orders, apparent demand collapses because the supply chain now fills up with unwanted beer. The problem is that nobody is managing the supply chain as a whole, and each is perfectly responding to the problem as they see it. The result is *both* retailer and brewery lost sales because of shortage of product and large stocks of unwanted product.

Sources: Goldratt and Cox (1993); Senge (1990).

perverse and negative. Sub-process optimisation leads to total process sub-optimisation because of the bounded rationality of sub-process decision-makers. Both Senge and Goldratt are advocates of *systems thinking*, or the ability to see the wood for the trees, and use these games as didactic tools. The fundamental insight of systems thinking is the same as that of sociology as an academic discipline – that system-level effects structure the behaviour of decision-makers in patterned ways, and that the only way to understand the behaviour of individual decision-makers is to understand their function in the system. Further examples of the insights provided by systems thinking will be presented in section 11.6.

7.9 Clustering the supply chain

The organisation of the resource bases within the construction supply chain has changed little since the emergence of the trade system described in Case 2. In the trade system the resource bases were defined around the manual skills required to work on different materials – carpenters and wood, masons and stone, and so on. Although construction technologies have developed enormously over the past 500 years, this basic principle of supply chain organisation has not. As technologies changed, and new ones emerged, they were simply added to the list of specialist trades, thereby creating an increasingly fragmented set of resource bases. This is an important source of inefficiency in both production and transaction cost terms. How, then, might things be organised differently?

A central organisational principle when decision-making under uncertainty is to group together responsibility for those tasks which have the greatest interdependence and hence greatest requirement for information processing between them[23]. This is because they will have the most intensive information flows as they collectively solve the problems they are assigned. This solution to the problem of interdependency under uncertainty has recently been developed as an alternative basis for the organisation of the resource bases in the supply chain – known as supply chain clusters[24]. One of the early implementations of this principle was on the Building Down Barriers project presented in Case 7 which has had a considerable influence on the UK's Defence Estates organisation as a client in recent years.

The organisational principle of construction clusters is presented in Fig. 7.4. It shows how a core team is responsible for the definition of the facility as a technological system, developing a budget and a strategy for its realisation. An important part of the realisation strategy is the definition of the clusters. The resource bases are clustered around technical sub-systems of the facility such as its structure, envelope or services for the processes of developing the detail design and execution on site. Where appropriate, the cluster leaders can also be members of the core team. The aim is to gain the maximum of information processing within the clusters, so as to seek efficiencies in production costs while clearly defining the interfaces between the clusters, thereby minimising the amount of information processing required between clusters so as to seek transaction cost efficiencies. Because the principle of clusters is, in essence, about the supply chain rather than

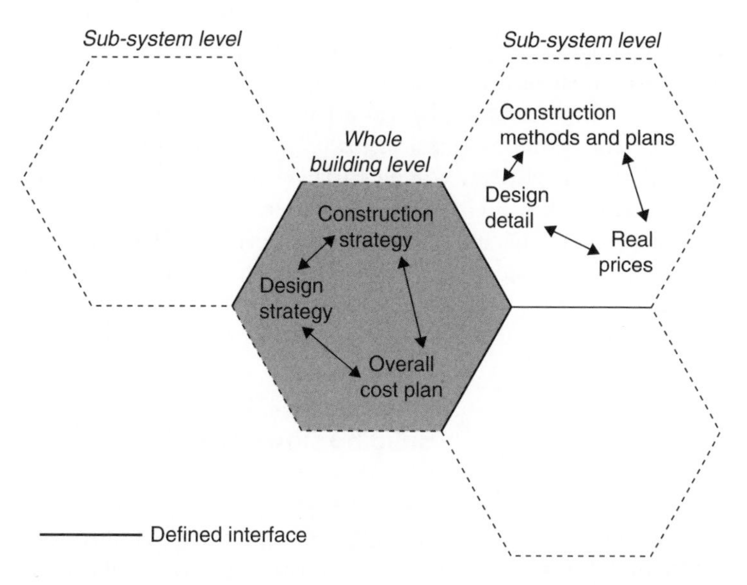

Fig. 7.4 The principle of construction clusters (source: Holti *et al.*, 2000).

the project chain, it can be implemented within either mediated or integrated project coalitions. It is currently diffusing within UK public sector projects in the form of an integrated coalition structure known as prime contracting, which draws on the experiences reported in Case 7.

7.10 Summary

In this chapter we have moved on from the project chain to investigate more closely the management of the supply chain that delivers construction resources to the project. Again, the basic problem is one of information – how do principals clearly express their requirements and motivate suppliers to meet those requirements under uncertainty? Where certainty is high – for relatively simple facilities, or relatively late in the project life cycle – it is possible to write clear and binding contracts and rely on the market to ensure performance. However, certainty is rarely high enough for this on construction projects. More elaborate arrangements are required that allow contracts to be adjusted during project realisation. So far as the supply chain is concerned, the main choice to be made is between subcontracting for the resources or using in-house resources – in particular employees. The uncertainties inherent in the construction process would encourage most firms to use in-house resources, but the low levels of transaction frequency make this difficult. The response has been to set up tiers of suppliers, with relationships varying from the one-off to the continuing.

In 'mobilising the resource base', we have posed the principal/agent problem as the central one in construction procurement – how to ensure that suppliers of resources to the construction project are competent and trustworthy and effectively motivated to play their part in realising the project mission. Many of the first-tier agents in the project chain then mobilise a further set of resource bases in the supply chain, creating chains of principal/agent dyads. These chains end in the employment relation, where the operative workforce is deployed on the material flows which are the physical act of construction. In order to help with the management of the principal/agent problem, clients often turn for help to third parties, creating relatively complex triad arrangements. The resultant project coalition structures are often complex, chartered by standard forms of contract that turn market into hierarchy. This part has been aimed at identifying the principles by which appropriate project coalition structures can be formed to allow effective governance of the project process. We will return to this issue of governability in Chapter 13 when we address the issue of risk management.

Case 7
Building Down Barriers

Influenced by its success in reforming defence procurement using the prime contracting concept for weapons and material, the UK Ministry of Defence decided in the mid-1990s to develop a new form of project coalition for its constructed facilities. They piloted this new form in the Building Down Barriers (BDB) experimental projects. The Army Land Command required new sports facilities at Aldershot and Wattisham, and these two projects were chosen for the experiment with new approaches to procurement. Additional funding from the UK government allowed a research team from the Tavistock Institute to work with the two project coalitions on an action research basis to facilitate the process and to capture their experience for wider dissemination. There are a large number of features of this project that are of interest – not least the concept of prime contracting in a construction context; however, this case will focus on the innovative approach to the management of the supply chain and its mobilisation in the reduction of both capital and operating costs.

The BDB project went through three distinct phases.

(1) *Concept (January 1991 to early 1999).* The first step was to appoint two prime contractors – Laing and AMEC. They then worked with their respective architects and Land Command to work up the design to beyond stage D of the RIBA plan of work. During this phase, decisions were taken as to how the clusters of resource bases were to be organised and their members selected. As appropriate, these cluster leaders and other cluster members participated in the design process. The outcome of this phase was the concept phase submission (CPS), which was evaluated by the Defence Estates Organisation and Army Land Command, together with fixed prices for executing the work from the two prime contractors – £10.8m for Aldershot and £4.2m for Wattisham.

(2) *Detail design and execution (early 1999 to mid-2000).* In January and February 1999, the construction contracts were signed – AMEC at Aldershot agreed its price on a GMP basis with a gainshare formula, while Laing at Wattisham offered to discount the fixed price submitted with the concept phase submission by 5% in expectation of savings being found during execution. Both buildings were handed over early – Wattisham in February and Aldershot in June 2000.

(3) *Proving (mid-2000 to 2002).* The prices submitted at the end of phase 1 included the running of the two facilities by the prime contractors for periods of 15 and 24 months respectively, from handover.

The expectation was that this new form of project coalition would be able to achieve the 30% cost reduction challenge specified in the Latham Report while also improving the whole-life cost performance of the facility. Whole-life costing is discussed in further detail in section 9.7.3. The key tool in establishing these savings targets was the historic reference cost (HRC) model for each building. This was an artificial construct generated by pricing from existing records the capital and whole-life cost of a facility designed to the normal MoD standards. The HRCs were accepted as being very tight. As Table 7.1 shows, the Latham target was not achieved; nevertheless, important savings were made against a tough benchmark by making capital investments (capex) in a higher quality of specification to achieve greater savings on running and maintenance costs (opex) over the facility life cycle; both projects show roughly a 100% return on the investment in higher quality of specification. Both projects were handed over ahead of time,

Table 7.1 The evolution of the budget for building down barriers.

	capex		opex (on NPV basis)	
	HRC	budget	HRC	budget
Phase 1: Concept phase submission				
Aldershot	9.7m	10.7m	16.6m	14.7m
Wattisham	3.8m	4.0m	7.2m	6.7m
Phase 2: Midway (October 1999)				
Aldershot	9.8m	10.3m	16.7m	14.3m
Wattisham	3.8m	4.0m	7.1m	6.7m
Completion (final account)				
Aldershot	9.8m	10.3m	16.7m	14.3m
Wattisham	3.8m	4.0m	7.2m	6.2m

HRC, historic reference cost.

with excellent safety records and no disputes. User satisfaction with both facilities is high.

All members of the two project coalitions are reported to have achieved good margins between 8% and 24%, and some achieved more because of the effectiveness of the organisation of the works on site. Those firms that did not achieve their budgeted margins attributed it to poor estimating. Client-initiated changes to specification after CPS were minimal, accounting for a £100k increase in the HRCs at Aldershot.

The budget at CPS figure was already the result of significant effort in design development involving formal value engineering, and was fully owned by the prime contractor. For instance, a significant decision was to invest in a relatively expensive combined heat and power unit so as to attack more effectively whole-life costs. There is a general agreement that, on both projects, the fit and finish of the buildings is good compared to what would have been expected through traditional design and build. The differences between the CPS figures and the outturn costs predicted midway through phase 2 in October 1999 were largely a result of intensive formal value engineering exercises on both projects. AMEC and Land Command thereby shared savings of £400k. The Laing figure is the discounted fixed price and cannot therefore yield further savings for the client against budget. The savings on Aldershot are intriguingly close to the 5% discount offered by Laing on Wattisham. A significant source of savings during phase 2 on both projects came from collaboration within the groundworks clusters between the structural engineer and the groundworks contractor on the foundation works.

The most difficult clusters to manage were the finishes clusters. These are inherently diverse, and consist of large numbers of very small firms. Many of these firms simply did not have the managerial capacity to engage with the collaborative approach within the clusters. Here, relationships between the trade contractor and the prime contractor tended to be much more 'traditional' in terms of the industry recipe described in section 2.2 than with the larger firms working on the main trade packages. The additional managerial effort required by the prime contractors for the finishing trades cut into their margins.

The proving period worked extremely well at Wattisham – Laing put a caretaker on site who handled minor defects and those that he could not handle were swiftly dealt with by Laing or its cluster members. The facility at Aldershot was more complex, and also its use patterns changed from those expected at scheme design. This led to some upgrading of facilities, but also means that the facility is not fully fit for its current purpose.

A notable feature of the two projects is a sustained attempt to implement the principles of clustering the supply chain. The overall structure of the project coalition is illustrated in Fig. 7.5. Around the prime contractor a project core team was formed which included the designers – architects and engineers – and the cluster leaders. While there were slight differences in implementation, the same basic clusters were used on both projects, as illustrated in Table 7.2. The table shows how the design work was divided between the designers working for the prime contractor who were responsible for facility definition – see section 8.5 for the terminology used here – and those within the clusters responsible for facility description and execution. It is notable that those cluster members who made the most of the collaborative arrangements and tended to make higher margins were those that had their own design capability – this was particularly important for the early identification of coordination issues with other clusters.

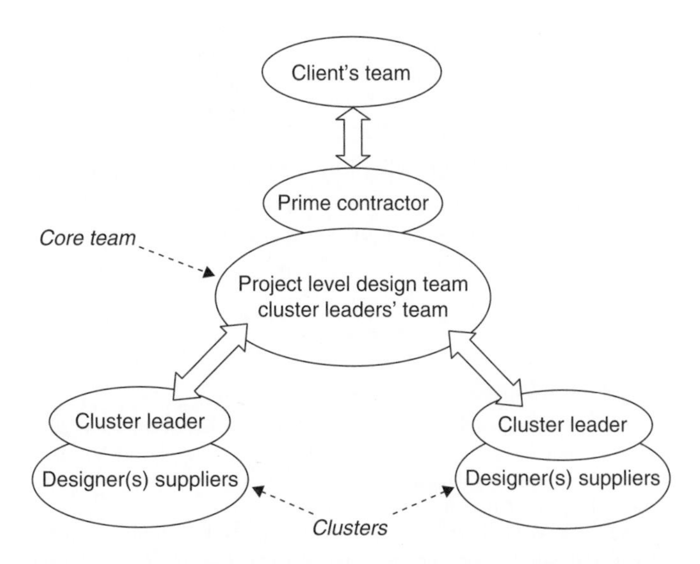

Fig. 7.5 The clustered project coalition structure for building down barriers.

Table 7.2 Organisation for building down barriers.

Wattisham pilot		Aldershot pilot	
Cluster leader	Design responsibility	Cluster leader	Design responsibility
Groundwork: Civil engineering contractor	Scheme and detail design by core team consultant advised by cluster leader	*Civils and groundwork:* Civil engineering contractor	Scheme and detail design by core team consultant advised by cluster leader
Frame and envelope: Steelwork fabricator	Scheme and detail design by core team; consultant advised by cluster leader	*Dry envelope:* Steelwork fabricator and roofing and cladding contractor	Scheme by core team consultant; detail by cluster leaders
Swimming pool: Pool contractor	Scheme and detail by cluster leader	*Water treatment:* Water treatment contractor	Scheme and detail by cluster leader
Internal finishes: Architect	Scheme and detail by architect, in consultation with other cluster leaders	*Blockwork and finishes:* Building contractor	Scheme and detail by architect, in consultation with cluster leader
Mechanical and electrical services: M&E contractor	Scheme by core team consultant; detail led by cluster leader	*Mechanical and electrical services:* M&E contractor	Scheme by core team consultant; detail led by cluster leader
		Sports equipment: Sports equipment contractor	Scheme and detail by architect in consultation with cluster leader

Sources: Holti et al. (2000); Nicolini et al. (2000, 2001); Building Down Barriers Evaluation Reports 1–4 (London, Tavistock Institute). See also Cain (2003); Fisher and Morledge (2002). I am very grateful to Richard Holti for his help in the preparation of this case.

Notes

1 Fraser (1969, p. 139).

2 Cacciatori and Jacobides (2005) argue that there has been significant integration on the supply side, but do not give case examples and largely rely on self-defined images of firms for evidence of this development. They also identify competing definitions of what 'integration' means between different types of firm in the sector.

3 It was Karl Marx (1976) who first identified this distinction, and it is the fundamental underpinning of what economists call X efficiency, and the rest of us call management.

4 See Goffee and Scase (1982).

5 Lumley (1980) indentifies some of the problems, and Cammock (1987) explores one solution through a pre-site agreement. See also Frenkel and Martin (1986).
6 The classic case here is electricians who are both relatively highly skilled and need to be scheduled late in the project. Both the Heathrow Terminal 5 and Wembley National Stadium projects suffered such problems – the electricians on Heathrow were outside the project agreement negotiated for building workers (http://www.building.com accessed 22/08/08).
7 According to Milgrom and Roberts (1992, p. 233).
8 See Winch (1986) for a review of the literature on construction labour markets.
9 Milgrom and Roberts (1992, Chapter 10).
10 Compare the account in Price (1980) with that in Winch (1986, 1998a).
11 Braverman (1974).
12 Sykes (1969, p. 30).
13 Interview, West London, 17/03/92; see also Green (2006) for a view from 'below the waterline'.
14 Hounshell (1984).
15 See Ball (1988).
16 See Eccles (1981a).
17 Dyer *et al.* (1998).
18 Eccles (1981b).
19 Cox and Townsend (1998).
20 Clark and Ball (1991).
21 The usage of the term joint venture in the construction industry is not in accord with the wider use of the term in economics and management. Consortia are, in essence, temporary, while joint ventures usually share equity participation and are often indeterminate in length. The classification used here follows mainstream practice, rather than construction practice.
22 The term comes from Tommelein and her colleagues (1999) who have developed Goldratt's matchstick game into a construction supply chain simulation called Parade Game.
23 Thompson (1967) was the first to articulate this principle explicitly.
24 The concept was developed independently in the UK and Finland – see Gray (1996) and Lahdenperä (1995).

Further reading

Cousins, P., Lamming, R., Lawson, B., and Squire, B, (2008) *Strategic Supply Management: Principles, Theories and Practice* Harlow, FT Prentice Hall.

Lorange, P., and Roos, J. (1992) *Strategic Alliances: Formation Implementation and Evolution.* Oxford, Blackwell.
An overview from the leading analysts of managing joint ventures and consortia.
An authoritative text on the principles and practice of supply chain management.

Torrington, D., Hall, L., and Taylor, S. (2008) *Personnel Management: HRM in Action* (7th ed.). Harlow, FT Prentice Hall.
The leading authoritative text on human resource management.

Part IV
Riding the Project Life Cycle

Riding the project life cycle[1] is the process by which the project mission is defined at greater and greater levels of detail through time, and different types of resources are mobilised as the level of detail becomes finer through a process of progressively *structured sensemaking*. The fruits of the intense intellectual effort early in the project life cycle start to become physically manifest on site until at completion, all the information required for the project is embodied in the constructed asset. This part will identify both the managerial approaches to managing this information flow over the life cycle as a whole, and the large number of techniques that have been developed for managing different parts of the life cycle. What all these different techniques have in common is that they are all coping with the fundamental problems of reducing uncertainty in the project information flow, and they all consist of decision-making cycles as specified in the generic information loop shown in Fig. IV.1.

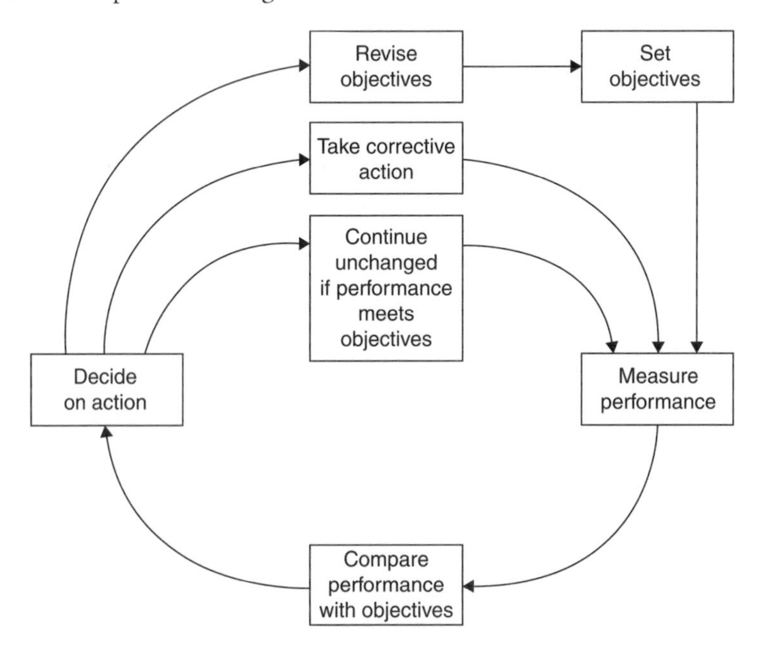

Fig. IV.1 The generic information loop.

The information loop starts with the setting of the objectives; this may be at the level of the project mission as a whole, or the work for the following day as appropriate. Indeed, establishing the appropriate *wavelength* for the control loop is one of the principal project management tasks. The principle usually adopted here is the 'rolling wave', illustrated in Fig. IV.2. Longer term objectives are set in broad terms, while nearer term objectives are set more precisely. This is because more information is typically available for what is to be done next month than for what is to be done next year. As uncertainty is progressively reduced – particularly in the design and execution phases – planning at a greater level of detail becomes possible and control loops become more intense. We will see many examples of both the information loop and the rolling wave in the following chapters.

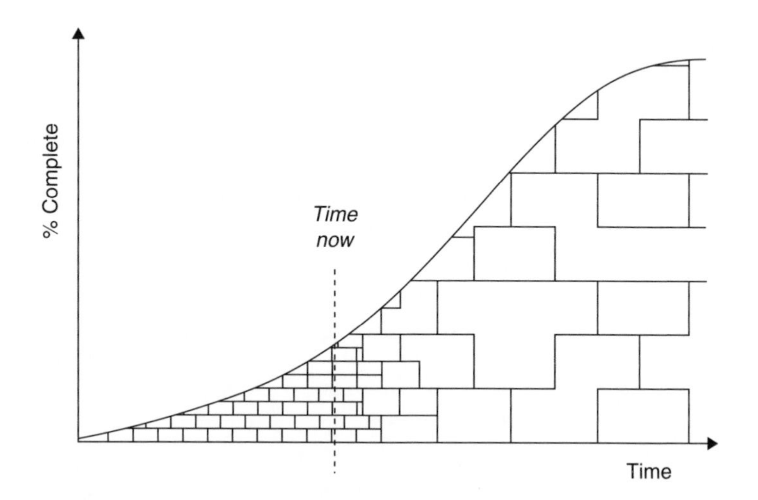

Fig. IV.2 The 'rolling wave' concept (source: Morris, 1994 figure 60).

The next step is to measure performance and then compare that performance with the objectives set. This comparison yields one of three options for action:

- continue with the task, or proceed to the next task, if performance is in line with objectives;
- take corrective action with task execution if objectives and performance are moving apart – do remedial work, supply more resources, change the specification and so on;
- reassess objectives if the original ones are no longer achievable because of new information becoming available.

These loops look very different earlier in the project life cycle compared to later as options close down and the loops become tighter. For instance, changing the specification is a widely adopted corrective action during the early stages of design, but is rarely used during finishing works. Many of these loops are already well known under 'brand' names – *plan, do, check, action* from total quality management or *last planner* from project planning for instance – but there is an underlying unity to the process. It is the principles that matter, not the branding of the particular technique.

Some information loops will have very long wavelengths – the longest are, by definition, as long as the life cycle of the project itself[2]. The shortest, such as managing a concrete pour, will be measured in hours rather than in days. Many of the loops are nested inside each other with daily cycles at the lowest level of management, weekly cycles higher up the hierarchy and quarterly cycles for the most senior participants in the project.

The process is essentially one of feedback and control, and the central project management problem in riding the project life cycle is to establish and maintain appropriate information loops for each task on the project. If things are going wrong on the project, then the failure of an information loop is almost always at the heart of the problem.

- New and surprising information becomes available and objectives have not been adjusted to cope.
- Tasks are not on budget or schedule, but opportunistic behaviour hides this information.
- Measurement is not taking place, so nobody knows the level of performance against objectives.
- Objectives are so poorly defined that it is unclear whether they have been met or not.

This part of the book will start by presenting in Chapter 8 an overview of the dynamics of the project process over the life cycle as a progressive reduction of uncertainty through time, and will show the way in which this dynamic can be managed through a series of filters or *gates*. Chapter 9 will turn to the management processes associated with setting project objectives – particularly briefing and scheme design. Chapters 10–12 focus on the challenges of delivering those project objectives in terms of budget, schedule and conformance to specification. Chapters 13 and 14 are more integrative in that they analyse issues that are pervasive in the management of construction projects. Chapter 13 focuses directly on the problem of uncertainty through exploring risk management. This part then closes with a more detailed examination of the use of information technology to manage the project life cycle in Chapter 14. However, before starting to investigate these issues, it is worth looking at two cases – one in New York and the other in London – which show the diversity of challenges facing those responsible for riding the project life cycle.

IV.1 The rock star of architecture in New York[3]

David Childs of Skidmore Owings and Merrill (SOM), completing his first major project in New York, was feted thus by the New York architectural community for his post-modern Worldwide Plaza building in 1989. He led the SOM design team in conceiving a 47-storey skyscraper clad in granite, brick and copper, capped by a crystal lantern. He had fought throughout the project for his vision of what the final building should look like, no doubt spurred by a very complimentary review of his initial concept in the *New York Times* in 1985. In delivering this vision, Dominic Fonti of the construction managers HRH faced many challenges, including:

- A running argument between Childs and one of the partners – also trained as an architect – in the client company ZCWK Associates who preferred a more pink shade of brick. This argument went on for several months and was only resolved when Childs backed down as Fonti insisted that the bricks be ordered so as to keep up with the schedule.

- Bricklayers who were not allowed – because of union rules – to use a plumb line which made the vertical alignment of the specially shaped bricks designed to give a vertical ridging to the façade very slow and difficult.
- Stonemasonry trade contractors who did not have adequate design capability to engineer the prefabricated granite cladding system, and whose owner spent much of the project incarcerated in a prison hospital in Atlanta because of his Mafia connections.
- Lack of competition in the tenders for both the supply of the bricks from Glen-Gery and the copper roofing by Dahnz because of their idiosyncratic specification.
- The ordering and delivery of three different colours and over 50 different shapes of bricks – all special order items – to meet the architect's visual intentions.
- An attempt by HRH to value engineer the copper roofing to increase prefabrication, which backfired because of a number of unforeseen consequential changes required to install the prefabricated roof components. An earlier attempt to suggest to SOM that full copper roofing over 700 ft in the air was a waste of money, and that the same visual effect could have been achieved with a coated material was so vehemently opposed by the architects that HRH were told not even to mention such a possibility to the client.
- A misalignment between the top of the brickwork and the eaves of the copper roof because of inconsistent information on drawings led to expensive refabrication by Dahnz. Whether this was a result of SOM's inadequate drawing coordination or HRH's inadequate drawing management remained to be settled at the end of the project.

Worldwide Plaza was delivered 10% over budget and 4 months late on a 24-month schedule. This building was let to prestige clients such as Ogilvy and Mather, then one of the world's largest advertising practices, and Cravath Swain and Moore, a long-established New York law practice. The permanent mortgage of $533m to replace the construction loan was successfully negotiated in May 1989, and overall the project was considered a financial success but it was believed by both parties to be unlikely that Zeckendorf, the principal partner in ZCWK, would use HRH in the future. Yet most of the problems on the Worldwide Plaza project which caused the overruns had their origins in design decisions by SOM to achieve visual effects. In particular, the choice of specialist cladding materials made HRH's estimating task very difficult. The resulting choices – beyond the granite at street level – were felt by many on the project to be pure luxury. The building is impossible to take in as a whole from street level – the advertisements for the building in the *New York Times* pictured it from the air. Moreover, the expensive detailing on the brickwork that clads some 44 floors and the roof is impossible to see from the ground. The copper roof and associated lantern are purely ornamental, weatherproofing being achieved by a conventional flat roof internal to the pitched copper.

David Childs went on to design some of the most prestigious buildings in New York and Washington, DC, for SOM. Currently he is the project architect for the Freedom Tower on the site of the World Trade Center in collaboration with Daniel Libeskind, which is due for completion in 2012. Its construction is being managed by Tishman Construction, which built the original World Trade Center.

IV.2 Power into art[4]

By 1981, the Bankside power station on the south bank of the Thames in Southwark was derelict, a victim of the changing economics of electricity generation. The National Lottery launched by the UK government in 1992 provided a source of funding for cultural projects, and was seized upon by Nicholas Serota, the Director of the Tate Gallery. The Tate had always played two roles in British cultural life:

- the repository of British art;
- the repository of international modern art.

The Tate found itself unable to fulfil these two functions in its existing gallery on Millbank, and was seeking a dedicated modern art gallery to match the Centre Pompidou in Paris and Museum of Modern Art in New York. The National Lottery's Millennium Fund allowed it to purchase Bankside and turn it into a 'world-class' modern art gallery – the Tate Modern. With a total budget of £130m and a construction budget of £50m, this was one of the most expensive of all the UK's millennium projects.

Herzog & de Meuron of Basel, Switzerland, won the competition for the design and developed their vision of a hard, industrial space that both echoed the history of the building and complemented the art on display. Between 1995 and 2000, the Tate Modern project coalition managed power into art to produce one of the most successful projects funded by the National Lottery. In achieving this within budget and with minor schedule overruns, a large number of challenges had to be addressed by the project management team from Stanhope and the construction management team from Schal (now part of Carillion):

- The site overlapped with that of another millennium project – the Millennium Bridge – and the design team of Norman Foster and the sculptor Anthony Caro had a very different idea of how the bridge should land on the south bank from Jacques Herzog and his Swiss landscape architects Keinast & Vögt. Attempts at compromise proved difficult and in the end it was left to project architects rather than design stars to come to an arrangement.
- The residents of a row of three seventeenth-century houses adjoining the site proved determined in asserting their view of the project, and even went so far as to find property developers willing to provide additional housing on the site before they could be sidelined by the project management team.
- The power station was sold to The Tate, warranted free of asbestos, but asbestos was still found, disrupting work considerably and costing £1.5m to remove, which was charged to contingency, and losing 2 months on the schedule. The project ended with Nuclear Electric and The Tate still arguing about responsibility.

- The rooflights over the turbine hall – consisting of glass blocks set in concrete – were a major feature which all wanted to retain, but closer inspection once access was available showed them to be beyond repair. Replacement was prohibitively expensive, yet the pitched glazed rooflight proposed by the construction manager did not offer the same quality of light. The suggestion of a separate light diffuser to achieve the same effect offended the architects' modernist principles, where form and function cannot be separated – 'we don't do that', stated the architects. A more expensive compromise was reached where the diffuser doubled as an access platform for maintenance, thereby uniting form and function.
- The structural steelwork package – tightly fought by the bidders – turned out disastrously for the trade contractor, Rowen Structures. There were three main problems. Firstly, Rowen did not realise how much detail design work was required to prepare working drawings, having assumed repetition where there was none, and they fell behind in that part of the schedule. Secondly, the specification called for extensive remedial works to the footings of the new structure that would support the gallery spaces on the existing concrete retaining wall, and Rowen did not have the experience required, leading to mispricing and considerable additional man-hours. Thirdly, the difficulties of erecting steel *within* an existing building were badly underestimated. The Tate negotiated a reduction of £100k on the budget price of £6.1m for the schedule overrun of 100%, while it is estimated that Rowen lost £1.5m on the contract.
- During a pour of the slabs, the concrete pump hired by Birse, the trade contractor, broke down and backup arrangements failed. As the slab would be understrength if poured in two halves, the proposal was to chip out the poured concrete from the reinforcement and to start again. However, the structural engineers designed a special system to tie the two half-slabs together.
- The architects made an error in the design of the grand staircase which meant that on one landing there was a small ledge in the adjoining wall. The project architect – Harry Gugger – was so mortified by his mistake that he tried to get the whole stair realigned at a cost of £25k and a 2-week delay. Even Serota, Director of The Tate, normally accused of indulging the architects, baulked at this, and so Gugger had to live with his shame.
- The company providing the glass covers to the main lighting – Bug of Austria – proved incapable of scaling their sample panels up to the 4.5m required on the project, and the lighting tubes could be seen through the glass, instead of a diffused glow. No solution was fully satisfactory, and the architects were forced to accept sandblasted glass at considerable expense to Bug.
- All the doors had to be stained because the veneer used was taken from too near the centre of the tree, producing an uneven effect. However, the factory contracted to do the work burned down one weekend, killing two sheep and destroying The Tate's doors.

The project finished within budget, but with a delay on schedule of some 9 months to handover. However, the Tate Modern successfully opened during 2000 to widespread acclaim – London finally had a world-class museum of modern art. Perhaps the single largest item on the agenda of design review meetings was colour. The architects – dubbed the 'princes of darkness' – had a penchant for 'industrial' black, while the client team had an aversion to gloom. This attention to detail by the client was, arguably, what made the project such a great success. Project leadership from the Director of The Tate – Serota – and an intimate involvement from the Tate's Director of Gallery Services – Peter Wilson – meant that the client's desires infused throughout the project, generating a strong project culture. If it tended to favour the whims of the architect, to the annoyance of the construction management team, the culture was strongly orientated towards the quality of conception rather than realisation. For its part, the construction management team was not highly regarded because of its inability to keep project directors in post – it got through four during the project life cycle, while all other senior members of the project coalition broadly maintained continuity of personnel.

The Tate project has been a remarkable success. Undeterred by injuries sustained by visitors after a 167 m long crack had been installed in the concrete floor as part of an exhibit, the UK government committed a further £50m to an extension to the main building in late 2007, again designed by Herzog & de Meuron. Opening is planned for 2012, but funding difficulties as a result of the 2008 financial crisis for the balance of the £215m budget may cause delay. Herzog & de Meuron went on to design the stadium for the 2008 Beijing Olympics and receive the RIBA Gold Medal in 2006. Schal were replaced by MACE for the management of the extension project.

IV.3 Riding the life cycle at the process level

As argued in section 1.8, the process level is shaped by the institutional and governance levels, and in turn shapes those levels. Figure IV.3 reprises the process level from Fig. 1.5 and shows how, for a given project mission, riding the project life cycle is a dynamic interplay between routines, tasks and teams[5].

- *Routines* are the learned practices developed within the industry recipe that are carried from project to project and then adapted to meet the needs of particular projects. Routines thereby provide the *cognitive resources* which teams will use to decide how to execute tasks; they also generate the *representations*[6] by which disparate teams relate to each other within the project coalition. Routines therefore specify the *how* of riding the project life cycle.
- *Teams* are the *human resources* allocated from the resource bases mobilised on particular projects, providing the *who* of riding the project life cycle. These issues will be explored further in Part V.
- *Tasks* are the *what* of riding the project life cycle – the set of tasks that have to be completed in order to realise the particular project mission typically captured in the work breakdown structure.

The importance of routines in economic activity was first identified by Richard Nelson and Sydney Winter who argued that they define the characteristics of an organisation (its DNA to use a genetic metaphor) and are the basis of its

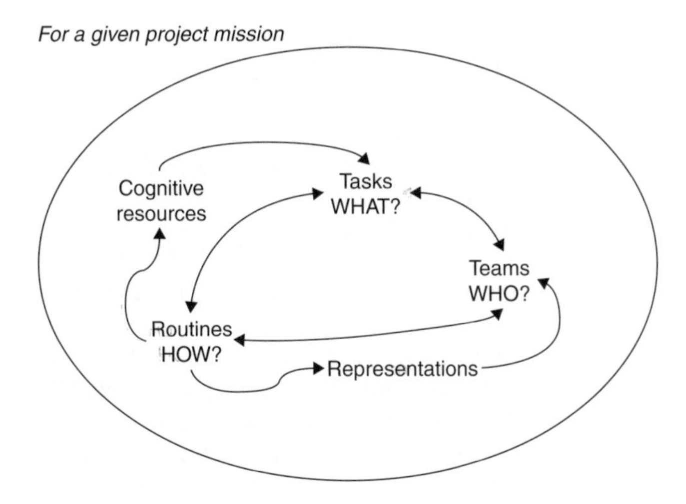

Fig. IV.3 The process level.

distinctive capabilities in the marketplace[7]. Routines have a number of important features:

- They act as the repository of know-how in the organisation.
- They provide the rules of engagement of the members of the organisation.
- They provide heuristics which economise on information processing and thereby increase efficiency.
- Their use in particular contexts requires improvisation, and routines are thereby a source of change as well as stability.
- They provide legitimacy for particular outcomes through perceptions of due process.
- They imply both a procedure and a division of labour, and it is the latter feature which distinguishes them from individuals' habits.

Routines can vary from the formally mandated when they take the form of rules, through management tools and techniques, to informally negotiated and accepted ways of getting along. They provide the articulation of the industry recipe – see section 2.2 – at the process level, and have to be continually adapted to meet the needs of particular projects in interaction with the tasks to be executed and the teams allocated to the project. Routines are open to both formal interventions through improvement programmes and subject to autonomous evolution. For instance, the dynamic of adversarial relations discussed in section 6.7 was the unintended outcome of routines that specified supplier selection should be on the (apparently reasonable) basis of lowest price competitive bidding, yet the Latham programme of reforms in the UK is a deliberate attempt to diffuse new routines for supplier selection which emphasise process integrity issues. Thus the notion of 'best practice' in project performance is diffused through the adoption and adaptation of recommended routines on particular projects.

Mediating representations are those produced by routines that allow the development of mutual understanding between different teams which need to co-ordinate their activities[8] – the BIM presented in Case 14 is an excellent example of this. Different tasks require different skills and information processing styles from the teams that execute them, and many of these skills are more tacit than explicit; translation between information processing styles is required. Boundary objects may take the form of:

- *repositories* – providing explicitly shared reference points such as international standards and regulatory codes;
- *standardised formats* – providing mutually intelligible ways of presenting information such as the Standard Method of Measurement for bills of quantities;
- *sketches and models* – communicating design intent across teams responsible for design tasks, with stakeholders, and through to execution task teams[9];
- *interface maps* – providing visualisations of the interfaces between task teams such as responsibility matrices and Gantt charts.

As we will see in the following chapters, effective use of mediating representations is particularly important in managing projects because the principal responsibility of the project manager – see Chapter 15 – is the coherent co-ordination of task execution teams to realise the project mission, rather than task execution itself.

Routines, tasks and teams are negotiated and renegotiated for a particular project chartered by its project mission. As the project moves through the life cycle, the tasks change, and hence different teams are mobilised which deploy different routines. However, prior choices of routines also shape which teams are selected by which criteria and which tasks are deemed to be in scope to the project. The co-ordination routines used by project managers for task execution teams need to be continually adapted to the needs of the particular project while retaining enough overt good practice to serve as a legitimation for the actions of the project manager. Thus the project process is indeed a negotiated order in which 'the bases of concerted action (social order) must be reconstituted continually; or . . . worked at'[10], and routines provide the raw material for this work in the context of governance and institutions.

IV.4 Identifying different types of projects

The dynamic interaction of routines, tasks and teams will vary according to the type of project, and a basic principle of the information processing approach to managing construction project is that the dynamics of those processes are contingent on the context of the project. In practice, knowing what kind of project is being managed is a first step in managing that project effectively. Some organisations may only build one kind of building and so need not have such a system, but many organisations will want to distinguish between prestige head offices with important symbolic value and functional buildings designed to house routine back-office functions; between new build and refurbishment; and so on. Case 5 illustrates how Marks & Spencer's project strategy changed as the type of stores it was developing changed.

Different organisations use different criteria for classifying their projects, and it should be remembered that construction projects are not the only types of projects that organisations invest in. The UK government might be the largest client for construction in the UK; it is also the largest client for IT projects. Such organisations will typically wish to standardise their formal project routines across the whole portfolio, not just their construction projects, and to transfer best practice between project types. Thus we have seen the advocacy of the PRINCE 2 routine for use on construction projects as well as the IT projects for which it was developed. Other organisations, such as oil and gas companies, may be highly project orientated but focused on large-scale engineering projects with construction as only a small part of their portfolio while wishing to standardise project routines across their entire portfolio. Some clear analytic principles are required for classifying projects so that the implications of a particular classification can be understood and worked through. We need a categorisation (classification based on analytic principles) rather than a taxonomy (classification based on descriptive characteristics), while bearing in mind that the simpler the classification system, the more likely it is to be useful in decision-making.

A number of researchers have considered this problem, and two frameworks in particular have been influential. One is the goals-and-methods matrix[11], which classifies projects on how well the goals of the project are known and how well the methods for achieving those goals are known. The principal problem with this framework is that the methods for achieving goals are largely a function of those goals, and so the two dimensions are not at all independent of each other. A related issue is that the framework is applied to identify the differences between projects in different sectors (e.g. oil and gas versus IT), rather than within one sector, suggesting its discriminatory power is rather crude. A second is the novelty, complexity, technology, pace (NTCP) model[12]. This has four dimensions:

- level of *technology* on a four-point scale varying from no new technology being deployed to key technologies not even existing at project inception;
- level of *complexity* distinguishing delivering discrete technological entities through functionally self-contained assets that create value in distinctive ways through to networks of systems functioning to achieve a common purpose. An example of the *assembly* might be an elevator; of the *system* might be a house; and of the *array* might be an urban transit system;
- *pace* identifying whether the project can proceed at a regular pace, is urgent or is a crisis response;
- *novelty* in terms of whether the resulting product is a derivative, a new platform or a breakthrough innovation.

As the authors admit, the typology is based around the technological characteristics of the asset being created and cannot handle projects where 'soft' aspects such as stakeholder complexity and the delivery of outcomes rather than outputs are important issues in project organisation design. Arguably, the NTCP framework is

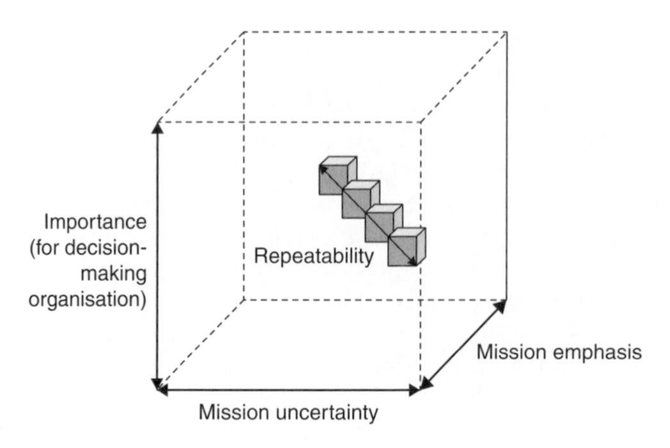

Fig. IV.4 Construction project classification model.

more appropriate – particularly on the novelty dimension – for new product development (NPD) projects and it could be argued that it is not clear what the difference between an array and a *programme* of system and assembly projects might be.

Figure IV.4 presents a typology which can be developed to provide a strategic project classification system. The two dimensions apply the concepts of *mission uncertainty* developed in section 1.3 and *product integrity* developed in section 3.8, and are elaborated to give a dimension of low to high mission uncertainty and whether the criteria for mission emphasis lie more towards specification or conception. It can be suggested that these two dimensions capture the relevant dimensions of the NCTP framework for asset acquisition projects. However, other factors need to be taken into consideration. One is *importance*, or the extent to which the project matters to the client, its supply chain and external stakeholders. A number of elements combine to contribute to the level of importance. One is simply the scale of the project – one definition of a mega-project is that it is a project that affects its environment[13]. A second is the proportion of overall assets at risk on the project – a £100m project may be absolutely critical for most organisations, but is relatively small change for a global corporation such as BP. For instance, it can be suggested that many of the difficulties in managing risk around the Wembley Stadium project presented in panel 6.5 are because it represented a very high proportion of assets at risk for both the principal contractor (PC), Multiplex UK, and the client, the Football Association. A third is profile. A project may be large but largely unnoticed beyond the project team, while a smaller project may have a very high profile for various reasons. Thus the Wembley Stadium project was not particularly large, but because of the high profile of the stadium in the UK sporting calendar it has received considerable outside attention, particularly from the tabloid press.

The fourth factor that needs to be taken into account in strategically designing the project organisation is *repeatability*. The extent to which the project is clearly a one-off in terms of its mission and the means to achieve that mission, or whether it is similar in its mission to a number of other projects is a crucial variable. Repeatability offers the possibility of organising the project as part of a programme, and thereby opening up the opportunities offered by programme management discussed in section 15.3. Figure IV.4 presents the construction project classification system proposed here as a four-dimensional framework. The classification system is to be read as a three-dimensional model for the single project, with a fourth dimension if it forms part of a programme.

Notes

1 I am grateful to Eunice Maytorena for her comments on the introduction to this part.
2 Of course, the project life cycle itself is embedded within the broader product life cycle through to deconstruction.
3 Sources: Sabbagh (1989); Wikipedia (accessed 27/08/08).
4 Sources: Sabbagh (2000); http://www.building.com (accessed 20/08/08). See also Moore and Ryan (2000).
5 The ideas underlying this framework are multiple, but most immediately they derive from the work of Manning (2008).
6 There is a considerable literature around the role of objects (boundary or epistemic), artefacts and the like in processes of knowledge management. We have chosen the term 'representation' here because, with René Magritte in *La Trahison des Images*, we would argue first that drawings (and other documentation) are representations of the finally achieved object, not the object or artefact itself, and second that the terminology of objects and artefacts implies a fixity in the fluidity of knowledge creation and dissemination.
7 See Nelson and Winter (1982); see also Feldman and Pentland (2003), Becker (2004) and Bresnen *et al.* (2005) for further insights.
8 The argument here is derived from the work of Star and Griesemer (1989) and Carlile (2002) on boundary objects.
9 See the fascinating special issue of *Building Research and Information* (**35**, 2007) on visual practices.
10 Strauss *et al.* (1971, p. 104).
11 Turner and Cochrane (1993).
12 Shenhar and Dvir (2007).
13 Ed Merrow, in a presentation at the *European Academy of Management* conference 2006, Oslo.

Chapter 8
Minimising Client Surprise[1]

8.1 Introduction

> 'The completed work, when constructed in accordance with my designs, will not only be the greatest bridge in existence, but will be the greatest engineering work of the continent, and of the age. Its most conspicuous features, the great towers, will serve as landmarks to the adjoining cities, and they will be ranked as national monuments. As a great work of art, and a successful specimen of advanced bridge engineering, this structure will forever testify to the energy and enterprise and wealth of the community which shall secure its erection'.

This chapter will explore the inherently fragile tension between aspiration and achievement in managing construction projects, presenting a perspective on the project life cycle which places the minimisation of client surprise at the heart of the project management problem. The contention is that the high levels of uncertainty at project inception mean that, inevitably, there will be changes from the original plan – some options will prove not to be viable, while other opportunities will open up. The essence of the challenge of managing construction projects is to manage this dynamic through time so that the client, or indeed any other project actor, is not surprised by the achieved conception, specification or realisation of the facility.

The chapter will start by exploring two very different ways of thinking under high uncertainty in the early phases of the project. Future-perfect thinking is about how order is created under high uncertainty to provide a focus for action, while options thinking focuses on how different potential outcomes can be assessed and evaluated. Turning to managing client surprise and the gap analysis approach to the management of service businesses will be developed and applied to the construction project case. This will then be linked to the project life cycle model using ideas from the world of NPD where the stage/gate approach is providing much greater discipline to design project management. Attention will then turn to the issues raised by the differences in the nature of the information flows between different stage-gates, and the implications this has for the nature of the management tools and techniques that can be deployed at the different phases of the project.

8.2 Projecting a perfect future[2]

A moment's thought suggests how remarkable an achievement a construction project is. Moving through the life cycle shown in Fig. 1.3 is inherently difficult, and the greater the mission uncertainty, the more difficult the achievement is. The future is inherently unknown – the implications will be explored in detail in Chapter 13 – yet each year half the wealth in the UK is invested in ventures into the unknown world of the future by investment in the construction of facilities. These facilities will only start to yield their value 2, 3, 5 or even 10 years hence. For instance, London's Crossrail project was funded for construction at £16bn in October 2007, with the expectation that the first trains would run in 2017, 28 years after the first government report recommending its construction was published in 1989. How is this effort sustained over such long periods?

The epigraph to this chapter gives us a clue. John Roebling[3] championed New York's Brooklyn Bridge in the late 1860s and died of a site accident during its construction. What Roebling is doing is projecting an idealised notion of a future state – what *will have been* achieved at project completion. This is known as *future-perfect thinking* in which we conceive of what we will have done by a certain point in time in order to motive action in the present. For some[4], future-perfect thinking is seen as a lateral thinking exercise to stimulate creativity; however, for Alfred Schutz, the Austrian philosopher, such future-perfect thinking is fundamental to all proactive human action in which 'the planned act bears the temporal character of pastness'[5] in distinction to reactive human behaviour. Schutz is careful to distinguish such 'protentions' from mere fantasy – the future-perfect state is believed to be realisable and the intermediate steps towards it identifiable.

Schutz frequently frames his arguments about action in terms of projects, and, indeed, occasionally uses the metaphor of a construction project to make his point. More recently, these concepts have been applied to the management of a major construction project as shown in panel 8.1 where a future-perfect strategy was developed through working back from the desired end state to current activity. To quote[6] the Chair of the UK promoter company for the Channel Fixed Link talking about putting together the deal in the mid-1980s:

> 'If I was to sum up the overriding ethos which governed the directors ... it was the unarticulated faith, difficult to define or explain, but an abiding faith that we would get there in the end'.

We can suggest that future-perfect representations are very important on projects because they provide collective resources for ordering the inherent chaos of the future. As John Dewey[7] argues, imagined futures give meaning to present activity, and in the context of projects they do this in the following ways.

- By providing *orientation for action*. From the chaos of all possible future states the imaginative act of future-perfect thinking pulls out one, or a narrow range of possible states. These can vary from extensions to the house to bridges across the Strait of Messina. Engineering and architectural effort can then be mobilised to test whether the idea is fantasy or not. Proposals for the Channel Fixed Link veered between fantasy and future-perfect thinking for over a century before the current technical solution began to crystalise in the 1950s. Through that time a number of remarkable entrepreneurs from Watkin and d'Erlanger, through Davidson and a younger d'Erlanger, to the team under Henderson who put the final deal together invested considerable effort and frequently a lot of their own money into nudging the project forward.
- By providing a *point of mobilisation* for resources. Projects are not built on enthusiasm alone; they require large amounts of finance. Even small projects – such as the extension to the house – require large amounts of capital relative to existing assets. Sources of finance have to be mobilised through convincing investors that a particular project merits their attention. This is done not only through narrative of the benefits of the facility in use, but also through many different kinds of representations such as artists' impressions, scale models and, most recently, 3D visualisations. Similarly, the human resources that will be required to realise the project can thereby be enthused.
- By providing a *focus of debate*. Stakeholder key players need to be convinced of the merits of the project, while potential opposition can be flushed out and accommodated. Alternative means to realise the project mission can be rehearsed, debated and evaluated taking into account stakeholder perspectives.

From a sensemaking perspective, future-perfect representations provide *enactments* of the potential future states so that further information can be gathered about the potential viability of those states. The representations thereby provide the cognitive resources for the definition of the project mission. The process can go awry from a point of view of the efficient allocation of resources, as the discussion in section 3.7 indicates, but the rationality of investment appraisal is inherently constrained because of uncertainty. Future-perfect thinking provides a narrative that shines a light into the shadow between the conception and the creation, and thereby reveals information that reduces that uncertainty.

Schutz argues that future-perfect thinking provides us with 'empty horizons' that require to be 'filled in' to become meaningful. Indeed, he suggests that what distinguishes future-perfect thinking from fantasy is the ability to identify the intermediate steps in the process of filling in. The many different routines that will be discussed in Part IV provide the tools for this filling in. However, as Dewey emphasises, the filling in of the future-perfect state through the identification of intermediate stages does not imply creating a plan in an attempt to determine the future but using 'foresight of the future' to orientate present activity. We will explore these issues again in Case 17, but now turn to the filling in of intermediate steps, starting with a broad phasing of the project life cycle before moving on to more focused activities within the rolling wave.

8.3 Strategies for imagining the future: options thinking

The basic principles of discounted cash flow (DCF) as an investment decision-making tool were presented in section 3.6. Three important features of the tool are:

- the assumption that the investment being appraised is indivisible and cannot be broken down into separate but linked elements;
- the assumption that the investment has to be made now or never, and that there is no value in delay;
- its inherent conservativism because downside risks are mitigated by raising the discount rate, but there is no parallel mechanism for assessing upside risks (opportunities).

These three features mean that DCF has great difficulty in handling uncertainty, and tends to undervalue potentially high risk/high return investments. While the impact of identified risk events can be understood through sensitivity analysis and simulation, uncertainties which, by definition, are unquantifiable cannot be so handled. Over the past 30 years a sophisticated way of improving investment decision-making has been developing by borrowing an analogy from the world of financial trading. An *option* is a right, but not an obligation, to invest (or divest) at some point in the future which may or may not be fixed. Options are typically purchased for a fraction of the overall investment required, referred to as the *premium* for the option. *Real options* are investments made to keep the option of a future investment opportunity open, and the most important insight is the understanding that this option itself has value[8]. At the point of exercise of an option, the *payoff* needs to be in excess of the cost of the *exercise price* of making the full investment *plus* the value of retaining the option which expires upon exercise; investments which meet this criterion are said to be 'in the money'.

There have been some important developments in using real options for property valuation[9], but this work focuses mainly on the timing of the decision to go ahead with the project. From a project management perspective, six types of options can be identified which are important to take into account in developing the strategy for the project[10].

- *Stage* in which a project is divided into distinct stages and continuing from one stage to another is contingent on the reassessment of the value of the next stage of the investment. The successful completion of each stage means that uncertainty has been reduced because more has been learned about the viability of the total invest on either or both the income and the expenditure sides of the investment appraisal equation.
- *Abandon* in which the 'exit strategy' is built into the project should events not turn out as expected.
- *Defer* in which investments are made which keep options open until more information is available and uncertainty thereby reduced.
- *Growth* in which a baseline investment creates opportunities for further investments once more has been learned about the use of the asset.

- *Scale* in which the asset being invested in can be scaled up or down as more is learned about the use of the asset.
- *Switch* in which the asset can be more easily switched to alternative uses should the expectations about the use of the asset not be met.
- *Select* in which a number of distinct options are developed simultaneously so as to provide a range of choices once conditions are better known.

It can be seen that what unites all these different types of options is flexibility, so in a very important sense options are a measure of the value of flexibility for the client on the project and hence what premium is required to make to retain the option of flexibility. It is also clear that the types of options overlap – for instance, one way of retaining an abandon option is to stage the project. The formal valuation of real options is highly technical and presents considerable difficulties, particularly in the case of compound options where investment in one option creates a further option. However, *options thinking* can be used to understand many aspects of the creation of the built environment, and its application yields some interesting – indeed counterintuitive – insights into construction project management.

Perhaps the three most interesting types of options thinking from a project management point of view are *stage*, *select* and *defer*, and we will explore each in turn. Large property and infrastructure investments are often sequenced into phases, which is the most obvious application of the stage strategy. However, there is another phasing that is usually implicit through the project life cycle from inception to completion. If we take the 'problems' shown in the stage-gate process in Fig. 8.2, we can see that there are a number of stages shown from briefing, concept design, detail design, through to completion. The work done at each stage prior to execution on site can be seen as an investment that acquires more information about the project and hence reduces uncertainty. The process of the client and its advisers working up a brief deepens the understanding of the client's requirements, eliminating some options and perhaps opening up some others. Some projects stop here as this understanding leads to a realisation that an investment is not appropriate; others are deferred awaiting more information. Moving to concept design allows further learning about the challenges of the project in technical and organisational terms, and, in particular, engagement with regulatory bodies which may constrain the options available to the client. Investment in concept design can be thought of as a premium which leads to rapid reductions in uncertainty. This investment may also create of itself an asset with a market value because a site with a permit for construction is typically worth more than a site without one. From a real options perspective, the client should at this point decide clearly whether to exercise the option of going forward with further investment or realising the asset created through concept design and obtaining regulatory approval by disinvesting.

This approach to staging the project life cycle clearly supports the development of stage-gate approaches to the project life cycle discussed in section 8.4 but it also has another implication. The approach suggests that integrating the process under the leadership of a single supplier could lead to the early closing down of options and losses associated with exercising options too soon. In options thinking, the total cost of the project is the price for the actual works *plus* the price for the options foregone in doing those works. Thus the benefits of freezing design early and moving speedily to execution are partly offset by loss of new options that may have arisen had the decision-making been deferred. On this basis there is an argument – at least on projects facing relatively high levels of uncertainty – for clearly separating contracts for design and execution so that options are fully evaluated sequentially through time and there is not a bias towards execution because those responsible for execution have already been hired. Efficiency gains from having an integrated team may be outweighed by the value of the options foregone thereby.

Architectural competitions – *concours* – described in section 5.3 are normally thought of as a procurement mechanism for selecting the architectural team with the best ideas for the project. These can also be thought of as an investment in *select* options as shown in panel 8.2. Inviting multiple design teams to develop concepts for the project has the advantage for the client of rapid learning about alternative ways of tackling the design challenges posed by the project, thus deepening knowledge about the project. The client then has the option of selecting the preferred option, an option that would not be so readily available if a single architectural team had been hired from the outset. While the explicit borrowing of ideas from losing teams' proposals is usually forbidden by the rules of the *concours* and intellectual property rights, the client has still learned from the rejected designs and this learning can be used to shape the successful one. It is on this basis that the reimbursement of architects who enter *concours* can be justified on the grounds that the client has acquired an asset while it is not normal to reimburse tenderers for construction works. A similar pattern appears to be emerging in complex PFI projects where payments to unsuccessful bidders can be considered as premia to keep select options open during concessionaire procurement. The pricing of such options is tricky, and includes interesting questions such as whether the third-placed tenderer should be paid as much as the second.

Turning to *defer* options, there are always the benefits of waiting until further information is available before proceeding with a project, but waiting too long can lead to the option expiring. However, *within* a single project options can be taken to make small additional investments which allow decisions to be deferred regarding certain aspects of the facility – this is known as *safeguarding*. For instance, a certain technology may be immature at the time of deciding to go ahead with the project, so provisions are made in the design so that the technology may be incorporated relatively easily at a later date. There may be grounds to believe that requirements might change in the future, but investment to meet those requirements is not yet warranted. For instance, student rooms in Wentworth College at the University of York – see Case 3 – were initially double rooms shared between two students, but designed to be easily converted to single rooms by paying a small premium to ensure that each room could have a washbasin installed and a dividing wall built easily. The growth in the vacation conference and holiday home market increased the value of exercising this option and so the university went ahead and made the additional investment to convert all rooms to single rooms. While the total cost of doing things this way was greater, this cost was offset by the value of the option of waiting until the market had matured before the additional investment was made. Panel 8.3 shows how safeguarding was used strategically on BAA's Terminal 5 project, which is discussed further in Case 12.

The value of safeguarding options varies as a function of the level of uncertainty and the level of modularity in the design. Where there is low uncertainty that an option will be exercised, its value is increased so that a greater premium for safeguarding is warranted. Where systems are modular, safeguarding tends to have a lower premium cost, so again its value is increased. The level of modularity

Panel 8.1 No Business as usual in Sydney Harbour

The project mission was to clean up the water in Sydney Harbour in preparation for the 2000 Olympic Games by building a storm drain to relieve the main sewage system which tended to back up during heavy tropical storms. The scope consisted of approximately 20 km of tunnels in sandstone, and associated treatment plants and other installations. This project was delivered on time and was slightly over the target budget. The project was organised as an experiment in project alliancing on an open-book basis between three contractors and the client. However, the espoused 'no-blame' culture tended to become a 'no-responsibility' culture.

Project performance was measured through five Key Performance Indicators (KPIs):

- schedule – immovable at 31 July 2000 because of the Olympics;
- budget – negotiated around $A380m;
- community – particularly the affluent stakeholders who would be affected by the installations above ground;
- occupational health and safety;
- ecology, particularly the marine life of Sydney Harbour.

A formal statement of project culture was developed by the project management team, and Business as Usual (BAU) levels of performance were identified to provide the baseline which the KPIs could be set to exceed. Performance rewards were available against all the KPIs, which could not be traded off against each other. Benchmarks for non-financial KPIs were developed by the project team and externally audited. The project organisation was self-consciously innovative in procurement terms, striving for excellence in rejection of the BAU mentality. The collaborative environment facilitated high levels of innovation and value engineering in order to meet the KPIs. The project was managed by a Project Alliance Leadership Team (PALT), and collaborative working was supported by team-building consultants. The research showed that future-perfect thinking was deployed by the PALT in three different ways.

- *Endgaming* was used specifying what was expected to happen stating when particular tasks will have been completed by.
- *Incentives* were clearly linked to the achievement of the KPIs.
- *Representation* of the project mission took the form of a large, strategically placed fish tank symbolising the project mission – clean water in the Harbour – and emphasising the importance of the fifth KPI in the co-located project team offices.

Sources: Clegg et al. (2002); Pitsis et al. (2003); Clegg et al. (2006).

Panel 8.2 Generating select options for the Tate Modern

> 'Without wanting necessarily to plunder the ideas of the hundred and forty nine architects who submitted, having been obliged to look at [such] material . . . causes you to rethink what you're doing. It's tremendously easy in a project of this kind to start narrowing options at too early a stage'.

Sir Nicholas Serota, the Director of The Tate, commenting on the experience of reducing the initial entries which were responses to the formal advertisement of the competition in the summer of 1994 down to a long list of 13, is using intuitive options thinking. Each architect on the long list was then given £4500 and a month to come up with proposals, a total premium of £58 500. A further round of assessment reduced the 13 to 6 who were then given more time to develop their proposals and interact with the client team so that the client could learn more about their intentions and the compatibility of their working style with that of the client team. The Tate finally exercised the option of moving forward to the next stage of design with Herzog and de Meuron, although the contribution of Renzo Piano to the client's thinking behind the design is also acknowledged. Had the client not made the investment in the *concours*, learning and hence uncertainty reduction would have been less rapid; from a real options perspective, all the failed entrants also supplied the client with architectural services.

Sources: Moore and Ryan (2000); Sabbagh (2000); http://www.tate.org.uk/modern/transformingtm/ (accessed 24/04/07).

is typically a function of technological issues – for instance, electrical systems are fairly easily modularised by using bus systems, but this is relatively difficult to achieve for mechanical systems. Safeguarding can be active or passive. Passive safeguarding typically has minimal real cost, and the premium of the option is largely the opportunity cost of not using the asset. Thus the route for the Metrolink to East Didsbury in Manchester was passively safeguarded even though there was no date fixed for the construction of the link by preventing new housing and other developments encroaching on the route, and its cost is the lost housing that could otherwise have been built. Active safeguarding involves making more or less large investments to purchase the option as shown in panel 8.3.

Panel 8.3 Safeguarding at Terminal 5

> 'Safeguarding is not so much about keeping my design solution generically flexible. It is about saying I can see a potential future use. It is not there now, but it will be very expensive to implement it when it comes unless I do a few things now which will have limited cost. It is about playing it safe. The trick is about how to safeguard while at the same time recognising that safeguarding can cost money. So it is about how to prudently stop waste but actually keep open that flexibility as needed'.

The Head of Design and Development on BAA's Terminal 5 project is using intuitive real options thinking in that options are purchased by doing a small amount of additional work now in order to defer the construction of particular elements of the terminal while minimising the costs of exercising that option at some point in the future. There are a number of examples of this on the project.

- Structural reinforcement above existing requirements at a cost of £250k, which will allow the addition of a mezzanine floor in terminal 5a to provide a *defer* option for additional retail and lounge space with minimum disruption to operations.
- Investment in the infrastructure that will allow the extension of the inter-terminal train and baggage handling systems easily to terminal 5c at a cost of up to £100m – a case of active safeguarding. This option had been created by an earlier *defer* option that designed terminals 5a and 5b to allow this to happen easily. This premium was paid before the decision to build terminal 5c was made because a tunnel would have to be bored under active aircraft stands if terminal 5c were eventually constructed at six times the cost. However, the extension of the train to terminal 5d was only passively safeguarded because of the higher level of uncertainty about whether it would be built. Terminal 5c will now be completed in 2010.
- Investment in the *switch* option of multi-access ramp stands that can service either a very large aircraft such as an Airbus A380 or two smaller aircrafts.
- The *growth* option of an additional 1000 m^2 of space in the baggage hall.

Source: Gil (2007).

The application of real options analysis and thinking in property and construction is in its infancy, but the examples of applications here suggest that it has potential. Sophisticated quantitative analysis is likely to remain restricted to the valuation of investments in property because of the lack of appropriate data sets to support other decisions. Applications of real options thinking such as safeguarding have a broader potential and could well be of considerable help in value management and through-life costing. For instance, investments made during initial construction, which reduce disruption during future possible extensions of the facility, could well be of benefit and the formal valuation of that option will facilitate the case for making such investments. While options thinking in construction project management will remain based largely on expert judgement backed by simple cost-benefit analysis, the rigour it brings to thinking about the value of flexibility in the future suggests considerable advantages for both clients and their construction suppliers.

There are, however, a number of issues in the implementation of real options analysis which should not be underestimated:

- Real options incur significant information costs associated with monitoring the movement of the underlying values so as to ensure that the options are exercised efficiently. These costs may outweigh the value of the option. In particular, principals and agents may have differing views regarding when an option is in the money.
- Where the exercise of options requires variations to contractual terms, additional costs of exercise because of opportunism by suppliers may be incurred for clients. Part of the premium may be the transaction costs of writing more flexible contracts.

- The tendency of projects to escalate, discussed in section 10.10, means that there are biases against the timely exercise of abandonment options.
- Optimism bias could be reinforced because downside risks are mitigated by only the premium being at stake. Overinvestment in more options than can ever be physically realised would be a misuse of resources.
- The actual analysis can be extremely complex, and although people with appropriate skills can be hired to do this work, unless decision-makers understand the implications of the results, errors are likely to be made.

We now turn to how routines for one of the most important form of options thinking – *staging* – are currently being developed in the construction sector.

8.4 Moving from phase to phase: gating the process

Having broken the process down into clearly defined phases with identifiable outputs, we need a clear set of review points where progress against the project mission is assessed. In effect, these routines are the measurement points in the overall project-level control loop and are known as stage-gates[11] which are pre-defined review and decision points in the project information flow where progress is assessed against predetermined criteria by those actors who can contribute positively to such an assessment. Pre-definition is important here because without it, mission drift and schedule slippage may not otherwise be noticed. Thus the key criteria for planning the sequence of stage-gates over the project life cycle are the *who, what* and *when* criteria:

- *Who* should be involved in each review?
- *What* criteria need to be met for the project to go on to the next phase?
- *When* should the reviews be held?

The answers to these questions will vary over the project life cycle. Early phase reviews will tend to have a higher level of client involvement, and review progress in fairly broad terms, reflecting the higher levels of uncertainty under which decisions are being taken, while later reviews will have less client involvement and will be able to measure progress much more precisely. These differing levels of detail in criteria are explored in the Glaxo project described in Case 8. However, at all gates the following questions need to be addressed:

- Is the project still on course to deliver the project mission?
- Is the project process well managed?
- Are the tasks for the next phase of the project clearly defined?

Stage-gate processes have been diffusing amongst a number of clients as they try to grapple with the problems of strategic misrepresentation and optimism bias inherent in the definition of the project mission. In the public sector – at least in Norway and the UK – this is driven by the finance ministry concerned by accountability for public expenditure. In the private sector, corporations such

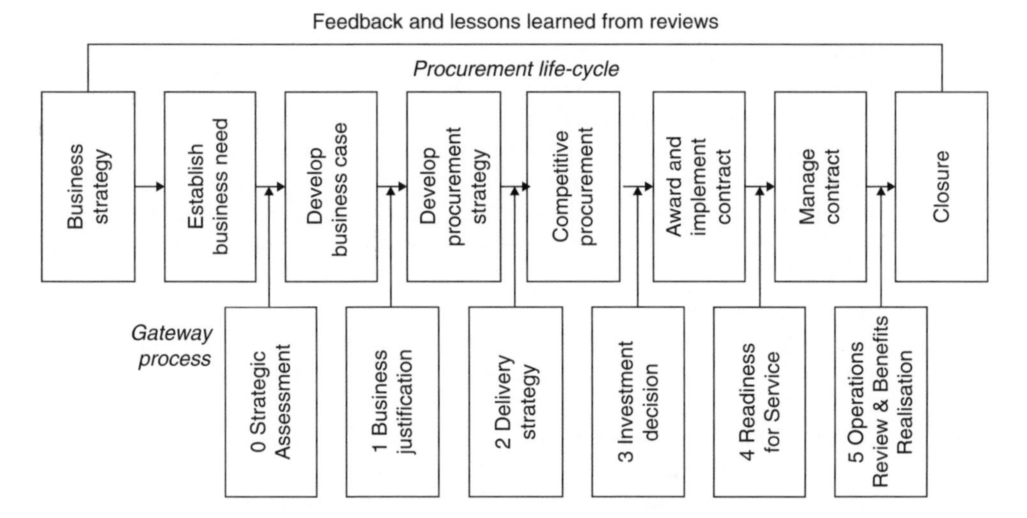

Fig. 8.1 The Office of Government Commerce's Gateway process.

as BP have their capital value processes to achieve the same objectives and reassure shareholders that capital expenditure is done wisely. The key insight here is that ill-conceived projects can be very difficult to terminate before they move to the high-spend planning and execution phases[12]. Figure 8.1 shows the Office of Government Commerce's Gateway process which is mandatory for all central government capital projects, and Fig. 6.8 shows the similar process within the ProCure 21 framework agreement. While the gateway approach very much provides a formalised routine for the management of the project, by combining it with service marketing concepts in the gap analysis approach, it can be developed into a more proactive tool for riding the project life cycle.

8.5 The gap analysis approach

Construction is essentially a service industry. What is sold to the client is not a product but a capacity to produce. As discussed in section 6.2, procuring construction services is qualitatively different from buying a melon, or, indeed, a car. When buying a melon, there is no interaction between producer and customer; when procuring construction services the interaction is intense. Buying a melon is a timeless spot contract; buying construction services is a process through time, and this is true of all service industries. Indeed, it is their defining characteristic.

Three basic dimensions of service transactions[13] can be identified, illustrated by the example of a meal at a fine restaurant:

- *Intangible* in that there is no melon to be squeezed or car to be test-driven. Both the quality of the service delivery (realisation) and the quality of the final

product (conception and specification) are difficult to measure on important parameters. Judgement and the 'feel good factor' remain important elements in evaluating service quality. The enjoyment of a meal is a combination of the quality of the cooking and interior design, where tastes are fickle and the judgement of what is excellent is open to contention.
- *Heterogeneous* in that performance varies from client to client and staff member to staff member. The celebrity chef may be ill and his understudy be working a double shift as a result; the diner may have won or lost on the horses that afternoon.
- *Inseparable* in that there is little opportunity to buffer the problem by control audits post production. Service production and service consumption are simultaneous – the meal must be eaten when it is ready.

The key to managing service quality in such a situation is the management of perceptions, or rather the various gaps of perception in the service delivery process. There are five such gaps in the delivery of consumer services in general:

- *Gap 1* between consumer expectations and management's perception of consumer expectations;
- *Gap 2* between management's perception of consumer expectations and management's translation of those perceptions into service quality specifications;
- *Gap 3* between service quality specifications and actual service delivery;
- *Gap 4* between actual service delivery and information provided to the consumer about the service;
- *Gap 5* between actual service delivery and consumers' perception of the service delivered.

It can be seen from this that there are plenty of opportunities for things to go wrong between the customer deciding to place the order and the completion of service delivery, as all of us who have spent a disappointing evening in an expensive restaurant will testify. The opportunities for things to go wrong on a construction project are exponentially greater.

The gap analysis approach, together with its service quality measurement instruments, was designed for business-to-consumer (B2C) services such as restaurants and medical care. For project-orientated business-to-business (B2B) services such as construction, the approach requires development along the lines presented[14] in Fig. 8.2. The model distinguishes four problems to be solved if the minimisation of client surprise is to be achieved in terms of the size of the *project performance gap* between what the client thought it was going to get and the perception of the facility it received. The smaller the performance gap, the smaller the level of client surprise and hence the greater the level of client satisfaction. In order to minimise the project performance gap, the previous four gaps must be minimised through the consummate solution of the problem at hand, and there is little overall benefit from solving one problem very well and failing on the others.

The *briefing problem* is the process of turning the client's desire for a facility into a clear *brief*. In other words, it is the process of defining the project mission – as

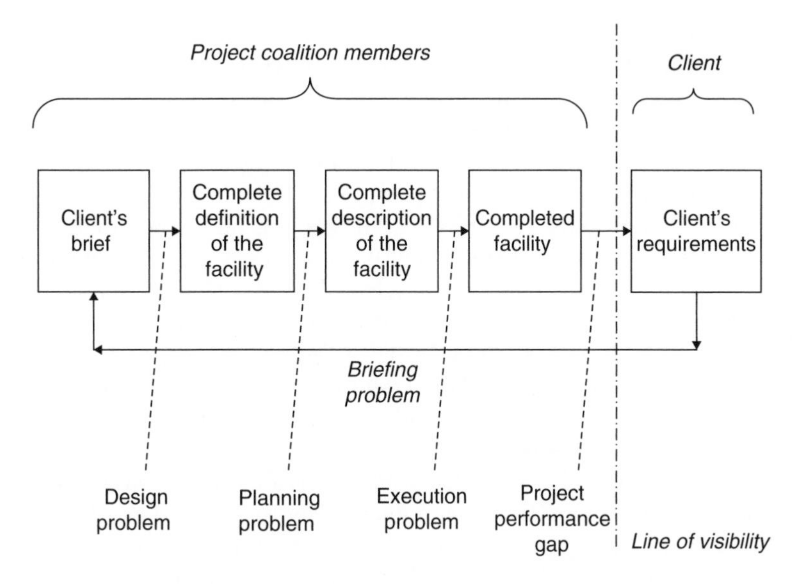

Fig. 8.2 The gap analysis approach.

discussed in Part II – and turning it into a brief against which resources can be mobilised. On some projects, this process can be iterative between briefing and design while where *concours* are used, there is a clear breakpoint between solving the briefing problem and moving on to the design problem. On complex projects, this phase can last for years – or over a century in the case of the Channel Fixed Link. A successful outcome to this stage is vital for the overall success of the project – there is little point in delivering the facility to budget and schedule if it is the wrong facility in the wrong place.

The solution of the *design problem* moves on to exploring particular avenues and options to arrive at a complete *definition* of the facility. This is an inherently innovative phase of the project – unless mission uncertainty is low, it is likely that there will be at least some aspect of the design which will require a novel solution. Here the needs of the client as expressed in the project mission need to be balanced against the possibilities afforded by the technologies available and the constraints imposed by the regulatory stakeholders. Further research into the needs of the facility users may reveal additional requirements and constraints. The quality of conception can come into tension with the quality of specification and realisation during this phase. There is no compelling reason why the judgement of peers and the judgement of clients with respect to quality should be in agreement here – these become matters of legitimate debate between the project actors, and often more widely as the reactions to the Scottish Parliament building presented in section 3.5 show.

Despite a clear brief, much can go wrong during this phase – commonly called concept and scheme design. Unless the process is carefully managed, it is easy for the designers to favour conceptual quality, supported by a rhetoric of professional integrity or the public good, rather than the quality of specification and realisation. For some, the latter is simply not a priority here – as Amanda Levete of Future Systems put it on the Tate Modern project, 'well, you can't design an idea to a budget'[15] – no matter what the implications for the project mission. The inherently innovative nature of the design process means that it is impossible to define the outcome of this phase in advance; what can be done is to identify clearly the criteria that the complete definition of the facility must meet in order to pass to the next phase of the project. The responsibility of the construction project manager here is to challenge those resource bases which threaten the project mission by refusing to accept budget and schedule constraints.

The *planning problem* is that of turning the definition of the building into a complete *description* – this is the phase that encompasses what are commonly known as detail design and project planning. Detailed and working drawings have to be prepared, calculations made and components specified. Plans for the next phase in terms of schedule and budget are laid, while the documentation required for the procurement of trade contractors is also developed. The issues are, by now, much more clearly defined and uncertainty is greatly reduced. Frequently, responsibility for this phase is passed to a different project actor from the definition phase – indeed this has probably been the biggest shift in relative responsibilities between architects and contractors in the UK over the past 10 years or so. There should by now be few surprises in store for the client.

Once the facility is completely defined and planned in terms of schedule, budget and the quality of intention for the physical artefact, what remains is to solve the *execution problem* by realising the completed facility on site so that it conforms to the criteria set out in the complete description. This is the moment of truth for the plans laid in the earlier phases. For instance:

- assumptions made regarding ground conditions may prove to have been optimistic or based on limited data;
- design compromises may prove to be unworkable in practice;
- the client's decision-making context may change leading to a mission redefinition.

One way of handling the last problem is, effectively, to split the planning and execution problems into two parts – create the structure and major installations in a first phase, and then fit out the spaces thereby created in a second phase. This approach is used in both shell-and-core speculative property development and the open building approaches to social housing – see section 9.7.3.

The project performance gap is defined in terms of the gap between what the client thought it wanted at inception and its perception of the completed facility. As there would inevitably have been changes as the project mission was developed in briefing, and then defined, described and executed on site, the key criterion of success is not the match to the original statement of the project mission in the brief, but whether the client is surprised by the difference between the mission and the completed facility. If the client has understood and approved the series of decisions as dynamic uncertainty is reduced, then the level of surprise should be low. If the client has not understood why these changes have taken place, then the level of surprise is likely to be high. It follows that the *line of visibility* of the client into the project process is an important variable in effective project management. Where clients are actively involved in the project process and possess the in-house capabilities to be active, it is easier to minimise surprise than if the client is distant from the process. The partnering and alliancing described in sections 5.6.4 and 6.8 provide new opportunities for deepening the line of visibility for the client into the project process.

8.6 What do we mean by project success?

The conventional approach to defining project success is in relation to the time/cost/quality project performance model – a successful project is on time, below cost and conforms to specification[16]. However, this is a rather limited notion because it takes as given the solutions to the briefing and design problems and ignores the differing interests of the project stakeholders; it is an execution-based approach not a total project life cycle approach. We need to develop a more sophisticated definition that allows for the differing interests of stakeholders and places the project mission at the heart of the definition of success. In section 3.8 a conceptual framework for *product integrity* in construction was presented, and summarised in Fig. 3.2. This is our starting point for assessing project success, and the chapters in this part of the book will tackle the various dimensions of achieving that success.

There are two distinctive challenges in minimising client surprise and thereby achieving project success:

- *Appropriate intention* – managing the process of briefing and definition as an increasingly precise definition of the project mission, which will be discussed in Chapter 9.
- *Predictability of realisation* – managing the process of realising that mission through planning and execution on site, which will be discussed in Chapters 10, 11 and 12.

These two challenges are visualised in Fig. 8.3. The first part of the figure is taken from Fig. 3.2, while the second part re-articulates the traditional time/cost/quality triangle in terms of objectives for budget, schedule and conformance to intention, providing a definition of *process integrity* in the predictability triangle to complement the product integrity defined in the intention triangle. Process integrity also includes the achievement of the generic element in all project missions of a safe working environment and minimum environmental impact; conformance therefore includes conforming to safety, health and environmental regulations and ensuring that the project is accident free, disruption to local residents is minimised and there is no longer term damage to either the health of the participants or the surrounding environment. These issues will be explored further in Chapter 12.

The rhetoric of project success typically revolves around *process integrity* issues. Commentators in the press, embarrassed politicians and chief executives, and thought leaders in the industry all castigate its performance on process performance – usually adherence to budget and schedule. Interestingly, individual

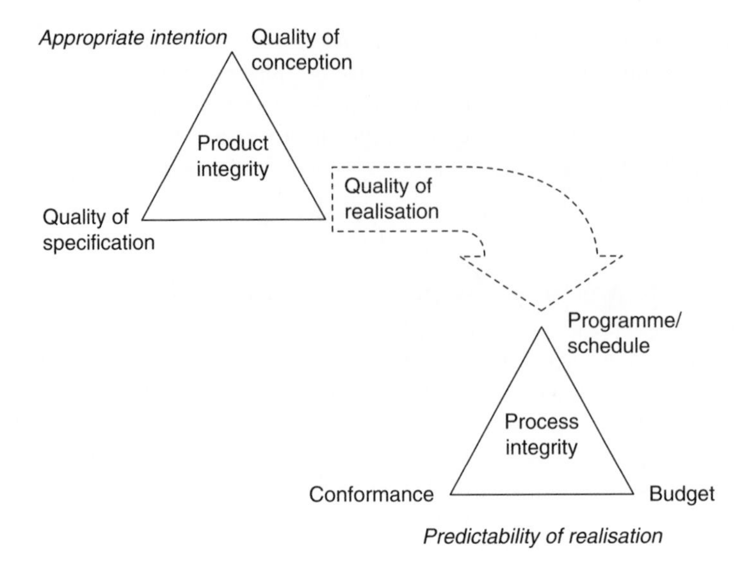

Fig. 8.3 Product and process integrity in construction.

projects are rarely faulted on conformance issues – a notable distinction from other project-based sectors such as IT[17] – although the overall performance of the sector on safety is improving only slowly. In the longer term, product integrity usually trumps process integrity in the definition of project success. Many projects that were notorious in their day have left lasting and much appreciated legacies; other projects that were celebrated in their day as innovative solutions to pressing problems have demonstrated their dysfunctionality in hindsight. The high-rise apartment blocks of the 1950s and 1960s fall into the latter category, while the Sydney Opera House presented in panel 8.4 falls into the former. The mobilisation of the Scottish architectural establishment behind the Scottish Parliament evinced by the award of the Stirling Prize to that building at a ceremony in Edinburgh suggests a concerted attempt to shift it from the process disaster category into product success; time will tell.

Panel 8.4 Ars longa; vita brevis: the case of the Sydney Opera House

The Sydney Opera House opened in 1973, 6 years late, at a cost of $A102m against an estimate in 1957 of $A7m. The Danish architect Jørn Utzon won the architectural competition for the project located on a spectacular site which shows all the signs of strategic misrepresentation for political purposes by the client. After enormous levels of friction between Utzon and the client (by then the New South Wales government), Utzon resigned in 1966 and the project was completed by local architects. Utzon also fell out with the engineers – Ove Arup – who had done so much to enable his vision to become a reality through innovative structural engineering of the shell roof. The project was castigated in the press at that time, and formed one of the cases in Peter Hall's book, *Great Planning Disasters*.

Following completion views began to change. The building became an international symbol of a culturally vibrant Australia, and Utzon's achievements began to be recognised more widely. Although he never built on any scale again, Utzon was awarded the RIBA Gold Medal in 1978, and the Pritzker Prize in 2003. He was reengaged for the upgrade programme on the House in 1999 which was designated a UNESCO World Heritage Site in 2007. Arguably the returns to the Sydney, and indeed the broader Australian economy, for the money spent have been very large in terms of tourism, branding and its contribution to the panorama of Sydney Harbour. Visiting it is an uplifting experience. With the benefit of hindsight it is seen as an outstanding success, rather than a notorious failure. The issue is whether effective definition of the project mission and leadership through the project life cycle could have achieved the same ends without the trauma of construction.

Sources: Hall (1982); Murray (2004); Watson (2006); Wikipedia (accessed 27/08/08).

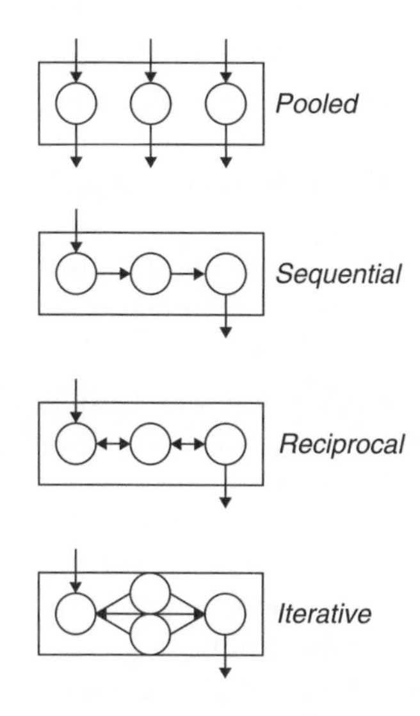

Fig. 8.4 Four generic types of information processing (source: developed from Van de Ven and Ferry, 1980).

8.7 The nature of information flows in problem solving

Problem solving in each of the phases is an information processing activity as possible solutions are sought, evaluated and chosen. Figure 8.4 shows the four basic types of information flows that make up a business process[18].

- *Pooled* information processing occurs where there is no requirement for information to be passed from one task to another, and all they share is a common participation in a project. For instance, there is rarely a need for the roofers to give or receive information from the electricians.

- *Sequential* information processing occurs where information is passed from one task to another, but there is no requirement for feedback between tasks – information flows in one direction downstream. For instance, the architect may specify the type of brick to be laid and does not require feedback from the bricklayers on that choice.
- *Reciprocal* information processing occurs where clearly defined feedback loops occur between tasks. While information basically flows in one direction, information from the execution of downstream tasks is required for the completion of upstream tasks. For instance, final confirmation of the foundation design may await exploratory excavations on site.
- *Iterative* information processing occurs when there is a requirement for intensive and unstructured feedback between tasks – in essence, it is very difficult to define some tasks as upstream and others as downstream. The intense interaction between architectural and structural designs while working out solutions to difficult structural problems requires such intensive iteration.

As might be expected, the more intensive and unstructured types of information processing are more common under greater levels of uncertainty, while the more sequential information processing is found when task definitions are more clearly defined and sequenced. Relating this model to the gap analysis model in Fig. 8.2 and the life cycle model in Fig. 1.3, the predominant mode of information processing at each phase can be defined in the following terms[19]:

- *Briefing problem* – here information processing is both intensively iterative and *divergent*, looking outward in a broad search for solutions to the briefing problem to solve a very *ill-structured* problem.
- *Design problem* – here information processing is still iterative around an ill-structured problem, but becomes much more *convergent*, looking inward to develop, evaluate and choose solutions to the design problem to achieve a complete definition of the facility.
- *Planning problem* – here information processing is more reciprocal than iterative as the facility definition is developed into a complete description. The problem becomes *well structured* and susceptible to the application of management science techniques such as critical path analysis, described in Chapter 11.
- *Execution problem* – here the problem is well structured with information processing tending more to be either sequential or pooled, but important reciprocal information flows still remain in high uncertainty aspects of the on-site works, such as foundation works, and structural works on refurbishment projects.

The most important implication of these different types of information processing in the four phases is that – as explored in Chapter 15 – the most appropriate forms of organisation and management also vary from phase to phase. Management styles and tools developed for processing information sequentially under high certainty are unlikely to be the same as those that are most appropriate for iterative information processing under high uncertainty. Table 8.1 presents

Table 8.1 Tools for decision-making (source: developed from Simon, 1977, Fig. 1).

Types of decisions	Traditional tools	ICT enabled tools
Well structured 'routine, repetitive decisions. Organisation develops specific processes for handling them'	• habit • standard operating procedures	• management science, e.g. linear programming, critical path analysis, fuzzy logic • computer simulation
Ill-structured 'one-shot . . . novel, policy decisions. Handled by general problem-solving processes'	• judgement, intuition and creativity • rules of thumb • selection and training of staff	• heuristic problem-solving techniques, e.g. cognitive mapping, soft systems thinking

ICT, information and communication technology.

an overview of the relevance of different decision-making tools for the different phases. The following chapters will explore thoroughly these differences.

8.8 Process representation

The fundamental proposition of this book is that the construction project is an information process through time – an information flow that stimulates and controls a material flow. However, such processes are intangible, and so what is needed is a method of representing those processes so that they can be described, communicated, analysed and redesigned as routines. These methods also require a level of standardisation so that the different parties understand each other. Two sorts of representation are commonly used[20]:

- *a process map*, which describes how the process *is*;
- *a process protocol*, which specifies how the process *ought to be*.

8.8.1 Process mapping

Process mapping is a well-established approach to visualising business processes, but it has been given a new lease of life over the past 10 years by new software applications which make it much easier to prepare and revise maps[21]. Process mapping has its origins in method study. The aim of method study is to find the most appropriate flow for materials on the shop floor in a manufacturing context using *flow process charts*. Here standardisation is limited to the symbols used to represent flows identifying where components are actually being transformed (i.e. value-adding activities), where they are waiting for transformation or are in storage, or being transported. Visualisation in flow process charts allows those responsible for the particular tasks being charted to analyse, discuss and thereby identify possible improvements.

The development of the computerisation of business processes such as accounting, stock control and order processing led to a more urgent requirement to map routine information flows. While normally subject to procedure manuals, the operation of such flows typically included levels of tacit understanding which had to be captured if computerisation were to be successful. This led to the development of standards for *data flow diagrams* to describe such flows accurately enough for computer programs to be written to perform them.

However, while the library of symbols for flow process charting and data flow diagrams is fairly standard, each group of people undertaking process representation tends to follow its own methodology in preparing the actual visualisations. For these reasons, a much more rigorous methodology was developed to support new developments in agile manufacturing for the procurement by the US Air Force of new aircraft and other weapons – this is the well-known Icam DEFinition (IDEF) suite of tools. In a sense, they combine the concerns of production engineers with process flow charting, and software engineers with data flow diagrams.

All three of these approaches to process visualisation – and the review is by no means exhaustive – share a concern to describe the process in enough detail so that production or software engineers could analyse and improve the process. One result of this *engineering* approach to process mapping is that the resulting maps are frequently complex and difficult to read and hence easily misunderstood. In many situations a more *management* approach to process mapping is appropriate. Here maps are drawn using a simple symbol library in a matrix representing the phases of the information or material flow along the horizontal axis, and the participants in the flow along the vertical axis. Within the body of the map, tasks are represented with their relationships. In a management context, such maps of 'the white space on the organisation chart'[22] provide a valuable basis for management discussion on process improvement, and training those who have to execute the tasks in how they operate – they act as mediating routines as defined in Part IV.3.

A major limitation of process mapping tools is that they are limited to mapping sequential and reciprocal information flows; iterative information flows remain too unstructured to be captured by such techniques. As the most important mission-defining and mission-developing processes in construction are iterative in nature, this places important constraints on the extent to which it is possible to use them to help solve the briefing and design problems. The whole development of design methodologies from the 1960s onwards attempted to structure the ill-structured problem of design, and is widely agreed to have failed[23]. More recent attempts to improve the effectiveness of the solution to the planning problem – described in section 11.5 – have proved, so far, impossible to extend upstream to the more iterative design processes.

8.8.2 Process protocols

Process protocols can be thought of as information route maps of how the process ought to work. These protocols are normative, while maps are descriptive; a map only tells you where A and B are, not how to go from A to B which a protocol does. As competition has moved in manufacturing away from costs towards design-led features such as functionality, there have been a number of attempts to improve performance in NPD. These attempts typically involve the formalisation of the NPD process into a firm-specific protocol which lays down the phases and decision points in the process, often in a stage-gate format, as discussed in section 8.4.

The use of process protocols in construction has tended to take a different development. Perhaps in response to the relative fragmentation of the industry, the aim of protocols has not been to improve the internal performance of individual firms, but to provide a national framework for the industry. Protocols thereby form meta-routines for the project process. Most countries have one or more national process protocols for visualising the process. In France this protocol is specified in the *Code des Marchés Publics*, and in the UK it is enshrined in the RIBA Plan of Work. These fulfil a number of important functions:

- They provide the basis for contractual arrangements between clients and suppliers of design services.
- They provide a model for the representation of the process in textbooks, industry guidance documents and debates about reform of the process.
- They provide a language by which participants can communicate their understandings of their roles and responsibilities in the process.

While these national protocols perform a number of different functions and are a reference point for a wide variety of actors, their origin in the interests of particular stakeholders tends to be obscured. For instance, the origins of the French code are essentially legal. The code plays a central role in the body of administrative law which governs the relationship between the French state and its suppliers; it is therefore prescriptive in its application to public procurement contracts, and is essentially an instrument of public procurement policy. The RIBA Plan of Work, on the other hand, was developed by the principal professional institution representing architects in the UK, is essentially aimed at formalising design processes and is less prescriptive. However, in many subtle ways it reinforces the role of the architect in the process, particularly in its assumption that the project will be managed by the architect. Neither of these two protocols pays much attention to the client processes of evaluating and promoting a project, and they are also silent regarding on-site material flows. A further weakness is that neither is capable of facilitating the development of ICT tools for design and management because of their lack of detailed specification of information flows.

8.9 Knowledge management and learning from projects

As argued in section 9.4, learning from the facility in use is vital for future projects so that clients can more precisely define the mission of future projects. Learning from the project process is also vital for the resource bases so that their effectiveness is enhanced on future projects. It is this process of organisational learning by the resource bases that has become known as *knowledge management* (KM)[24]. Construction projects are inherently innovative, yet frequently the learning generated in solving project problems remains with the individuals and task teams concerned, and is not diffused more widely in the organisation. Thus KM, defined as organisational learning from the project process experience, is a vital element of the project life cycle. However, before going on, it is worth clarifying the difference between 'information' and 'knowledge', because there are no clear definitions of these two terms that are widely accepted.

For the information processing approach taken in this book, there are important differences between the two concepts, which are analogous to the economist's distinction between stocks and flows.

- *Knowledge* is a stock of information – a resource that has the potential to be mobilised on the project to create the constructed product held by the resource bases. Knowledge can be *tacit*, in that it is not fully expressed and is closely related to the personal skills of an individual and only meaningful in specific contexts, or it can be *explicit* in that it is fully communicable between any two or more individuals. Knowledge can be traded internationally in the form of *intellectual property*, or it can be held in secret as *private* knowledge to gain competitive advantage. Whatever the form in which it is held, knowledge is a resource which only creates value for its owner and others once it is situated[25] – that is mobilised on projects to meet particular client needs. To extend our river analogy from section 1.3, knowledge is a reservoir that is used to supply the river in order to maintain water flow levels, or to provide a head for greater pressure.
- *Information* is knowledge in use – a resource mobilised to create new values in the manner shown in Fig. 1.1. The learning generated during this mobilisation has the potential to enhance the existing stock of knowledge, which is why learning from projects is so important – it is a central means of maintaining and developing the resource bases that feed the project information flows. A book in the library is knowledge – only when mobilised by a reader for learning does it become information.

There are a number of ways of capturing learning from projects so as to renew the resource bases and develop their capacity to realise projects more effectively in the future in the construction industry.

- Conducting post-project reviews shortly after project close. Key participants are brought together to debrief and identify main learning points from their experiences on the project.
- Using intranets within the resource bases to capture and diffuse learning points, as described in panel 8.5.
- Establishing centres of excellence in the competencies desired by clients.

It is now broadly accepted that there are two main approaches to KM, depending on the level of codification of the information to be captured. Table 8.2 presents

Panel 8.5 Knowledge management at COWI Consult

The key to KM at COWI Consult is the 29 'professional networks' covering 7% of staff (figures as of 1999) in areas ranging from project management to advanced engineering technologies. These networks are given funding centrally, and enabled through the COWI intranet with their own web pages. A major activity for these networks is the identification of best practices through peer review of innovations made by the network members. Once accepted, the best practices are documented and placed on the intranet where they constitute a searchable database. As of 1999 there were 612 best practices focused on technical issues.

A knowledge manager reports within corporate services, and his main role is to prepare the Intellectual Capital Report. The first one, for 1998/1999, was prepared manually and used to establish the appropriate metrics. The 2000/2001 report used the SAP system shown in Figure 14.4 to calculate the IC metrics. This took considerable programming effort because of the distinctiveness of the COWI metrics and associated algorithms. The reporting is at departmental level and can be used to benchmark different departments against each other, and to evaluate their managers.

These metrics include:

- market share – measured by turnover by region and type of client;
- client satisfaction – measured through a 30-point satisfaction survey;
- professional profile – measured through numbers of professional publications and public presentations by staff;
- staff profile and turnover – by age, sex and educational attainment;
- staff training – measured by total length of education, discounted through time as the relevance of knowledge acquired diminishes;
- international experience – measured by staff travel days outside Denmark;
- project management capacity – measured by number of projects managed per member of staff.

By the publication of the 2007 report, there were 55 metrics for performance drawn, unaudited, from COWI's management systems, integrated throught the COWIPortal.

Sources: *COWI Intellectual Capital Report 1998/1999*; Interview, Jesper Hjerrild Rild, IT Systems Manager, COWI, 10/04/01; http://www.cowi.com (accessed 23/08/08).

the results of work conducted at Harvard Business School on management consultancy projects, which shows the differences between the codification and personalisation strategies in managing knowledge. Where resource bases aim to craft their solutions to specific clients' needs, the personalisation KM strategy is more appropriate, with an emphasis on hiring highly skilled people and developing them through intensive coaching, and developing communities of reflective practitioners. Where the aim is to deliver standardised solutions to clients, less skilled people are hired and there is a greater reliance on IT-based KM systems and less individual attention in personnel development. Those resource bases that tend to operate when dynamic uncertainty is high and to mobilise on high mission uncertainty projects will favour the personalisation KM strategy, while those that tend to operate later in the project life cycle and to mobilise on lower mission uncertainty projects will tend to favour the codification KM strategy.

Table 8.2 Knowledge management strategies (source: developed from Hansen *et al.*, 1999).

Codification KM strategy	Personalisation KM strategy
'provide high-quality, reliable and fast information-systems implementation by reusing codified knowledge'	'provide creative, analytically rigorous advice on high-level strategic problems by channelling individual expertise'
Standardised solutions – invent once and reuse	Crafted solutions – solving unique problems
People-to-documents learning, capturing and codifying knowledge	People-to-people learning, developing networks for sharing knowledge
IT-based KM systems, to make codified knowledge widely accessible	People-to-people KM systems, facilitating conversations and the sharing of tacit knowledge
Hire graduates, who can be taught the standardised solutions	Hire postgraduates, who can solve problems creatively and tolerate ambiguity
Group training, to facilitate solution implementation	One-to-one mentoring, to nurture problem-solving skills

KM, knowledge management.

While KM is agreed by everybody to be a good thing, and many companies have now established KM strategies, many of these remain ineffective. There are a number of reasons for this.

- *Time*. In many project-based organisations, completing the current project and winning the next is the overwhelming priority. The time required to take a day for the post-project meeting, to add to the intranet database and to make presentations at professional conferences is all time that is not directly fee-earning.
- *Incentives*. The aim of KM is to turn individual learning into organisational learning. This means that incentives are required to encourage people to share their expertise. Some of these issues were addressed in section 7.3. Where employment relationships are collegial, the organisational culture can play an important role in generating incentives; where employment is casualised through sequential spot contracting, there are positive disincentives to share personal learning organisationally. Panel 8.6 describes Vinci's approach to formalising incentives.
- *Centralisation*. There are important trade-offs between the centralisation of knowledge capture within the organisation so that diffusion is maximised, and its decentralisation so that more intensive team learning is favoured. In many centralised KM systems, senior managers become arbiters of what is

Panel 8.6 Knowledge management incentives at Vinci

In order to stimulate innovation and capture learning from projects, Vinci runs a biannual innovation awards scheme in which staff are encouraged to present their innovative ideas to a jury of senior managers and external experts. One hundred and twenty-one proposals were received in the 1995 round, and 265 in the 1997 round. Ninety eight of these in the first round, and 215 in the second, were judged worthy of dissemination. These are judged by experts on the basis of: importance of innovation; personal merit; the impact on efficiency; other impacts and benefits; and the innovation process. The best proposals are then submitted to the jury of 17 experts who award between one and three prizes in each of four categories: product innovation; product adaptation; process innovation; and process adaptation. Prizes amount to more than 1 month's salary for each member of the team concerned. The jury also votes on which proposals should be awarded a grand prize. All the innovations are collated and diffused throughout the Vinci group through internal publications, such as the research compendium, of which 5000 copies are printed. The large rise in proposals between the two rounds is put down to growing awareness of the scheme, particularly through the benchmarking of different operating companies in the Vinci group on the basis of proposals and prizes per 1000 employees.

Source: Cousin (1998) and company documentation.

appropriate learning, and KM becomes seen as a specialist function. The ultimate parody of the centralised knowledge manager is Jorge the blind librarian in Umberto Eco's novel *The Name of the Rose*, who would rather commit murder and burn down the library than allow ordinary monks access to the knowledge it contained.

8.10 Summary

'You did well, Bernardo, in lying to us about the expense involved in the work'.

The words of Pope Pius II to his architect Bernardo Rossellino, on the handover of the cathedral and papal palace of Pienza[26], suggest that he was an unusual client in responding to a 500% cost overrun. In Renaissance Italy, clients tended to take very seriously their patronage of architecture and few clients of any period like to be surprised by the outcome of the projects they finance. More commonly, the project life cycle is a co-learning process as the client and its suppliers made sense of the project mission through the project life cycle; effective co-learning means that clients are not surprised if the realised facility does not match the future-perfect representation that gave meaning to the project from its inception.

Construction projects represent enormous cognitive challenges as we try to give meaning to present activity by imagining a future state through the process of future-perfect thinking. However, as John Dewey would argue, such a state is not determined by prediction but created through purposive action. In complement, real options thinking provides a structured way of evaluating alternative possible actions, and encourages delaying decisions until the time is right to make them as information becomes available.

Thus the project life cycle is an information and materials flow that starts with learning and finishes with learning. This chapter has taken an overview of the project life cycle as a sensemaking process, presenting a gap analysis approach to the cycle and showing the ways in which the process is gated by project reviews. These gate reviews are, perhaps, the most important project management tools for managing the life cycle, particularly for solving the briefing and design problems as they structure the flow of information towards the future-perfect state. The role of process mapping and protocols was then reviewed before issues in KM were briefly explored. The subsequent chapters in Part IV will examine in more detail the project management of different elements of these information and materials flows as the future-perfect state is 'filled in', starting with the resolution of the briefing and design problems.

Case 8
Riding the Life Cycle on the Glaxo Project

The Glaxo Group Research (GGR) campus is located on a 30ha site in Stevenage, and houses the Glaxo Group's main UK pharmaceutical research activities. Completed in 1995, it contains 140 000m^2 of laboratories and accommodates 1500 scientists. This £500m research and development centre consists of five main facilities arranged around an open, landscaped courtyard aimed at promoting staff interaction – see section 3.3 for why this is important – centralisation of building services and efficiency of operation. This case focuses on the project process – for the project organisation structure, see Case 15. An overview of the project process, identifying the types of information outputs prepared at each phase, is offered in Fig. 8.5, while Fig. 8.6 shows the overlaps in responsibility of the coalition members through the life cycle and identifies the principal stage-gates in the process. Figure 8.7 shows the gap analysis model from Fig. 8.2 applied to the Glaxo project, identifying the management tools used to ride the project life cycle.

The *briefing problem* was addressed in the master planning phase by the master planner (MP), during 6 months in 1990. This process used client reviews, user interviews and the collection of technical data for the compilation of the final brief. The *project mission* was defined in the brief under two headings:

Scientific objectives

- removing technical and statutory restrictions on research;
- consolidating UK research on one site.

Design and construction objectives

- provision of facilities for GGR to meet its commitments under the worldwide research strategy;
- cost effectiveness;
- security and safety of GGR personnel and property;
- functional efficiency of the facilities;

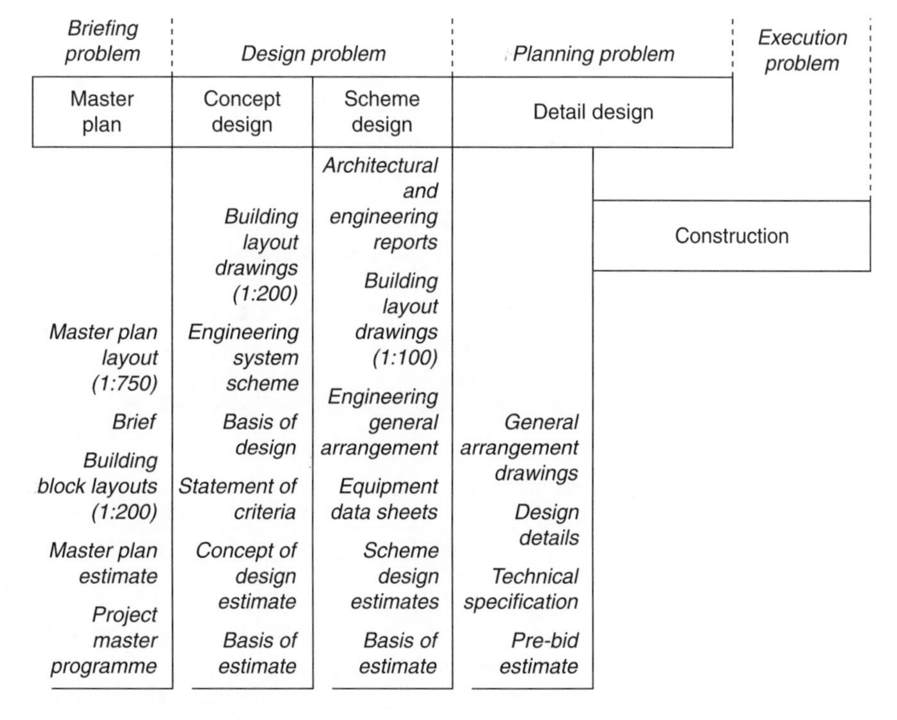

Fig. 8.5 Glaxo project information outputs.

- completion of the facilities within the agreed schedule;
- encouragement of communication and interaction between scientific personnel;
- creation of a humanistic and motivating environment;
- flexibility of growth;
- responsiveness to site and environment;
- creation of a strong corporate image;
- accommodation of ongoing construction works.

The MP then moved on to the *design problem* and took a further 9 months on the concept design phase to prepare a set of documents known as the *control documents*. The control documents can be regarded as the development of the brief by the MP for communication to the principal architect–engineer (PAE) for scheme and detailed design. The control documents were prepared by using four sources of data: GGR's accumulated experience as an operator of pharmaceutical research facilities; the MP's experience as a designer of similar facilities in the past; interviews with operational and user groups; and relevant UK codes and standards. Separate control documents were prepared for each building on the campus.

A typical control document contained both qualitative and quantitative data for the individual spaces to be accommodated within the structure. Detailed analysis of information was carried out by each discipline, such as architectural, structural,

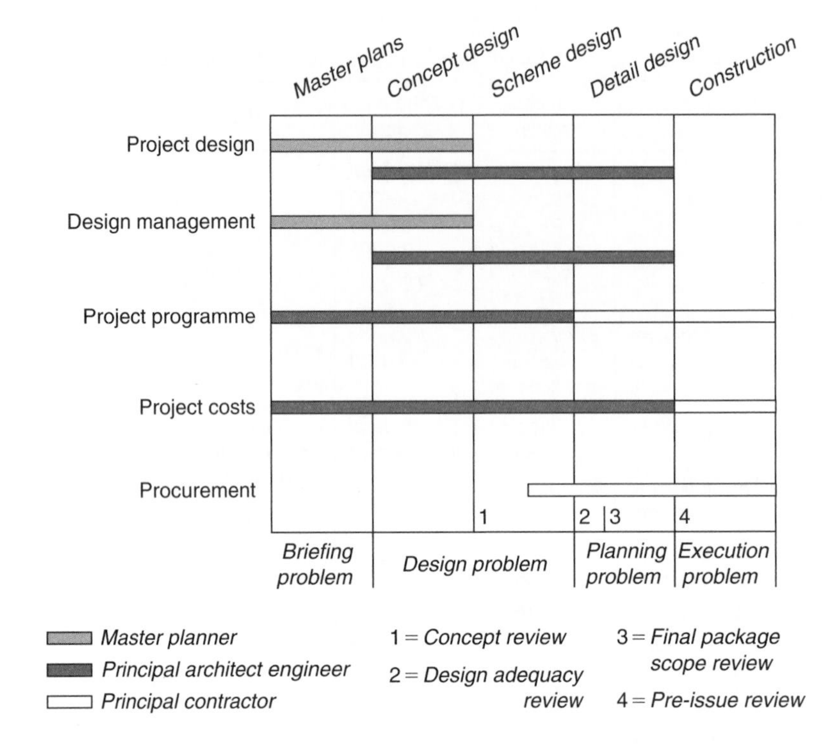

Fig. 8.6 Project management responsibilities.

HVAC, electrical, lighting, plumbing and fire protection. Control documents also contained technical information, recommendations of the type of systems to be used, relevant sketch drawings to explain a system and a cost estimate for the concept design stage. The control documents were divided into two sections:

(1) *Statement of criteria (SOC).* This statement listed criteria for specific buildings as well as standards which were applicable to all buildings on the campus. At each stage, Glaxo was invited to comment and detailed review meetings were held. This was, therefore, an agreed document with complete client involvement at each stage. The SOC could again be divided into two sections.
   (a) *Design criteria.* This section primarily described the engineering objectives and constraints, according to which the proposed facility had been designed. For example, the architectural criteria included functional space programme, building design concept and exterior and interior materials to be used. Similarly, the HVAC criteria listed standards of indoor design, temperature, humidity, sound and ventilation. In addition to these general criteria, specific details were also provided for the chemistry building. For example, the provision of fume cupboards in this section of the campus was set as an essential criterion for the safety of the scientists who would be working in this building.

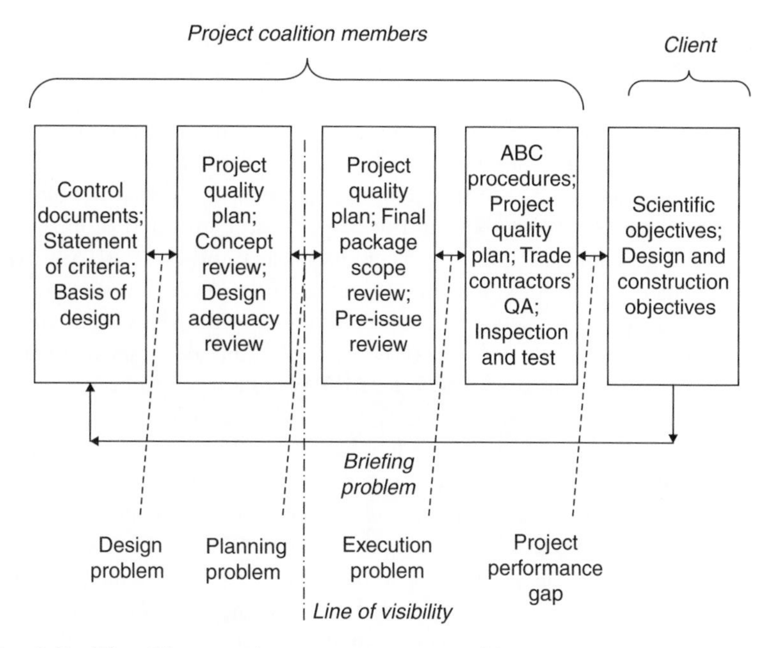

Fig. 8.7 The Glaxo project management problem.

(b) *Space criteria.* At the end of the SOC, detailed tables were provided summarising specific criteria for each campus building. This formed the basis for room data sheets. For example, under the architectural section, finishes for the floor, walls and ceilings were listed for separate rooms, and ceiling height and door types were also determined and presented in the form of a table.

The SOC was the document which summarised all the relevant data and information that the MP had collected from all its sources. This document systematically listed the relevant information after considering the client's requirements. The SOC also gave a clear picture of what the design teams needed to aim for in terms of standards and equipment. In short, the SOC was a document which defined the *project scope*.

(2) *Basis of design (BOD).* The SOC was then used to develop the second section of the control documents known as the basis of design. The design criteria from the SOC document were used to recommend technical systems which were to be used in the building. For example, heating loads were calculated in the SOC document whereas heating systems were suggested in the BOD document. In addition, drawings were provided where necessary; for example, heating system flow diagrams were included to explain the system in detail. Therefore, the BOD was a much more detailed and explanatory document and identified those specifications which were considered by the client to be vital for the success of the project.

The quality management system – see section 12.2 for a definition – within the PAE was codified in the project quality plan (PQP). The PQP contained two types of information. Firstly, it contained *procedures* for the assurance of quality at various stages of the process. Secondly, it defined the arrangements for *reviews and audits* of the process. A design review was defined as 'a formal examination of the design to evaluate the design requirements and the capability of the design to meet these requirements and to identify problems and propose solutions'. Design reviews took place on a project team basis and were of two types. Final design reviews occurred at specified stages of the project. These reviews were organised by the project managers and undertaken by directors from the design executive with the participation of GGR staff. MP and PC staff also participated in these reviews. Interim design reviews took place as required within the PAE and could include GGR representatives. At each review a quality review record was completed. The project manager for that particular team then ensured that action was taken on those comments to the satisfaction of the client.

For the PAE team the *design problem* was to translate the control documents into the more detailed information identified in Fig. 8.5, which provided the base for the detail design phase. The basic procedure for managing the design problem was the *design review* as defined above, because the PQP emphasised that the most appropriate method of assuring quality while solving the design problem was peer review with the participation of the client. The first two formal reviews were the concept review and the first final review, the design adequacy review. These acted as stage-gates in the project process.

The concept review was earned out by the PAE team when the MP handed over the control documents to it. This gate marked the point where the PAE formally took over the responsibility for the project; therefore, a complete understanding of the design solutions recommended by the MP was vital for the development of a positive attitude between all the parties involved. This review included a value engineering exercise, and the PAE team identified a 'few areas where some kind of a change was needed'. In order to maintain a high level of communication between the two parties, the MP was partially involved during scheme design, where it continued to comment on the development of design by the PAE. This two-way communication continued until the end of scheme design when the design adequacy review took place. Its aims were to ensure that GGR's objectives as defined in the control documents were met in the complete definition of the buildings, and that the design packages had been clearly determined.

The *planning problem* is that of turning the design definition into a complete description of the building ready for construction. The PQP identified two final reviews for the detail design phase. The first was final package scope which took place two months after detail design had started on each design package. Its objectives were to ensure that the building was dimensionally co-ordinated across the design disciplines, that the scope statement for each package was clearly defined and that interface requirements between packages had been clearly defined. The second was pre-issue which was undertaken one week before the release date of each design package. The issues here were the formal co-ordination of drawings, the checking of drawings and specifications to ensure completeness and accuracy, and the documentation of interface requirements with other packages.

A number of procedures codified in the PQP for ensuring quality control during the detail design phase helped to solve the *planning problem*. The more important procedures were as follows.

- *Design control and co-ordination* which described procedures for the control and co-ordination of the design-related activities in order to ensure conformance to GGR requirements. This procedure specified the manner in which design inputs and outputs were documented, monitored and reviewed. This section also dealt with the verification of the design to ensure the design output conformed to the design input requirements.
- *Design documentation control* which defined procedures for the following:
  - *Document approval and issue* – the process protocol is shown in Fig. 8.8.
  - *Document changes/modifications* – all design changes were to be requested in writing, and their review and issue to relevant personnel documented.
  - *Document control scope* – this included the types of communication media to be documented such as verbal, computer discs, microfilms and so on; it also listed the points of control such as circulation control, issue and authorisation of originals, recording of document status and withdrawal of superseded documents.
  - *Project-specific requirements* – this included procedures to be followed to provide information regarding published reference data such as standard specifications and national codes.

Adherence to these procedures was controlled through design audits. A design audit was defined as 'a separate internal examination to determine whether design

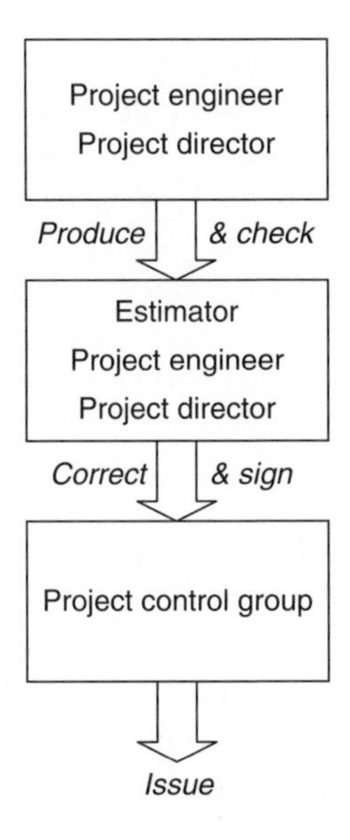

Fig. 8.8 Drawing quality control process protocol.

activities and related results comply with planned arrangements and whether these arrangements are effective and are suitable to achieve given objectives.' While these audits were internal to the PAE team, GGR staff could also participate.

The *execution problem* identifies the gap between the detailed description of the project and what is actually built on site, and is largely the responsibility of the PC. However, the PAE was responsible for ensuring the accurate transmission of design information and approving design work undertaken by trade contractors. The PQP also identified procedures for the interface with such trade contractors, the most important of which was the 'ABC Procedure' which classified the level of approval of trade contractor and supplier drawings by the PAE, and timescales for the process. The quality management system for this phase is illustrated in Fig. 12.7.

The *project performance gap* is the difference between the characteristics of the completed building and client's requirements. It is this gap that interests both the client and the members of the project coalition, for it defines whether the project can be considered a 'success'. Considerable effort went into moving the *line of visibility* further into the project coalition's sphere of activity, thereby narrowing the gap between the client's expectations of the project outcome and the completed facility in various ways:

- Formal *design reviews* at pre-defined stages of the project. These had the critical role of providing feedback loops where project performance could be compared back to project objectives. The formality of the procedure, the prior definition of review objectives and the predictability of the timing all served to focus attention on reducing potential gaps in the project process. In addition, the participation of GGR in these reviews made a considerable impact on the line of visibility. These reviews were particularly important at the earlier phases of the project, where more conventional techniques were considered to be inappropriate, and reliance upon peer review was argued to provide the most effective means of managing quality. These design reviews were central to tackling the briefing and design problems and made a major contribution to the planning problem. These project reviews were complemented by a formal *value engineering* exercise – see section 10.5 for a discussion.
- *Quality assurance procedures* acted as *blueprints* for project procedures, so that the client could see what was expected of the project coalition. These procedures pre-defined what sort of documents were to be produced and how they were to be managed and changed once they had been produced. They were complemented by *quality control procedures* for the verification and issue of drawings.

Overall, the *line of visibility* of Glaxo Group Research into this project was relatively deep. The 100-strong-client project management team had the technical capabilities to ride the project life cycle collaboratively with the project coalition and to take the initiative in areas such as value engineering.

Sources: Usmani and Winch (1993); Edkins (1998); see also Winch *et al.* (1998).

Notes

1 I am grateful for the comments of Nuno Gil on this chapter.
2 The ideas in this section are currently under development in collaboration with Kristian Kreiner of Copenhagen Business School.
3 Cited in Shapira and Berndt (1997, p. 339).
4 For example, Weick (1979); Tassoul (1998).
5 Schutz (1967, p. 61); see also Schutz (1973).
6 Henderson (1987, p. 15).
7 Dewey (2002).
8 Dixit and Pindyck (1995) and Howell *et al.* (2001) provide general overviews of real options analysis.
9 Patel *et al.* (2005) provide a review of work on property, while Miller and Lessard (2000) and Ford *et al.* (2002) propose its application to major projects.
10 These are developed from Fichman *et al.* (2005); the concept of a select option is derived from the work of Sommer and Loch (2004) on selectionism.
11 Cooper's work on new product development is seminal here in identifying the importance of stage-gates – for example Cooper (1993).
12 Royer (2003).
13 The original formulation of this concept is provided in Parasuraman *et al.* (1985). See also Zeithaml *et al.* (1990) and the subsequent debates in the marketing research literature.
14 See Winch *et al.* (1998).
15 Cited in Sabbagh (2000, p. 26).
16 See Barnes (1988) for a classic statement, and Morris (1994) and Winch *et al.* (1998) for critiques.
17 The notorious 1994 Standish Group Report which is widely cited in research on IT project performance singles out the construction industry as one to be admired thus: 'bridges do not fall down'.
18 These were developed from the work of Van de Ven and Ferry (1980), and first presented in Winch (1994a).
19 The terminology used here comes from Guildford (1959) and Simon (1973). See also Sidwell (1990) for a similar application to the construction project life cycle, and Ivory *et al.* (2006) for a perspective in terms of sensemaking.
20 See Winch and Carr (2001b).
21 Hunt (1996) provides a useful overview; MS Visio is one of the most widely used process mapping software tools.
22 The citation is taken from Rummler and Brache's comprehensive guide to process mapping (1995).
23 See the contributions to the debate collected in Cross (1984).
24 The problem of knowledge management as organisational learning from projects was initially formulated by Nonaka (1994), and has been taken up by many authors since that seminal article. A special issue of *Management Learning* (2001) **32** (1) was devoted to the problem of learning from projects. See, especially, Ayas and Zeniuk (2001), and Keegan and Turner (2001), upon which parts of this section draw.
25 Understanding situated learning and knowledge is a major research topic – Bresnen *et al.* (2004) and Ewenstein and Whyte (2007) provide analyses in a construction context.
26 Cited in Hale (1993, p. 400).

Further reading

Geary A. Rummler and Alan P. Brache (1995) *Improving Performance: How to Manage the White Space on the Organization Chart* (2nd ed.). San Francisco, Jossey-Bass.
A comprehensive management guide to process improvement in organisations.

Richard J. Boland and Fred Collopy (eds.) (2004) *Managing as Designing*. Stanford, Stanford Business Books.

A wide-ranging reflection on the nature of the design process stimulated by the experience of working with Frank Gehry on the Peter B. Lewis Building at Case Western Reserve University with contributions from a wide variety of disciplinary perspectives.

Robert G. Cooper (2001) *Winning at New Products: Accelerating the Process from Idea to Launch* (3rd ed.). Reading, Perseus Books.

One of the most comprehensive guides to using stage-gates to manage the new product development process.

Chapter 9

Defining Problems and Generating Solutions

9.1 Introduction

> 'In Architecture, as in all other Operative Arts, the End must direct the Operation. The End is to Build Well. Well Building hath three conditions; Commodity, Firmness and Delight'.

These words of Sir Henry Wotton – first published in 1624 – have echoed over the last four centuries throughout the anglophone world as the problem definition for all architects. He had just returned from a posting as English ambassador to the Venetian Republic, where he was exposed to the latest architectural ideas of the day. The three conditions of well building are his translation of the words of Vitruvius, who wrote during the first century BC in Rome, that good architecture consists of *Utilitas*, *Firmitas* and *Venustas*. These writings had been rediscovered, and then popularised by Alberti as part of the Renaissance revaluation of the classical heritage in the fifteenth century. During the sixteenth century the same ideas were popularised in France by Philibert de l'Orme. For at least 500 years, then, the commonly accepted definition of good architecture in Western culture has not changed; what has changed is our interpretation of these words.

This chapter will focus on the management of the processes by which architects and engineers go about 'well building' through defining the problems that the client is trying to solve and generating solutions to those problems through briefing and design. It will start by examining the briefing and design processes themselves, for we have little hope of managing effectively a process we do not understand. It will then discuss separately the particular issues in managing the briefing and design problems. The regulatory interest in 'firmness and delight' will then be discussed as the institutional constraints to 'well building'. In the latter part of the chapter, a number of different routines will be discussed which have been developed in order to facilitate and improve the process.

9.2 Tame and wicked problems in the project process

The construction project process is, in a profound sense, the 'conception and planning of the artificial'[1]. This conception of the artificial is a difficult process, for it entails grappling with the articulation of a desired future state under high levels of uncertainty. These difficulties have encouraged some of the more thoughtful analysts of the process to see design as the process of solving a *wicked* problem[2]. This is distinguished from a *tame* problem in both the challenges it poses and the process of solution, as illustrated in Table 9.1.

Table 9.1 The characteristics of tame and wicked problems

Tame problems	Wicked problems
Solution set describable	Problem definition incorporates solution
Determinate solutions	Indeterminate solutions
Optimised solutions	Satisfied solutions
True solutions	Good solutions
Solution achievement definable	Solution can always be improved by further work

Tame problems are those which are amenable to mathematical modelling in some way or other because the bounds of the problem can be set and the process of solution is rule-bound. Thus a game of chess – while sorely stretching the cognitive capacity of the human mind – is a tame problem, because it is clear where the bounds of the problem are, there is an explicit set of rules which each of the players must follow, and it is crystal clear who has won and who has lost. On the other hand, designing a new boutique to sell the latest fashions is wicked, because there is no way of determining what will catch the imagination of fashion-conscious shoppers next year; whether the choice of the extensive use of black is a challenging statement or just plain depressing is a matter of taste; and the success of the shop may be a result of its location at a key intersection and have nothing to do with the efforts made in design. The outcomes from the interior design process are the matter of legitimate debate and opinion, while Grand Masters at chess can be ranked precisely.

An important feature of dynamic uncertainty in the project life cycle is that problems shift from being wicked to being tame over time as information becomes available and the problem definition becomes more precise. While defining the structural configuration of a facility may well be a wicked problem, sizing a beam to support a given load is a relatively tame one. This suggests that the management of the solution to wicked problems will require a very different approach compared to the management of the solution of tame problems.

In particular, much greater responsibility will need to be given to the problem solver – the designer – because there are few clear criteria by which the client can audit or otherwise determine the appropriateness of the solutions offered. Unless the client trusts the designer to solve the problem, progress is impossible. Even once the facility is in use, it may be extremely difficult for the client to determine precisely why things are not as expected – those disputes that end up in litigation are typically the failed solutions to tame problems, not dubious solutions to wicked ones.

9.3 Solving the briefing problem

At the heart of the briefing problem is the dynamic between the client's desire for a constructed asset to achieve its broader objectives, and the set of possible realisations of that desire, given technical and regulatory constraints. The difficulties in addressing these problems revolve around the client not having fully articulated the project mission for itself and not being fully aware of the range of options open to it, while the designers may have difficulty in understanding what the client actually wants. As information is flowing between the parties under very high levels of uncertainty, the probability of misunderstandings occurring is high and can only be mitigated by frequent iterations around the problems.

These dilemmas can be captured visually in the so-called Johari Window, shown in Fig. 9.1, which classifies the information available on the project and its distribution between the client and its designers (or indeed any other agents).

- *Shared* information is both available and commonly understood by both parties, that is it is shared and not opportunistically manipulated.
- *Private* information is known by the client but is not communicated to, or understood by, the design team.

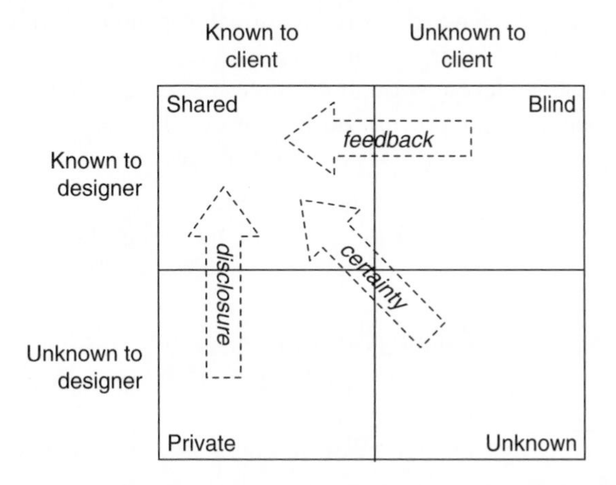

Fig. 9.1 The Johari Window (source: developed from Barrett and Stanley, 1999, Fig. 3.2).

- *Blind* information is known by the design team but not communicated to, or understood by, the client.
- *Unknown* information is not available to any of the parties – this is the zone of uncertainty.

There are a number of formal briefing models available, but these only operate in the realm of shared information[3], and so are inadequate for grasping the full nature of the process. Unknown information is the zone of uncertainty, and as the briefing process takes place at relatively high levels of dynamic uncertainty, it will inevitably tend to be large compared to the other three zones. What can be managed in order to solve the briefing problem are disclosure and feedback. *Feedback* is the process by which the design team communicates with the client regarding potential design solutions, and hence shifts the client from the blind to the shared state as indicated by the horizontal arrow. *Disclosure* is the process by which the client releases information to the design team, allowing it to share the information privately held by the client, as indicated by the vertical arrow. Disclosure and feedback are likely to be more difficult where there is a high level of mission uncertainty on the project, or where the client is inexperienced in procuring constructed facilities. As unknown information becomes available, dynamic uncertainty is reduced, indicated by the diagonal arrow. An example of the dynamics of the briefing process using disclosure and feedback is presented in panel 9.1.

The client may hold information privately for a number of reasons:

- it did not appreciate that the design team needed the information;
- internal disagreements within the client body mean that it cannot arrive at a clear position and restricts disclosure to hide this state of affairs;
- it does not give its representative at briefing meetings the authority to make decisions, and so the disclosure process is slow and insecure and decisions taken are overridden later by senior management;
- it does not have the organisational capabilities to communicate its needs clearly to the design team;
- it does not devote enough resources to being a client;
- it is behaving opportunistically towards the design team, that is information is impacted because it does not trust its designers.

The client may be blind as a result of the design team withholding information because:

- the design team thought that the client did not want the information;
- the design team is not capable of clearly communicating the possible range of design solutions;
- the design team may be searching for ideas, and needs more time to bring them to maturity;
- scarce resources are being deployed on other contracts;

Panel 9.1 The Houston Mosque and Community Centre

In 1984 a group of families belonging to a Muslim sect commissioned a local architect to design a complex of housing, community and religious buildings on a 12-ha site outside Houston, Texas. The two parties were unknown to each other, but a rapport was immediately formed between the two representatives of the group of families and the architect. The architect's office researched traditional Muslim architectural forms and presented some ideas to the two representatives. They in turn showed the architect pictures of a much more modern mosque. Thus the two parties' initial conceptions of the project were moved into the shared sphere and compromised. However, the mosque required the blessing of the sect's leader in India, and particular requirements regarding arches and windows had to be met. But the leader was uncommunicative on the precise requirements, leaving both client and architect in the unknown sphere on these points. The architect told the client that internal columns would be required for structural reasons, although this was not strictly the case and the architect was actually seeking approval for his preferred solution, leaving the client in the blind sphere. Alternative design solutions were presented to a raucous meeting of the whole community who argued the merits of each for some 3 hours. As this was all in the community's native language, these debates remained private to the client, although the architect could see that the women and men favoured different solutions. In the end, the community representatives announced that the community had agreed on a further compromise solution, the rationale for which the architect could not see. The architect publicly agreed to this compromise, aiming to try to shift the design towards a more coherent solution during design development. However, final approval still rested with the sect's leader in India, and nobody in Texas had a clear idea of what his requirements would be.

Source: Cuff (1991).

- the design team is behaving opportunistically towards the client, that is information is withheld because it does not trust its client.

There are a number of ways of stimulating the feedback and disclosure processes. The first, and for many clients the most important, is to work with the same design team as on the previous project. This allows a greater level of trust between the parties, while ensuring that many of the basic communication problems in terms of understanding the ways in which the other party addresses problems, and the range of acceptable solutions, are resolved through learning. This is one of the most important reasons why, as discussed in section 5.3, designers tend to be appointed rather than being subject to competitive tendering. Where the *concours* is used, clients are often tempted to use these as 'fishing expeditions' for good ideas, as shown in panel 5.4. As the director of the Tate put it: 'without wanting to plunder the ideas of the hundred and forty-nine architects who submitted, having been obliged to look at [such] material . . . causes you to rethink what you are doing'[4]. Even rejected design solutions reduce uncertainty because they close off options, while nearly acceptable solutions provide the basis from which to work towards the most appropriate one.

Visualisation techniques can be used. Typically, the client expresses its requirements through a description in text form, backed by appropriate figures for expected schedule and budget. One of the first tasks of the designer is to turn that description into a visual representation in a sketch, so as to pose the question 'is that what you meant'? Such sketching processes may be made a little more user-friendly by hiring specialists to create artists' impressions or to build scale models of possible solutions. Sketches have the enormous benefit of being very quick and cheap to produce, and they are deeply integrated into the thought processes of most designers[5]. Thus they are a very easy tool for the design team to use, but many clients find it difficult to make the mental transformation between a 2D visualisation and a 3D reality. More sophisticated methods are becoming available, using 3D graphics in virtual reality systems, particularly to create a 'fly-thru' for the proposed solution, as discussed in section 4.7. These are by far the best ways of communicating design intent to a client. However, they are expensive and require dedicated large-screen facilities for their most effective use.

Much knowledge of desires and possibilities is tacit – on both sides. One way to make this tacit knowledge explicit is to visit other facilities which resulted from similar missions. Architects and engineers maintain an enormous repertoire of possible design solutions, not only from their own past experience and that of their colleagues, but from the constructed heritage of their culture. The briefing process often involves visits to a subset of known earlier solutions, either independently or jointly. Joint visits offer the enormous added advantage that the informal interaction between client and design team during the visit allows trust to be built up,

Panel 9.2 Joint visits

Two types of joint visits were held by the Tate Modern client team. Firstly, a trip to Switzerland was made to see various buildings by Herzog & de Meuron. This visit allowed the team both to reassure themselves of the merits of their choice of architects – see panel 5.4 – and also to gain a deeper understanding of the ways in which Herzog and de Meuron had solved earlier design problems. They came away impressed with the economy of their designs and the quality of their detailing. Secondly, a Tate team visited two newly opened museums – the Guggenheim at Bilbao (designed by Frank Gehry) and the Getty at Los Angeles (designed by Richard Meier) – which focused more on the operational features of the buildings. The issues addressed in these visits varied from hanging the spaces, through signage, to the number and locations of toilets. Complemented by their deep professional knowledge of the principal competitors to the Tate Modern, the visits helped the client's operational team to specify more closely what they wanted from the new facility.

Source: Sabbagh (2000).

and a greater awareness of each other's needs to develop. Client teams may also visit similar facilities to gain insight into how their own facility might work. Such visits were used extensively on the Tate Modern project, as illustrated in panel 9.2.

A further approach – and one that is widely advocated if less frequently practised – is to involve building users in the process. Clients frequently do not know enough about their own business processes to fully articulate the needs of the users of the building; those that have to work in a school, an operating theatre or an open plan office will often have a much better idea of what is required in the brief than those who are responsible for making investment decisions. There are also important change management reasons for empowering users in the briefing process, as this aids acceptance of the organisational and environmental changes that the new facility brings. For these reasons, a number of methods have developed for involving users in the briefing process, ranging from questionnaires to focus groups to actual involvement in design. On some social housing projects in the UK, tenants have been given an active role in developing and selecting design solutions. However, one of the most important aspects of such user consultations is the management of expectations; budgetary and schedule constraints, and the fact that different user groups may have conflicting interests in the new facility, mean that the ability of the users to have a real impact on the design is, in most cases, very limited.

While the client/user gap identified in Fig. 9.2 is well known, it is important to take into account another gap internal to the client – that between the procurement function and senior management. Senior management is, quite rightly, concerned with the core processes of their business; the facility being procured is simply the means to the end of ensuring that those processes are efficient and effective. However, those charged by the client with procuring their facilities – be they the procurement function, facilities management function or development function – have more focused objectives. These are to procure the facility within the framework set by senior management. To put the problem in the terms of Fig. 8.3, senior management tends to place the emphasis on appropriate intention, while procurement executives tend to place the emphasis on predictability of realisation. These problems are compounded when the client itself is not in full control of the project, such as when permission from a landlord has to be sought for adaptations to the facility – see panel 10.8. This difference in perspective between senior and procurement executives tends to generate two types of problem for the management of the project:

- senior management are unwilling to delegate decision-making authority to procurement executives;
- new information becomes available which changes senior management's perception of the mission and tempts them to change that mission in spite of its implications for the predictability of realisation.

Whatever methods are used, it is clear that briefing is a process through time. Although the formation of the brief may be considered to be complete once all the internal client groups have signed off the design proposal as it affects them, the acquisition of additional information through the project life cycle may mean that the brief has to be revisited as costs escalate, regulatory constraints are imposed and initial design ideas prove not to be viable.

9.4 Client organisation for briefing and design

Solving the briefing and design problem requires organisational effort by clients. Clients need to understand enough about their own business processes to articulate their requirements for facilities. This understanding needs to range from how the facilities in the existing facilities actually work to how the symbolic values of the client as an organisation should be expressed in the new facility. Thus the processes of creating and exploiting the facility are symbiotic. We will discuss here two different aspects of client organisation for briefing:

- *facility management*, generating learning about how existing facilities work;
- *design championing*, generating a commitment to high standards of conception and specification on new projects.

Projects create new facilities to be exploited, and exploitation of those facilities generates requirements for the new facilities to be delivered by projects, as illustrated in Fig. 9.2. The only way to create better facilities in the future is to learn from the exploitation of existing facilities so as to enable the more precise definition of the project mission, as shown in panel 9.3. Yet, remarkably, the systematic collection of data on the ways in which constructed facilities perform in use is typically very limited. Learning tends to be restricted to what can be

Panel 9.3 Learning from facility management for the Tate Modern project

Peter Wilson, head of Gallery Services at the Tate, was actively involved in the Tate Modern project throughout its life cycle as a member of the design review team. It was his job to make sure the new gallery worked, both in terms of the visitor experience and the curatorial process, effectively and efficiently. He paid particular attention to these issues during his visits to Switzerland and the Getty Museum to explore how others were achieving these goals – see panel 9.2. He made an important contribution to saving costs because he could challenge the architect's assumption that column-free spaces would be required in the galleries. He knew that, in fact, little advantage had been taken of the column-free spaces at the existing Tate over the previous 20 years, and so the more complex engineering required to achieve this could be avoided. Similarly, he supported floor grilles over more expensive air handling solutions proposed by the architects, despite pressures from the curators worried about locating sculptures, because he knew that the problems could be worked around in practice.

Source: Sabbagh (2000).

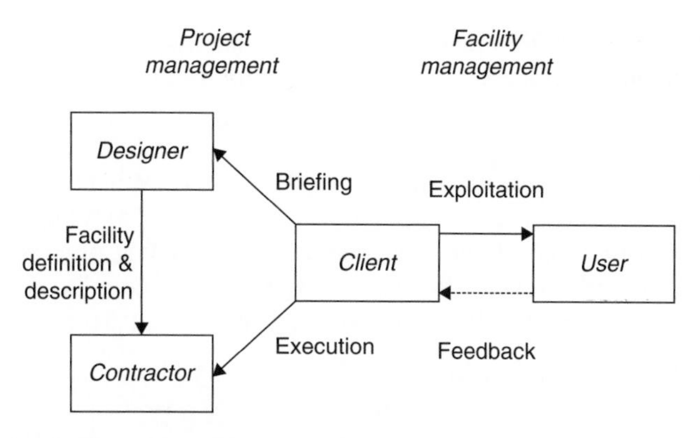

Fig. 9.2 The facility management feedback loop.

easily measured, such as energy consumption, rather than what is important to the value of the asset, such as the ways in which it enhances employee performance or attractiveness to customers. The contribution that such analysis can make to client business processes was explored in Chapter 3.

The first responsibility of the client in defining the project mission is to articulate the needs of the users, be they future customers, existing employees or sitting tenants. Where the users are an identifiable group at project inception – such as in the refurbishment of social housing – then more sophisticated methods can be deployed of getting user feedback on the existing and proposed facilities. However, on most projects such a group cannot be identified, and so the client must rely on past experience and the resource bases it hires for this expertise. Learning from the past experience of exploiting facilities, and, thereby, the identification of ways of improving exploitation in the new ones, plays a very important role during the briefing process. However, as Case 3 showed, few client organisations have in place effective learning loops to provide the kind of experience-based inputs that this process needs.

Developing the facility management feedback loop requires an organisational capability – preferably in-house – that enables to develop a more proactive engagement in briefing. This can be broken down into three distinctive phases[6]:

- *Operational briefing* which determines the needs of users in the day-to-day operation of the facility, and feeds into:
- *Strategic briefing* which takes a longer-term (2–5 years) view of the facility requirements of the organisation, which, if a need for a new or refurbished facility is identified, feeds into:
- *Project briefing*, which draws on the strategic and operational briefs to identify to the design team the requirements for the new facility, which feeds back to:
- *Operational briefing*.

While the facilities management profession is becoming well established and can address indoor environmental quality issues, and the nascent commercial management profession[7] is starting to place oversight of the project realisation process on a sound footing, client arrangements for articulating needs in terms of symbolic and spatial quality remain *ad hoc*. Where senior managers within the client are architecturally literate – such as on the Worldwide Plaza and Tate Modern projects – their engagement in the briefing process and oversight of the design process can be effective. However, this is rare. The UK's Commission for Architecture and the Built Environment argues that:

> 'Ultimately, the responsibility for delivering high-quality projects rests with the client. It is not the procurement process itself that determines the outcome. The essential ingredients are a committed client, with the right skills and an adequate budget, focused on whole life costs, with a quality designer as part of the procurement team'[8].

The recommended way forward – at least by CABE and the OGC – is that clients should appoint a 'design champion' to ensure good standards of design[9]. However, the suggested remit of this champion, their role in procuring design services and their relationship with the design team during the project life cycle is unclear. For instance the OGC[10] suggests that this person need not have experience of construction projects, but they also argue that the advice of externally appointed design professionals may also be required. In either case, it is not at all clear that such a person would understand how buildings add value for clients – even many trained design professionals are unaware of this and still treat buildings as artefacts. This is an area that clearly needs more research[11].

9.5 Solving the design problem

All human artefacts are designed to the extent that someone has thought about their form, function or aesthetics prior to commencing physical work. This is the burden of Karl Marx's famous dictum that 'a bee would put many a human architect to shame by the construction of its honeycomb cells. But what distinguishes the worst of architects from the best of bees is that the architect builds the cell in his mind before he constructs it in wax'[12]. The process of preconception of a future state which we defined in section 8.2 as future-perfect thinking is an elusive and ethereal process – we know that it exists but we cannot easily grasp how it is done. The nature of this process, and the extent to which it can be formalised into one or more methodologies which can provide the basis of a more rigorous approach to its management, have been the topic of intense debate for the past 40 years.

In essence, two perspectives on the process have emerged[13]:

- the *linear model*, illustrated in Fig. 9.3, which suggests that good design is produced when certain key steps are followed in a reciprocal manner and

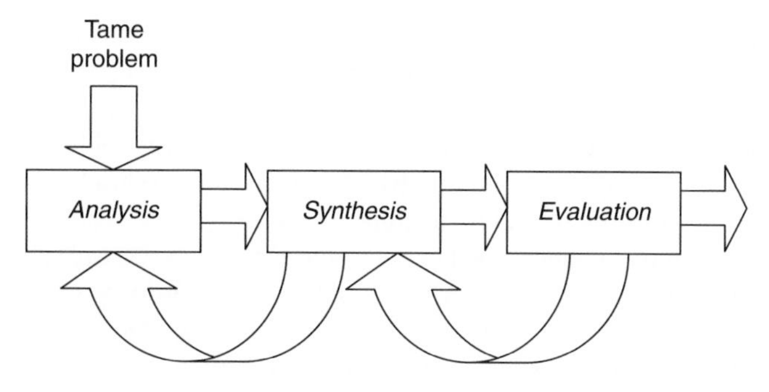

Fig. 9.3 The linear model.

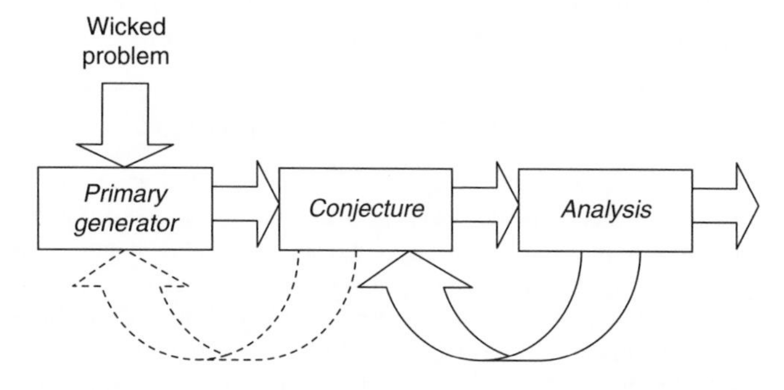

Fig. 9.4 The conjectural model.

- the *conjectural model*, illustrated in Fig. 9.4, which suggests that designers iterate around a number of cycles.

In the linear model, the first step in the design process is to analyse the problem completely, look at it from all sides and get a clear idea of the problem that the artefact is required to solve. The next step is to match the problem definition to the range of possible solutions, as constrained by available technologies and regulatory requirements, through a process of synthetic thought. Finally, the outcome of that phase in a proposed solution needs to be evaluated to ensure that it meets the original requirements. The logic of such a process is faultless, and it is difficult to gainsay the benefits from working in such a way. However, in their very nature wicked problems cannot be solved in this linear way, for they defy comprehensive, or even adequate, analysis prior to developing solutions. The cognitive range required for complete analysis is too great, and the process inevitably stalls before it has got going – paralysis by analysis is the result.

A combination of conceptual critique, and research into how architects actually work, has led to the development of a different model of the process that is much more in tune with the challenges of solving wicked problems – the conjectural model. This suggests that faced with a wicked problem, designers (and indeed all other decision-makers faced with such problems) start with an initial hunch as to what the solution might be.

This hunch comes from prior experience, some initial scoping of the problem and examination of solutions to similar problems by others. This hunch forms the primary generator, which then drives the rest of the process. The process then proceeds through a dialectic of conjecture in which ideas are put up as possible solutions and analysis in which they are critiqued and tested for viability. The feedback loop from conjectures to the generator is weak – only infrequently do designers abandon their original generator. If they cannot convince the client of its merit, they are more likely to be relieved of their contract than to be inspired by a completely new primary generator. The conjectural model is, then, a model of future-perfect thinking in design.

Although the linear process tends to be favoured by engineers and is widely seen as more rigorous[14], in fact the conjectural model is much closer to scientific method than the linear one; see panel 9.4 on the position of the last century's leading philosopher of science, Karl Popper. The primary generator is the big idea – the cry of 'eureka' – that drives the design process, and indeed the design process is often managed in order to

Panel 9.4 Scientific method

Karl Popper was concerned to understand how we know, and what counts as, a scientific theory. He criticised the ideas that truth was obvious to the enlightened and that knowledge comes from careful observation, arguing instead that it came from learning from our mistakes. He argued that a scientific theory was the result of an inspired conjecture, which was then tested by experimentation. Empirical observation of the data is impossible in the absence of a theory which guides us to what to look for in the data. Scientific theories are testable theories, and a proposition that cannot be tested experimentally is not a scientific proposition. According to Popper, we can never accept a theory to be true, but simply hold it as one that has not yet been shown to be false. Thus, argues Popper, we learn from our mistakes through cycles of conjecture and refutation. The scientist then is both intuitive and sceptical – intuitive because he or she can formulate testable propositions that explain puzzling phenomena, and sceptical because he or she is always questioning those theories through designing tests that would refute them.

Source: Popper (1969).

encourage big ideas to be formed early in the process. The architects for some of the most challenging – and hence wicked – projects are often procured through the *concours* system described in section 5.3.3, which does not favour analytic refinement, but grand concepts that capture the jury's imagination. It is around the generation of these big ideas that the oft-mentioned 'arrogance' of the architect revolves. The task of generating and driving through a solution to a wicked problem requires enormous self-confidence as it is assailed from all sides by 'lesser mortals'; the appropriateness of the solution cannot always be rationally defended simply because it is a solution to a wicked problem. The image of Howard Roark as the lone voice against mediocrity in the book and film *The Fountainhead* is a tempting one for the architect, and pervasive in stereotypes held by other actors within the project coalition. However, as Bill Hillier argues[15], the process is inherently a social one, and architectural design is a reasoning art – mobilising analysis where there are data, and conjecture where there are not. The arrogance of conjecture needs to be coupled with the humility of refutation.

9.6 The cult of wickedness

Are all design problems wicked problems? Some commentators suggest yes, as they place the emphasis on the uniqueness of each project mission. Yet many projects have very similar missions – designing one out-of-town shopping centre is, arguably, very much like designing another. The client is not looking for major innovations – simply one as successful as the one built in the adjoining town. Whether designing constructed assets is a wicked problem or not surely depends on the project mission; it is, in essence, a function of the level of mission uncertainty on the project mission. Projects where the mission emphasis is on specification and realisation are more likely to be tame than those where conception is paramount. Projects that can deploy well-established engineering solutions with incremental innovation will be tamer than those which demand new techniques to create longer bridge spans or solve complex geotechnical problems. However, even here, regulatory constraints can turn a tame problem into a wicked one, as Case 9 illustrates. The problem is to identify which missions are at the tame end of the scale and which are more wicked, and to develop the management approach accordingly.

Architectural practices, like other professional service organisations, have basically four strategic options[16]. Perhaps this strategic management model can be adapted to address this problem in the manner indicated in Fig. 9.5. The two dimensions are the level of mission uncertainty that the project faces, and the mission emphasis on a scale between specification and conception, as defined in Fig. 3.2. Where mission uncertainty is low, with an emphasis on specification, it is likely that solutions can be largely borrowed and there need be only *incremental adaptation* to the problem at hand. Where mission uncertainty is higher but the emphasis remains on specification, then *pushing the envelope* of available technologies will be required. If the mission emphasis is more on conception rather than specification, yet mission uncertainty is low, then the *bright idea* will suffice to

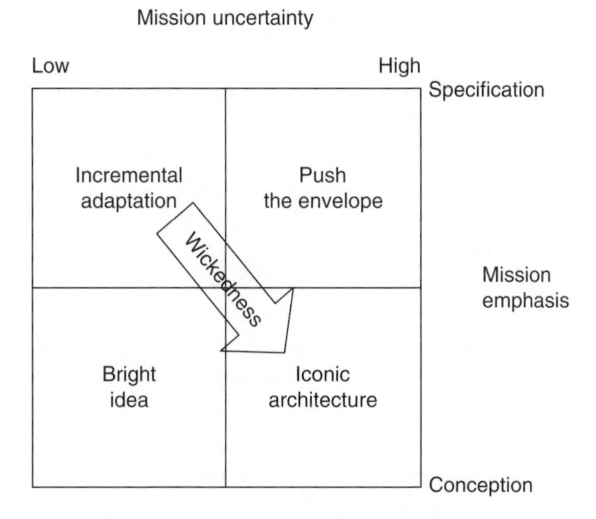

Fig. 9.5 Tameness and wickedness in the design problem.

meet the needs of the client. Where both mission uncertainty and the conception orientation are high, then *iconic architecture* will be required. This shift from bright ideas to iconic architecture can be seen in the careers of many architects famous for their iconic architecture whose early commissions were for private houses. These are projects with low levels of mission uncertainty but with demanding patrons emphasising originality and flair in the design solutions – in other words looking for bright ideas.

In Fig. 9.5, wickedness increases as the design problem moves from the incremental to the iconic. The challenge in managing the briefing and design problems is to know where in the model the problems lie. Where the problem is incremental, then the *total design* approach advocated by Stuart Pugh[17] is viable, with its reliance on the linear approach and emphasis on defining the product design specification at inception. Where the problems are at the wicked end, then a much more intensive briefing process is required as clients engage with the primary generator presented by the architects. The traditional form of design process management, where architects are procured through *concours* and then left to get on with it, is most appropriate for the realm of bright ideas. The cult of wickedness lies in insisting that all design problems are wicked problems, without first understanding the uncertainties inherent in the project mission and the aspirations of the client in relation to conception and specification.

9.7 The management of design

How, then, is this process to be managed? The answer needs to be at two levels. The first is the process internal to the architectural studio working on a given project. Typically, the source of the primary generator – the partners – leaves the conjecture and analysis process to other, more junior staff. This is then managed through a process of *propose* and *dispose*[18]. The junior staff makes design proposals (conjectures) which are then interrogated by the partners as to their fit with the primary generator, and the developing understanding of the brief. This process may be on a one-to-one basis, or more openly in a *crit*. The second level is the process of interaction with the rest of the project coalition. Proposals that survive the *crit* process are fed back to the client as part of the briefing process, forming a visual conversation with the client. In response, the client discloses further information to the design team, enabling them to gain a deeper understanding of the problem. As options are narrowed down and particular solutions become favoured, the emphasis shifts more from briefing around the primary generator to the cycle of conjecture and analysis through design.

Once the design problem has been tamed, a number of design management techniques become available[19]. Six that have found applications in construction and will be discussed here are – quality function deployment (QFD), functional analysis, whole life costing, prototyping, design quality indicators (DQI) and sustainability charrettes. All six have in common the use of scoring systems of specified alternatives against agreed criteria in a group decision-making context and a commitment to the involvement of users in the design process. The requirement to specify alternative solutions makes them difficult to use for tackling wicked problems. It is also notable that the benefit of such techniques comes not from the slavish following of any specific methodology, but from the fact that they provide a structured environment for arguing about possible solutions to the problem[20]. Busy people need to be given 'permission' to get around a table and discuss the issues, instead of getting on with producing 'deliverables'. An important element of the benefits of these techniques is to give such permission, while capturing the outcomes of the discussions in standardised ways that can enter into the project information flow.

9.7.1 *Quality function deployment*

Quality function deployment was first developed in the Japanese shipbuilding industry to improve the articulation of the relationship between the client's needs and the technological possibilities, and it has more recently found considerable favour in new product development in the motor and other volume industries. The essence of the approach is to create a matrix of customer requirements against design parameters. The former are ranked in order of importance for the customer, while the latter are scored on their ability to meet those customer requirements. An important element in QFD is benchmarking the proposed design solution against competing offers in the marketplace – such analysis frequently involving the reverse engineering of competing products. Design parameters are also costed to ensure that the proposed solution is within the target costs for the proposed market segment. Panel 9.5 illustrates the use of QFD in the design of nursery facilities.

Panel 9.5 QFD for nursery design

Under the UK's Sure Start programme, private sector companies were encouraged to invest in the provision of new nursery facilities – 2500 are planned for 2008. A construction company with a contract for two new nurseries proposed to use a QFD methodology to elicit design requirements from which the design team could work. Government recommendations were the start of the process, producing a list of 180 design criteria; a focus group formed from stakeholders at an existing nursery was used to pare these down to 20 elements, which were then rated for importance. Competitor nurseries were then identified and managers asked to rate the 20 elements on 5-point scales to produce the *customer requirements*.

A set of *technical characteristics* was then generated to meet these customer requirements. Trade-offs within the 'roof' of the house of quality were then identified, and costed, and characteristics that were deemed important to the customer but were also expensive such as car parking were reduced in priority in this process. Later value engineering exercises then led to attention to standardisation across other nurseries within the group. The process is reported to have been a success, and important features from the customer point of view such as daylight were identified.

Source: Delgado-Hernandez *et al.* (2007).

9.7.2 *Functional analysis*

The essence of the approach is to break down the facility in terms of the business processes which it is to house, through *functional analysis*. This is typically done

> **Panel 9.6 Functional analysis in action**
>
> The project for the new 'world class' laboratory was going wrong – the client was deeply dissatisfied with the design proposals from the architect, and the architect complained that different parts of the client organisation were inconsistent with each other. A one-day value management meeting was organised, with an experienced value manager in the chair, consisting of the representatives of the design team, client and the cost consultants. Each participant was asked to present what they believed to be the key objectives of the project, and it became clear that there was a wide diversity of views, both within the client and between the client and the architectural team. The chair then facilitated a discussion in which the value tree shown in Fig. 9.6 emerged. Once this was agreed, brainstorming addressed some of the issues of concern for the client with the existing design proposals, and some of these ideas were taken forward for development. At a subsequent one-day meeting some weeks later, the validity of the value tree was confirmed and weights were given to its various elements. These were then used to score the five different design proposals produced by the architects, and a clearly favoured option emerged.
>
> Source: Green (1992).

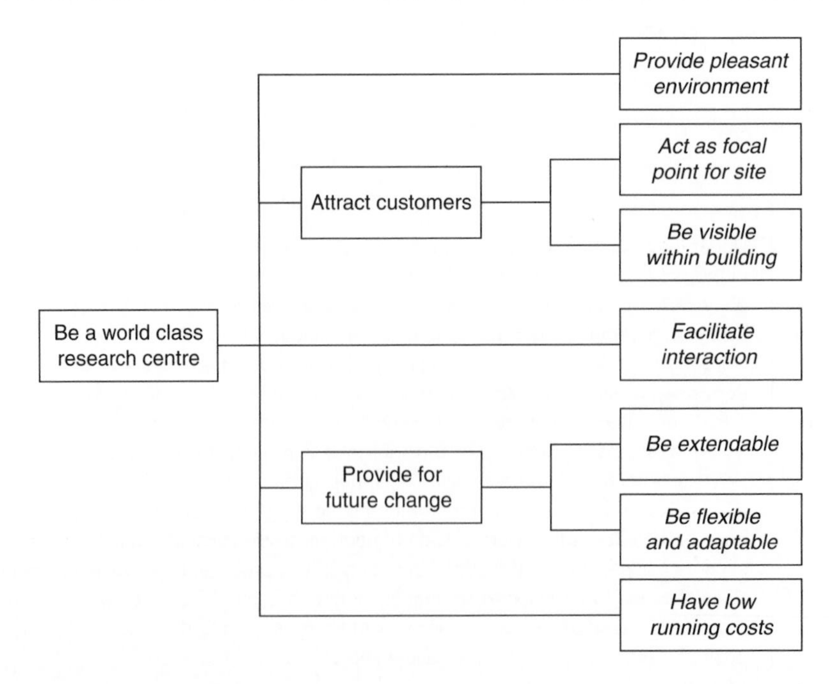

Fig. 9.6 Project mission for a pharmaceuticals research centre (source: adapted from Green, 1992, Fig. 5).

using verb/noun couplets in the manner shown in Fig. 9.6, which shows a functional analysis, or value tree, for a scientific research centre for an international pharmaceuticals company; see panel 9.6 for further details. Figure 9.6 shows the project mission broken down into a number of sub-missions. Once established, the functions can be rated in order of priority, to provide criteria against which to make future design decisions. The most important aspect of such meetings is the dynamic between the parties, rather than the details of the methodology which 'provides a framework around which the professional team can think and communicate [and] also ensures that the decision-making process is explicit'[21]. The importance of functional analysis in defining the project mission is illustrated by the failure to define properly the mission of the Channel Fixed Link (Case 1) – to provide transport – which led to management effort being focused on the tunnel itself, rather than the transportation systems the tunnel was to house.

9.7.3 Whole life costing

The fundamental idea behind whole life costing is that investment in the quality of specification now can reap returns later in lower operating costs, and less frequent repair and maintenance of the facility. Conceptually, it is based on cost–benefit analysis as presented in section 3.6 and in the terminology used in Case 7, design decisions that increase capital costs (capex) may reduce life cycle costs (opex) to provide a positive NPV. Although, as Table 3.1 shows, opex costs are relatively small in the overall budget – typically amounting to 3% per annum for office buildings – there are still benefits to be had by reducing opex through more sophisticated investment appraisal[22]. In addition, and probably much more important than direct opex savings, there is the lower level of disruption to the client's business processes caused by maintenance operations. Closing a floor for redecoration means disruption for office staff, or lost sales area; compromised customer movement because a lift or escalator is out of order means lost sales for a retailer. In all cases efficiency and effectiveness suffer.

There is, however, a considerable reluctance on the part of clients to fully appraise whole life costs, for a number of reasons:

- The client decision-maker may not be responsible for repair and maintenance, because the building may be sold or let at completion, or simply because capex and opex are the responsibility of different departments within the client.
- It is in the interests of project promoters to minimise capital cost in order to have the project funded. Once funded and completed, the owner is locked into any life cycle expenditures – this is a special case of escalating commitments, explored in section 10.10.
- Whole life costs are incurred in the future, about which we have very low levels of certainty, while capital costs are incurred in the (relative) present and have (relatively) higher levels of certainty – costs discounted 20 years to the future will always look trivial.

While the theory of whole life costing is well developed as a special case of the investment appraisal techniques discussed in section 3.6, the serious problem with the reliability of data on what costs-in-use of constructed assets actually are, hampers effective analysis. There are two basic sources of data:

- The facility management learning cycle discussed in section 9.4;
- Component manufacturers' reliability analysis, drawing on the reliability engineering concepts presented in section 12.2.

The practice of whole life costing is presently coming nearer to the theory, thanks to a number of recent developments:

- One of the benefits of developing concession contracts – as described in section 2.6 – is that they create stronger incentives for the collection of actual costs through the facility life cycle – in effect, the capex and opex budgets are held by the same decision-maker, the concessionaire.
- Systematic efforts to improve data sets are starting to reap rewards.
- The development of open building concepts is generating a better understanding of the expected life cycle of various building sub-systems, and hence more precise periods for NPV analysis. The DEGW 7 S framework for office buildings[23] is one such approach:
  Site – which is timeless, and provides the context for the facility.
  Shell – the structural frame of the building which may have a life of 50–75 years.
  Skin – the cladding to the building, which may be expected to last around 25 years before the frame of the building is reclad. However, where the building is high on conceptual quality, skin and shell may have the same life cycle.
  Services – these typically last 10–15 years, before mechanical and electrical unreliability becomes a problem.
  Scenery – which is the basic interior decor of the building, lasting some 5–7 years, and will need to be regularly refurbished, if only to maintain staff morale and customer appeal.
  Systems – ICT systems typically only last 3 years before they require significant upgrading.
  Settings – which are the everyday moveable elements of the buildings such as furnishings and decoration and, if required, partitioning.
- The regulatory pressures to improve sustainability, as discussed in section 4.4.4, are particularly associated with energy consumption by facilities in use.
- The facility management profession is growing in capability and influence.
- The structural shift of Western economies to a lower inflation environment means that discount rates in NPV calculations are lower, and hence future expenditures appear greater.

9.7.4 Prototyping: simulations, models and mock-ups

It is often argued that one of the major problems in managing design in construction is that all projects are prototypes. To an important extent, this is an expression of the cult of wickedness, as discussed in section 9.6. The use of the standardisation and pre-assembly techniques presented in section 12.10 is one way of shifting the project mission towards the incremental category in Fig. 9.5. Even for the most challenging project mission, though, there is much that can be done to move on from the prototype to allow particularly crucial elements of the facility to be prototyped *before* commitment to actual construction is made.

Panel 9.7 Models and mock-ups

Physical models and mock-ups were used in three different ways on the Worldwide Plaza project:

- Two 1:500 scale models were tested in a wind tunnel, complete with a replica of the surrounding buildings to simulate any turbulence in local wind conditions. One in wood tested deflection; the other in perspex tested wind loadings. As a result, changes were made to the structural design.
- Full-scale mock-ups of key elements of the building were tested in a different wind tunnel for air and water penetration. A segment of the curtain walling to the 35th floor leaked like a sieve because of the use of the wrong mastic and incorrect fitting of the flashings to the windows. It took 2 months to rectify the problems by closely defining the construction sequence and using the appropriate materials. Similarly, a segment of the roof was tested which also failed its penetration test and displayed signs of structural failure in the wind buffeting. Changes to the fixings were made as a result.
- A full-size furnished mock-up of an office for one of the major tenants – who would be occupying a number of floors – was built so that its staff could get a much better feel for their new working environment, and thereby staff resistance to the move would be reduced.

Source: Sabbagh (1989).

The development of computer-aided design (CAD) since the early 1990s from a 2D draughting tool to a 3D modelling tool has allowed the development of sophisticated simulation packages to allow the performance of various components and sub-systems of the facility to be analysed prior to execution; these issues are developed further in section 14.4. The implementation of virtual reality techniques as the interface to the 3D model pushes this potential even further, and the virtual protyping of the completed facility is becoming a viable possibility[24].

Even without CAD, there is much that can be done using models and mock-ups both to understand the performance of the facility in use and to rehearse its execution on site. The issue of the guardrail in Case 9 was resolved using a mock-up which demonstrated what could be less easily seen from the drawings – the best height for the rail. Scale models – particularly when used at design review meetings – allow all the parties to better understand the proposals and their implications. There are a wide variety of uses of mock-ups for management construction projects, as panel 9.7 illustrates.

9.7.5 Design quality indicators

The DQI framework modernises the language of Vitruvius and Wotton. It uses a basic triangle of build quality, functionality and impact, breaking them down into subcriteria as shown in Fig. 9.7. A standardised questionnaire is then used to rate the building against each of these criteria on a FAVE basis (fundamental, added value and excellence) in terms of what the building is intended to achieve[25]. Does a high score on the DQIs mean that a building is 'good design'? As a first step to

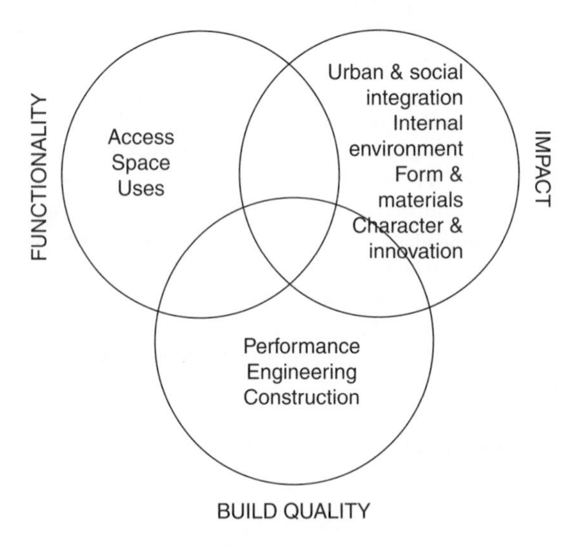

Fig. 9.7 The design quality indicators (DQI) criteria (source: http://www.dqi.org.uk, accessed 10/04/07).

answering this question, it is helpful to place their development in the context described in Case 2 of the UK reform programme. One important aspect of this programme was an increased emphasis upon the measurement of the performance of the project process. Architects and engineers whose concern was more for the performance of the constructed product became concerned about this emphasis upon the process. As Sunand Prasad, a member of the committee that oversaw the development of the tool who later became President of the Royal Institute of British Architects, argued:

> 'The decision to devise DQIs might be taken then to contain all the intellectual rigour of the position: *if you can't beat 'em, join 'em*. But it would be more accurate to describe it as an act of appropriation. We have to distinguish between measurement and the uses to which it is put. To fear measurement is no more rational or less logical than to fear technology. Those with a critical and indeed sceptical attitude to measurement are best placed to make good use of it. *Measurement is too important to be left to the measurers*'[26].

Michael Dickson, the chair of the DQI development committee and a structural engineer, concluded that

> 'Initial feedback suggests that the DQI tool is robust and readily understandable, applicable generically to all building types. It is designed to be used by professionals and lay people alike and is intended for all who wish to see the quality of the built product improve.. . .It can be used by all parties at any stage of the design procurement and occupation processes to inform on the compromises that need to be made in the process to achieve affordable, enduring quality'[27].

The team from the Science Policy Research Unit at the University of Sussex that did the development work of the DQI tool view the contribution of the tool to the design process a little differently:

> 'As a tool for thinking about design. . .the DQI . . . is most useful as a starting point for discussion. It cannot provide an absolute measure of the design quality of a building but can be used to articulate the subjective qualities felt by different stakeholders in the design process and thereafter in the use of a building. Tools for thinking aim to elicit and represent knowledge about design in order to initiate conversations about client and user priorities, design possibilities and consequences. This is possible because results from different members of the project team and user groups can be compared and contrasted during design and subsequent evaluation processes'[28].

A major weakness, however, of the DQI process is that it is still based on a view of facilities as artefacts as defined in section 3.2, and not assets because it relies on descriptors of the facility rather than analysis of how it adds value for clients and other stakeholders.

9.7.6 *Sustainability charrettes*[29]

The development of sustainable facilities presents new challenges for the briefing and design processes. These include:

- The necessity to adopt relatively unfamiliar and sometimes unproven technologies ranging from ground source heat pumps to green roofs which can

have unexpected interactions with each other in completed facility and hence present additional risk to process and product integrity.
- The increased requirement to model analytically the proposed facility as a system, particularly with respect to thermal performance.
- The desire to meet external accreditation targets such as LEED and BREEAM to validate publicly the claims made for the building.
- The broader commitment to community that underlies the philosophy of sustainability implies much greater effort to involve building users and other local stakeholders in the design process.
- The importance of 'ownership' of the commitment to a sustainable building, particularly where specifications do not provide returns on a cost–benefit analysis basis, but on a broader ethical basis as defined in section 4.6. This ethical commitment can be at risk during value engineering exercises – see section 10.5 – founded on cost–benefit analysis.

Charrettes are typically organised as workshops over 4–7 day periods facilitated by an experienced charrette facilitator and are expected to iterate through three phases of presenting alternative proposals, refining proposals and agreeing on a proposal to take forward into design development. They typically take place at or near the proposed construction site, and should include potential opponents as well as proponents – see section 4.3. The aim is that the intensive interactions during the charrette generate greater commitment to the green aspects of the conception and specification than ordinary consultation processes would.

9.8 Summary

> 'One of the hardest roles you have to learn is to accept the eighty per cent rule: that it's better to settle for eighty per cent of what you wanted and get it delivered when you want it, than to wait for a hundred per cent delivered at some indeterminate time in the future that might be never'.

Just as process integrity is a negotiated order, quality in the constructed product is also a negotiated order[30] as the Gallery Manager of the Tate suggests above[31] and Case 9 demonstrates in greater detail. The level of quality appropriate to a client's needs as articulated in the project mission is the outcome of a long negotiation process between articulated desire, technological possibility, regulatory consent and financial viability as a process of future-perfect thinking. Achieving all the elements of the project mission is rarely possible in full; what is important is that the client knows why they have not been achieved, while the facility meets the base criteria laid out in the project mission.

There is a very important sense in which the process cannot be 'managed' in the conventionally accepted way. The indeterminacies inherent in solving a wicked problem mean that attempts to manage the process directly are likely to be counterproductive. The sources of the primary generator and the information flows through the conjecture/analysis cycles are much too tacit to allow direct influence by an outside manager; the designers at this stage of the process need to be trusted to manage themselves. What can be managed, however, are the stage-gates in the process, discussed in section 8.5. Design reviews allow the opportunity for progress to be reviewed in a collaborative manner, and for stops to be placed on the cycles on the agreed grounds that the design is good enough to pass to the next stage of development. Thus design reviews, with a clear definition of the answers to the who, what, when questions, are the key to managing the briefing and design processes, as illustrated in Case 8. The tools and techniques introduced in section 9.7 can be used to support this process.

Once the briefing and design problems have been solved, the project moves on to the planning and execution problems, although returning to the design problem is not unusual if the client changes its mind because of unexpected developments or new opportunities. The next three chapters are devoted to ensuring process integrity in planning and execution.

Case 9
Designing the Sheffield Arena

By the mid-1980s, Sheffield had suffered the severe and permanent decline of its traditional industrial base, leaving large areas of the inner city derelict. Like many provincial British cities, it sought to revitalise its economy and urban fabric through investment in cultural and leisure facilities. In 1986, it saw an opportunity to do this by offering to host the Universiade XVI (16th World Student Games) in June 1991. This, it was argued, would provide the investment opportunities and lever the capital grants that would lead to the desired regeneration. However, Sheffield City Council received no support from central government for its efforts, at a time when relations between Sheffield and some other Labour controlled cities and the Thatcher government had reached a nadir.

A number of projects were involved – the Ponds Forge swimming pool, the Don Valley Stadium and the conversion of the semi-derelict Hyde Park flats into a games village. An additional requirement for an 'events hall' was identified in order to house gymnastics, basketball, volleyball and ice sports. As funds were not available from central government, and there were many other demands on the resources of the city, it was decided to seek a private sector partner to design, build, finance and operate the facility. It was also clear that the facility could not be commercially justified on sports uses alone. Therefore a multi-use facility was proposed, which could also host concerts and exhibitions, to be called the Sheffield Arena.

The overall programme for the games was managed by two companies set up by Sheffield City Council. Sheffield for Health Ltd would build and run the facilities, while Universiade GB Ltd would run the actual games. Both were companies limited by guarantee, with Sheffield City Council as the first guarantor. Sheffield for Health would borrow in the open market, while Universiade GB would live off sponsorship money. During 1989 neither strategies were raising adequate funds, and both companies were running up serious debts; the Universiade was simply not attractive enough to the international television and advertising groups which typically sponsor such events.

By the middle of 1989, design for the Arena was well under way and enabling works had started on site. In addition to the enormous time pressures generated by the immovable deadline of 1 June 1991 for the opening of the games, an event completely unconnected with the project had an enormous impact – the disaster in Sheffield's Hillsborough football stadium in April 1989 in which 95 people were killed when spectators were crushed during crowd surges in the stands.

Because of difficulties with putting together the funding package, the funds were not in place to finance the project during 1989, and the architects and engineers were asked by the client to work on the project on a 'goodwill' basis in order to keep things rolling, with a promise of payment when funding finally materialised. As a client representative said to the architects and engineers, 'if you do not show faith in us at this critical period, then we will have to think twice about retaining you for our other projects which we are in the process of finalising'.

Sheffield City Council was the *promoter* of the project, fulfilling what it saw as its democratic mandate to regenerate the industrially devastated Don valley in spite of intense opposition from central government and other political opponents. It saw the Universiade as a way to achieve its regeneration objectives, and to provide the people of Sheffield with a lasting resource. Its interests in this respect were represented on the project by its Land and Planning division; these interests were to ensure the success of the Universiade and through that, stimulate the regeneration of key areas of the city. The Council was constrained by lack of support from central government, shortages of capital, mounting cost overruns on other projects in the Universiade programme and the fear of a revolt by local Council Tax payers if the Council's guarantees were called in. *Mowlem Management Ltd* was appointed by Sheffield City Council as *project manager* to oversee the progress of the project on behalf of the promoter and to chair project review meetings. Mowlem was also working on other construction projects in the overall games programme.

Spectacor Management Group (SMG) was to be the operator of the facility. It successfully tendered to Sheffield City Council to design, build and operate the facility, and came to an arrangement where it would help the promoter to put together a financial package from the private sector. Based in Philadelphia, SMG manages sports facilities on behalf of public and other authorities throughout the USA. It became the *client* for the actors in the project coalition. SMG's objectives were to ensure that the costs of the facility were low enough, and the quality and capacity of the facility were high enough, to ensure a revenue stream that would allow the financiers to recoup their investment and SMG to run the facility at a profit. As its operations manager stated, 'give me as many comfortable seats with good sightlines as the Arena auditorium will accommodate. Remember more seats mean more revenue, more revenue means more profit'. SMG's principal constraints were that it would be liable to Sheffield City Council if the facility were not ready in time for the games, and that it lacked an understanding of how British local government worked.

The design team was led by *Hellmuth Obata Kassabaum's* Sports Facilities Group (HOK), part of one of the world's largest US architecture/engineering firms. Specialists in designing sports facilities, HOK had worked for SMG many times.

HOK was the *concept architect* for the facility, appointed on a fee basis. Its objectives were to satisfy a long-standing customer while retaining its reputation as the leading architects in sports facilities. *Lister Drew Haines Barrow* (LDHB) was a leading British architectural specialist in sports facilities. It entered into a 50/50 consortium with HOK and undertook most of the detail design for the facility as *project architect* on a fee basis. Its objectives were similar to those of HOK with whom it was also in consortium on the Docklands Arena, and to build up a reputation in sports facilities. *Oscar Faber & Partners* was the *engineer* for the facility, responsible to the client for engineering both the structure and the mechanical and electrical services inside the building. *Poole Stokes & Wood* was appointed as *quantity surveyor*. It was responsible both for advising the client on the costs of the various design options, and preparing the bills of quantity that would be used as the basis of competitive tendering for the trade contractors. Any proposals for changing the design, from the architects or engineers, would have to meet its approval from the point of view of costs.

R.M. Douglas successfully tendered as *construction manager* responsible for selecting and co-ordinating the trade contractors which actually executed the 46 separate works packages for the construction of the Arena. Working on a fee basis, its objectives were to realise the client's mission. Douglas' principal constraints were that it could do little until design issues were resolved, and that the National Coal Board found a seam of coal at the eastern end of the site and exercised their statutory right to exploit it as an open-cast mine during 1989, not releasing that portion of the site until the end of the year. The contract form used was JCT 87.

Sheffield City Council was also responsible as *statutory authority* (SA) through its Building Control, Fire and Civil Defence division. These responsibilities on the project consisted of ensuring that the Arena complied with the Building Regulations – the statutory codes which provided for the integrity of the structure and the safety of its users. Its objectives were to fulfil its statutory obligations; its principal constraint was the strong local sensitivity following the Hillsborough disaster. Indeed many of the staff concerned with the Arena project were appearing before the inquiry into the Hillsborough disaster during 1989, when the key design decisions described in this case needed to be made to ensure that the project remained on schedule.

During 1989, three main issues came to a head in the design process which demanded speedy and cost-effective resolution. These were the means of emergency escape; the tiered seating gradient; and the guardrail height. The relationship between architects and engineers on the one hand and statutory authorities on the other is one of propose and dispose. The design team proposes solutions that will meet client needs, while the SA comments on these design solutions, indicating the points at which they are unsatisfactory from a regulatory point of view. Redress is available in the case of dispute through an appeal to central government in the shape of the Secretary of State for the Environment.

The specific issues were as follows:

- *Emergency escape*. The SA insisted on a 2400 mm wide cross-aisle to be used for the evacuation of the Arena in the event of fire. This would result in the loss of 900 seats.

- *Tiered seating.* The SA said that the tiers were too steep, and that the rake should be lowered to ease the climbing of the stairs. This would have spoiled sightlines for the spectators, and as the raking beams were already under construction it would have had severe implications for budget and schedule for rework.
- *Safety rails.* The SA requested 1100-mm high safety rails in the auditorium, which complied with the prevailing regulations. However, these would have obscured the spectators' view and 800 mm was preferred by the architects.

In sum these instructions from the SA would have the effect of:

- obliging structural works already under way because of the very tight schedule to be scrapped and redone with obvious implications for the schedule and budget;
- reducing the number of seats with good sightlines, and hence the profitability of the operation of the facility;
- reducing the comfort of users of the building, and hence its attractiveness to potential customers.

SMG and their architects HOK took the view that they had designed and managed large numbers of similar buildings in the USA which had operated in perfect safety for many years. These were designed and operated in accordance with standards approved by the American Standards Association. Similar buildings in Birmingham and London designed by HOK had been accepted by the relevant SAs. As one SMG representative argued, 'it is evident that the [SA] have no idea about what is involved. We have been appointed by Sheffield as experts in our field. Surely they must rely on expert knowledge and experience'.

The statutory authority, on the other hand, took the view that full compliance with the British Building Regulations was required in the context of high public concern following the Hillsborough disaster. However, no regulations for this type of multi-purpose arena existed, and so the SA were obliged to adopt those for theatres, cinemas and football stadia. This led to considerable room for interpretation, and argument.

In an attempt to resolve problem 1 – the emergency escape – HOK and LDHB jointly commissioned Warrington Fire Research Consultants (WFRC) to provide specialist advice on fire safety. They provided a comprehensive review of the latest research in fire safety, and appropriate building codes from other countries. Their report questioned the technical basis of the SA's position. The report was presented to the SA, which merely noted it and continued with its insistence on the cross-aisles. A similar WFRC report had been accepted by the London Borough of Newham for the Docklands Arena, but that was prior to the Hillsborough disaster.

A second strategy was to demonstrate to the SA how SMG had effectively managed similar facilities for many years without incident, by offering to take its representatives over to the US and show them the operations there. However, this would have contravened local government staff guidelines, and in the words of one SA official, would have been seen as 'as trying to unduly influence the local government staff, in order to obtain a favourable outcome to the detriment of the health and safety of Sheffield's ratepayers'.

The next strategy was to try to play off the two parts of Sheffield City Council against each other. SMG tried to place the Land and Planning representatives in a situation where they feared for the future of the games and would put pressure on Building Control. This backfired as the basis of the SA's position was statutory, and its officers could be personally liable for failing to carry out their statutory duties. They dug in their heels even further.

Appealing to the Secretary of State was ruled out because time was too short. When the schedule was about to be seriously disrupted because of the lack of a decision on the cross-aisle, SMG gave in and instructed their architects to include it in the design.

In order to resolve problem 2 – the tiered seating – the independent expert strategy was tried again. This time the promoter was persuaded to commission a report from Hughes Associates Inc. of Maryland USA, which had also advised other British authorities. The report argued that there was nothing inherently dangerous in the HOK/LDHB design, and that in the absence of clear guidelines in the British codes, 'the lessons learnt in the design, approval and usage of the Sheffield Arena can be shared more widely as part of a broad reconsideration of the UK regulations, standards and outlines'. This position was accepted by the SA.

In order to solve problem 3 and to bridge the gap in the height of the safety rails, a mock-up was constructed so that everybody could see what the effects of the different heights were on the spectators' amenity. It was then clear that 1100 mm would obscure the view, and a compromise between 800 and 900 mm was arrived at.

The Taylor Report into the Hillsborough tragedy was published in January 1990. Its principal recommendation with regard to stadium design was that all spectators should be seated, conclusions which had no relevance for the Arena project.

During 1990, SMG withdrew from its position of client for the project, retaining the role simply of operator of the completed facility. This left Sheffield for Health with no alternative but to step in. This effectively transferred commercial risk for the Arena to the ratepayers of Sheffield, as SMG receive a guaranteed management fee which is offset against the profits from events held at the Arena.

In the end, the project was completed on time and the Arena was opened by the Queen on 30 May 1991. It won a number of local design awards for that year, and did in the end receive £4.75m of central government funding under the urban programme. Between August 1989 and the opening of the games, the overall costs of the sporting facilities rose from £111m to £147m, or by 32%. Sheffield Council Taxpayers presently pay over £20 per year each to cover the losses made on the Universiade programme as a whole.

In defining the project mission, the principal constraint on all the actors was quality of realisation in terms of the schedule – the deadline was immovable and so dominated decision-making. As the project manager said, 'You cannot say to 6000 eager runners, gymnasts, jumpers and throwers: could you all come back next month, we are a little behind programme'. After that, quality of specification tended to be favoured because very few of the actors were cost-sensitive – only the promoter and client were. Most of the other actors were appointed on a percentage fee basis so as costs mounted their fees mounted, or these issues were irrelevant as in the case of the SA. Thus the original proposals for the Arena facility in 1986 envisaged a building costing around £10m; the final outturn cost was £34m. This was partly because of the decision to build a multi-event arena rather than a sports hall, and partly because of inflation over the period, but it was largely as a result of a classic schedule/budget trade-off by the actors in the project coalition in project realisation.

Source: Developed from Lookman (1994). This material is also available as a role-play – see Winch (1999).

Notes

1 The definition is from Buchanan (1995, p. 12).
2 The concept was first introduced by Rittel and Webber (1973).
3 Barrett and Stanley (1999).
4 Cited in Sabbagh (2000, p. 18).
5 The special issue of *Building Research and Information* (**35,** 1, 2007) on visual practices explores these issues in some depth.
6 See Nutt (1993) and also Nutt and McLennan (2000).
7 See Lowe (2006) for an overview of the issues.
8 CABE (2006, p. 23).
9 OGC/CABE (2002).
10 Current Office of Government Commerce Guides.
11 See Winch (2008) for a more detailed evaluation of the current situation.
12 Marx (1976, p. 284).
13 See the papers collected in Cross (1984) – this volume is the best single source for research developments in our understanding of the design process.
14 Pugh (1990) and Ulrich and Eppinger (2008) provide excellent statements of the engineering approach.
15 Hillier (1996).
16 Winch and Schneider (1993); see Winch *et al.* (2002) and Winch (2008) for further empirical support for this four-fold model.
17 Pugh (1990).
18 The formulation comes from an unpublished paper by Bill Hillier. See Schön (1983) for case data to support this contention.
19 See Ulrich and Eppinger (2008) for a comprehensive review.
20 See Wheelwright and Clark (1992, Chapter 9).
21 Green (1992, p. 42).
22 See Boussabaine and Kirkham (2004) for an authontative overview.
23 See Blyth and Worthington (2001).
24 See various contributions to Brandon and Kocatürk (2008).
25 The DQI tool and a selection of application case studies can be found at http://www.dqi.org.uk/
26 Prasad (2004, p. 176); emphasis in the original.
27 Dickson (2004, p. 194).
28 Gann *et al.* (2003, p. 322).

29 This section draws extensively on Kibert (2005); see also Luck (2007) and Gifford *et al.* (2002) for very different analyses of some of the challenges of involving users in design, and http://www.charretteinstitute.org for more on charrettes in general.

30 The idea comes from Powell and Brandon (1984).

31 Cited in Sabbagh (2000, p. 114).

Further reading

Cuff, D. (1991) *Architecture: The Story of Practice*. Cambridge, MIT Press.
An ethnography of how clients and architects work together to solve the briefing and design problems, and its implications for architectural education.

Lawson, B. (1997) *How Designers Think: The Design Process Demystified* (3rd ed.). Oxford, Architectural Press.
The standard work on the design process, reviewing the research covering a number of fields, but with the emphasis on architecture.

Ulrich, K.T. and Eppinger, S.D. (2008) *Product Design and Development* (4th ed.). London, McGraw-Hill.
A normative, but authoritative, analysis of the design process from the point of view of engineers.

Chapter 10

Managing the Budget

10.1 Introduction

> 'Estimating is like witchcraft; it involves foretelling situations about which little is known'.

The budget is the most important single measure of project performance, and in the end, schedule and quality are reducible to budget on the expenditure side of the NPV calculation. The aim of this chapter is to explore the way in which the budget for the project is managed. The briefing and design processes discussed in Chapter 9 effectively set the budget; design decisions regarding conception and specification, as evaluated through the NPV calculation, determine the overall budget envelope. The task of budgetary management is to work within this envelope determining and redetermining the budget at an increasing level of detail through the project life cycle. The epigraph above from a former head of operational research at Costain[1] indicates how difficult this seemingly straightforward task is. However, by the end of the process, the exact price of the building is known or at least knowable on the condition that effective accounting systems are in place, but all that is then left for the client to do is to pay up.

This chapter will firstly review the differing levels of accuracy that are possible in the budget at differing points in the life cycle. Secondly, the difference between the product breakdown structure (PBS) used for early-phase budgetary planning and control, and the work breakdown structure (WBS) used during the planning and execution phases of the project, will be identified. A discussion of the principles of budgetary control using earned value analysis (EVA) will follow. The chapter will close with an investigation into the problem of escalating commitments, or the strong temptation to throw good money after bad on the part of clients, and how this is used by project managers to ensure continued funding for the projects they are running.

10.2 Levels of accuracy in project budgets

The reduction of uncertainty through time has an important effect on project budgets; it means that budget estimates made early in the project life cycle are relatively inaccurate compared to those made later as more information becomes available regarding both the precise work to be done and the current market level of prices. Best practice in estimating budget costs is to provide a three-point estimate between a maximum (pessimistic), minimum (optimistic) and most likely (mode or central estimate) outcome through group elicitation techniques[2]. Such estimation is a progressive process through the project life cycle as more information is acquired about the challenges of delivering the project mission, and uncertainty is progressively reduced through time as shown in Fig. 10.1. Gateway reviews discussed in section 8.4 provide good opportunities for reviewing the successive accuracy of estimates by agreeing the formal reduction of the variance around the mode of the three-point estimate, and so a step-wise effect is typically generated in the convergence of the maximum and minimum estimates on the central estimate.

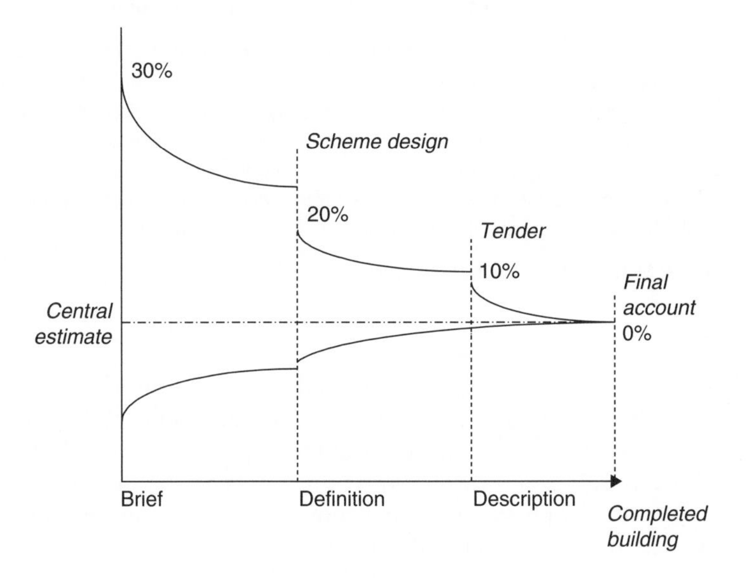

Fig. 10.1 The changing accuracy of estimates.

Table 10.1 summarises this successive reduction of budget variance and compares conventional budgeting rules of thumb to a data set derived from a study of estimating accuracy of pioneer process plant construction – see panel 10.1 – while the 10% figure is in line with data on contractor estimating accuracy[3]. In formal terms, the central estimate is the mean of a distribution with a 95% (or 99%) chance of the true cost lying within the upper and lower accuracy figures given. In practice all the evidence indicates that the distributions are skewed – underestimates of budget costs leading to optimistic outcomes are much more common than overestimates and so the optimistic outcome lies much closer to the mode than the pessimistic one.

Table 10.1 Budget-estimating accuracy at different project phases (source: developed from Merrow *et al.*, 1981, Table 4.1).

Project phase end (see section 8.3 for phase definitions)	Conventional wisdom on accuracy level ± (%)	Estimated cost as a proportion of actual cost at completion for 44 projects (%)
Brief	30	62
Definition	20	78
Description	10	83
Trade package	5	93
Completion (by definition)	0	100

Panel 10.1 The pioneer process plant study

In the late 1970s, the RAND Corporation studied the performance of pioneer process plants. Pioneer plants are those which have a significant level of technological innovation – they are the first of a type. So, mission uncertainty is relatively high for the sample of 44 projects benchmarked, compared to the population of all process plant projects. Data were obtained from 34 firms on the client and supply side. Complemented by the later desk study of megaprojects which included large civil engineering projects, this research is almost unique in the quality of insight it gives to project performance.

Sources: Merrow *et al.* (1981); Merrow (1988).

The final column in Table 10.1 shows the extent of underestimate of outturn costs at each phase; estimating errors are systematic, rather than random, and the outturn cost will nearly always lie above the budget estimate. Indeed, for the pioneer process plants, outturn cost lay entirely outside the expected range of estimated cost at each phase until completion. Some of the reasons for this lie with the strategic misrepresentation discussed in section 3.7; this chapter will assume that estimators are acting in good faith.

The very high level of inaccuracy in early-phase budget estimates is, of course, a major reason for the separation of the contract for design from that for execution on site. Typically, this breakpoint occurs when costs are down to around the 10% range of accuracy after completing the description of the facility. This breakpoint has two functions – it allows the switch towards a contract that shifts more of the remaining risk towards the supplier, and it provides a natural review and audit point for whether the project should continue. If either the capital costs prove to be at the higher end of the estimating range, or the income streams from the exploitation of the facility are now estimated to be smaller or less sure than originally thought, then the NPV may come into question, and the most appropriate thing to do may be to cancel the project before any more money is spent. From the options thinking perspective introduced in section 8.3 design development can be thought of as a stage option on the project.

Estimating accuracy may be improved significantly through better data; this reflects the importance of databases which provide tender prices for a variety of building types, such as the Building Cost Information Service (BCIS). Many clients also possess their own databases. The closer the match between the facility proposed and the sample of facilities in the database, the greater is the attainable accuracy of estimates. There are, however, limitations with such approaches including:

- sampling problems because of the data set not containing enough observations to provide a reliable estimate of the distribution of actual prices;
- the site specificity of each project introducing random variables into the data set;
- the fact that it is *prices* that are observed, not costs, so that the drivers of price changes are difficult to identify within the data set.

Budget estimates may be made on a top-down basis, using the experience and judgement of the estimator supported by price databases, or on a bottom-up basis using quotations for every resource required for the project. The latter approach is clearly futile during the early stages of the project, and the top-down approach is preferred; the theory of probability provides mathematical justification for this point, as discussed in panel 10.2. By the time uncertainty has been reduced enough for trade contracts to be let, at around 95% accuracy, a bottom-up process becomes viable. At this point, material prices can be known with great accuracy and most of the remaining uncertainties relate to labour productivity. For this reason, trade contractors prefer bottom-up estimating in their tendering.

Panel 10.2 Probability theory and estimating accuracy

The probability of an event occurring in repeated trials – where the event may be, for instance, a task running 10% over budget or schedule – is given by the binomial distribution. This shows that the percentage error of the estimate of the mean performance on the task as a ratio of the percentage error for the mean performance on all tasks in the WBS is inversely proportional to the square root of the number of tasks, for a discrete sequence of tasks. Thus for a 52-week project to be estimated at ±10% accuracy in budget, a task lasting a fortnight should only be planned with a ±50% accuracy because some tasks will exceed their budgets and others come in under budget. This calculation therefore defines the lowest planning unit for the WBS. The larger the number of tasks in the WBS, the less point there is in planning each task accurately, because positive and negative variances will even themselves out at the level of the project as a whole. This argument, of course, does not apply where there are systematic biases in estimating.

Sources: Turner (1999) and material supplied by John Kelsey.

10.3 Developing a budgetary system

As with most things, the devil is in the detail with budgetary estimating and control. A structure for the database is needed so that the information can be organised effectively. There are two basic options:

- the PBS, which is the proposed facility (product) broken down progressively into all its constituent *components* and
- the WBS in which the construction process is broken down progressively into all its constituent *tasks.*

Despite the contention of many authorities, the PBS and WBS are *not* the same, and the differences between them cause a number of problems which make the PBS to WBS transformation problem a difficult one to resolve satisfactorily. The PBS only contains those components that are incorporated into the final product, and does not include any components that are associated with preparatory or temporary tasks. For instance, the specification for a beam will define the product components – concrete and reinforcement – but will not include erection and demolition of formwork, and any falsework required to support that formwork. It is for this reason that the PBS is only the starting point of the WBS. Figures 10.2 and 10.3 show respectively, the PBS and WBS for the roof of the Centuria building, described in more detail in Cases 10 and 11[4]. These figures were prepared using Uniclass[5] – in particular, tables D (facilities) and G (elements for buildings), and tables L (construction products) and J (work sections for buildings) respectively. Comparison of the two figures illustrates how the PBS is *product* orientated and the WBS is *process* orientated. A further complication is that two separate tasks in the WBS may be performed by the same trade. For instance, both the roofing works and insulation works in Fig. 10.3 would normally be performed by the same gang. Within the WBS, tasks need to be unambiguously assigned to trades so as to enable full cost control where they form different subcontract packages. In effect, the WBS is a function of the PBS and the structure of the project coalition as analysed in section 5.4, sometimes referred to as the organisational breakdown structure (OBS) described in section 15.8. The differences between PBS and WBS also apply to design processes; tasks such as mock-ups and testing models in wind tunnels, described in panel 9.7, and research to solve particular problems, are all off the PBS yet are vital for the progress of the project.

In the early phases of the project, it is natural to use the PBS for development of the budget – after all, it is the product that is of interest to the client, not the process. However, as the PBS is developed, it starts to become clear what the task content of each component is, and a switch needs to be made to the WBS for structuring the budget data. This switch is usually made at the point where the contractor takes over from the designers. This is relatively unambiguous and unproblematic where tenders are on complete design, with the bill of quantities performing the translation role, but even here the bill of quantities has been

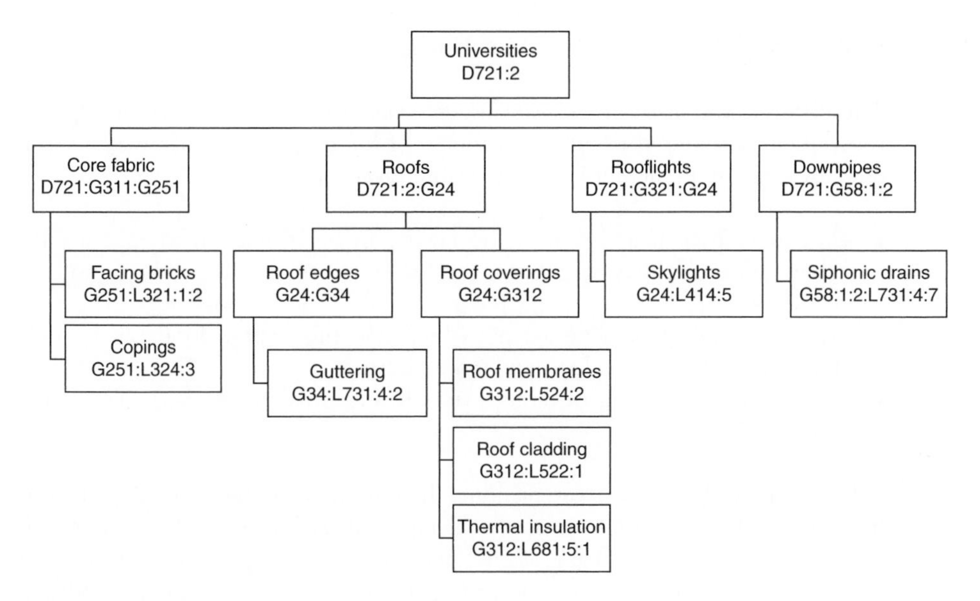

Fig. 10.2 Centuria product breakdown structure.

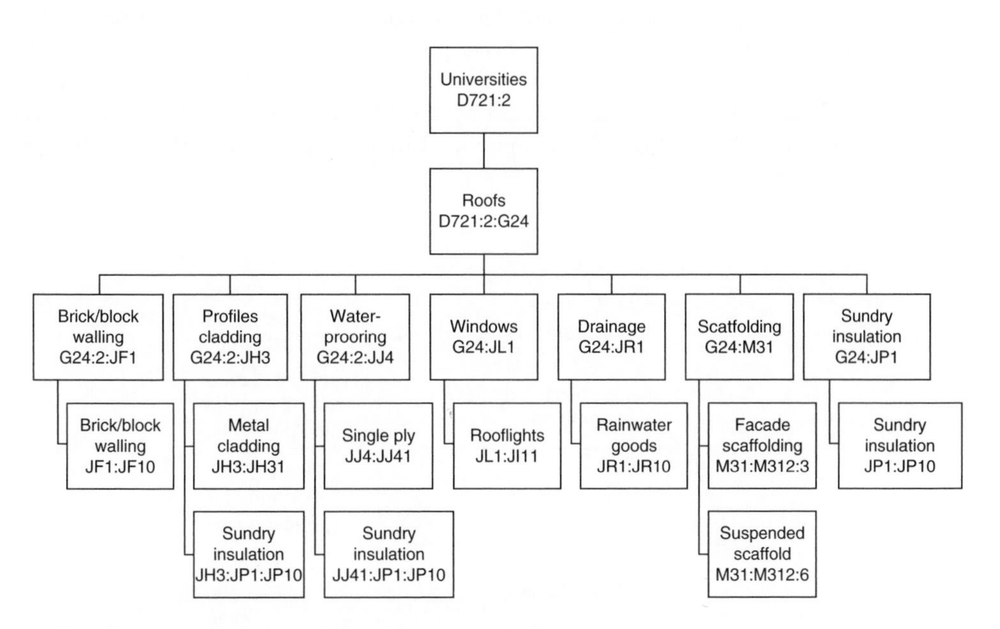

Fig. 10.3 Centuria work breakdown structure.

criticised for its lack of relationship to processes on site. However, with attempts to integrate across the design/construction interface, the switch can cause problems – for instance in the development of 4D planning, described in section 11.5.5.

10.4 Using the PBS to control the budget

The PBS is at the heart of *cost planning and management*. The cost planning process is essentially one of setting up a cost model of the proposed facility, which can be used for three different, but very important, purposes:

- allowing the estimated cost of the facility to be benchmarked against either other similar facilities or industry-level data sets such as the indices produced by the BCIS;
- creating a budget baseline against which the information loop can be made during later design development and execution;
- enabling the definition of target prices in incentive contracts, as discussed in section 6.3.

Figure 10.4 shows the cost planning hierarchy, with the familiar graduation from simple, broad-brush estimates to detailed breakdowns. In the earliest phase of the project, the benchmark of cost per unit is used. Frequently, the rates are the benchmarks within which the project budget must stay – an approach particularly favoured by the public sector with its cost yardsticks. As the design is developed, more precise figures on a metre square or other appropriate measure can be developed. As more design detail becomes available, the information can be broken down into the elemental cost plan by broad PBS headings. In the UK, a standard form of cost analysis is available to enable comparison with industry cost databases known as the Standard Method of Measurement. The lowest level of detail available in PBS terms is provided by the bill of quantities, which is, in essence, a measure PBS.

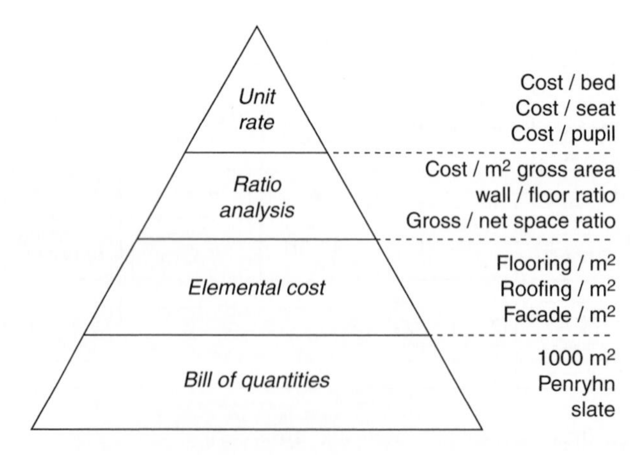

Fig. 10.4 The cost planning hierarchy.

The cost planning process is an essentially reactive one. Proposals from designers are benchmarked against available data sets through an iterative process[6]. However, the lack of active budgetary management inherent in such an approach has encouraged the development of new approaches. One derived from the car industry turns the problem around and drives the design process by the estimation of costs – a technique known as *target costing*. The starting point of the target costing process is what the market will bear – in other words, setting a price point and then on no account exceeding it. At one level, the construction industry does have target costs in the public sector cost yardsticks, but these tend to include adjustment factors for various differences between sites, and tend not to be reduced over time so as to drive out costs from the product. The true target costing approach sets non-adjustable targets and then uses learning curve principles to reduce cost over time. The challenge in target costing is to provide the maximum value for the customer at the given price point. The main techniques for doing this are value engineering (VE) and cost management.

10.5 Value engineering and cost management

As architects and engineers grapple with the design problem, it may become apparent that the early estimates on which the NPV was based were optimistic. The aim of VE process is to identify 'unnecessary cost' and thereby bring the facility back on estimated budget, thus ensuring the viability of the project. Thus VE is essentially an audit approach, providing third-party review of the design to date[7]. The classic methodology is the 40hour VE workshop, where a team of specialist value engineers and independent designers review the design at around the completion of sketch design. The format is one of brainstorming followed by evaluation of the ideas generated over the 5-day period. The arguments for such an approach include the following:

- A fresh perspective can sometimes identify the wood for the trees – designers who have been struggling with a problem for a long time may have missed something by being too close to the problem.
- The client can have additional confidence that it is getting value for money.

Against these advantages, the disadvantages are as follows.

- The process can be very demotivating for the design team as the solutions over which they have worked so hard are picked over by those who may not have fully grasped the challenges of the brief.
- The process takes 3–4 weeks out of the design schedule.
- It is an additional cost to be borne by the client.

It is perhaps notable that the techniques have been most enthusiastically adopted by clients which tend to require buildings strong on specification rather than conception. Where the brief is more challenging, effective VE might be more difficult. However, the greatest limitation of a VE approach is the lack of reliable cost information. If cost estimates at definition (scheme design), as shown in Table 10.1, are accurate to ±20%, then a claimed saving of 20% has a 27% chance of being entirely spurious[8], especially as the smaller the element under consideration, the higher the potential variability in the estimate compared to outturn. Thus VE can only ever be a broad-brush approach, and is unlikely to warrant the investment in the 40hour workshop for most projects. A more viable approach would be to provide incentives for the design team commissioned to do the work in the first place to think through the problem with a broader value management perspective. The difficulties inherent in developing VE further towards a target costing approach are well illustrated by the Building Down Barriers project in panel 10.3.

Panel 10.3 Target costing and value engineering on Building Down Barriers

One of the innovative aspects of the Building Down Barriers project presented in Case 7 was the attempt to use target costing techniques derived from manufacturing. Instead of the cost-plus approach traditional in construction, the aim was to design the facility to a cost set by the price the market was prepared to pay for the value generated by the asset less profit and overheads. This approach relies on intensive value engineering activity, which in turn relies on the ability to cost alternative design proposals accurately. However, it was quickly discovered that the prime contractors were not able to do this. The problem is that they subcontract for all resource inputs, and therefore do not possess adequate information on costs. They rely instead on prices from their suppliers, who in turn typically subcontract for many of their inputs, and so on. The lack of basic knowledge on the construction supply side regarding costs and their drivers means that a target costing approach is not viable in the current state of the industry, and, more generally, attempts at value engineering are vitiated.

Source: Nicolini *et al.* (2000).

A broader critique might be that the necessity to mount a full VE exercise is, arguably, the result of a failure of incentives within the procurement process because of the use of fee-based rather than incentive contracts; see section 6.5 for a discussion. While budgets can mount on any project as a result of the inherent uncertainties of the process, and need to be brought back under control when more information becomes available, the necessity to use such techniques on every project suggests that designers are being inappropriately motivated. If the client requires an absolute cap on expenditure, then the procurement of designers should provide incentives for them to meet this cap[9].

10.6 Constructability

Constructability is the process of designing to enable the most efficient deployment of resources on site with the aim of reducing waste, improving conformance quality, increasing productivity, removing safety hazards or reducing schedule. Achieving constructability requires that those whose expertise lies in realising the project be involved in the design of the facility. This poses considerable problems for procurement using fixed-price contracts, and is one of the main arguments in favour of switching to incentive contracts as discussed in section 6.5.3.

Under incentive contracts, the constructor has the motivation to propose more effective or efficient ways of realising the project. As uncertainties have reduced by this stage to around ±5%, it becomes possible to make some fairly precise calculations regarding where benefits may be found.

There are two main aspects to constructability. One is a more finely tuned version of VE. In Anglo-Saxon systems this expertise is often provided by a construction manager, rather than trade contractors. Where incentives are appropriately aligned, this sort of constructability approach can become the norm rather than a specific exercise. In the French system, construction contractors know that they will need to develop *variantes* in order to win the project, while contractual conditions often make this approach difficult in the UK[10]. Italian steelwork fabricators routinely win major trade contracts throughout Europe, including T5, by re-engineering the design to add value for the client by obtaining for themselves larger trade packages, while reducing significantly the overall price of the project – see panel 10.4.

Panel 10.4 Competitive advantage and value engineering

During the 1990s, Italian steelwork fabricators were winning work throughout Europe – such as on the Severn Bridge, described in panel 2.9, and Friedrichstadtpassage 207, described in panel 10.7. Some suggested that this was because of their access to cheap steel, but in fact it was because of their ability to re-engineer the project. On two projects – the Storabælt East Bridge and the Grand Canal Maritime Bridge at Le Havre – Italian firms won the contracts through providing the client with savings on other trade packages. In Denmark, CMF proposed upgrading the specification of the steel deck to the approach viaducts to the cable-stayed bridge, which allowed spans between piers to be increased from 168 m to 193 m, thereby reducing their number and hence cost. In France, Cimolai proposed v-steel supports founded on the banks, as opposed to the concrete piers founded on the river bed proposed by the design team, which would have involved the construction of caissons. Again, construction costs were saved on other trade packages through finding elegant engineering solutions.

Source: Micelli (2000).

The second aspect of constructability is the development of process capability, as defined in panel 12.3[11]. Here the issue is very much one of trade-offs, as illustrated in Fig. 10.5. Higher process capability requires investment, and there comes a point when the returns on this investment – either from higher incomes from asset exploitation or from savings in whole-life costs – start to diminish. The greatest return on investment is achieved at the point where the distance between the two curves is greatest. The appropriate level of capability for a process will depend on a number of factors.

- *The project mission in terms of appropriate intention*, defined in Fig. 8.3. Clients may be prepared to pay for higher levels of workmanship (i.e. greater process capability) in sensitive areas for aesthetic or operational reasons.

- *The robustness of the technology.* For instance, enclosure technologies such as traditional brickwork are relatively robust in that they remain perfectly functional at relatively low levels of process capability, and typically have to be *fitted* by craft-trained operatives. Others, such as cladding systems, are less robust and need to be *assembled* to tight tolerances by achieving higher levels of process capability.

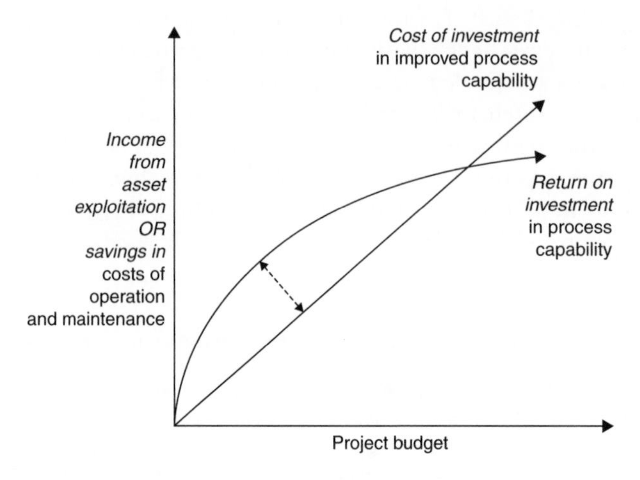

Fig. 10.5 The process capability trade-off.

There are various ways of enhancing process capability in construction:

- investing in the training of operatives and site management;
- investing in better quality management systems, as discussed in section 12.2;
- investing in standardisation and pre-assembly, as discussed in section 12.10;
- investing in better machinery, capable of working to higher tolerances;
- working more slowly, to reduce the effects of haste on conformance.

However, all these imply additional costs, and hence reduce the efficiency of the process. There are diminishing returns to the greater investment in process capabilities. In other words, it is also a form of waste to achieve a higher level of process capability than the project mission requires. Workmanship that would be acceptable for the rear of a back office would not be acceptable on the facade of the corporate head office.

10.7 Controlling the budget

The WBS, at its full level of detail, assigns an identifier code to each task – these can be derived from Uniclass or can be proprietary. The WBS tends to be developed iteratively as the project progresses in the rolling wave illustrated in Fig. IV.2.

The first level of the WBS is the trade package level, unless it is worthwhile splitting the project into sub-areas of the site. The next level might be whole trade tasks such as formwork. Below this level, the WBS might be broken down into particular elements, such as work on a particular beam. The finest level of detail – which is often left to the subcontractor – is the weekly one. This level is required because operatives tend to be paid on a weekly cycle, and this is the planning cycle recommended by advocates of the last planner technique described in section 11.5.3.

Each element in the WBS will be broken into the three classic elements of all accounting systems – the costs of the labour, materials and plant required for task execution – in what is known as the cost breakdown structure (CBS). Under activity-based costing schemes, project overheads – such as site installations – would also be allocated to the task. However, the relatively low proportion of project overheads in construction costs means that this is rarely worthwhile[12]. The matrix of the CBS and the WBS generates the database categories for the project at their highest level of detail. Combined with the OBS, this matrix provides the *cost control cube* in which every cost element can be unambiguously allocated to a cell determined by the task, the type of cost and the member of the project coalition responsible.

Costs are expenditure at a point in time, but what point in time? The costs of inputs purchased from other suppliers are in fact prices – they include the supplier's mark-up. They are also booked as costs in three different ways:

- as commitments, when the purchase order is placed or the work is carried out;
- as expenditures, when the invoice for the delivery of the good or service is received;
- as cash flows out, when the invoice against the purchase order is actually paid.

It is commitments that project managers need to manage; once a commitment is made, that amount is no longer available within the budget for other costs. However, expenditures can be lagged by months against commitments, especially for long-lead items, because most corporate accounting systems record expenditures not commitments. Project managers typically have to maintain their own accounts, rather than relying on their accountants. A further problem is that payments against invoices may be lagged because of corporate cash flow management policies, and this can result in non-delivery of ordered goods and services if payment terms to suppliers are not being observed.

The budget is broken down into the elements indicated in Fig. 10.6. The project budget is the budget after profits, as amended following agreed (i.e. invoicable) variations. Within that budget, there will be contingency sums allowed for the unexpected, and these are removed to establish the project budget baseline against which performance is measured. This performance measurement baseline (PMB) is then broken down against the WBS into cost accounts (CAs) which form the active units of managerial responsibility for the budget. Where work is subcontracted, a CA will typically coincide with the contract for the subcontract for a group of tasks to be executed. Within the CAs, some will be at the stage of

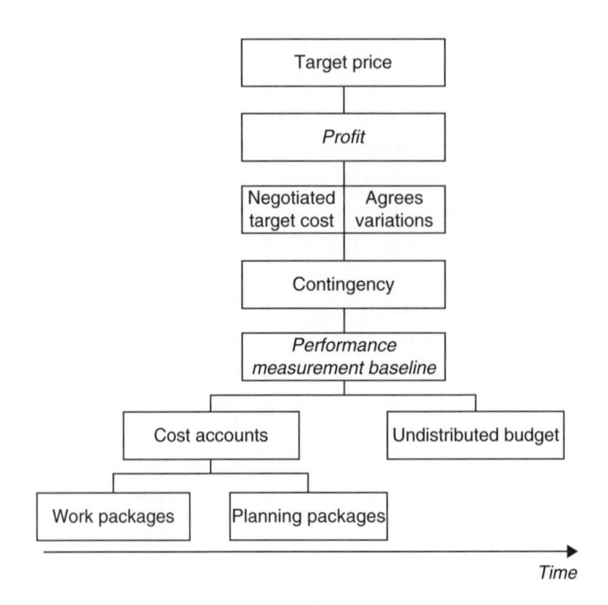

Fig. 10.6 The performance measurement baseline (source: adapted from Fleming and Koppelman, 2000, Fig. 8.9).

Table 10.2 A cost account plan (CAP) for architectural services (source: RIBA).

Project milestone (end of stage)	Percentage of total fee
ABC	15
D	20
EFG	40
H	25

active work packages, while others will be in planning/negotiation stages. Work not yet formulated into CAs is undistributed budget against the PMB.

Expenditure within the CA is determined by the cost account plan (CAP). Where the CA is based on a lump-sum subcontract with no interim payments, the CAP simply defines payment upon completion. More frequently, the CAP will designate a schedule of payments against milestones, or allow for monthly valuations of the work completed. Table 10.2 shows a CAP for a package of architectural work, using the milestones from the RIBA Plan of Work. Within a CAP, the manager responsible for the CA may well need to take the WBS to a finer level of detail, depending on the nature of the task. For the reasons described in panel 10.2, there is little point in planning tasks with a high degree of uncertainty, such as design, on a daily basis, but a fast-track fit-out of a shop may well require planning on an hourly basis.

The level of detail of the WBS, and hence the cell sizes within the cost control cube, is a matter of some debate. Too many cells will lead to information overload, yet, as will be discussed in the next chapter, projects need to be managed on a weekly cycle so as to ensure effective task execution. This implies that the lowest level of the WBS should have a typical duration of a week. However, this is too small for most project reporting purposes; a monthly reporting cycle is more typical. This means that the information systems used must have the capability to be aggregated easily to the most appropriate level of detail, depending on who is to receive the reports.

10.8 Earned value analysis

Simply comparing actual expenditure against the budget to identify variances is of limited use for proactively managing the project. The critical variance is between what has been spent at a point in time, and what is planned to have been spent by that time as defined by the PMB. This variance is measured by EVA because proactive control of the budget implies that it must be related to the schedule. This can only be done if the schedule and budget have been prepared using a common WBS. EVA combines three main measures, which are defined formally in panel 10.5.

- *Earned value*[13] (EV) which is the budgeted cost of the work completed at a defined point in the schedule. Its name derives from the fact that it is the cost basis for the amount (also including overheads and profit) which can be invoiced to the client against the contractual master schedule.
- *Actual cost* (AC) which is actual expenditure to date to be compared with earned value to identify budget variance (AV) at a defined point, which can be expressed in the cost performance index (CPI).
- *Planned value* (PV) which is what ought to have been spent by the defined date according to the budget. This can be compared to earned value to identify the schedule variance (SV) at a defined point, expressed in the schedule performance index (SPI). The sum of the PV for all tasks is the project PMB.

Panel 10.5 Earned value analysis

$$CV = EV - AC \qquad SV = EV - PV$$

$$CPI = \frac{EV}{AC} \qquad SPI = \frac{EV}{PV}$$

$$BAC = AC + \frac{(PV - EV)}{\Sigma CPI}$$

Purchase costs on a commitments basis should go against the WBS upon delivery of materials, and against subcontract values should any include retentions held back, or the AC will be systematically underestimated. The relationship between

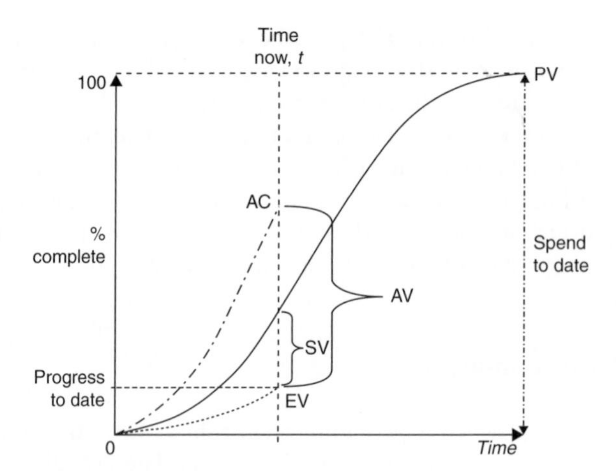

Fig. 10.7 Earned value analysis (actual cost, AC, should be read against the right-hand axis).

these three metrics at point *t* in the schedule is given in Fig. 10.7, which shows a project that is both behind schedule (SV) and over budget (AV).

These measures can also be related through ratios to give the CPI and the SPI, where a figure of 100% indicates performance to plan, and a figure below that indicates a negative variance. CPI is also termed – erroneously – 'productivity' by many authorities, because the numerator and denominator are often taken from time sheets and measured in labour hours. This is erroneous because it measures planned inputs against actual inputs, while true productivity is a ratio of inputs to outputs. Two different tasks could both have 100% CPI (i.e. EV = AC) yet have very different productivity rates because the plan may call for each worker to add twice as much value on one task compared to another. It should also be noted that SV is a monetary value, not duration. It uses the budget variance as a proxy for schedule slippage measured in time units, and it needs to be verified against the schedule tracking Gantt chart – see Case 11. By projecting these analyses forward, it is possible to predict their impact on the overall predicted budget variance at completion (BAC). The variance between the PMB and BAC can be estimated most simply by taking the work remaining (PV at completion less EV), and dividing by the *cumulative* CPI (ΣCPI) on the assumption that existing levels of performance will continue, thereby producing a steadily increasing variance over time against the plan. The evidence is that ΣCPI does not change by more than 10% once the project is 20% complete; indeed it only gets worse as attempts are made to bring the project back on budget and schedule. This calculation gives an optimistic view of overrun; dividing by the multiple of ΣCPI and ΣSPI gives a more pessimistic outcome prediction.

The point of this deterministic extrapolation of past performance to the future is to provide an information loop assessing what needs to be done to bring the project back on to the original plan. If this is not possible, it identifies how the plan now needs to be changed, and what additional resources would be required for this change to have effect. Where the problems with schedule and budget can be isolated to particular completed packages, it is not necessary to use CPI and SPI – the variance can simply be added to PV. However, once a project is running late, all sort of knock-on effects make it very difficult to regain the original schedule and budget, which is why cumulative CPI and SPI are preferred in estimating final outcomes.

10.9 Mitigating optimism bias

The issue of optimism bias was introduced in section 3.7, and the UK's HM Treasury accepts that 'there is a demonstrated systematic tendency for project appraisers to be overly optimistic. To redress this tendency appraisers should make explicit, empirically based adjustments to the estimates of a project's costs, benefits and duration'[14]. The recommended way to address this problem is to apply standard uplifts to the estimates provided in the development of the business case shown in Table 10.3. These uplifts are to be applied to the present value of the costs of the project by multiplication using the upper bound, effectively adjusting upwards the distribution of the three-point estimate. It is recommended that further business case development should then be used to reduce the optimism bias factor from the upper bound to the lower bound, at which point the project can be given authorisation to go ahead. The uplifts are based on the analysis prepared by Mott MacDonald of data[15] on completed projects.

Table 10.3 Recommended adjustment ranges for optimism bias (source: HM Treasury, 2004a, Table 1).

Project type	Optimism bias %			
	Works duration		Capital expenditure	
	Upper	Lower	Upper	Lower
Standard buildings	4	1	24	2
Non-standard building	39	2	51	4
Standard civil engineering	20	1	44	3
Non-standard civil engineering	25	3	66	6

This methodology, which is laid out in more detail in panel 10.6, raises a number of important issues, for it is not at all clear that it is getting to the root of the problem of optimism bias.

- In effect, the recommended adjustments are large contingencies and will lead to the establishment of relatively generous overall budgets for projects. The availability of contingencies in the budget can itself cause a moral hazard problem in the relationship between project actors and project funders. If the funds are there, they are more likely to be spent than if they are not.

- Many of the sources of optimism bias identified by HM Treasury would also feature in any risk register, but there appears to be a complete disconnection between recommended processes for the adjustment of optimism bias and those for risk management. For instance, the Orange Book on risk management procedures[16], published after the supplement to the Green Book, does not mention optimism bias.
- It ignores the organisational institutional context in which optimism bias becomes strategic misrepresentation through encouragement to 'play the game' to make investment appraisals stack up.

Panel 10.6 Optimism Bias Mitigation Procedure

Step One – Decide which project type(s) to use

Projects need to be categorised so that the correct adjustment factors can be applied. A project is considered 'non-standard' if it (a) is innovative; (b) has mostly unique characteristics or (c) involves a high degree of complexity and/or difficulty.

Step Two – Always start with the upper bound

Use the appropriate upper bound value in Table 10.3 as the starting value for calculating the optimism bias level.

Step Three – Consider whether the optimism bias factor can be reduced

The upper bound can be reduced to the extent to which the contributory factors have been mitigated by applying a mitigation factor with a value between 0.0 and 1.0 where 0.0 means no mitigation and 1.0 means complete mitigation.

Step Four – Apply the optimism bias factor

The present value of the capital costs should be multiplied by the optimism bias factor resulting from step 3 and then be added to the total net present value of the project cost to provide the adjusted Base Case.

Step Five – Review the optimism bias adjustment

Clear and tangible evidence of the mitigation of contributory factors should be observed and independently verified before reductions in optimism bias are made; procedures for this include the Gateway Review process.

Source: HM Treasury (2004a).

Another aspect that needs to be taken into account here is that there is also some evidence that optimistic decision-makers are better motivated to perform than pessimistic ones[17] – optimistic targets also act as stretch targets. A little optimism bias could be a useful element in the effective management of projects. Flyvbjerg does not rely upon formalistic uplifts on estimates for minimising optimism bias. He emphasises the importance of 'the outside view' in the preparation of estimates, and also cites what Frank Gehry calls 'the organisation of the artist' as a way of ensuring that budget targets are met[18]. However, as Gehry himself points out, he first ensures that the client is prepared to pay for the type of building he wants to design and he can make 'that budget become real'[19].

10.10 Budget overruns and escalating commitments

Why do budgets overrun? Is it simply the result of bad estimation, or are there other dynamics at work? Overrun is here defined as excluding any change in the project mission. Any such changes – and they are not at all unusual – will (or should) be the result of a new NPV calculation, and hence be covered by newly identified or augmented income streams. They are, therefore, by definition, unproblematic for the client. We also exclude from this discussion projects on which the NPV has been subjected to strategic misrepresentation as discussed in section 3.7 – for obvious reasons such projects are doomed to budget and schedule overrun.

Our focus here is on cost overruns for an unchanged project mission. The analysis will deploy data from the RAND study of 47 megaprojects (>$500m in 1984), including a number of civil engineering projects, described in panel 10.1. The use of this data set can be defended on the grounds that such megaprojects are the most likely to flush out the issues, and that no similar analysis is available for other more typical projects. Firstly, RAND found that errors in project execution (defined to include detail engineering design) were not a major problem: 'to our knowledge, no systematic analysis of capital projects has ever been made which concluded that blunders by project managers in executing projects were even an *important* source of cost growth or schedule slippage, much less a dominant one'[20]. Having excluded planning and execution problems, RAND identify budget-estimating errors as drivers of both cost growth and schedule slippage. The sources of these drivers can be grouped into two main categories:

- unexpected regulatory constraints and
- innovation in new product components or construction processes.

The first of these is the most important and indicates the importance of the management of the stakeholders, discussed in Chapter 4. The principal problem with regulatory systems is the extent to which they are negotiable. Where they are unambiguous in principle and rigorous in application, there is no excuse for regulatory problems delaying the project. However, even experienced project managers can be let down by the failure to understand how different foreign regulatory systems can be from their own, as the experience of the French company CBC in Berlin indicates – see panel 10.7. Where the regulatory system is more open to negotiation – such as urban planning in the UK, or structural integrity in France – there is more room for surprise. In such situations, extensive negotiations with the regulatory authorities can form an important element in the definition of the project mission. Alternatively, clients take a risk with obtaining regulatory approval and commit to the project in order to reduce the schedule duration, sometimes getting it wrong; this is frequently done, for instance, in retail development where opening in time for the Christmas season is vital – see panel 10.8.

One of the more common ways for projects to fall foul of regulatory authorities is to innovate. The fact that there were no regulations in place to cover a multi-use

Panel 10.7 CBC in Berlin

The fall of the wall dividing Berlin in 1989 created enormous new construction opportunities, which attracted many foreign – particularly French – companies. CBC won the contract to construct the new Galeries Lafeyette store at Friedrichstadtpassage 207, designed by Jean Nouvel with innovative structural elements. CBC organised the work in the French way – in particular, it assumed that it would be responsible for structural design, because, as shown in Fig. 2.2, the French regulatory system is performance-related. Unfortunately for CBC, Fig. 2.2 also shows that the German system is very prescriptive. The regulations are enforced by the representative of the local authorities, the Prüfingenieur, who has to sign off the structural calculations, the structural drawings and the working drawings showing the layout of the reinforcement and formwork. CBC would normally use their own *bureau d'études* for structural design, but it was unfamiliar with the German codes. The first German consulting engineers hired did not produce work acceptable to CBC, and so it turned to a second firm and the work again had to be checked by the Prüfingenieur. Design changes led to further cycles of approval; CBC changed the supplier of the HVAC system, and the client initiated further changes to the layout, both of which meant that structural elements had to be redesigned. As delays accumulated, stopping work on site for several weeks, CBC went ahead with construction without formal approval. As a result, it was forced to tear down completed columns, thereby losing even more time and incurring considerable additional costs. The accumulated delays and additional costs on the project resulted in CBC losing its independence and being absorbed into its parent company – now Vinci.

Source: Syben (1996).

arena greatly exacerbated the design management problems on the Sheffield Arena, described in Case 9. On other occasions, clients are simply unlucky, as the case of NATM on the Jubilee Line extension, described in panel 10.9, shows. As well as generating regulatory risks, innovation also increases the level of mission uncertainty in the project and is the second most important source of cost growth and schedule slippage. 'Doing something in a different way reduces the amount of information available to the cost and schedule estimator. The effect of innovation on cost and schedule estimates is thus similar to that of poor project definition.'[21] Innovation may yield important benefits, but the additional investments required to attain those benefits are frequently underestimated.

As information regarding additional costs comes in, clients are faced with a dilemma. In formal terms, they need to recalculate the NPV for the project – can the additional capital required be covered by the expected revenues or can additional revenues be identified now that the mission is more clearly defined and the facility is nearer delivery? However, all the evidence is that clients rarely take this disciplined approach after the very early stages of the project, and very rarely once the project has started on site[22]. This is the problem of *sunk costs* – or the tendency to throw good money after bad. Once significant capital has been expended, it becomes extremely difficult not to authorise additional expenditure in response

Panel 10.8 Marks & Spencer and their landlords

Marks and Spencer frequently takes a lease on retail space within a large commercial complex, and so has to negotiate with the owners of the complex – typically a property company – for permission to make alterations to its stores. In such cases, the performance of M&S retail units is dependent upon the overall attractiveness of the complex as a whole. This poses a number of problems for M&S Store Development Group, with a tendency to disrupt the smooth management of the project:

- During the redevelopment of the Milton Keynes outlet, the complex was sold to another property company. This meant that the project had to be put on hold while the alterations were approved by the new owner. As construction contracts had already been let, these contracts had to be renegotiated to allow for the later start date and the storage of equipment already ordered for installation. Significant budget and schedule overruns resulted.
- In Versailles, M&S did a deal with the complex owner to expand the floor area of their store. These plans required two small retail outlets to be moved. The owners proved unable to keep their side of the bargain because these outlets refused to move. In the end the project was abandoned and the construction contract determined.
- In Marseilles, the footfall in the Grand Littoral commercial complex was much lower than expected. This meant that M&S sales predications were now too high. Senior store operations management therefore insisted that the floor area be cut in half, although this would not save any money.

M&S sometimes goes ahead before all regulatory approvals are obtained because of the absolute necessity to meet store opening deadlines, particularly linked to the Christmas season. The risks of budget overruns are less than the risks of missing the seasonal retail peak periods.

Source: Carr and Winch (1999).

Panel 10.9 NATM on the Jubilee Line Extension

The collapse of the tunnel on the Heathrow Express project in October 1994 had a major impact not only on that project – see panel 16.5 – but also on the works at London Bridge station for the Jubilee Line extension project to the London Underground. The Costain-Taylor Woodrow joint venture proposed using the New Austrian Tunnelling Method (NATM) for the excavation of the station tunnels for the new Jubilee Line, and also for the refurbishment of the existing Northern Line tunnels and platforms. Cautious, the client requested a pilot project to validate the construction method, which was proven without incident. Shortly after the implementation of NATM on the main tunnelling works, work was stopped while the implications of the Heathrow collapse were assessed through a major design review and approval was sought from the responsible regulatory body – the Health and Safety Executive. For the Northern Line tunnels – which were more critical in schedule terms – the project team largely reverted to traditional methods, while NATM was retained for the Jubilee Line because of the benefits it still offered. However, a more expensive and time-consuming three-phase enlargement approach was used instead of the side-drift method which had caused the problems at Heathrow and was in use at London Bridge.

Source: Field *et al.* (2000).

to surprises, and the problem grows, or escalates, as the project nears completion[23]. The dynamic of escalation occurs for a number of behavioural reasons:

- cancelling now is an admission of failure, while continuing gives the possibility of mitigating the problem;
- investment decisions involve credible commitments to competitors in the case of the private sector, and voters in the case of the public sector;
- there is always the hope that additional revenue streams might turn up;
- a half-completed facility would generate no revenues, while a complete one will generate some revenues.

Because of the fundamental transformation – defined in section 6.3 – project managers are able to behave opportunistically by exploiting this escalation process so as to minimise the chances of the client cancelling the project in the face of surprises. The tactics in such a strategy will include the following.

- Sinking as much capital as possible as early as possible which is why the project manager of the Storebaelt project 'had to have the concrete on the table in a hurry'[24]. This may even be at the cost of the overall effectiveness of the management of the project, and may itself generate further cost increases, as in the case of the Channel Fixed Link in Case 1.
- Withholding or underestimating the budgetary and schedule implications of the surprise from those to whom the project is accountable – this is particularly common on projects where the NPV was based on strategic misrepresentation such as the Boston Central/Artery described in Case 13, the Channel Fixed Link presented in Case 1 and the Scottish Parliament[25]. One tactic here is to remove all contingency from estimates submitted for approval to reduce their total.

The dynamic of the escalation of commitment also gives us insight into why strategic misrepresentation is an effective strategy, and optimism bias such a serious problem. Typically, the poor quality of estimates becomes apparent relatively early in the project life cycle, and in principle, an effective gateway process should prevent them from continuing. However, the psychological investment in the project as it is developed can make it very difficult to kill the project[26], and there is a strong temptation to explain away apparently bad news and claim that everything is now under control. These escalation dynamics are well illustrated by the case of the London 2012 project for the Olympic Games presented in panel 10.10.

10.11 Summary

There are no facts about the future[27]; at best we have estimates of what those facts might become the day that the final account is done. To the extent that we judge that the past is like the future, we can base estimates for the current projects on

Panel 10.10 Budget Escalation for London 2012

London's bid to host the 2012 Olympic Games was accepted amidst great jubilation in July 2005 with a total budget of £3.3bn. The London Organising Committee for the Olympic Games (LOCOG) established the Olympic Delivery Authority (ODA) in September 2005 as client for development of the facilities. In turn, the ODA appointed the CLM consortium (consisting of Laing O'Rourke, MACE and CH2M Hill International) as delivery partner to manage the construction programme in September 2006 using the competitive dialogue process discussed in section 5.6.2. By March 2007, that budget had tripled to £9.3bn as shown in the table below – the deadline for completion of the project in July 2012 remains immovable. Is this escalation because of bad luck, optimism bias or strategic misrepresentation? From the evidence presently available publicly, it would appear to be a combination of all three factors as indicated in the table. Whatever the answer, the rational thing would have been to cancel the project in early 2007, but it possesses a classic escalation driver – the UK government has made *credible commitments* to the International Olympic Committee and the global sporting community. Cancellation was unthinkable because of the damage to the perception of the UK internationally that would result.

Budget line	**2005 £m**	**2007 £m**	**Comment**
ODA Core Cost, including venues, site security, . . . transport infrastructure and public sector contribution to the Olympic Village	2992	3081	This increase of £89m includes an additional £554m on the programme management costs of the ODA itself and of hiring CLM netted against £600m of savings identified through value management by the ODA supported by KPMG. It allows for an inflation rate of 5%. The original budget for programme management of £16m can only be described as naïve.
Policing and non-site security	0	600	From the point of view of the programme this can be considered bad luck – the day after the announcement four bombs exploded on London Transport murdering 52 people.
Value added tax	0	836	Although a cost to the budget this is simply an accounting transfer within government and should probably be excluded from the headline escalation.
Programme contingency	0	2747	This figure was established through the optimism bias mitigation process presented in panel 10.6. Omitting contingency is a classic symptom of strategic misrepresentation, but this judgement is tempered by the provision within the original ODA core costs of individual project level contingencies between 10 and 23.5%. £360m of this contingency was released to the ODA in June 2007.
Other (non-ODA)	0	388	These are mainly for contributions to training programmes for athletes.
Regeneration	1044	1673	This would appear to be because of scope creep as legacy objectives are still in definition.
Private sector contribution	−738	0	Optimism here was apparently unfounded.
Total	3298	9325	The figures presented in this table come from the summary presented in the Financial Times dated 17/03/07.

The figures above do not include the total cost of the Olympic Village, which was expected to be largely funded by the private sector, but the property crisis that broke in 2008 placed such optimism in doubt and the ODA is presently obliged to take all the risk. Any additional public funds required would be drawn from the programme contingency. Also excluded from this budget is the actual cost of running the games by LOCOG, which is expected to be revenue neutral.

The official position in 2008 is that the revised budget will on no account be exceeded, but the spectacular success of the 2008 Beijing Olympics increased the pressure on the London authorities not to penny-pinch and the collapse of the housing market threatens to undermine the private sector funding for the Olympic Village. Energy costs rose spectacularly in 2008, but the rapid easing of construction demand will result in input lower prices for individual projects. Schedule outcome is known – how much it will cost to meet that outcome remains unknown.

Sources: National Audit Office (2007b); *Building* 05/09/08; *Financial Times* 16/03/07.

data about past – this is the essence of reference-class forecasting using public (e.g. BCIS) and proprietary data sets. However, the alignment of the current project with the reference classes in the databases remains a matter of judgement, and as soon as predictability is reduced by factors such as innovation, volatility in markets and lack of clarity in project mission, the potential variance in those estimates is increased. Yet budget is one of the three main performance criteria in effective project realisation even though construction projects often surprise their clients by exceeding the budget derived from the NPV calculation. This is sometimes a result of the systematic biases in estimation derived from the desire of project promoters to get their project funded as discussed in section 3.7, but more often because of the inherent difficulties in predicting the future.

To provide a single-point estimate of the budget for a facility to be delivered years hence is, literally, nonsense. Yet specious certainty dominates the budget-setting process, and when the supply side points out the optimism of the estimates, it is encouraged to keep quiet to ensure that those providing the capital – be they the board, bankers or the taxpayer – remain willing to fund the new facility. Sometimes project managers are sacked for speaking out honestly on the real prospects for budget and schedule[28]. Yet, it is by being honest about these issues – as HM Treasury is trying to be – that we can start to address the problem of strategic misrepresentation, optimism bias and the inevitable variability around estimates. Budget setting is an inherently social process as Lichtenberg recognises, and all clients need to be prepared to take on responsibility for that social process and its inherent uncertainties and manage them through time as BAA and other sophisticated clients do – consultant cost engineers and quantity surveyors can only provide the outside view; they cannot make the key decisions on behalf of clients.

Even where such biases are not distorting the definition of the project mission, managing the budget on construction projects remains challenging. The principal way to control the budget is through design, yet there are considerable difficulties in doing this in reality; construction contractors have such poor understanding of the drivers of their costs that they frequently find it difficult to offer reliable advice on costs at an adequate level of certainty to inform decision-making, even when they are brought into the design process, as Case 7 shows. When compounded with the lack of incentives for the members of the project coalition to minimise cost growth, discussed in Chapter 6, and the very human tendency to throw good money after bad in the hope that things will get better, it is not surprising that budget escalation is so common. However, we have indicated some tools such as the successive refinement of budgets in the context of gateway reviews and value management techniques which can help to mitigate such escalation.

Managing the budget is a classic case of the control loop shown in Fig. IV.1. The NPV calculation provides the targets against which the cost planning process manages the budget within the PBS. As the project moves into planning and execution, the development of the PMB provides a detailed set of targets against which to manage costs within the WBS. The principal difficulty in estimating at the task level is the poor quality of contractors' budget estimates, which, at base, is the result of low levels of process capability within the construction industry. We will return to this topic in Chapter 12, and in the meantime we will explore the second main performance criterion in project realisation – schedule.

Case 10
The Centuria Project Budget

The Centuria Project is a £5.25m guaranteed maximum price contract, with a 50:50 incentive split between the principal contractor, AMEC, and the client, Teesside University, to provide new facilities for the university's School of Health. The project was planned to run for 49 weeks from 2 August 1999 to 10 July 2000. This case study will focus on one particular parade of trades – roofing[29]. It is assumed for the purposes of this case study that they are the responsibility of a single trade contractor[30], with the exception of the provision of scaffolding. The roof plan is shown in Fig. 10.8. While the project is a real one, the actual performance on the project has been changed to allow key points in this chapter to be illustrated; the material in Cases 10 and 11 is in no way a reflection on the performance of the real actors in the Centuria project coalition. The analysis will also be based on labour costs because this is the major source of budget variances in project task execution.

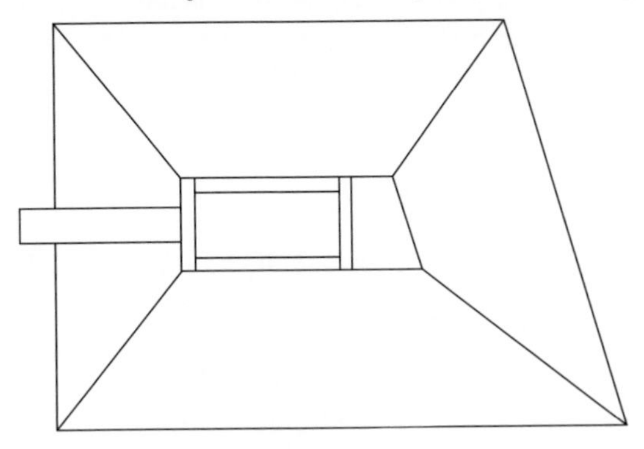

Fig. 10.8 Centuria building roof plan.

Absent unexpectedly high rates of waste or theft from site, the other elements in the CBS do not vary significantly from budget during task execution.

The works, as shown in the PBS in Fig. 10.2, consist of:

- Kalzip standing seam aluminium roof cladding on a 6° pitch over insulation boarding;
- single-ply covering to the central flat roof areas over insulation boarding;
- a patent glazed atrium rooflight with a central valley to the central area;
- siphonic rainwater systems to the eaves with associated downpipes;
- brick parapets to the flat roof area and walkway over the stairwell.

In addition, the WBS in Fig. 10.3 shows that task execution requires temporary works consisting of providing an additional lift to the brick-laying scaffold to form a roofing scaffold, and a birdcage scaffold to the atrium area to allow installation of the rooflight.

The total price for the roof works[31], excluding the scaffolding, is £100 637 (ex VAT). An extract from the bill of quantities for the single-ply roofing system to the flat roofs is shown in Fig. 10.9. Table 10.4 shows the baseline budget and realised budget for labour for the Centuria Building roof. As can be seen, the labour budget overran by over 12%, leading to a total budget overrun for this trade package of 5%, just at the maximum of the normal range of budget estimates for trade packages shown in Table 10.1. So, this was a poorly performing project in terms of budget. This was largely driven by labour cost overruns in placing the insulation to the north and south slopes of the roof, and the roof covering to the northern slope, and on the installation of the rooflight because of additional resources required to complete work without a serious SV. We will see in Case 11 how these overruns occurred, and how they could have been avoided with better programming of the trade package.

An important aspect of the cost planning of a project is the cumulative cash flow – defined in commitments terms. Figure 10.10 presents the total trade package cash flow, excluding scaffolding, derived from the cost plan as embodied

Table 10.4 School of health roofing labour cost baseline and outturn.

Task	Budget baseline (£)	Budget outturn (£)
Raise scaffold	1499	1799
Guttering (north)	5097	5097
Guttering (south)	5097	5097
Insulation (north)	2249	3598
Roof covering (north)	3748	4497
Insulation (south)	2249	3598
Roof covering (south)	3748	3748
Birdcage scaffold	5097	5097
Brickwork lobby roof	1350	1350
Insulation (flat)	900	900
Covering (flat)	3598	3598
Rooflight	6746	8096
Total labour cost	41 378	46 475

BILL ITEMS

Schedule: SCHOOL OF HEALTH ROOF

Bill: ROOFING

ID	Description	Unit	Quantity	Rate	Total £
	Standing seam aluminium roof cladding; 'Kalzip' Hoogovens Aluminium Building Systems Ltd or other equal and approved; ref AA 3004 A1 Mn1 Mg1; standard natural aluminium stucco embossed finish				
	Roof coverings (lining sheets not included); sloping not exceeding 50 degrees; 305 mm wide units				
A	0.90 mm thick		2108	25.85	54 491.80
B	extra over for		0	0.00	0.00
C	raking cutting	m	132	6.43	848.76
	Eaves detail for 305 mm wide 'Kalzip' roof cladding units; including high density polythylene foam fillers; 2 mm extruded alloy drip angle; fixed to 'Kalzip' sheet using stainless steel blind rivets				
D	40 mm × 20 mm angle: single skin	m	205	6.54	1340.70
	Accessories for roof coverings				
E	semi-rigid insulation slab; 30 mm thick; tissue faced	m2	2108	5.78	12 184.24
	Ridge detail for 305 mm wide 'Kal-zip' roof cladding units				
F	duo ridge including natural aluminium stucco embossed 'U' type ridge closures fixed with stainless steel blind sealed rivets; 'U' type polythylene ridge fillers and 2 mm extruded aluminium alloy support Zed fixed with stainless steel blind sealed rivets; fixing with rivets through small seam of 'Kalzip' into ST clip using stainless steel blind steel rivets	m	66	28.32	1869.12
C	abutment ridge including natural aluminium stucco embossed 'U' type ridge closures fixed with stainless steel blind steel rivets; 'U' type polythylene ridge fillers and 2 mm extruded aluminium alloy support Zed fixed with stainless steel blind sealed rivets; fixing with rivets through small seam of 'Kalzip' into ST clip using stainless steel blind sealed rivets	m	73	14.19	1035.87
	'Trocal S' PVC roofing or other equal and approved				
H	Coverings		200	15.79	3158.00
I	Skirtings; dressed over metal upstands not exceeding 200 mm girth	m	108	12.21	1318.68
	Insulation board overlays				
	Dow 'Roofmate SL' extruded polystyrene foam boards or other equal and approved				
	To collection £				76 247.17

Fig. 10.9 Extract from the Centuria building roofing bill of quantities.

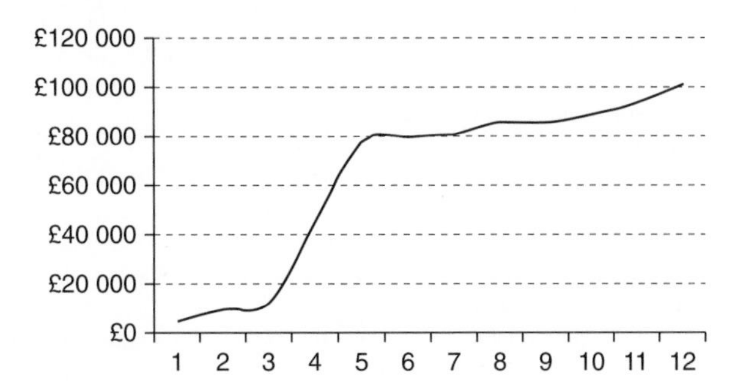

Fig. 10.10 Centuria Project planned value (total costs basis).

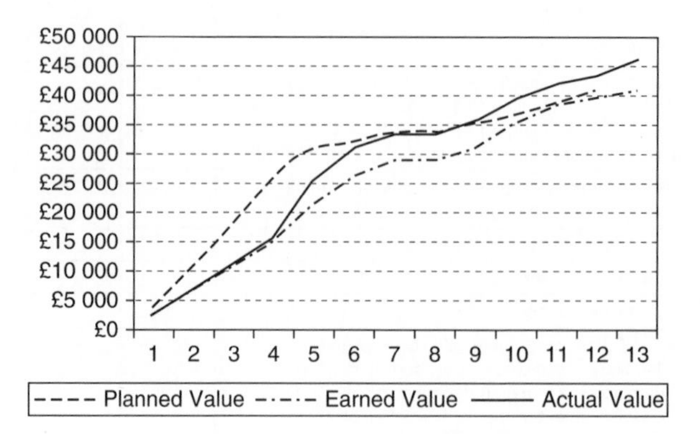

Fig. 10.11 Centuria Project earned value analysis (labour costs basis).

in the bill of quantities. It follows the familiar *S*-curve, but with a flat for the Christmas break, when no work was performed. Figure 10.11 presents the EVA for the project, on the basis of labour costs. It shows that by the end of week 3 (commencing 22 November 1999), the project was still on budget (AC = EV) but was already behind on schedule (PV > EV) with an SPI of 0.61. It then went over budget as well (AC >EV)[32]. This yields for the end of week 10 a CPI of 0.9, and an SPI of 9.6, with a BAC of £45 126, progress having been made in pulling back against the SV. Figure 10.11 also shows the usefulness of EVA, because overall baseline budget was not exceeded until week 9. EVA is a more responsive information loop than conventional monitoring of budget variances, giving an earlier warning of problems. The actual BAC was worse than the week 10 estimate because the last task in the schedule also overran.

However, the type of *S*-curve presentation of EVA shown in Fig. 10.10 is limited in the amount of information it can communicate because of its 2D character. Using a 3D presentation of the data offers a much richer way of communicating project performance information, especially if combined with the use of colour.

Notes

1 Brian Fine (1975, p. 221).
2 Lichtenberg (2000) and Goodpasture (2004) argue that such three-point elicitations can be usefully approximated to continuous distributions. Lichtenberg stresses the importance of the successive nature of estimates through the project life cycle and the importance of the thoughtful design of the estimate elicitation process. However, Fortune and Weight (2002) present evidence that, at least in the UK, deterministic single-point estimating as opposed to stochastic three-point estimating still predominates practice.
3 Brian Fine (1975).
4 All the examples of the Centuria project are developed from materials supplied by Nash Dawood and his team at the University of Teesside.
5 Uniclass is the UK's Unified Classification for the Construction Industry, issued in 1997 in accordance with ISO Technical Report 14177 and compatible with ISO 12006.
6 See Kirkham (2007) for a discussion.
7 Kelly *et al.* (2004) provide an authoritative review of value engineering and value management practice.

8 This is the area underneath the overlap of the original and revised cost probability curves, that is, there is a 27% chance of the outturn cost being within the original cost curve without the purported saving having been made. I am grateful to John Kelsey for this calculation.
9 A fee calculated in inverse proportion to the price of the construction contract might be one thing to try.
10 See Winch and Campagnac (1995) and Edkins and Winch (1999) for comparative research on this issue.
11 See also Dalton and Kenward (1975) on dimensional accuracy in building.
12 Cyril Tompkins, presentation, Bath University 04/07/00.
13 The terminology used here is that recommended by Fleming and Koppelman (2000) on whose book much of this section is based. For the British Standard guidelines on EVA terminology, see BS 6079.
14 HM Treasury (2004a, section 1.1).
15 HM Treasury (2002).
16 HM Treasury (2004b).
17 See Armor and Taylor (2002).
18 Flyvbjerg (2005, 2006).
19 Gehry (2004); the Dean of the Peter B. Lewis building for the Weatherhead School at Case Western Reserve University claims that he was forced to resign due to budget escalation on that Gehry-designed building, which escalated from an initial budget of $25m which became $40m when Ghery was hired and finished at over $60m (Carlson, 2001).
20 Merrow (1988, p. 23), emphasis in the original.
21 Merrow (1988, p. 25).
22 Northcraft and Wolf (1984).
23 Staw (1997) provides a synthesis of the literature on project escalation from a social psychological perspective; see also Conlon and Garland (1993).
24 Cited in Bonke (1998, p. 10).
25 Bain (2005).
26 Royer (2003).
27 Goodpasture (2004).
28 Although it is always difficult to determine the details from outside, this would appear to have been the fate of senior project managers on both the Boston Central Artery/Tunnel and the Scottish Parliament.
29 The principal reason for this is to allow the output from the planning software to be reproducible in the format of this book.
30 In fact, five separate subcontractors were deployed.
31 The works were priced from a take-off by the author using the Spon's 2001 *Architect's and Builder's Price Book* (126th edition), priced by the Estimating System software provided. Labour rates for earned value analysis were also taken from this source.
32 These graphs are prepared in Excel from the data presented in MS Project's report tables – MS Project 98 does not provide a graphics function for its reports.

Further reading

Fleming, Q.W. and Koppelman, J.M. (2006) *Earned Value Project Management* (3rd ed.). Newtown Square, Project Management Institute.
The standard reference on earned value project management.

Lichtenberg, S. (2000) *Proactive Management of Uncertainty Using the Successive Principle*. Copenhagen, Polyteknisk Press.
An alternative approach to project budgeting and scheduling derived from a Bayesian approach to statistics to support the progressive reduction of uncertainty through time.

Turner, J.R. (2008) *The Handbook of Project-Based Management* (3rd ed.). London, McGraw-Hill.
A standard reference and comprehensive text on all aspects of project-based management.

Chapter 11
Managing the Schedule

11.1 Introduction

> '...to watch the time, and assign things by time, to devote oneself to business and never lose an hour of time ...'

Leon Battista Alberti, although he only wrote one book on management[1], compared to ten books on architecture, knew the essence of managing a schedule. The rising Florentine bourgeoisie, whose views Alberti articulated, were very concerned about time and money, and because of this they were able to finance the emergence of a new role in the building process – the architect. This chapter will investigate the role of time in managing the construction project, showing how it has become central to the discipline, before relating it back to the management of money discussed in the previous chapter.

Managing the schedule is, for many, the core competence of the discipline of project management, yet, as Peter Morris points out, it has become associated with project management as a middle-management, operational discipline, rather than the strategic discipline that its challenges require – it has become *the* project management routine. Or, as Eli Goldratt puts it, there has been nothing new in 40 years[2]. This comment refers to what some see as the heyday of the discipline when it armed America and sent men to the moon. However, this is starting to change, so the task of this chapter will be to identify current good practice with the *critical path method* before moving on to investigate some of the more recent developments which are starting to diffuse – most notably critical space analysis (CSA), *critical chain*, *last planner* and 4D planning.

Our focus in this chapter is on the tools and techniques of scheduling, but it should be emphasized that construction planners do a lot more than schedule tasks through time[3]. Planners start their work during the tender process and have to create an overall master schedule that will become contractually binding. Given the time constraints during the tender period, the resulting schedules can only be at a low level of detail and rarely involve a WBS. Planners are typically expected to produce method statements along with their schedules, and may also be asked to produce site layouts, procurement schedules and assurance plans for quality, environment and safety – see Chapter 12 for more on these.

11.2 Critical path method

In the previous chapter, we identified the role of the WBS in establishing the budgetary management system for the project. The WBS is similarly at the core of the management of the schedule. The WBS identifies the tasks that need to be completed for the project to progress, and these tasks have *dependencies* upon each other. Certain tasks have to be completed before others can be started. These dependencies may be created in a variety of ways.

- *Technical* – a beam needs to be placed on supporting columns; therefore, its construction depends on the completion of the relevant columns.
- *Organisational* – a floor covering can be fixed before work to the ceiling is completed, but such a schedule greatly increases chances of damage to the floor covering after it is fixed.
- *Spatial* – while the plumbing and electrical systems are technically independent, their installation typically occupies the same space and so one needs to follow the other.

In order to cope with the complexities of these dependencies upon even a relatively small project, a technique now known generically as *critical path method* (CPM) was developed separately during the 1950s by DuPont and the US Navy. The essence of the method is to array graphically all the different tasks in temporal sequence, forming a network of task dependencies; it is essentially 'a method of thinking and also a method of presenting information'[4].

There have been various conventions for graphing such arrays, but the one most commonly used today is the precedence method (activity on the node) because this is easier to use in computer-based applications. CPM is one of the management techniques that have been transformed by the desktop and laptop PC. Only the smallest projects can be drawn by hand – a 1000 activity network would take one person one week to calculate manually, and any change to the network would imply a recalculation[5]. So, this was one of the earlier computer applications, but the diagram had to be drawn up in tabular form, entered on to IBM punch cards, sent to the computer department and returned to the planner. As late as 1988, Laing were sending schedules and schedule revisions around London by courier to and from their central computing facility at Mill Hill; as a result, this was considered 'too much bother' by construction managers on large projects such as the British Library[6]. Today, PC applications such as MS Project and Primavera allow calculations to be done instantly, and options to be evaluated effortlessly. Arguably, it was during the 1990s that the true potential of CPM as a project management tool began to be realised.

For CPM, the WBS needs to be turned into a set of dependencies. Each task in the WBS is identified using a task descriptor of the type shown in Fig. 11.1. The descriptor typically identifies[7]:

- the task, by name and/or code[8];
- its duration, typically in days;
- the earliest day on which it can start, depending on when those tasks that need to be completed beforehand will be completed;
- the earliest finish day, as a function of duration and earliest start;
- the latest finish day, depending on when those tasks that rely on the completion of this task need to be started;
- the latest start day, as a function of duration and the latest finish day;
- slack (float), being the difference between earliest finish/start and latest finish/start in days.

Early start	Duration	Early finish
Task name		
Late start	Slack	Late finish

Fig. 11.1 Critical path task descriptor.

When arrayed according to dependencies, as shown in Fig. 11.2, the shortest possible duration for the project is given by the *longest* sequence of tasks where the earlier and latest finish times are all equivalent, and there is, therefore, no slack. This is known as the *critical path which* should be the focus of management attention to ensure that the project is completed in the shortest possible time. Typically it is protected by using slack to create feeding buffers where non-critical task outputs feed into tasks on the critical path, and by using a project buffer to protect the project end date, which is usually contractual. One of the important debates in determining the critical path is whether it should be planned forward or backtimed. Backtiming – setting the target date for completion and then fitting the planned sequence of tasks to that date – is more intuitive but can lead to a compromised plan from the outset. Some prefer the project to be scheduled from the expected start date, and if the result leads to completion too late to meet client needs, then it can be rescheduled. This way, the tradeoffs made and risks accepted to provide a shorter critical path can be more clearly identified. The choice will depend on how important the schedule is for the project mission. Where the design process has also been scheduled, the critical path can be used to determine the *last responsible moment*[9] for making a design decision, thereby maximising the opportunity to make late changes to the design without comprising project progress.

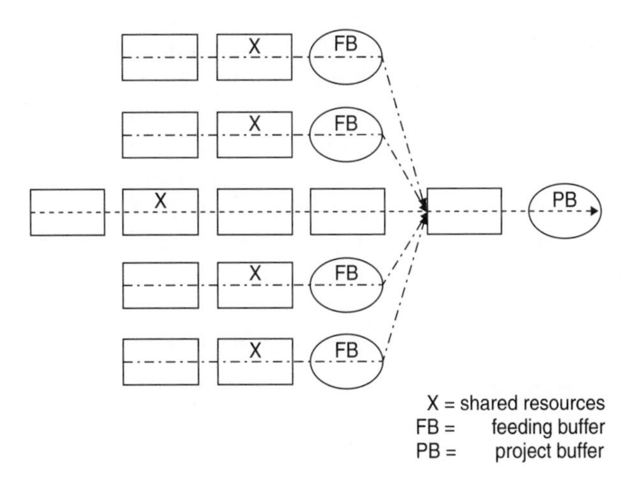

Fig. 11.2 Critical path with buffers.

Once the basic network is established, different options can be tried out for the sequencing of tasks in order to identify:

- the shortest achievable schedule;
- where sequence constraints exist which can be mitigated through management initiative;
- risks to the schedule from tasks where durations are particularly uncertain;
- opportunities to reschedule tasks off the critical path to achieve other objectives, such as resource utilisation;
- the implications of new information being acquired which changes task durations, task dependencies or latest start dates.

A fundamental assumption of the method is that task durations are known accurately to at least the planning unit used for the network (weeks, days, hours) – the method is deterministic in allowing a single optimal solution. There are two sources for the data required:

- measuring performance on the same or similar tasks on earlier projects entered into a database;
- asking those who are charged with executing the task how long it will take.

These data tend to be less reliable where task objectives are more uncertain, such as during design. CPM is, therefore, most commonly used for execution on site, yet even here, important uncertainties remain regarding task durations. Apparently random fluctuations in task execution times have been noted by a number of observers – unpredictable events such as the weather, a flu epidemic or material shortages can extend task durations, while managerial failures to adequately plan task execution also take their toll.

One solution to these problems is to accept that task durations are uncertain, and to use probabilistic methods. The two best known of these are Program Review and Evaluation Technique (PERT), developed for the US Navy in the 1950s, and Monte Carlo simulation, which has been the focus of more recent research. PERT allows task durations to be estimated as three values – most likely duration, worst case duration and best case duration. The distribution between these three estimates for a task is assumed to a beta distribution, and therefore a probability of achieving the planned duration of the project as a whole can be calculated – see panel 11.1.

Panel 11.1 PERT and the beta distribution

The expected duration (*d*) of a task is given by MacCrimmon and Ryavec (1964) as:

$$\frac{d = (t_0 + 4t_m + 3t_p)}{6}$$

where t_o = most optimistic outcome (1% probability of occurring)
t_m = most likely outcome (mode)
t_p = most pessimistic outcome (1% probability of occurring)
However, if Murphy had retrained as a statistician, he or she would have reformulated the famous law – 'what can go wrong will go wrong' – as 'things are more likely to go wrong than to go right'. In other words, the distribution is typically skewed so that the most frequently occurring outcome (mode) has a less than 50% probability of occurring.

In order to handle the inherently skewed nature of the distribution of durations, Turner (1999, p. 232) proposed the following formula:

$$\frac{d = (t_0 + 4t_m + 3t_p)}{8}$$

while Locke (1996) proposed:

$$\frac{d = (t_0 + 3t_m + 2t_p)}{6}$$

The only real way to move this debate forward is to move beyond simulation and obtain task data from a sample of real construction projects – research that awaits to be done.

Monte Carlo simulation techniques take this probabilistic approach further by generating the distributions of durations of tasks randomly, as is indicated by the reference to the casinos of Monaco. This is done by providing two estimates – a worst case and a best case duration – and then randomly allocating either one or the other to each task through a large number of calculation cycles. This simulation produces a normal distribution of the probability of a task or the whole project (last task in the network) being completed on a particular date. A cumulative curve of the distribution can be used to identify the probability of the project being completed on or before a specified date. However, probabilistic methods have not tended to be very popular among construction planners because the results tend to be difficult to interpret, they are more demanding of computational resources, and no additional information is provided by the technique. Deterministic CPM remains the standard practice of construction planners.

Networks soon become large and difficult to read. They are an analytic tool for specialist construction project managers. For reporting and control purposes they tend to be presented as bar (Gantt) charts, as shown in Fig. 11.11. Such charts are excellent visualisations of the project schedule, allowing all the participants to quickly ascertain how they fit into the schedule and how they are performing against the schedule. However, they are not management tools in terms of allowing the proactive response to uncertainty, because they do not allow the assessment of implications and the evaluation of options – only the underlying network can do this. However, on simpler projects they are often the tool for managing the schedule, and are prepared directly without an underlying network. The summary level can also be used to identify the major milestones of project progress, against which payments can be made and project review gate meetings organised.

On many projects, particularly larger ones, there is more than one plan. Figure 11.3 shows the interrelationships between the different plans for a major London hospital project[10]. The client's project manager develops its own schedule, which drives the schedule for the procurement of the resource bases, and also the architect's design schedule. The contractually binding agreement between the client and the

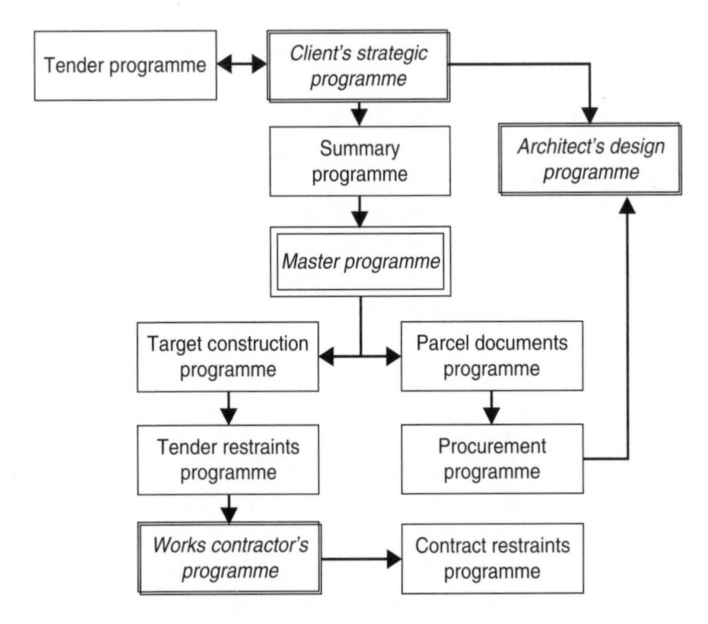

Fig. 11.3 Planning schedule hierarchy (source: Laing Management interview 22/05/89).

construction manager is the *master schedule*. The construction manager's project managers then prepare the *target construction schedule* which guides the procurement of trade packages, and the *parcel documents schedule* which drives the production of drawings by the architect. This last schedule is non-contractual, but can be used in claims for delays against the client caused by the non-delivery of drawings. Within the target schedule, trade contractors are given 'windows' for the execution of their responsibilities on site and these are formally agreed in the *works contractors' schedules*. Within those, the trade contractors then schedule their task execution at the level of WBS that suits them. In order to gain more bargaining power, construction managers may not reveal the master schedule to the trade contractors, so as to buffer the completion date to ensure a satisfied client even if the works contractors' schedule slips[11].

11.3 Resourcing the project

Task execution requires resources; it is this that generates the cost of task execution. It is therefore imperative that the use of resources is as efficient as possible. However, standard CPM approaches in construction tend to assume that resources are infinite – resourcing is a secondary issue if it is considered at all. There are two interdependent problems here:

- Resources need to be deployed on projects for the maximum proportion of their available time (i.e. not *underloaded*), for this is the only way that they generate value; undeployed resources are waste for which the client typically pays in one way or another.
- Resources need to be deployed so that they are not *overloaded* by being expected to perform two tasks at the same time, or accommodate more than one work gang. Overloading generates additional costs in a large number of different ways, such as premium payments for out-of-hours working, inefficient working through failure to provide adequate space, inefficient working because of the deployment of operatives and plant of marginal quality, and start delays to dependent tasks as a result of late completion of precedent tasks.

The resources available on a project are of three types.

- *Plant* – either access and transportation equipment such as hoists and cranes, or installation equipment such as piling rigs, which are constrained in their output rate and inability to be in two places at once.
- *Labour* – skilled workers are typically at a premium, and overtime is the usual response to this constraint with inevitable consequences for productivity. Labour, too, finds it difficult to be in two places at once.
- *Space* – technically independent tasks may require the same space for execution so as to maintain effective working conditions; erect falsework, thereby closing off workspaces; maintain safe working around plant or overhead working; or allow the transportation of materials to workspaces.

Where resources can be unambiguously allocated to tasks, such as a carpentry gang or a piling rig, there are facilities now available within most CPM software applications for *resource levelling*. However, where resources are effectively overheads shared by more than one task at a time – cranes are the classic example here – other planning methods need to be used. Similarly, spatial resources are not accommodated in existing software applications.

Resource levelling for a single project is done by taking the network and identifying all the resources required on the project for a particular period – such as week by week – categorised by type. Thus the number of bricklayers, carpenters and so on required on site each week can be found. By charting those requirements in a histogram with time on the x-axis and number on the y-axis, the resource loading of the project can be identified. Underloading occurs when there are troughs in the requirement for a particular type of resource from one week to another. Overloading occurs when the peaks of requirement for a particular resource are greater than the amount available, implying either overtime working at additional cost or delays in execution of the task. A level resource plan is one without peaks or troughs.

The basic method is to take those tasks with the greatest float, and to shift the start dates so as to reduce peaks and troughs. Typically, this is only partially successful in providing level resource use. For a *schedule-constrained* project, this is the best that can be done. Where the project is *resource constrained* it may be appropriate to extend the project duration to reduce the waste of resources inherent in the troughs of the resource plan, or the costs inherent in overloading. This is the essence of the schedule/budget trade-off in project realisation. Where additional resources are deployed in order to reduce task durations from the original plan, this is known as *crashing* the project. It should be noted that reduced execution times will reduce expenditure on overheads, so these need to be brought into the calculation as well. Once a project is running behind schedule, however, there is often little that can be done by adding additional resources because adding additional resources to a project that is running late will make it even later because of the disruptive effect of introducing those unplanned resources.

Resources of labour and plant do not necessarily need to be levelled within a single project, although space, by definition, is not shareable between projects unless they are physically contiguous. Levelling can take place between projects – lending resource to fill troughs and borrowing resource to reduce peaks – across the portfolio of projects illustrated in Fig. 1.4. This is, of course, a much more difficult problem which needs to be solved at the level of the firm rather than the project. Subcontracting is one way of resource smoothing – indeed the production cost efficiency argument for subcontracting discussed in section 7.6 relies as much on this resource smoothing effect as the learning curve effect.

Construction planners tend to ignore resource constraints associated with labour and plant because of their reliance on spot contracts, discussed in section 7.5. The competitive market for plant hire means that most types of plant are readily available at short notice, while casualised labour markets tend to mean the same for operative skills. However, the levelling problem has simply been shifted and not removed. Peaks and troughs in the demand for resources are now signalled in the spot market through price movements, rather than directly in costs. Where resources lie idle in a trough this is still waste – plant that is scrapped early or workers that are unemployed are waste, even though these costs have been externalised from the project. From a stakeholder point of view, however, such externalisation might not be acceptable. For these reasons, governments have often invested in construction projects to reduce unemployment in particular, and stimulate the economy in general.

11.4 The limitations of the critical path method

For the past 40 years, both academics and leading professionals have advocated the use of CPM, particularly in its PERT application, yet many construction project managers stubbornly refuse to move beyond the bar chart. Why? The answer lies in both limitations with the technique itself and organisational problems with implementation.

There are a number of problems with the technique. First, it requires extensive computing power to do any sort of analysis and evaluation of a network, and unless this computing power is easily accessible, the network will not be kept updated as the project progresses. Moreover, early software tended to ignore the rolling wave and insisted that the plan was developed at the highest level of WBS for any planning to be done. It was only during the 1990s that computing power has become cheap enough to be distributed to site and used as an everyday tool by planners. The introduction of user-friendly CPM software such as MS Project, which allows rolling wave planning, is even more recent. These limitations are likely to disappear.

A second problem is that project managers actually have remarkably little time to plan on most construction projects. The overall project schedule is formed during tender and becomes enshrined in the master schedule, as illustrated in Fig. 11.3. Short tender periods mean that this master schedule is very much a guess at what the actual task durations might be, which means that subsequent schedules developed for actually managing the project tend to be constrained by decisions made in haste during tender. However, this is changing. The development of new forms of procurement described in section 5.6 means that projects have longer lead times and project managers have more time to plan their projects. Complemented by some of the new techniques that are discussed later in this chapter, construction project planning is coming back on to the agenda.

However, four significant problems remain. The reliance on deterministic programming approaches is heroic and tends to engender a site culture of ignoring the schedule on the grounds that it is always wrong. The schedule becomes decoupled from the realities of the weekly management of the site, a problem that *last planner* addresses, as discussed in section 11.5.3. The central proposition of this book is that the management of construction projects is an inherently uncertain process, and any purported management tool that fails to take that uncertainty into account is unlikely to be very useful. The probabilistic methods are an advance on the deterministic, but their dependence on the unweighted beta distribution presented in panel 11.1 means that they tend to underestimate project duration. The same argument applies to deterministic durations to the extent that they are more optimistic than the worst case scenario. The only solution currently available to this is to embed additional slack in feeding and project buffers into the schedule to protect the critical path and near-critical path tasks.

The organisational limitations in the implementation of CPM are also serious. Partly because of the difficulties in updating the network due to the technical problems identified above, CPM has tended to become a historically orientated control tool, rather than a future-orientated management tool. Clients typically require the development of a network – on either a deterministic or PERT basis – which then becomes the baseline against which progress is measured and payments are made. This does not encourage its proactive use and any client 'requirement for the use of a management tool is certain to reflect unfavourably on that tool, whatever its intrinsic merits'[12], a phenomenon also visible with quality assurance and value engineering.

Panel 11.2 The planning fallacy

A team of Canadian researchers has been investigating why we make inherently optimistic assessments of task durations. By sampling various groups of students, asking them how long it would take them to complete assignments and comparing the answer to the actual durations, the team identified chronic optimism in task durations, which appeared to be a cross-cultural phenomenon. Their experimental work suggested that the reason for this optimism bias was that people focused on the task at hand, rather than reflect on the experience of themselves and others in completing similar tasks before. When respondents were asked why their tasks had overrun, they tended to blame external factors such as computer malfunction, but when asked to comment on the failings of others, they tended to suggest that the other person had time management issues. The team also investigated ways of debiasing task duration estimates, but they did not identify any that had a significant effect. For instance, when people were asked to visualise worst case and best case scenarios, they regarded the best case scenarios as much more plausible than the worst case ones and therefore weighted them more in coming to their conclusions. They conclude that the planning fallacy is a robust and pervasive phenomenon which is difficult to overcome. Later research has shown that where groups plan tasks together, the group dynamic tends to increase optimism and thereby worsen the effects of the planning fallacy.

Sources: Buehler *et al.* (2002, 2005).

A third problem is the padding of task duration estimates. The most common way – indeed the way recommended by many authorities – of obtaining these is to ask those who are responsible for executing the work for appropriate estimates. In a context where those people – be they at the operative level or their managers – will then be held to account for meeting those estimates through either their pay packets or their bonuses, the temptation to behave opportunistically is enormous. Operatives will wish to give undemanding times to allow room for shirking, while managers will similarly collude in such estimates to ensure that they can meet their targets. It should be noted that this padding of durations is not the same as slack. Slack is a function of the relationship between tasks and is transparent in the network – padding is hidden within each task duration. A fourth problem is linked to the discussion in section 10.10. Psychologists – see panel 11.2 – call optimism bias in scheduling the *planning fallacy* to capture our hopeless optimism in planning task durations even when we are not being consciously opportunistic.

Many planners simply do not know the actual distribution of achievable task durations, as the information they receive on which to base their estimates has been opportunistically padded or optimistically underestimated by various parties along the way. There are two complementary solutions to this problem – to build up databases of task duration through extensive benchmarking and workstudy, or to shift to the type of 'no blame' culture encouraged by the shift from the CPM to the critical chain method (CCM). In the meantime, planners rely on intuition and experience as much as on accurate data on previous project performance.

11.5 New approaches to project scheduling

These limitations have prompted the development of a number of enhancements to the CPM, which are typically developed as plug-ins for standard project scheduling software such as MS Project. *Critical chain* addresses the interaction between resources and the schedule; *critical space analysis* formalises planning for spatial constraints; *last planner* directly addresses the sources of excessive task execution variability which undermine project scheduling; and *dependency structure matrix* analyses dependencies in reciprocal, as opposed to sequential processes as defined in Fig. 8.4. The review is completed with a brief discussion of schedule visualisation techniques.

11.5.1 The critical chain method

The critical chain method (CCM) has evolved from the theory of constraints presented in panel 11.3. It addresses two of the key problems of CPM – the inherent uncertainty of task durations and associated opportunistic behaviour in establishing the true duration of tasks, and the resourcing of tasks. In CCM, the *critical chain* is the *longest* resource-constrained path through the network, theorised as a constraint to be elevated, as illustrated in Fig. 11.4 which should be compared to Fig. 11.2. Thus a critical chain looks like a critical path, but it includes resourcing in the dependencies; plug-ins for MS Project are available to allow this to be done, such as ProChain used in Case 11.

The conceptual shift from critical path to critical chain by including resourcing issues in the latter might be considered simply a technical development, moving on from the resource levelling approaches which are well established. However,

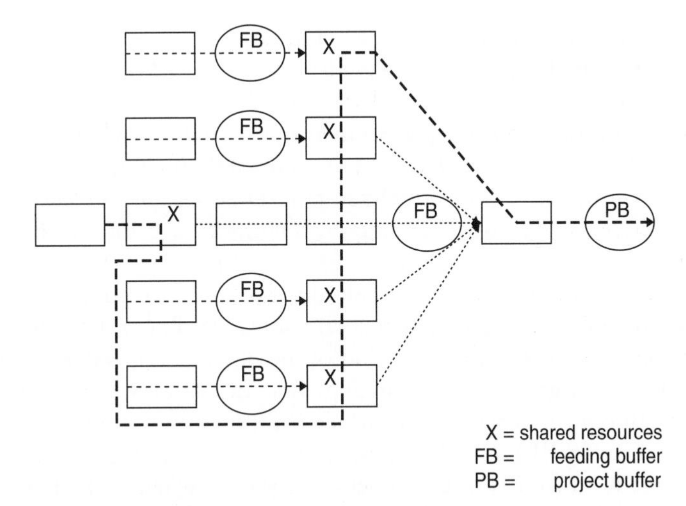

Fig. 11.4 Critical chain with buffers.

Panel 11.3 The theory of constraints

Eli Goldratt has created a remarkable body of work over the past 20 years. Schooled in systems thinking, he has developed an approach to operations management that is more broadly applicable than the lean manufacturing concepts that have only really proved themselves in very high volume industries such as cars. His approach is that of a storyteller – but a storyteller schooled in the Greek philosophy of Socrates. His ideas are expounded through dialogue between teachers such as university professors and management consultants, and managers who have customers and board directors to please. His ideas are then turned into practical tools through the Goldratt Institute – http://www.goldratt.com/

The core idea – developed in his first novel – is the *theory of constraints*, which has been applied to problems in areas such as marketing, strategic management, operations management and, more recently, project management. The starting point of the theory of constraints is that any production system is constrained in its performance – in a project context that constraint is the critical path. The process that embodies the theory of constraints is a five-step one:

- identify the constraint;
- exploit the constraint;
- subordinate to the constraint;
- elevate the constraint;
- identify the new constraint.

Sources: Goldratt and Cox (1993); Goldratt (1997).

the elevation of the constraint introduces a much more radical aspect. This elevation starts from two observations:

- actual duration for any planned task is unknown, but can be assumed to be normally distributed around expected task distribution;
- existing estimates of expected task durations are padded because they include *safety time*, greatly extending the length of the critical path.

In a situation where actual task duration is uncertain, and managers are held accountable for meeting stated durations (deadlines), task duration estimates are going to be at the worst case end of the distribution, as defined in panel 11.1, otherwise managers risk overrunning their deadlines. In other words, the incentive structure for transaction governance stimulates opportunistic behaviour by managers – see Fig. III.1. CCM proposes instead that the average estimated duration should be used – there is a technical debate regarding whether the mean or the median of the distribution is more appropriate. An inevitable result of this is that managers will overrun their planned durations half of the time; this problem is solved by removing from managers the absolute responsibility to meet task-level duration deadlines, and instead providing incentives for early completion. This approach was tried on the A13 project with great success, as described in panel 11.4.

Panel 11.4 The theory of constraints in East London

Balfour Beatty on the 5.2 km A13 highway project opted for a decentralised project management organisation, pushing responsibility for performance down to area-based teams; see panel 15.4 for a similar choice in New England. However, performance did not improve, so the project manager applied the theory of constraints thinking described in panel 11.3. Instead of simply allocating schedule slack to each task manager, who then took the time allocated to complete the task, senior management took control over slack and used it to buffer the critical and near-critical paths. Combined with the implementation of a 'no blame' culture of management, performance improved so that 96% adherence to weekly work schedules was achieved, compared to 52% before the change. The most important step in achieving this improvement was to change the culture from a confrontational one to a more co-operative consensus-driven management style. Members of the project management team agreed that although improvements in performance had been made, this was only the start, and they were keen to try out the ToC ideas on their future projects.

Source: Barber *et al.* (1999).

What do project managers gain from not delegating responsibility for meeting task-level deadlines? They gain control over safety time, or as Allan Flanders put it 'management learns to manage by being forced to accept the full responsibilities of management'[13]. Having released safety time from the control of those executing the task, project managers can now allocate it strategically to form buffers to protect the critical chain. By elevating the constraint in this way, overall project durations are reduced so long as the durations of the strategic buffers sum to less than total safety time. A further advantage of this approach is that it reduces the impact of Parkinson's law, which states that 'work expands so as to fill the time available for its completion'. By reducing target times, the effect of Parkinson's law is reduced for that half of tasks where the actual duration is less than the estimated duration. One US architectural practice removed stated deadlines from task assignments; they found that the variability of task duration increased, but mean task duration, and hence overall project duration, fell[14].

11.5.2 Critical space analysis

Critical space analysis (CSA) addresses the issue that task execution creates not only the completed facility but also, temporarily, the spaces in which tasks are to be executed. Task execution space availability is therefore dynamic because different trades parade through the same space and may clash spatially, and because the spaces themselves change as, for instance, floors are laid creating work spaces, and walls are built closing off work spaces. Of course, experienced project planners *do* take into account the availability of task execution spaces, but this is typically based on intuition and rarely formalised beyond the most basic rules of thumb. CSA provides a way of analysing the spatial constraint during task execution on construction sites, allowing it to be treated strategically as a resource.

The space planning problem in construction has two main elements which are interdependent, but which require rather different approaches. These are the *space scheduling problem*, which is focused on the planning of task execution spaces, and the *site layout problem*, which is focused on the location of temporary facilities of various kinds. There is now a significant body of work on the site layout problem[15] but comparatively few researchers have turned their attention to the space scheduling problem which CSA addresses.

Resources require space for operation. For example, it has been reported that studies conducted by Mobil suggest that 19 m^2 per person is required and that 50% more man-hours are required when this declines to 10.4 m^2 which is an absolute minimum. For well-planned emergency labour-intensive short-term tasks, it is possible to manage with 9.4 m^2. Maximum productivity occurs at 30.2 m^2. Other studies confirm 28.3 m^2 as the desirable lower limit for effective task execution and evidence that work space congestion reduces productivity[16]. Similarly, equipment operation requires a working clearance plus a safety zone.

Any analysis of the spatial configuration of the construction process needs a set of definitions of space types which is provided in Table 11.1 and illustrated in Fig. 11.5[17]. *Total space* is that enclosed by the site boundary, and consists of *product space* taken by fixed product elements such as walls; *installation space* occupied by site installations, prefabrication areas and access platforms which are typically a function of the solution to the site layout problem; and a balance of *available space*. *Required space* is that needed for effective task execution. This can be the task execution space itself for equipment or operatives, materials storage spaces in support

Table 11.1 Space type definitions (source: Winch and North, 2006).

Definition	*Space type*
Total space	t
Product space	p
Installation space	i
Available space	$a = t - p - i$
Required space	r

Table 11.2 Critical space analysis definitions (source: Winch and North, 2006).

SpaceMan concept	*Definition*
Spatial loading	$s = (r/a)100$
Spatial overload	$s > 100$
Spatial slack	$a - r$ (where $s < 100$)
Critical space	$s = 100$

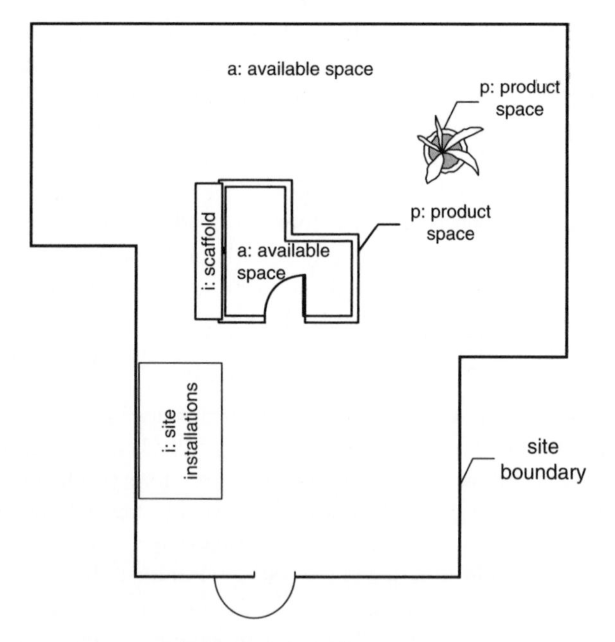

Fig. 11.5 Critical space definitions.

and movement paths for the supply of resources or the removal waste. Spatial loading is the ratio of required space to available space, where a ratio greater than unity means congestion, calculated as shown in Table 11.2. Critical spaces are defined as those where loading is at unity, that is, there is no spatial slack[18].

11.5.3 Reducing task duration variability – last planner

One of the main problems with managing by earliest start dates is that there is a strong tendency to start work even if not all the resources required for the completion of the task are available. This tendency – known as multi-tasking by Goldratt – has the inevitable result of increasing the variability of task durations by introducing greater uncertainty, and extending average task durations because of switching costs between tasks. These problems have been addressed in more depth by the advocates of 'shielding production' through the *last planner* technique which complemented by the kinds of process improvement tools presented in section 12.4 forms part of the Lean Project Delivery System (LPDS)[19].

The thinking behind the LPDS starts from the observation that construction project managers do not manage task execution but manage contracts; as discussed in section 7.5, the work of managing processes on site is not seen as the responsibility of managers in construction, but is delegated to self-organising gangs. The LPDS proposes that it is the responsibility of management to ensure that tasks are executed as efficiently as possible in a collaborative context, thereby regaining control by sharing it as in panel 11.4. Efficient task execution will reduce variability in task execution, and because the distribution of durations is skewed towards the pessimistic end, as shown in panel 11.1, it will also reduce average task execution times. The key to efficiency is shielding task execution so that tasks only start when precedent tasks have been completed, and all the resources are available. Such ready-to-start tasks are known as *quality assignments*. By making only quality assignments, managers can both reduce costs through increased efficiency and reduce durations by eliminating uncertainty. In order that resource utilisation is not reduced owing to delays in task commencement as a result of shielding, managers are expected to build up buffer stocks of quality assignments off the critical path, to which underloaded resources can be allocated.

Panel 11.5 4D modelling for the Walt Disney Concert Hall

The Walt Disney Concert Hall in Los Angeles opened in 2003 to a Frank Gehry design. The complex forms typical of Gehry's style posed challenging scheduling problems for the constructor Mortenson. With the help of the Center for Integrated Facilities Engineering at Stanford University, Mortensen used 4D modelling to improve scheduling on the project. The surface model of the building from CATIA was linked to the schedule created in Primavera using a proprietary interface which gave Mortensen much more analytic power in understanding the schedule. In particular, it used the 4D models for the following purposes:

- **Schedule creation**: to plan the laydown areas for steel erection; to visualise access routes at critical junctures in the project; to refine the scaffolding strategy; and to plan the installation of the complex ceiling of the main concert hall.
- **Schedule analysis**: to help identify several schedule conflicts which were not apparent in the CPA analysis including a wall scheduled too early while steel was being erected directly overhead conflicting with temporary shoring and creating a safety risk; an air handling unit scheduled for installation too late because access became closed off; and

issues with the scaffolding within the interior hall. As a result of the last analysis, the scaffolding contracts for the interior hall were consolidated from three to one contract.

- **Communication**: to support training sessions, and to enable collaborative review of the schedule in the Walt Disney Imagineering Virtual Reality Cave.
- **Team building**: to encourage trade contractors to discuss issues and solutions to problems or questions identified during the schedule review sessions and to focus the creative energy of trade contractors.

Sources: Haymaker and Fischer (2001); Wikipedia (accessed 27/09/08).

The approach is called last planner because making quality assignments is the last stage in the project planning process. The planning horizon is typically 1 week, and the decision-making process is delegated down to the level of first-line supervision. It is a tactical tool, not a strategic one. Although last planner and critical chain were apparently developed independently, they would appear to be highly complementary, with critical chain solving the strategic problem inherent in last planner – if the project ran out of quality assignments, progress would grind to a halt. The weekly decision-making cycle can be placed in the context of look-ahead planning on a monthly or quarterly cycle using critical chain. The difference between this approach combining critical chain and last planner and CPM-driven task allocation is shown in Fig. 11.6. Last planner could also be combined with CSA where spatial resource availability is one criterion for a quality assignment.

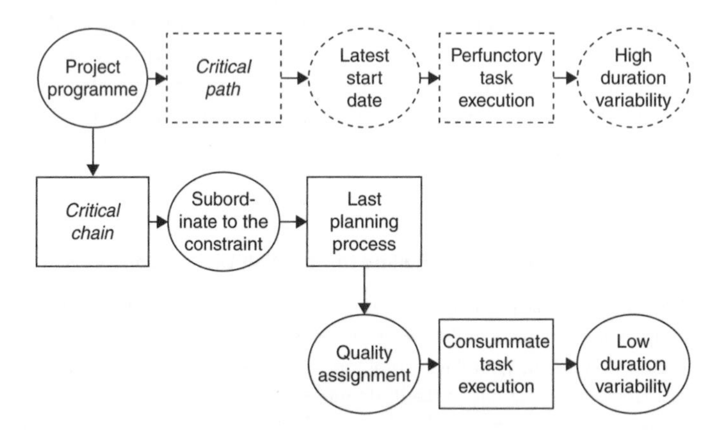

Fig. 11.6 Last planner and critical chain combined.

11.5.4 Scheduling reciprocal processes

A fundamental assumption of CPM and CCM is that task dependencies are sequential – that there is no requirement for reciprocation between dependent tasks, as defined in section 8.7. While viable for execution on site, this assumption is not viable for design tasks; design tasks have interdependencies as well as dependencies. A different approach is required for the programming of the design phases of the project. One way forward is to use a dependency structure matrix (DSM). This uses a matrix format to relate all design tasks to all design tasks in a time-related sequence, identifying those that are dependent on others by filling the boxes in the matrix[20]. Those below the diagonal of the matrix formed by the intersection of a task with itself present no problem – the precedents are all completed earlier. Those above the diagonal require information from tasks that start later – the information flows are reciprocal. The only ways for the task under such conditions to be completed are:

- estimating the information that will come from the later task, thereby increasing costs through overengineering;
- delaying task completion, thereby increasing durations.

By applying algorithms to the matrix, it is possible to identify groups of interdependent tasks, a process known as prioritising the matrix. This provides a much improved design schedule by identifying where interdependencies lie and programming all the non-interdependent tasks sequentially. The second stage of the analysis is to tear the matrix by placing those tasks which require reciprocal information flows as close together as possible, thereby reducing durations. It is also possible to identify those tasks where the costs of estimating the information required from later tasks can be traded off against the benefits of reducing durations.

This approach has been successfully applied to the detail design phases – moving from complete definition to complete description where uncertainty is already much reduced. It is not viable for the earlier stages of design[21] where information processing is not so much reciprocal as iterative. A second issue is that the DSM only identifies task dependencies and interdependencies; it contains no analysis of task durations or resources, and is not of itself a complete programming method. Work remains to be done before it can offer the same sort of programming capability for the detail design process that CPM – with all its limitations – presently offers for the on-site execution processes.

11.5.5 Visualising the schedule

Advances in computing are making it possible to visualise the project schedule, so that constraints and logic failures in the network can be more easily identified. Known as 4D planning ($x + y + z + t$), this approach takes the 3D product model of the facility created in the designer's CAD package and adds the capability for the different components to be added to the model in the sequence defined by the proposed schedule, thereby simulating the facility constructing itself in virtual reality on the screen. Figure 11.7 shows a 4D image of progress on the Centuria Project featured in Cases 10 and 11. These visualisations can significantly enhance the ability of the project management team to analyse constraints in the schedule as shown in panel 11.5. As one senior BAA project manager on put it, 'The 4D model saved Heathrow's T5 project £2.5m in the first 9 months of use'[22]. To enhance communication and debate, these visualisations can then be projected on to a large screen display and 'walked through' in the same way as other VR models and Walt Disney Imagineering has experimented with using an immersive *cave* for schedule visualisation as presented in panel 11.5.

Panel 11.6 Oaker Project and the butterfly effect

The project was carefully planned to allow on-site measurement at partially completed stages of the fabrication and installation of a cantilevered steel staircase over three floors and surrounding glazed steel box. The site measurement error by the steel fabricator was because of using recently hired staff without additional supervision. This led to rework on the staircase – the whole thing had to be craned out again and re-fabricated. This was completed willingly and quickly by the fabrication subcontractor – there was no doubt as to the responsibility for the error. However, the delay meant that the glazing subcontractor was unable to proceed and the installation slot missed. By the time steelwork had been reinstalled, the glazing contractor's only crane had been allocated elsewhere at various places around the UK including the BAA site at Heathrow Terminal 5. As autumn became winter, the weather deteriorated and planned installations were cancelled as a result of high wind. All this time, the glass was in store at the glazing contractor until an accident with a fork-lift truck – again a newly hired operative – resulted in all but one pane being smashed. The glazing contractor – who was understandably fed up about losing the slot in the first place – was now almost certainly losing money on the job and became increasingly uncooperative. It is notable that the glazing contractor was the only subcontractor who was not part of the main contractor's regular supply chain network. This was not a formally partnered network, rather a *fraternalist* one that is traditional for small, regionally based builders.

All this time, the house had an 8 m × 2.5 m hole in its gable wall. Remarkably, the temporary protection kept the weather out even in the winter gales, and progress continued with fitting out. However, out-of-sequence working was required which led to difficulties in maintaining the level of finish required, and made testing the heating system prior to the floor being laid an expensive business. Increasing desperation on the part of the client and builder increased the pressure on the glazing contractor. This was because of schedule pressure arising from the tenancy on the client's temporary accommodation drawing to a close – a one-month buffer on a six-month build schedule was starting to look completely inadequate. The builder's assurances that the client should not worry because it was entirely his problem became increasingly thin as homelessness loomed – a classic example of how risk is never really transferred by clients to contractors. Finally, the glazing contractor installed the glass two days before the removal men arrived with the client's furniture. Although the furniture was moved in, the client and family stayed in a hotel for three days at the builder's expense while he threw resources at the job resulting in some arguments between gangs on site, and at least one gang walking off site on the grounds that they could not fit balustrades on a staircase with people walking up and down it – ironically, they were from the steel fabricators. The costs to the client of all this were some sleepless nights and acceptance of some compromises in finish; the cost to the contractor was significant in terms of both sleepless nights and loss in labour productivity; two subcontractors probably lost money on the job. This was truly a butterfly effect where a small error was amplified right through the whole project. The system dynamic model allows us to analyse the situation more incisively and pinpoint the interaction between the two positive reinforcing loops. A full system dynamic model would assign values to the various elements which would allow simulation of the project dynamics and might even have found a better way forward than phoning or visiting the glazing contractor virtually everyday.

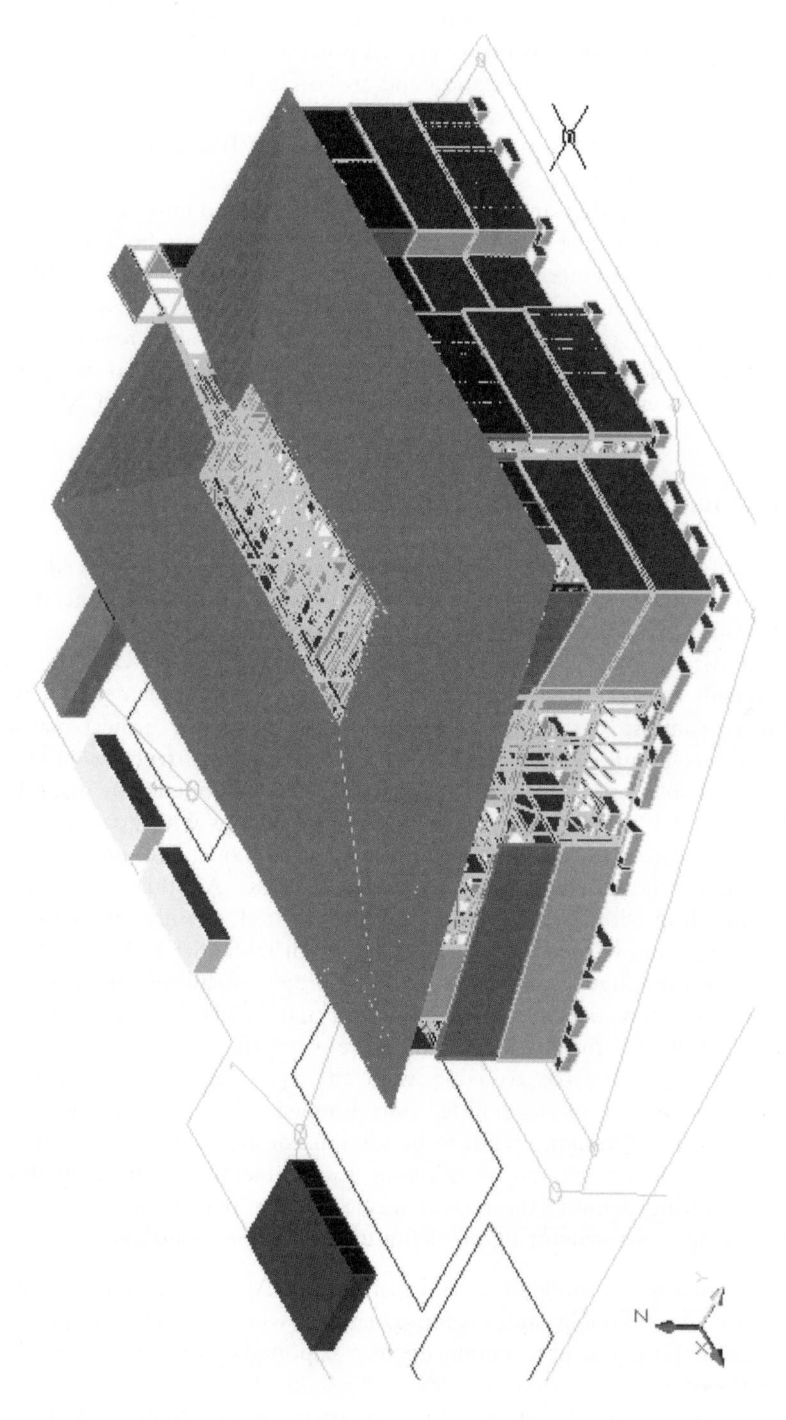

Fig. 11.7 Centuria Project: January 2000 (source: University of Teesside).

A further development of the thinking behind 4D planning is *virtual prototyping*, where the Building Information Model (BIM; see panel 14.4) is linked to scheduling software such as Primavera[23]. A process model can then be constructed of the project allowing visualisation and analysis at key stages. Parametric objects associated with temporary works are also generated thereby allowing the BIM model based on the PBS to be integrated with the WBS. Some of the benefits are illustrated in Case 14.

The potential of 4D planning to enhance CSA discussed in section 11.5.2 is obvious, although some proposals tend to be overelaborate. Apart from a few situations where overhead clearances for plant movement need to be planned, spatial planning on a construction site is essentially a 3D problem ($x + y + t$) because working at height effectively sterilises the space below for access or safety reasons. The ability of the latest generation of BIMs to treat space as a building element rather than just a residual of its enclosures should greatly ease the implementation of CSA.

11.6 The dynamics of the project schedule

As the information loop (Fig. IV.1) illustrates, thinking in terms of feedback loops in information flows is central to the effective management of projects, particularly while riding the project life cycle. The information loop is a basis application of *systems thinking* which holds that the interactions between the elements of a system have properties greater than the sum of those elements. It therefore stands in contrast to the decomposition inherent in many project management tools and techniques – most notably the work breakdown structure and tools that build on it such as critical path analysis. Typically, a distinction is made between *hard systems*, which are analogous to engineered systems with quantified variables and self-regulating features, and *soft systems* which are open and interact with their environment and usually lack self-regulating features[24]. Another way of expressing this basic distinction is between the well-structured and ill-structured problem, and the tame and wicked problem summarised in Tables 8.1 and 9.1 respectively.

Systems thinking has a long history in project management. Hard systems thinking underlay the success of the Atlas and Polaris weapons systems programmes as well as the Apollo mission – see panel 15.1. Such analytic approaches proved, however, unable to cope with the wickedness of the ill-structured problems of urban America and fell into disrepute. There are currently two ways in which systems thinking tends to be used in construction project management. The first is to use *soft systems* thinking as a problem-structuring method, particularly during defining the project mission. A number of different soft systems methodologies for structuring the definition process are available:

- Strategic options development and analysis (SODA) can be used to elicit and visualise trains of thought using cognitive mapping, which then allow further development of the argument in a group context supported by *Decision Explorer* software.
- Soft systems methodology (SSM) which involves capturing the real-world problem in a rich picture and then comparing it with frameworks for possible

solutions, which are then discussed through structure interaction and debate. SSM can be used both to understand the real-world problem, and also to structure the group process for coming to an understanding of that problem[25].
- Mixtures of these and other methods to explore different aspects of the problem in a structured and facilitated way. Soft systems methods can also be combined with harder approaches where appropriate data are available.

The second application is the development of *systems dynamics* analysis of the project process – see panel 7.6 for some background. Systems dynamics approaches typically focus on the limitations of critical path analysis in responding to disruptions and changes to the schedule, be they deliberate because of scope changes or unintended because of the discovery of additional work to be done[26]. The argument is that the ripple effects from such changes have much greater impact on the schedule and budget than simply the quantum of additional work to be done for a number of complementary reasons.

- The typical response to such events is to increase overtime working to bring the project back on schedule. However, once people are working more than 40 h a week their efficiency and effectiveness starts to tail off, and so errors increase and the amount of rework to be done also starts to increase as a direct result of the attempt to catch up. As actual working time as a proportion of total working time starts to drop off with increasing overtime, the marginal rate of pay starts to climb exponentially[27].
- An alternative option is to add additional resources, but then Brooks' law comes into play which states that 'adding manpower to a late software project makes it later'[28]. This law arguably applies to all design projects as existing staff have to stop work to induct the newcomers who take time to get up to speed and communication networks become more dense and time-consuming. A similar effect occurs with on-site execution as additional resources start to overcrowd the site and workers start to get in each other's way as discussed in section 11.5.2.
- Projects can reach a *tipping point* where the rate of addition of tasks because of scope changes and rework actually exceeds the rate of completion of originally planned amount of work to be done when the project becomes unstable and effectively starts to go backwards towards failure and possible abandonment[29].

Panel 11.6 provides a narrative of the ways in which a small error early in the project led to serious schedule escalation on a small house extension – the ability of apparently trivial events such as the flap of a butterfly's wing to generate major changes in weather systems is known as the butterfly effect and lies at the heart of chaos theory[30]. Figure 11.8 presents a systems dynamic diagram of the process that nearly led to (personal) project disaster.

11.7 Summary

Is Goldratt's claim that there has been nothing new in construction planning for 40 years, plausible? Even as late as 1995, the answer might have been yes but

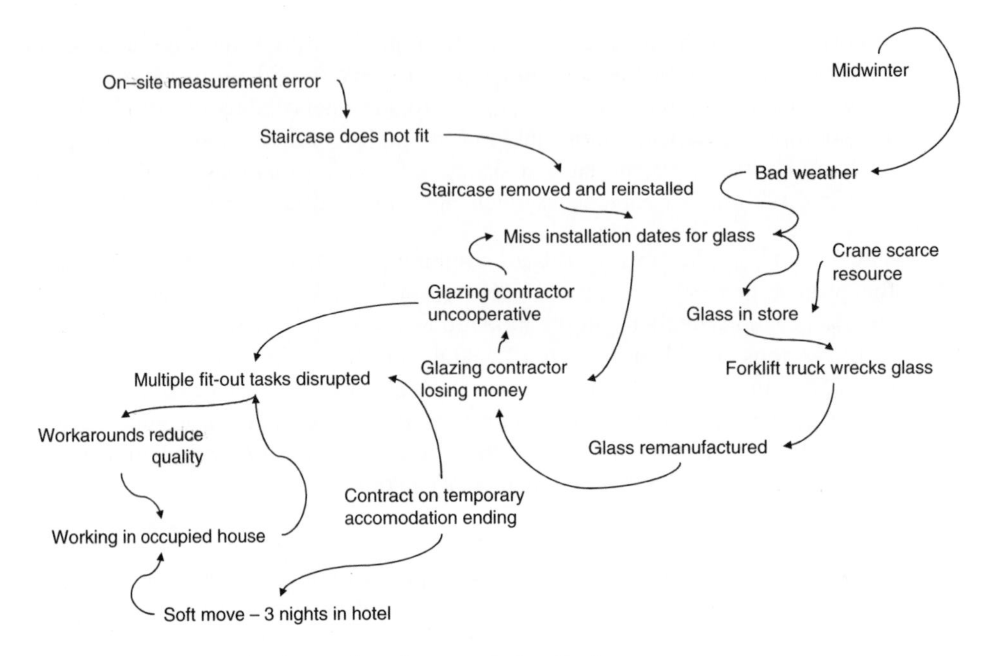

Fig. 11.8 The butterfly effect on Oaker Project.

this is clearly no longer the case. Last planner, DSM, systems dynamics, critical chain, critical space and 4D planning are now all coming to maturity with usable applications for the management of construction projects. There are a number of exciting new tools and techniques available, at the same time as changes in the ways clients mobilise resource bases are giving project managers much greater opportunities to plan the schedule, and link it more tightly with the budget. Further work is now required to integrate these new tools into a single information system to support riding the project life cycle.

Many of the issues in improving scheduling performance in managing construction projects are linked to those for budgeting – process capability. There is considerable variability associated with task execution durations, which make scheduling difficult and encourage the padding of durations. While last planner and critical chain do address the problem of variability, they do it by buffering task execution rather than addressing directly the problem of process capability; for this reason they cannot be considered full 'lean construction' tools. It is to this problem that we turn in the next chapter.

Case 11
Centuria Project Schedule

As presented in Case 10, the Centuria Project roofing package overran by 5% on budget. Instead of being completed with the achievement of the atrium watertight milestone by 20 January 2000, it was achieved on 27 January 2000, a schedule variance of 5 on 52 working days, or 9.6%. The first pass at the Centuria Project roofing schedule to achieve the atrium watertight milestone produced the Gantt chart and critical path network presented in Figs. 11.9 and 11.10 respectively. Week 9 is the middle of the Christmas break. The two parallel critical paths in Fig. 11.10 are highlighted in bold. This schedule used the common principle of start every task as soon as possible, which allows maximum slack to be stored up for (sometimes literally) a rainy day. A PERT analysis was performed on this by taking the pessimistically weighted average of the optimistic, expected and pessimistic task durations (1-3-2), generating a two-working-day (3.8%) variance on the roofing package as a whole, but with the building not watertight for 4 more days because of the weekend. This risk could be largely mitigated by buffering the rooflights by either starting the flat roofing earlier or putting more resources into that task.

However, this planned schedule contained resource clashes between various tasks:

- Scaffolding to eaves and birdcage scaffolding;
- Guttering to the north and south elevations;
- Insulation and Kalzip roofing to the north and south elevations.

Inspection of Fig. 11.10 shows that neither the birdcage scaffolding nor the guttering is on the critical path. This means that the birdcage scaffolding can be moved to follow on from raising the scaffold to the eaves, and the guttering can be rescheduled to flatten resource use without affecting the schedule. This *levelled* schedule was used as the basis of the budget management scenario presented in Case 10. However, this still leaves a problem with the roofers, where five roofers at 40 h gives a base working week of 200 h. While the additional hours required in weeks 2 and 3 could be covered by overtime – albeit at additional cost – this is simply not possible for weeks 4 and 5. The only way out is to provide additional resources, or let the milestone date slip.

During week 2, the scaffolders are hit by a bit of bad weather, losing 2 days, which delays the start of all dependent tasks. However, most critically, the roofers start to lose ground against the schedule because of resource constraints – there are simply not enough of them to cover both the north and south slopes, despite extensive, and expensive, overtime. Further problems with the snowy weather in mid-January mean that the installation of the atrium roof is delayed, and again, extra steel erector resources have to be used to prevent further slippage against the schedule once the weather clears. Figure 11.11 illustrates the performance actually achieved against the schedule baseline, using the Tracking Gantt facility. The upper bar shows achievement against the schedule baseline. For the guttering and scaffold, the difference between the two is the effect of levelling resources. For the insulation and roofing tasks, this is the result of overruns because of resource constraints.

Can the project be planned better? By enhancing the CPM with critical chain techniques it can, in the following manner[31].

Step 1 Critical path – as early as possible

Figure 11.10 shows that there are two parallel critical paths to the milestone running via tasks 1-4-5-11-12 and 1-6-7-11-12. This lasts 52 working days.

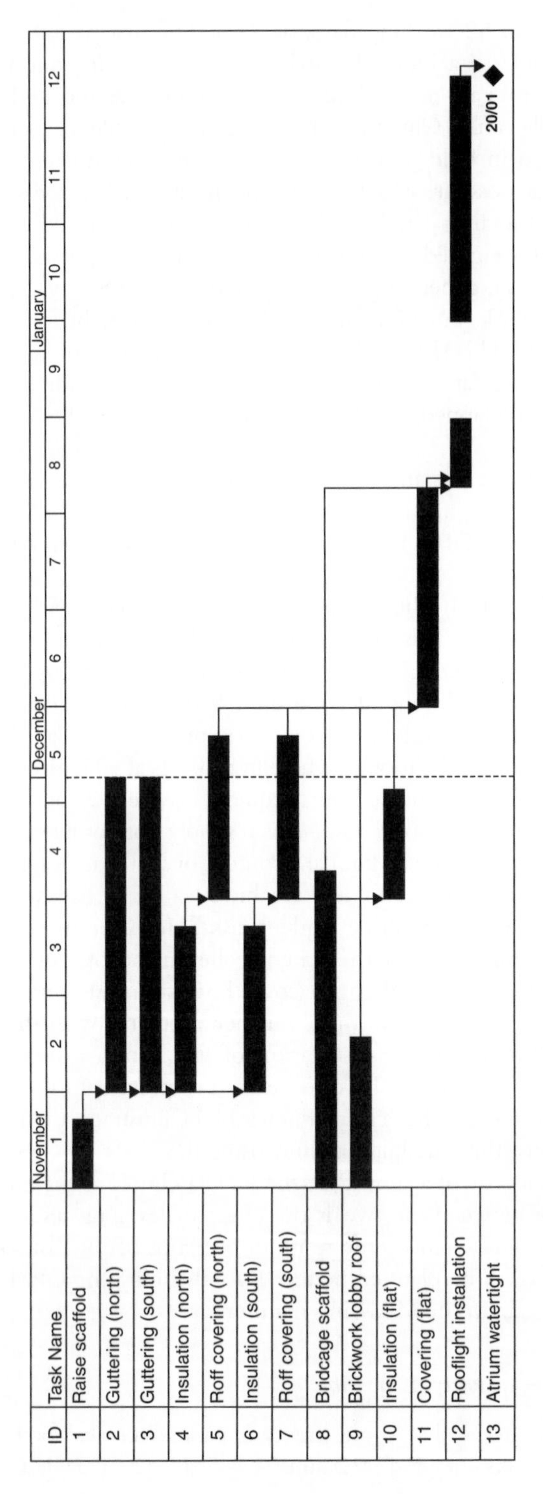

Fig. 11.9 Centuria Project planned Gantt chart.

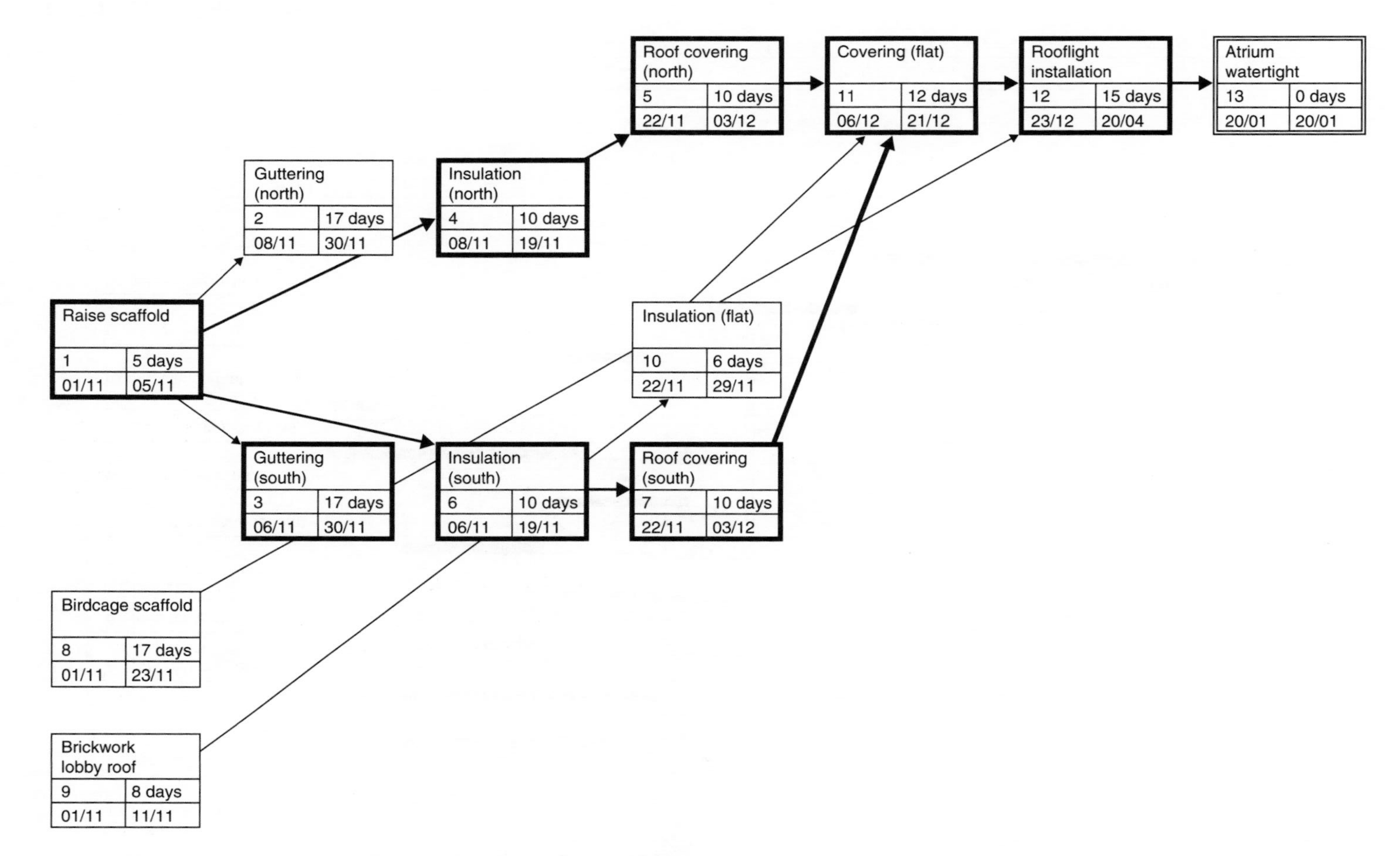

Fig. 11.10 Centuria Project planned critical path network.

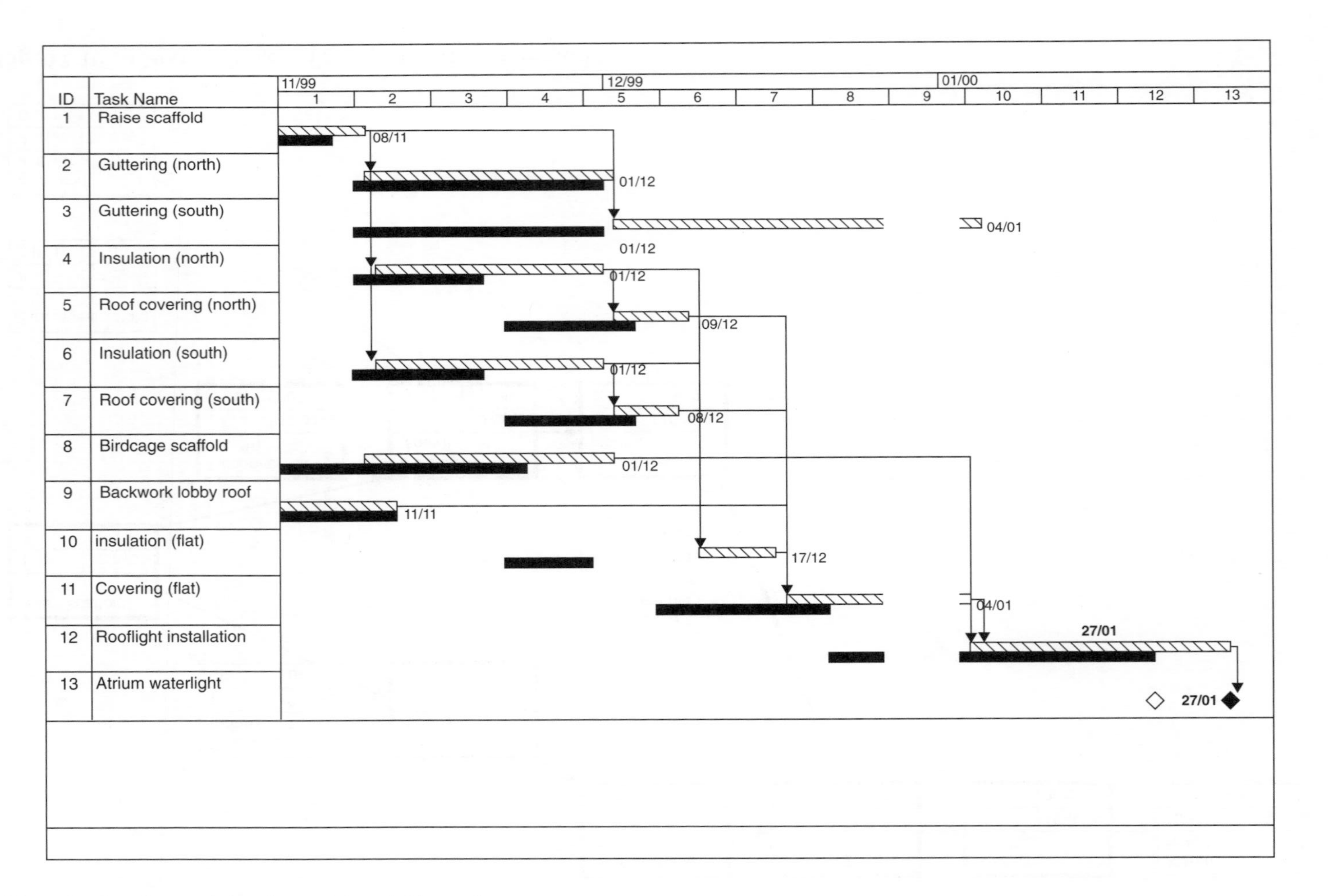

Fig. 11.11 Centuria Project achieved Gantt chart.

In addition, there is a near-critical path running through tasks 1-9-10-11-12 which takes 47 working days. Of the total budget baseline of £47 518, 37.8% is overtime. This figure should be compared to the figures in Table 10.4, which are without overtime. Clearly there is some room for improvement.

Step 2 Critical path – team enhancement

Here it is assumed that five scaffolders can do in 4 days what four can do in 5 days. Also, it is assumed that those roofing tasks which do not use all five roofers (fixing the insulation – tasks 4, 6 and 10) can be speeded up proportionately by throwing all roofing resources at them. This results in the two critical paths being shortened to 43 days and the near-critical one to 40 days. The budget baseline is slightly up at £48 117, of which 41.1% is overtime.

Step 3 Critical path – as late as possible

This is similar to the first pass except that the cash flow would be improved by having to pay costs as late as possible. Because the eaves scaffolding is on the critical path and the birdcage is not, we eliminate the need for scaffolder overtime and reduce the budget baseline slightly to £46 769, of which 33.6% is overtime. To maintain the logic, it is necessary to make milestone completion dependent on the guttering, although there is no technical reason for this.

Step 4 Critical path – levelled resources

In this case we start with the first pass and then heuristically reschedule tasks without amending the milestone date. In this case the two halves of the guttering and the two scaffolding tasks can be separated. There is no change possible on the roofers' overtime. The budget baseline comes down to £44 224, of which 18.3% is overtime.

Step 5 Critical path – resource constrained (as soon as possible timing)

In this case we relax the milestone date but disallow any overtime. The teams are not changed. The budget baseline – not surprisingly – comes down to £41 528 but the time taken gets pushed up to 72 days.

Step 6 Critical chain – identify and exploit the constraint[32]

Here we use the step 1 critical path but then take the padding of the duration for each task to try and represent what package contractors might do in real-time estimating. We assumed that each duration was padded by around 25%, adjusted to make a whole day. The budget baseline without buffers is £36 800, of which 38.5% is overtime. Figure 11.12 shows the ProChain critical chain. The constraint is clearly identified as the roofer resource, and task 10 (flat roof insulation) now appears on the critical chain, shown by the bold boxes while it was not on the critical path – compare with Fig. 11.10 – because it uses a critical resource. This resource is fully exploited through overtime working.

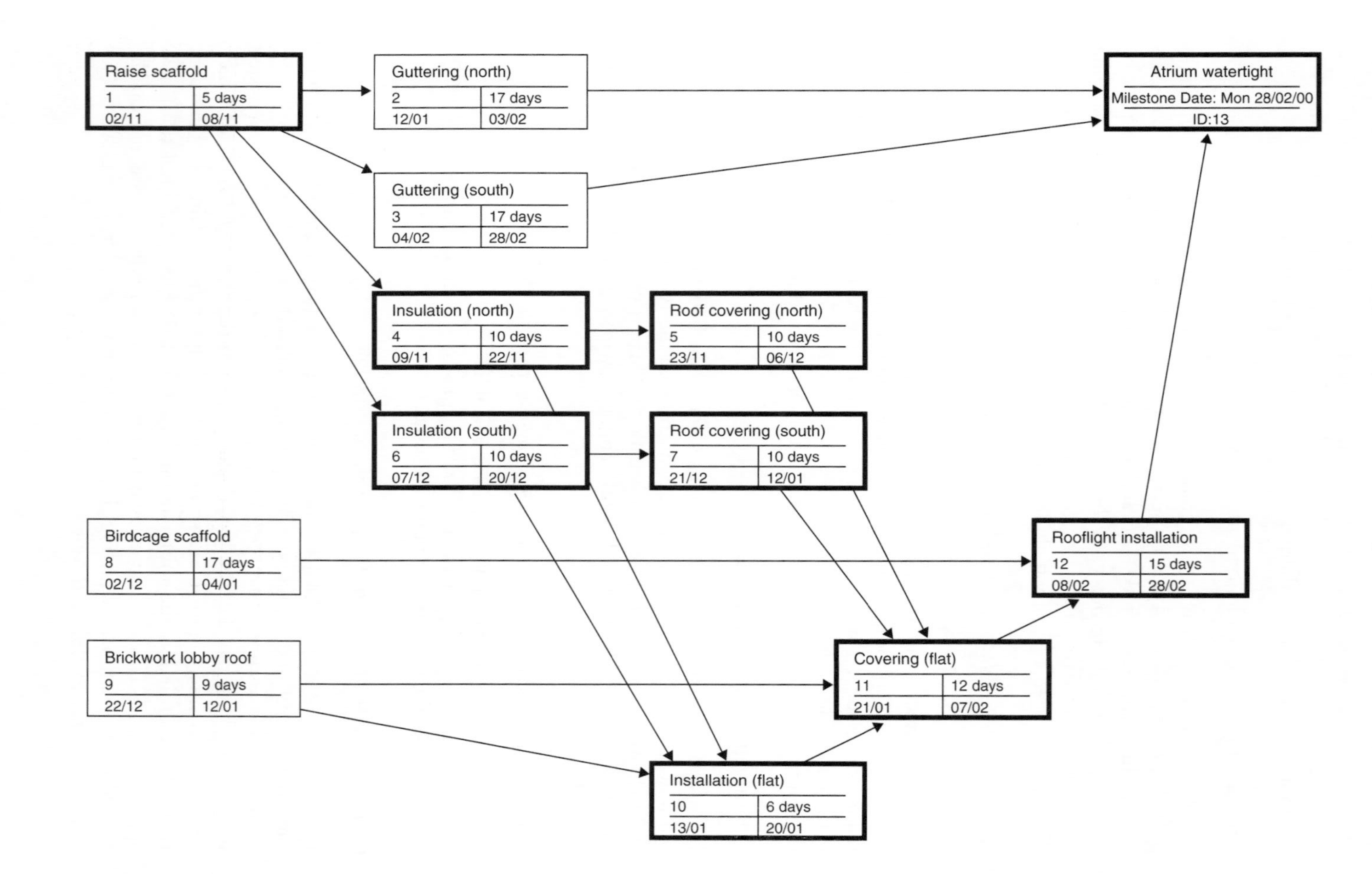

Fig. 11.12 Centuria Project critical chain network.

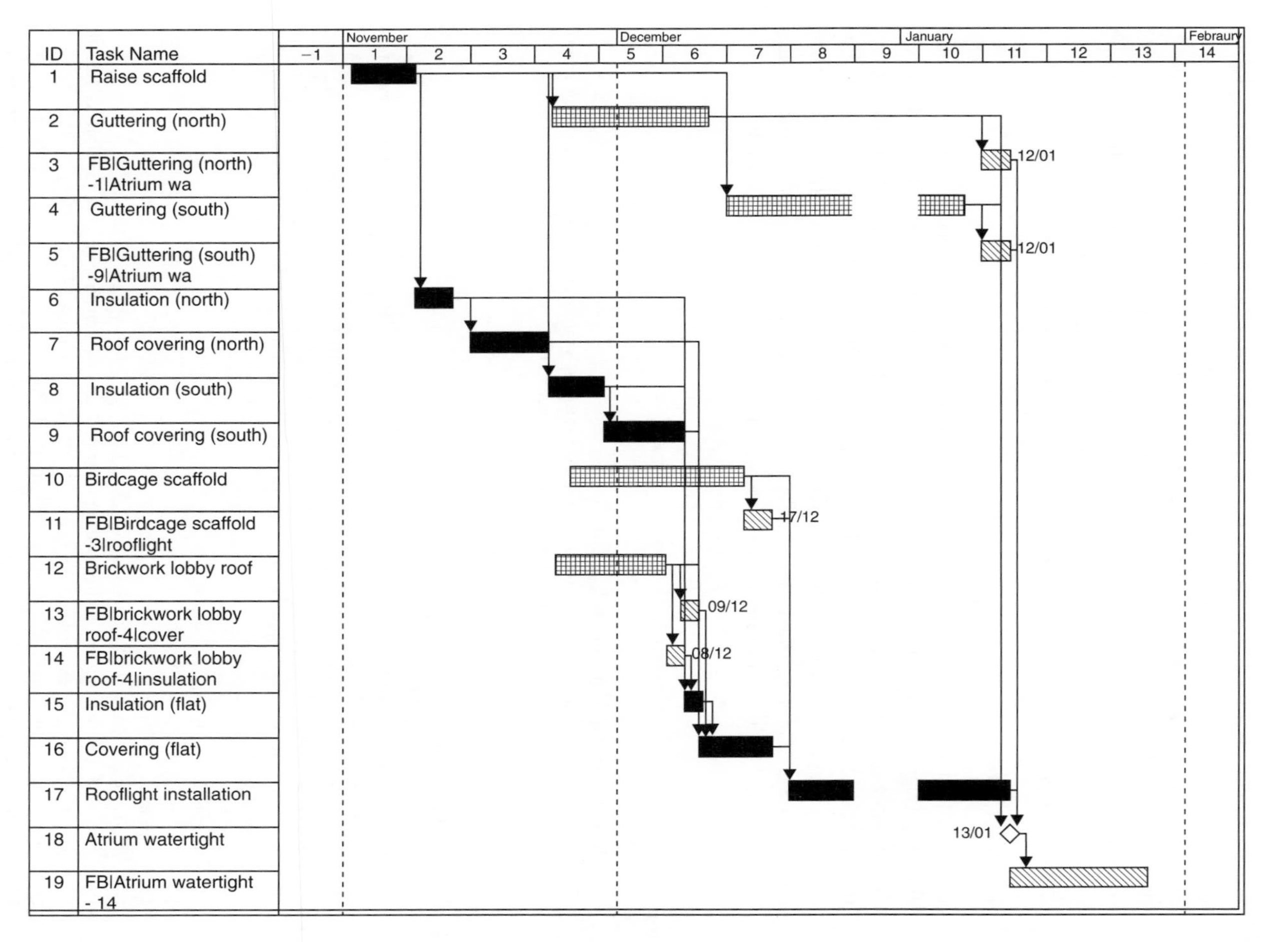

Fig. 11.13 Centuria Project critical chain Gantt (critical chain in solid, non-critical tasks in cross-hatch, and feeding and project buffers in hatching).

Step 7 Critical chain – subordinate to the constraint

Here we subordinate to the constraint by pushing the tasks back to their latest start times and constraining resources to disallow overtime. Here task 10 needs to be buffered from its precedent task which is not on the critical chain – task 9, lobby roof brickwork. The atrium complete milestone also needs to be buffered from the guttering tasks. Figure 11.11 shows the ProChain Gantt. The milestone completion date is pushed back to between 56 and 68 days – shown by the last bar – while the budget baseline comes down to £32 083, should the project be achieved in 56 days. The last bar in Fig. 11.13 shows the expected range for the project completion milestone – this forms the project buffer for the completion milestone.

Step 8: Critical chain – elevate the constraint

Here we add additional roofing resources by increasing gang-size by one roofer and allowing overtime working. This is now possible because we have reduced the baseline budget by removing padding from task durations and deploying it to protect the critical chain. The baseline budget rises to £37 260 but the completion date is reduced to between 43 and 55 days. In other words, we have slightly better than evens chance of delivering more quickly than the original master schedule, within a significantly reduced budget of 15% against the costs in the PMB in Table 10.4.

Notes

1 In 1433, cited in Landes (1983, p. 92).
2 Our hero in Goldratt's *Critical Chain* (1997) is a business school professor who saves the school by teaching an innovative approach to project management.
3 This discussion is derived from Winch and Kelsey (2005).
4 Handout from a course on critical path analysis attended by my father in the mid-1960s – Lockyer (1963).
5 Lockyer (1963).
6 Interview 05/12/88.
7 It will be seen from the networks presented in Case 11 that MS Project does not entirely follow these conventions.
8 Preferably derived from Uniclass, as discussed in section 10.3; this is an important finding from the VIRCON project.
9 The concept was developed by BAA in the late 1990s. I am grateful to Glenn Ballard for this information.
10 Interview 22/05/89.
11 Interview 30/09/89.
12 Cleland and King (1983, p. 412).
13 Flanders (1970, p. 61).
14 Meredith and Mantel (2000, p. 364).
15 The work of Tommelein and her colleagues is perhaps the best known contribution to the site layout problem, for example Tommelein *et al.* (1991), Zouein and Tommelein (1999).
16 Horner and Talhouni (1995), Thomas and Smith (1990).
17 The argument here is that of Winch and North (2006) which builds on that of Thabet and Beliveau (1994) and Riley and Sanvido (1995).
18 Winch and North (2006) present decision support tools for the analysis of spatial loading.

19 See Ballard and Howell (1998, 2003) Ballard *et al* (2002) and, more generally, the work of the Lean Construction Institute at http://www.leanconstruction.org/.
20 Austin *et al.* (1999); see also Ulrich and Eppinger (2008).
21 Baldwin *et al.* (1999).
22 Marcus Kapps at http://www.arup.com/projectmanagement/ (accessed 01/11/08).
23 See the work of the Construction Virtual Prototyping Laboratory at Hong Kong Polytechnic University – Baldwin *et al.* (2008), Huang *et al.* (2007), Li *et al.* (2008).
24 This review of system dynamics is based on Rosenhead and Mingers (eds.) (2001).
25 Winter (2006).
26 See Eden *et al.* (2000) and Lee and Peña-Mora (2007) for examples of this type of analysis.
27 Cooper (1994).
28 Brooks (1995, p. 25).
29 Taylor and Ford (2006).
30 See Gleick (1988) for a lucid exposition of chaos theory.
31 The analysis here was prepared by John Kelsey. It was prepared initially in Excel, and then verified using MS Project 98 and ProChain: http://www.prochain.com/.
32 See panel 11.3 for the critical chain terminology used here.

Further reading

Brooks, F.P., Jr. (1995) *The Mythical Man-Month: Essays on Software Engineering* (2nd ed.). Reading, Addison-Wesley.
This classic on software engineering has much that is of relevance to construction project managers. The author identifies one of the major problems of software engineering as being the lack of people responsible for the product as a whole, or, in his words, architects.

Meredith, J.R., and Mantel, S.J. (2006) *Project Management: A Managerial Approach* (6th ed.). New York, Wiley.
An excellent all-round project management text.

Williams, T. (2002) *Modelling Complex Projects* Chichester, John Wiley.
One of the first texts to take a systems dynamics approach to modelling projects

Chapter 12
Managing Conformance

12.1 Introduction

> Steel doesn't always fit and there are many reasons for it. A detailer can goof. The shop can goof. Even worse, Murphy can cause us to goof, and when I say Murphy, I mean an accumulation of tolerances, all added, like eighty-six basic steel tolerances, plus shop tolerances, plus the field tolerances. When they all add up to a misfit, then we have to address it. But we have troubleshooters on each job and they resolve them.

Gene Miller of Mosher Steel[1] identifies both the cost of non-conformance and the problems generated by the culture of broad tolerances; troubleshooters do not add value, and their resolutions of misfits can have knock-on effects elsewhere on conformance, budget and schedule. As we have seen in section 8.6, quality in construction has at least four different, but related meanings:

- quality of conception;
- quality of specification;
- quality of realisation;
- quality of conformance.

Chapter 9 was about defining product integrity as a function of the project mission. The dimensions of product integrity – the qualities of conception, realisation and specification – define the requirements that must be met in realisation. Chapters 10 and 11 turned to the aspects of process integrity in terms of budget and schedule. This chapter is about the fourth aspect of quality – conformance to requirements. However, this chapter also pushes beyond this product-related definition to embrace quality in processes. High-quality processes are those that minimise the risk of accidents to both people and the environment. The quality of intention on all projects includes, by definition, the avoidance of physical loss through accidents; this principle is, for instance, enshrined in the European Union health and safety directives.

Process integrity, therefore, means conformance to specification, which is the traditional quality management definition, and also conformance to the regulations on health, safety and the environment. It is this broader sense of managing conformance to intention that this chapter will address.

Many things can go wrong during realisation: drawings might be ambiguous; workmanship can be inadequate; standards and regulations may not be followed; details may be overlooked; and, most seriously, accidents can happen. The aim of the conformance management process on the construction project is to avoid these pitfalls so that the facility is delivered as intended, and central to achieving this is effective management systems. This chapter will start by examining some general principles of quality management before going on to investigate the various aspects of quality management systems (QMS), and some of the tools and techniques that have been developed over the past 50 years or so of formal quality management. Conformance quality is treated first, because it was here that management systems were first developed. The same principles are now being extended to health and safety, and to environmental conformance to form integrated management, or QUality, ENvironment, Safety and Health (QUENSH) systems. The concept of process capability is used to define more closely the objective of conformance management, and the reason for the creation of a culture of improvement. One way of enhancing process capability through standardisation and pre-assembly is then explored. The discussion in this chapter illustrates well the relevance of the process-level model introduced in section IV.3 – process performance improvement depends on the successful development and implementation of routines such as integrated management systems so as to enable teams to execute tasks which conform to requirements.

12.2 The principles of quality management systems

The four main components of QMS are illustrated in Fig. 12.1[2]. The first dimension in the model is whether the approach is a reactive one, picking up non-conformance after the event, or proactive, trying to prevent non-conformance occurring in the first place. The second dimension is the range from the search for blame in the sense that non-conformance must be somebody's fault, to empowerment through encouraging those doing the work to take responsibility for achieving high quality, and rewarding such achievement. The four basic approaches identified are:

- *inspection*, or the reliance upon physically checking that work has been completed satisfactorily;
- *quality control (QC)*, or the reliance on management control techniques to achieve high levels of quality;
- *quality assurance (QA)*, or the use of externally accredited procedures to ensure that quality management practices are rigorously followed;
- *total quality management (TQM)*, or the motivation of continuous process improvement to achieve higher and higher levels of conformance to intention.

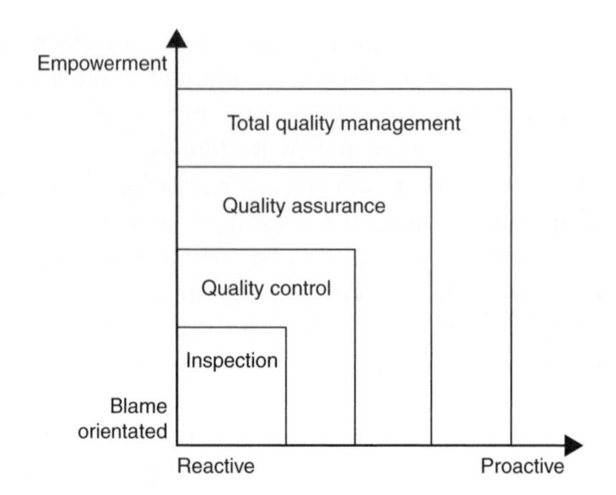

Fig. 12.1 Approaches to conformance quality management.

In the model all the different techniques are nested. Thus TQM includes elements of inspection, QC and QA, but QC can operate satisfactorily without QA or TQM and often does. Thus the different approaches are *not* alternatives to each other, but complements. The following sections will discuss each of them in turn, based on definitions taken from the standard quality vocabulary which defines *quality* itself as 'the degree to which a set of inherent characteristics fulfils requirements'[3]. This definition is enormously important for understanding how quality is managed on the construction project, for, as Chapter 9 showed, the process of identifying requirements precisely enough for design to proceed is, itself, difficult. However, it is the essential prerequisite for any effective conformance management in the sense described in this chapter.

What, then, is meant by stating that a component, or indeed an entire product, conforms to requirements? There are two aspects to this which might be considered as in-process conformance and pre-process conformance. *In-process conformance* is more familiar, and is usually defined in terms of whether the characteristics of the component as measured through inspection lie within the pre-specified tolerances. If the component lies within the tolerance band, then it is in conformance. Through sampling, it can be determined whether a batch of components reaches an acceptable quality level (AQL), where AQL is defined as the percentage of components to be in conformance. If the AQL is 100%, then the process is producing 'zero defects'. The essence of Crosby's argument is that achieving zero defects will pay for itself through the elimination of 'quality costs' such as wasted effort, scrap materials and warranty claims – 'quality is free'[4].

However, the tolerance-based approach to conformance through zero defects has been criticised by many. The main problem is that of tolerance stack-up[5], as illustrated in panel 12.1. The fundamental problem of defining tolerances is that process variations are random within the tolerance bands, and they can interact unpredictably to create non-conformance at the level of the product as a whole, which will only emerge when the product is in use. Thus a zero defects approach cannot ensure a working product, especially as it is possible to achieve zero defects by setting undemanding tolerance bands – for instance, on grounds of constructability as discussed in section 10.6. The losses implicit in failing to achieve quality targets are called the *quality loss function* (QLF), which is defined in panel 12.2. A more rigorous approach is always to aim for the target by steadily improving *process capability*, defined as the ability of the process to consistently deliver the conformance required in panel 12.3. These two different approaches to managing in-process conformance are compared in Fig. 12.2. Panel 12.3 demonstrates how far construction is from manufacturing (the analytic techniques also apply to services) levels of conformance – Case 12 shows that on T5 they were pleased with only a 10% defect rate.

Even though additional costs of investment in training and equipment may be incurred to achieve process capability, those investments will be returned through reduced losses from the product in use as given by the QLF. However, the QLF is not just a function of process capability but also of the design of the product. The components must be appropriately specified to achieve the level of performance

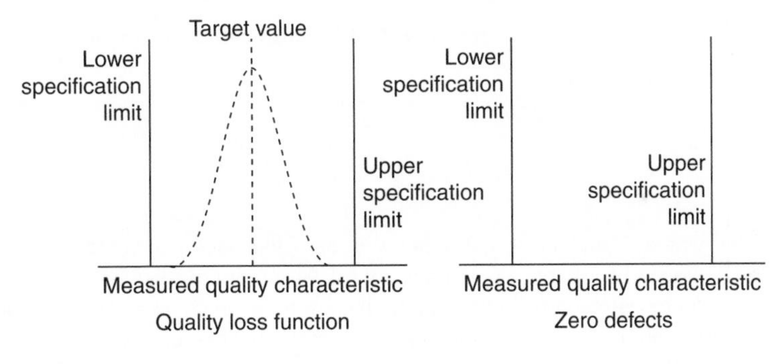

Fig. 12.2 Process capability and zero defects approaches to conformance.

Panel 12.1 Tolerance stack-up in steelworks fabrication

The Grand Staircase of the Tate Modern Gallery in London is a major feature. Yet, because of tolerance stack-up, remedial works had to be conducted to make sure that it aligned with the rest of the structure. The staircase consists of 22 flights of heavy steel plate, with wooden treads and handrails running through the steel structure supporting the gallery spaces. Two main problems arose as it was constructed. Firstly, the thickness of the finishes on the landings and the thickness of the finishes on the adjoining floors were different, so the whole staircase had to be jacked up to ensure that they met flush. Secondly, the whole staircase was a very tight fit as drawn, and tolerance drift in the concrete floor on which it stood meant that the head height required by the regulations was not met at one point; as the architect puts it, 'building is in centimetres . . . so you very easily lose two centimetres'. The offending flight had to be cut out and set back 15 cm at a cost of £60 000 and 3-week delay.

Source: Sabbagh (2000).

Panel 12.2 The quality loss function

When a product fails in use, the total costs of that failure are many times the costs of manufacturing that component. The costs include repair costs, the downtime costs incurred because of the loss of functionality of the product, the loss of the repeat business of the client which switches to a more reliable product of a competing supplier, and general loss of goodwill. These costs are measured by the QLF, where the loss (l) is given by

$$l = d^2 c$$

where d is the deviation from the specified production target value and c the cost of the countermeasure deployed to fix the problem.

While notional, the QLF does give a feel for the costs of non-conformance – think of the cost of a plumber taking 25% longer to make a connection, against the costs of breaking the joint out and remaking it when it fails in use.

Source: Taguchi and Clausing (1990).

Panel 12.3 Process capability and 6σ

Process capability, C_p, is defined as

$$C_P = \frac{T_U - T_I}{6\sigma}$$

where T_u and T_l are the upper and lower tolerance bands respectively, defining the tolerance band, and σ is the standard deviation of process performance around the mean. A process capability of 1 is where the specification range and natural variation of the process are equal; a figure less than 1 indicates that one or more of the observed parameters lie outside the tolerance range and the process is not capable. This formula means that sampled variability will be within $\pm 3\sigma$ of the mean 97.3% of the time resulting in 66 800 defects for every one million components produced. Six-sigma is a much more demanding target where sampled variability is within half of the $\pm 6\sigma$ specified variability about the mean, resulting in 3.4 defects per million components produced.

Source: Slack *et al.* (2007).

expected of them. This is the problem of *pre-process conformance* – how to ensure that the desired level of performance in use is achieved by choosing the appropriate design parameters, in particular, tolerances; in other words, how to achieve reliable operation.

Constructed facilities are remarkably robust products, and are almost always repairable and adaptable. Complete demolition usually only occurs when the NPV of constructing a completely new facility on the existing site is greater than the NPV of maintaining or adapting the existing facility. The issues, therefore, revolve around the reliability of the various sub-systems of the facility, rather than the system as a whole. One of the major design issues in the management of construction projects is that different component systems have different reliability profiles; the attempt to cope with these differences is one of the major reasons for the open building approach, as discussed in section 9.7.3.

The reliability of a component is given by the probability of its failure in use at a given point in time, and the overall reliability profile of a component system is given by the bathtub curve illustrated in Fig. 12.3. The two axes are time and the probability of failure at a given time. The overall reliability profile consists of three elements.

- Failures of components which do not achieve their design life. This curve follows a declining path – such failures are typically more common in the early phases of their life cycle. The aim of the commissioning process is to get as far down this bit of the curve as possible before handover.
- Failures of components because of external factors such as accidents and misuse, the chances of which are constant.
- Failures because of components having achieved their design life. Design life is specified through the whole-life costing process presented in section 9.7.3.

One of the most widely used ways of establishing the reliability profiles of a component or product system is the use of failure mode and effect analysis (FMEA). This is a tool for analysing the specification in terms of its reliability profile. There are three elements to an FMEA analysis:

- *occurrence*, or the likelihood of a specific failure occurring;
- *detection*, or the likelihood of the failure being detected during realisation;
- *impact*, or the severity of the failure should it occur in terms of the functionality of the component system in use.

Each of these is rated on a scale of 1 to 10, with a view to allocating to each component a risk priority number (RPN). Improvement of the design to ensure that it is in conformance with requirements is then prioritised by focusing on those components with the highest RPN. Panel 12.4 describes an FMEA for a cladding system.

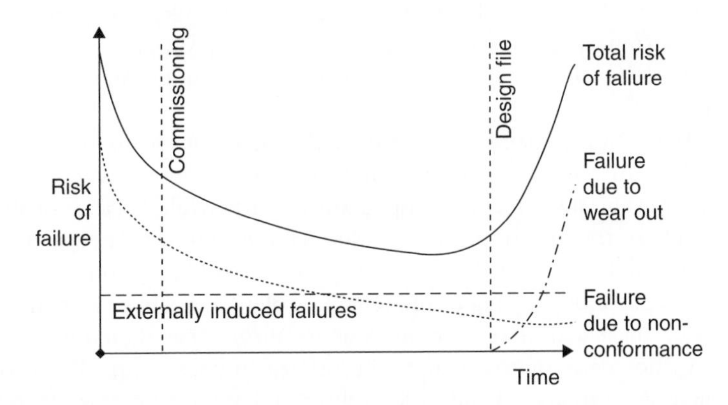

Fig. 12.3 The reliability profile (source: adapted from Dale, 1994, Fig. 8.1).

Panel 12.4 FMEA in cladding design

Research at Bath University has applied FMEA to a major building component – cladding. The first stage was comprehensive risk identification, and 16 different failure impacts in cladding were identified, with water penetration topping the list in terms of frequency. The cause of failure with the highest RPN was joint sealant. In terms of occurrence, it has a high probability of occurring. Its impacts are serious, including water penetration, increased air permeability, aesthetic losses and loss of thermal performance, with the most serious being structural failure. As the application of joint sealant is an inherently site-based process, its likelihood of detection is also fairly low. Potential risk responses identified included better attention to specification by designers, increased inspection during execution, designing out site applied sealants and better training of operatives.

Source: Layzell and Ledbetter (1998).

12.3 Inspection

Inspection is 'conformity evaluation by observation and judgement accompanied as appropriate by measurement, testing or gauging'. Inspection is usually carried out by persons other than those who have executed the work – usually some form of dedicated quality function – which makes an accept or reject decision. Non-conforming components – rejects – are then usually either reworked or scrapped. While inspection is capable of achieving high levels of quality – at least in physical goods – it is also expensive for two reasons:

- The inspection process requires dedicated resources;
- The rejected components generate non-conformance costs.

Because of its expense, inspection typically relies upon sampling methods, which immediately reduce its effectiveness. In any sampling method, because of the basic principles of statistical inference, there is a known chance of a faulty component not being rejected. This small proportion of faulty components will be picked up at some point either in later parts of the process or in use by the customer, and rectification will generate even further non-conformance costs through the QLF.

Inspection remains vital, particularly for components which are either safety-critical or the failure of which threatens the integrity of the whole product system. The investment in inspection is worthwhile because of the possible magnitude of the loss. Inspection is also vitally important for ensuring that the equipment used in construction remains process capable – in other words, ensuring that it is capable of working to the required tolerances. Thus plant and scaffolding need to be inspected at regular intervals to ensure that their functionality and safety have not deteriorated in use. The role of inspection in QC is to identify when there is a problem to be tackled. Higher levels of process capability will reduce inspection requirements for a given tolerance band.

12.4 Quality control

Quality control is the 'part of quality management focused on fulfilling quality requirements'. There are a number of operational techniques available – often virtually the same thing under different badges. Reference is often made to Ishikawa's seven QC tools[6]; this section will pick out some of those more applicable to the management of construction projects. These tools are often the basis of process improvement projects (PIPs) which will be discussed in section 12.7.

12.4.1 *Cause and effect diagrams*

These are often known as Ishikawa diagrams after their inventor, or fishbone diagrams. The idea is to work back from the manifest problem asking why it occurred, and then repeating the question for each of the underlying problems identified. This is a basic methodology in root cause analysis. Classifying the results of this process according to types of cause produces the classic skeletal shape, illustrated in Fig. 12.4. This shows the underlying causes of a fatal accident on a building site when a scaffold tower fell against a power line, electrocuting both the painter on the scaffold and the labourer excavating at its foot.

12.4.2 *Performance measurement*

Little progress can be made in understanding the roots of quality problems unless the outputs from the process are measured and analysed. Such data can be summarised in histograms, scatter diagrams or Pareto diagrams. These place the causes of problems in order of frequency of occurrence, and it is often noted that a small proportion of the causes is responsible for a relatively high proportion of non-conformance, typically on an 80/20 pattern. Where more than one parameter can be measured, the correlation techniques can also be used to further investigate the cause of problems. Figure 12.5 presents a Pareto diagram for non-conformance against quality on site. The data are taken from a study of 27 building sites, in which the team identified 501 quality-related events, 98 of which were serious enough to cause water ingress or structural instability. These data show that the biggest single cause of problems is 'unclear/missing design information', followed by 'lack of care'.

12.4.3 *Statistical process control*

As Ishikawa puts it, 'in every work there is dispersion. Data without dispersion are false data. Without statistical analysis, there can be no effective control'[7]. Output from a process will always be variable; the question is whether that output is within predetermined tolerance bands, defined as *control limits*. An important

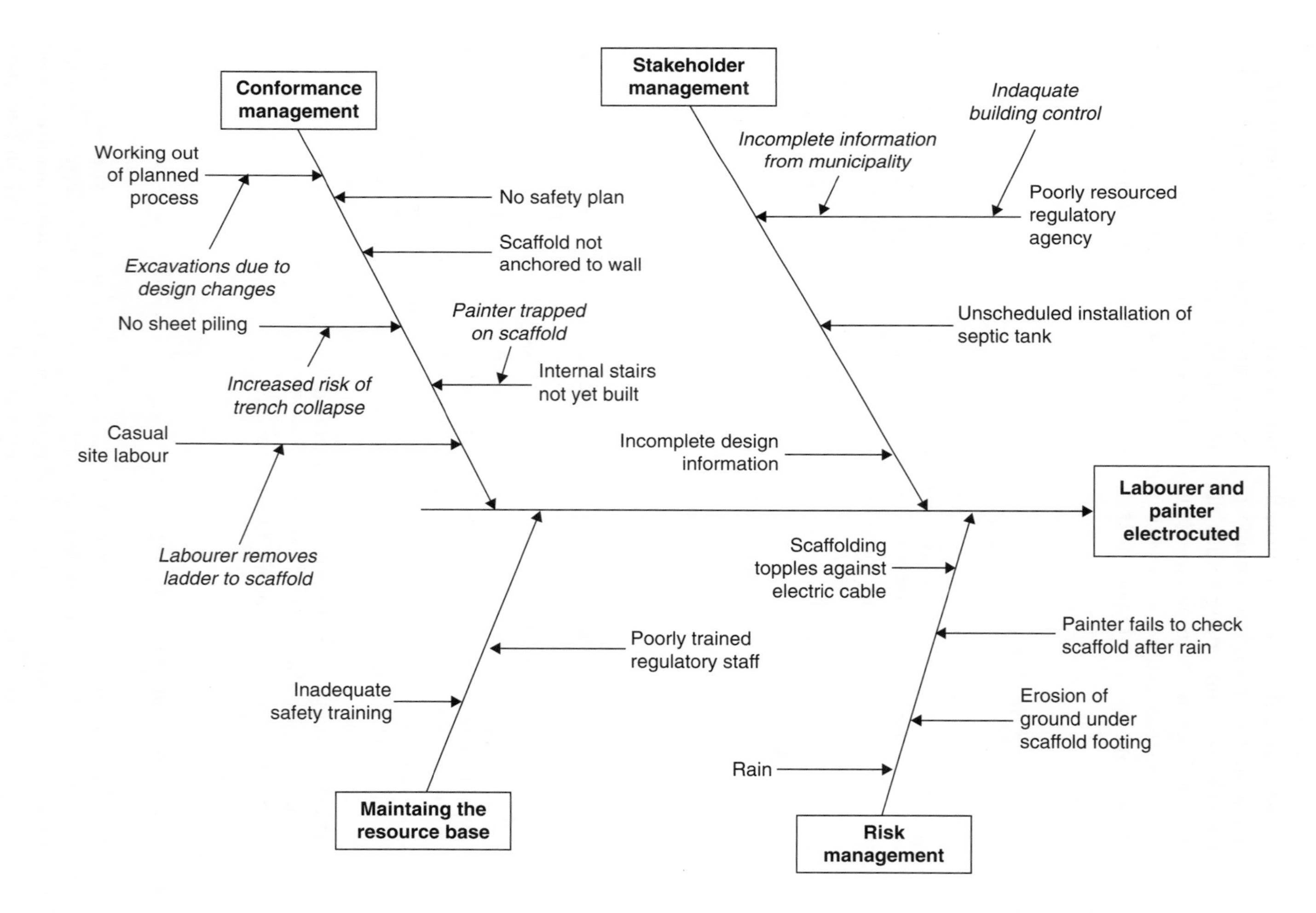

Fig. 12.4 The causes of non-conformance: safety (source: developed from material in *New Builder*, 11/02/94).

means of addressing whether variances are tolerable is to use control charts which present the data acquired from logging over a period in terms of its mean and standard deviation. Signs of trouble in such charts would include wide dispersion around the mean; trends over time shifting the mean towards a control limit; and any other discernible pattern in the data indicating that variances were non-random. The definition of process capability in panel 12.3 is derived from statistical process control and is central to six-sigma improvement processes. However, sophisticated statistical techniques are only really viable where data logging is to some extent automated, otherwise the costs of collecting enough data to identify trends can become prohibitive.

12.4.4 *Value-stream mapping*

Value-stream mapping is distinguished from the high-level process representions discussed in section 8.8 – known as SIPOC (supplier > inputs > process > outputs > customer) diagrams in six-sigma – by their focus on understanding discrete elements of overall processes at a detailed level. These representations capture flows of materials towards the customer and flows of information back from the customer. The 'as-is' map can then be used to redesign the process to eliminate steps that do not add value or generate waste. One difficulty in value-stream mapping in a project context is that a discrete value stream for the project may only form part of a much broader value stream for a resource base, making it difficult to redesign the value stream for a particular project. Value-stream mapping is also the essential prerequisite to observational methods of analysing process performance such as activity sampling[8].

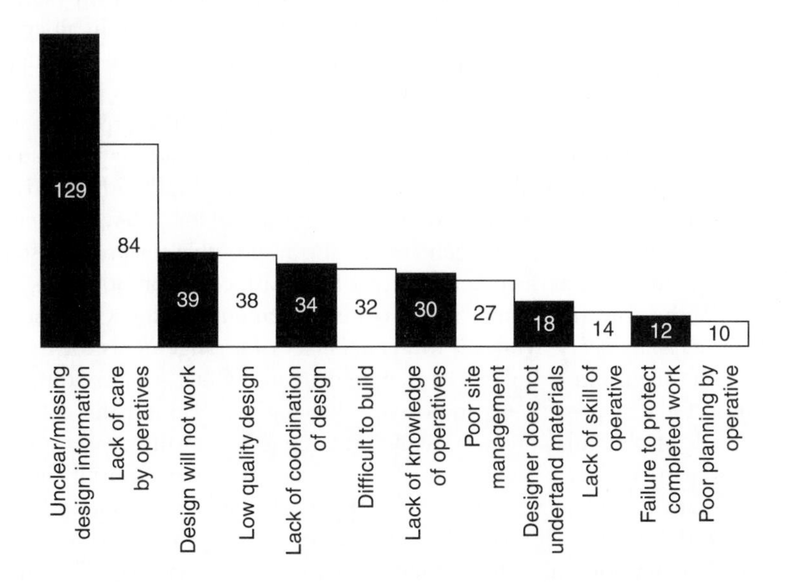

Fig. 12.5 The causes of non-conformance: quality (source: developed from data in Building EDC, 1987; 'other' category not shown).

12.5 Quality assurance

Quality assurance is the 'part of quality management focused on providing confidence that quality requirements will be fulfilled'. Its aim is to provide an overall set of procedures for the management of quality on the project. Thus it specifies what inspections should take place using which criteria, and which QC tools should be used when. Thus the QA system does not, in itself, identify conformance or non-conformance, but specifies the procedures for such identification. QA systems are of three basic types:

- first-party systems, which are the responsibility of the company concerned alone;
- second-party systems, where suppliers are accredited by buyers using proprietary standards – common in defence procurement and the volume manufacturing sectors;
- third-party systems, where the QA system is certified by an independent third party.

Third-party certification-based systems are now best practice and usually follow the international standards developed by the International Organization for Standardization (ISO) described in panel 12.5, particularly the ISO 9000 series, first published in 1987 and most recently revised in 2005. These specify the criteria that QA systems should reach, and as this is public knowledge, clients can be reassured regarding the capability of the firm to meet the agreed quality standards. The QA certification arrangements for the UK are shown in Fig. 12.6; as this is also an international standard, the basic system is the same in all countries adopting the standard. A firm applies for certification to one of the suppliers of such services, which are usually procured on a competitive basis. These certifying firms are, in turn, accredited by a national service under the procedures laid down in another standard and are responsible to the national industry ministry. Each country also has a national standards agency – such as the British Standards Institute – which is a member of the ISO and is responsible for policy and setting standards. These national standards agencies are also responsible for product standards, specifying minimum performance criteria for a variety of components.

Within the firm, there are two main elements to the QA system. Each firm has a *quality policy*, which is the general set of arrangements for managing quality in the firm. These policies are applied to individual projects through the *quality plan* for that project. An important challenge is adapting the quality plan to the needs of particular projects, while remaining within the procedures laid down in the firm's quality policy. Figure 12.7 shows the quality plan for the Glaxo project, presented in Case 8. Using BS 5750 Part II, the predecessor to ISO 9001 in the UK, the quality plans of the consortium partners (Laing and Morrison Knudsen) had to be combined to produce the project quality plan. This in turn had to be aligned to the procedures of the client (Glaxo Group Research) to produce the project procedures manual. In combination with the QA systems of the trade

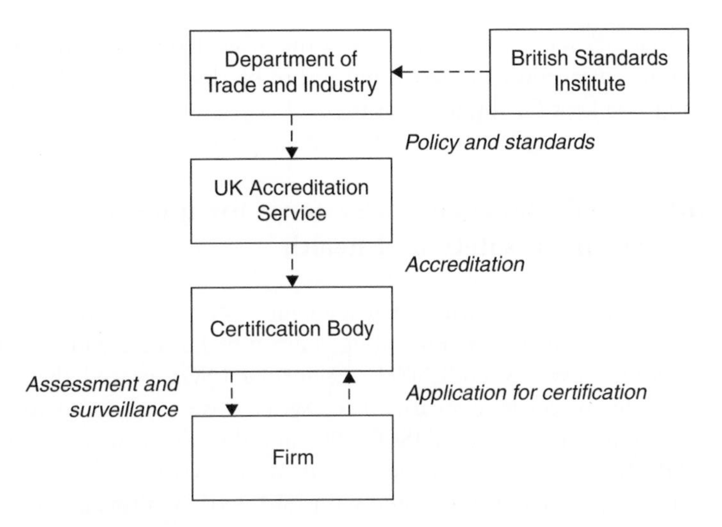

Fig. 12.6 The UK's QA certification arrangements.

Panel 12.5 The International Organization for Standardization

The ISO is a worldwide federation of national standards bodies from some 140 countries, based in Switzerland. ISO is a non-governmental organisation established in 1947 to facilitate globalisation by removing technical barriers to trade. The mission of ISO is to promote the development of standardisation and related activities in the world with a view to facilitating the international exchange of goods and services, and to developing co-operation in the spheres of intellectual, scientific, technological and economic activities. ISO's works result in international agreements which are published as International Standards. ISO is not an acronym, but derived from the Greek *isos*, meaning 'equal'. ISO standards are developed according to the following principles:

- *Consensus*. The views of all interests are taken into account: manufacturers, vendors and users, consumer groups, testing laboratories, governments, engineering professions and research organisations.
- *Industry-wide*. The aim is to develop global solutions to satisfy industries and customers worldwide.
- *Voluntary*. International standardisation is market-driven and therefore based on voluntary involvement of all interests in the marketplace.

National standards organisations – for instance, the British Standards Institute in the UK – may also issue their own standards to meet their particular needs, such as British Standards (BS).

Source: http://www.iso.ch

contractors and the requirements of the project, quality plans for each trade package, and inspection and test plans, had to be drawn up. Finally, verification records had to be kept for future client use.

12.6 Integrated management systems for quality, environment, safety and health

As the number of formal management systems has grown, there has been an increasing appetite for integrating them into business-wide routines which have become known as QUENSH[9]. As formal QMS proved their worth, the same principles of certified management systems were extended to the avoidance of environmental risks in the ISO 14001 and the management of safety and health in OHSAS 18001[10]. Both of these standards are written in such a manner that they can be used alone or integrated with ISO 9001; all provide for third-party certification and stress the importance of top management support for their implementation. Integrated management systems covering all three areas are now starting to be implemented in construction as recommended by ISO, particularly in the rail sector – see panel 12.6. Although BAA does not, apparently, use the term QUENSH, it is clear from Case 12 that there was a high degree of integration in their quality, safety and environmental management systems.

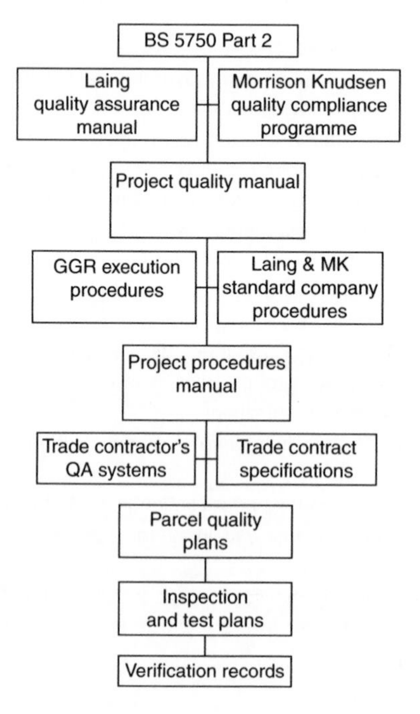

Fig. 12.7 Glaxo project QA plan.

Like ISO 9001, ISO 14001 does not specify performance levels but only how the firm should manage to achieve conformance with these performance levels. Similarly, OHSAS 18001 only specifies how defined levels of safety and health performance should be achieved, not what those levels should be. Yet, most within the industry would argue that not only is there a problem of conformance in the construction industry to the conformance standards that are currently set, but the standards that it is capable of setting for itself are inadequate. Continual improvements in process capability are required if projects are not to continually surprise clients by disappointing them regarding both the quality and the predictability of realisation.

Panel 12.6 QUENSH at Carillion

Carillion – formerly Tarmac – has taken the lead in developing integrated management systems for construction project management. Around 1995, it realised that its QMS was becoming bureaucratic and a hindrance to innovation. Carillion started bringing its systems together, a move which was given much greater impetus by the appointment of a director responsible for business systems. The vision was to develop 'management systems integrating the issues of safety, the environment and quality across the whole of Tarmac Civil Engineering. Existing practice is to be challenged and innovation encouraged in our aim to be more effective and efficient in everything we do'. Every project now has a single management plan, integrating 10 reports into 2, which is much more closely integrated with risk management. Separate management policies remain in place at corporate level. In 2008, Carillion took over Alfred McAlpine, another construction company with well-developed QUENSH routines, posing the challenge of integrating the two systems.

Sources: Interview 05/11/98 and *Construction Productivity Network* Workshop Report E9080, May 1999, Ekins (2006).

12.7 Creating a culture of improvement

Inspection, QC and QA – and QUENSH systems – all share two important limitations:

- In the terms of Fig. IV.1, they are all first-order information loops around static standards. Performance is 100% if zero defects is achieved.
- These systems are operationally orientated in that the measures are operational ones and do not, as such, take into account factors such as customer satisfaction.

Quality improvement is the 'part of quality management focused on increasing the ability to fulfil quality requirements', and the processes by which such improvement is generated are usually called total quality management. The aim of TQM is to tackle both these issues by encouraging continuous improvement in levels of conformance by enhancing process capability, and ensuring that the quality standards set for the process are the most appropriate ones for the customers of the output from that process. TQM is more an approach to managing the business than a set of quality management procedures and techniques. TQM requires a culture of improvement, and its principles apply equally to the improvement of performance on environmental, safety and health requirements. The project manager's role in generating an appropriate project culture will be discussed further in Chapter 16; here we need to note that the culture is essentially about the set of *values* that influences behaviour by mediating between the motivations generated in the work context and actions taken by individuals in response to those motivations.

The issues are most starkly put in the case of safety. Like all other risks, as will be discussed in Chapter 13, taking physical risks with life and limb has its rewards – the reward of the adrenaline rush in what might be called the bungee-jumping syndrome, or conviviality by driving drunk. Frequently – and most difficult to manage – by reducing risks in one place we generate them elsewhere[11]. Individuals continually risk accidents to gain rewards in line with their appetite for risk – what John Adams calls their risk thermostat. These dynamics are basic to the human condition, particularly for testosterone-laden young men, inherited from the days when survival depended on taking physical risks, and ability to take them successfully suggested high potential as a mate[12]. It is for this reason that induction to a new organisation may contain 'rights of passage' that entail demonstrating an ability to take physical risks[13].

However, on construction projects, the risk/reward profile is typically skewed. The rewards usually accrue to the decision-maker in terms of extra profits as a result of savings on costs associated with accident prevention, such as better access platforms or more secure storage of polluting materials. The risks, however, impact most directly upon other stakeholders such as operatives who are exposed to the greatest risk of accident, or users of the environment damaged who may have little other interest in the project or the facility it is creating. On top of the behavioural dynamic is, therefore, a predominant economic dynamic. The role of a culture of improvement is to channel these dynamics positively so that appropriate values mediate between the motivation and behavioural response. The key elements in creating a culture of improvement are empowerment, training, organisational learning, incentive alignment and senior management support.

12.7.1 *Empowering those doing the work*

Unless individual operatives take personal responsibility, the process capability cannot be improved; this is clear in the case of safety, but applies equally to environment and quality. For instance, no inspection system is foolproof; a rule of thumb is that any regime of 100% inspection will fail to reject 20% of non-conforming components. Where sampling techniques are used, an expected percentage of non-conforming components will get through. Moreover, such systems tend to encourage the passing of responsibility from the operative to the inspectorate. The person who first knows that a component is non-conforming or a process is unsafe is the person making it or doing it, so it is both cheaper and more effective to empower that person to do their own conformance control.

This requires a culture where operatives are not blamed for conformance problems, and do not lose pay by stopping task execution should a problem arise, as was implemented on T5 described in Case 12.

12.7.2 *Training is crucial for success*

Process capability is a key variable in achieving high conformance, and the skills of operatives are one of the most important – if not the most important – determinants of process capability. Unless operatives are properly trained to do their jobs, there is little chance of achieving acceptable levels of process capability. Operatives also need to be trained in the various tools of QC and how the QA system works so that they can take responsibility for identifying and removing quality problems, and follow procedures properly. Similarly, training in when and how to use safety equipment and safe working methods in hazardous environments is essential.

12.7.3 *Organisational learning is the aim*

A culture of improvement extends the single-loop and double-loop learning presented in Fig. IV.1 to form an overarching loop – learning how to learn. Knowledge management was discussed in section 8.9; a culture of improvement is the more tightly nested dynamic of organisational learning. The achievement of zero defects and the enhancement of process capability are slow and steady processes of incremental improvement, typically achieved through quality circles, using the QC tools in their forensic work, tracking down why components are non-conforming. Quality circles are teams of operatives and line managers working together to solve conformance problems, where the analysis of problems in a team context is as important as the identification of the appropriate solution to the problem, because it enables team ownership of that solution.

12.7.4 *Align incentives with desired performance*

The operational incentives in organisations frequently undermine their espoused objectives. Espousing conformance 'right first time', while rewarding volume of output, is an obvious contradiction, yet most construction operatives are paid on the amount they produce, not whether that output is in conformity with requirements. The operational metrics chosen for the organisation define that organisation, yet are often completely inappropriate for its espoused objectives. The current fashion for performance measurement and benchmarking could make this situation even worse, because metrics are often poorly chosen and so are either useless as incentives or encourage teams to blame other teams for poor performance, or motivate perverse behaviour[14]. Case 12 of BAA's Heathrow T5 project shows how important it is to align the incentive framework with the performance desired by the client.

12.7.5 Senior management commitment is essential

If the project culture is to change, if resources are to be invested in additional training, if appropriate behaviour is to be rewarded, if operatives are going to be empowered to take their own conformance decisions, then senior project management has to make it very clear that this is what the project coalition values, as the BAA team did on T5. Building a learning organisation is a senior management responsibility[15]. The test will come when a task takes longer than expected because component or safety arrangements are not in conformance, and those responsible receive thanks for taking action to resolve the problem, not blame for delaying the project.

12.7.6 Use formal process improvement projects

Management practice in conformance is increasingly focused around the role of formal process improvement projects (PIP). These combine the control loop thinking behind *kaizen*-type plan-do-check-action (PDCA) cycles with type of process improvement tools introduced in section 12.4. *Kaizen* is well established in quality improvement, and both ISO 14001 and OHSAS 18001 also have PDCA 'front ends' which emphasise the dynamic nature of achieving high environmental and safety performance.

One of the most widely diffused PIP routines is six-sigma[16] with the DMAIC cycle at its core.

- *Define* – identify and validate the problem, listen to the voice of the customer, create project team and define the end state desired.
- *Measure* – use value stream mapping to understand process and measure its performance.
- *Analyse* – identify root causes and prioritise for action.
- *Improve* – develop solutions removing root causes and validate through proposed value stream map, pilot solutions and develop implementation plan.
- *Control* – implement solutions, pay attention to training issues while continually measuring, finally completing and passing new process to the process owner.

Six-sigma PIPs are typically organised as stage-gate projects led by 'black belt' six-sigma practitioners. Panel 12.7 presents the use of a six-sigma process on the High Speed 1 project – notably it was used to address schedule rather than conformance issues. Six-sigma is very rigorous, and requires a tighly defined problem for solution to be most effective. The solution space also needs to be within the control of the six-sigma team; if it involves suppliers then they will need to be included in the team and this may not be a worthwhile investment outside a partnering arrangement. Arguably, few construction projects will be able to support the full six-sigma treatment, but a more modest programme of process improvement in the UK which uses the same basic concepts is showing considerable promise, as described in panel 12.8. Work of this kind is profoundly important if the average level of performance in the sector is to be improved, and the prestige projects of the kind described in Case 12 are not to leave the rest of the industry behind.

Panel 12.7 6 Sigma at St Pancras

Contract 105 for the construction of the extension to St Pancras station included extending the platforms in two parallel processes – the east and west decks. The east decks were behind schedule and a PIP was launched to address the situation. Led by a black belt, the PIP:

- *Defined* the problem as late delivery of the 15m raised platform beams, and the business case for the PIP was based on the liquidated damages, the additional direct and overhead costs, and the risk to the contract as a whole that would be incurred if construction continued at the present rate.
- *Measured* the problem using progress charts and identified an 8-week delay at present rates of activity.
- *Analysed* the root causes of delay as poor coordination with precedent trades, particularly piling, and inadequate amounts of formwork and falsework available.
- *Improved* by investing in more equipment and improving the piling process through its own PIP.
- *Controlled* the process by continuing to measure progress and noted both a reduction in variability of beam construction, and convergence over 100 beam pours towards the target cycle time of 8.5 days for the construction of one beam.

Source: Steward and Spencer (2006)

Panel 12.8 CLIP: Improving Process Capability

The UK's Construction Lean Improvement Programme (CLIP) was established in 2003 to provide support for construction companies aiming to make process improvements. CLIP provides government subsidised master classes led by CLIP accredited facilitators to work on a well-defined construction process problem in areas such as supply chain management, component manufacturing, on-site productivity and waste reduction. Master classes are also available in areas such as teamwork and leadership. CLIP typically works with firms at the medium to small end of the market – mainly general building contractors.
The CLIP process is based on a PDCA cycle using the standard kit of process improvement tools supported by two distinctive mnemonics:

- The 7Ws (wastes): motion, waiting, defects, transport, overproduction, unnecessary inventory and inappropriate work.
- The 5Cs: clear out; configure; clean and check; conformity; and custom and practice.

Source: http://www.bre.co.uk/page.jsp?id5355 (accessed 30/09/08)

12.8 Quality awards and self-assessment

In order to give greater focus to TQM efforts, a number of countries have developed quality awards such as the EFQM Excellence Model. The latest version was released in 1999, in slightly different versions for the public and private sectors. The model is divided into *enablers* and *results*, and firms are scored against each of the nine categories of performance; high scorers are deemed to be 'excellent' organisations. Figure 12.8 shows an adaptation of the basic model for project organisations, which is endorsed by the International Project Management Association (IPMA). The EFQM model is used in four ways[17]:

- As a framework which an organisation can use to help develop its vision for the future in a tangible, measurable way.
- As a framework which an organisation can use to help identify and understand the systemic nature of its business, the key linkages, and cause and effect relationships.
- As the basis for the European Quality Award (EQA).
- As a diagnostic tool for assessing the current health of the organisation through benchmarking and self-assessment. Through this process an organisation is better able to balance its priorities, allocate resources and generate realistic business plans.

The EFQM runs the EQA in four categories:

- large firms;
- public sector organisations;
- small and medium size enterprise (SME) subsidiaries of large firms;
- independent SMEs.

The national partner organisations of the EFQM, such as the Deutsche Gesellschaft für Qualität, provide TQM training and may also run national quality award schemes. Similarly, the Gesellschaft für Projektmanagement runs annual training for assessors for the Project Excellence Model. One of the most important uses of the model is to provide a framework for self-assessment and benchmarking. Self-assessment is a process internal to the firm, which scores itself against the model, thereby indicating areas of strength and weakness, while benchmarking studies allow more thorough investigation and assessment of particular aspects of performance. Most importantly though, the EQA and similar schemes such as the IPMA Project Excellence Award provide the framework for incentives that motivate a culture of improvement.

12.9 Conformance management in a project environment

Conformance management in construction still relies largely on the least sophisticated level – inspection. The question is whether this is inevitable, or whether

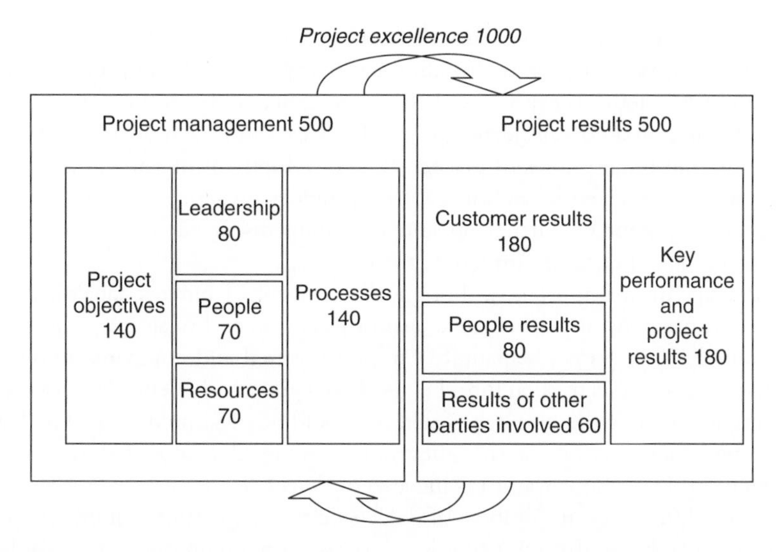

Fig. 12.8 The Project Excellence Model (source: IPMA private communication 04/08/09.

construction project managers can deploy the principles of conformance management discussed in this chapter to achieve the same sorts of remarkable improvements that have been seen in other industries. There are some important issues to address.

- A fundamental assumption of the literature on conformance management is that process capability is incrementally improved over time through the information loop. At each loop around, performance is slightly closer to target, and once zero defects is reached, the target can be made more difficult to hit with the aim of achieving even lower levels of defects. How can this apply to 'one-offs' in construction?
- Statistical approaches to QC demand repeated observations of the same process to detect defects, and the definition of process capability offered in panel 12.3 assumes repeated sampling. How can these principles be applied to one-off projects, where repetition levels are low and, as a result, there are high levels of reliance on inspection?
- Quality might well be free at the level of the project as a whole, but the returns on investment by the architect in ensuring that the detail drawings are error-free are often reaped by the contractor executing the work, so how can such benefits be shared equitably?
- Most PIPs are firm-based – they work on the horizontal dimension of Fig. 1.4, while projects flow on the vertical dimension. This implies that unless incentives on the vertical dimension are aligned across all the resource bases participating in the improvement, process improvement will be difficult.

The crucial realisation is that the sampling takes place at the component level. Even the most idiosyncratic building has many repeated components, and sophisticated QC methods can be applied to these. Secondly, to the extent that there is standardisation across projects, QC methods can also be applied to repeated processes within programmes of projects. The development of standardisation and pre-assembly – discussed in section 12.10 – provides greater opportunities here, while the development of partnering and alliancing discussed in sections 5.6 and 6.8 provide the context for equitable gainsharing.

Perhaps most importantly, though, the concept of process capability rests upon the measurement of performance against intention and measures relatively, and not absolutely. Thus process capability can be measured and compared between different processes; it can, therefore, be used to benchmark between processes where a simple count of the number of defects would be of limited value. The basic problem with defect counts is that they only measure degree of failure, not height of aspiration. The easiest way to achieve zero defects is to have wide tolerance bands, because differences in numbers of defects could be explained simply by differences in tolerance bands. In other words, a process with a high defect rate could actually be performing better than one with a low rate because the former has tighter tolerance bands. As the industry moves towards being an industry based on assembly rather than fitting on site, with greater levels of standardisation and pre-assembly, these issues will become more important. Traditional components with high tolerance bands which allow considerable room for fitting on site can, perhaps, rely on a zero defects approach, but the tighter tolerances required for assembling standardised component systems will require greater attention to process capability.

A further problem with the more sophisticated QMS is that they depend on the empowerment of the workforce. Empowerment is not the same as autonomy. Autonomy means simply being left to get on with it, and this is the typical way of managing operatives in construction. As a result, gangs take no responsibility for anything outside their immediate scope of work; if the problem is caused by poor design information, they 'work-around' the problem rather than addressing it. Empowerment means that there is a constructive dialogue between management and the operatives, and where operatives can see non-conformance they stop working and sort out the problem, rather than 'working-around'. In a word, it requires *trust*. Empowerment therefore implies a two-way commitment, and places human resource issues at the heart of effective quality management. However, the casual approach to the employment of operatives and many other members of staff in construction, discussed in section 7.5, makes the development of an empowered workforce virtually impossible, and encourages reliance upon inspection rather than more sophisticated methods. The achievement of TQM in construction will need a radical shift in the culture of the industry towards a permanently employed workforce.

12.10 Standardisation and pre-assembly[18]

One way of improving process capability is to use standardisation, pre-assembly, or a combination of the two:

- *Standardisation* is the use of standard components or modules, the development and production overhead costs of which are shared across a number of construction projects, and thereby benefit from the learning curve in repeated tasks. Many building projects are highly standardised at the component level, with components such as window frames, door sets and electrical fittings being bought off the shelf. Standardisation may also imply the definition of a standard dimensional grid for the facility.
- *Pre-assembly* is the prefabrication and/or assembly of components into sub-assemblies, either off-site in a factory or on-site in a dedicated facility, prior to final installation in place. The aim here is to provide a more controlled working environment to improve process capability.
- *Mass customisation*[19] is the combination of the two, where standardised modular kits are configured to meet particular project needs and pre-assembled for installation on site.

The aim with standardisation is to push up the volumes of identical components so that *economies of scale* in design and production can be achieved, on a model analogous to mass or lean production methods in manufacturing. The main limit to standardisation is the extent of the market for such components, where the quality of conception demands that every element of the building is crafted to the particular needs of the client standardisation is difficult, as the case of the bricks on the facade of the Worldwide Plaza, in section IV.1, shows. Pre-assembly is aimed at gaining economies through working in more controllable working environments away from the actual point of installation; the major limit to pre-assembly is usually the ability to transport the sub-assembly from the factory, or to place it in position in one piece on site. Prefabrication on site is very common on civil engineering projects, such as the segmental tunnel linings on the Channel Fixed Link described in Case 1 and the approach viaduct decks on the Severn Bridge described in panel 2.9, where specialist factories for casting the components were set up on, or adjacent to, the site. Complex steel fabrications are often pre-assembled and painted on the ground before erection. Panel 12.9 presents an excellent example of pre-assembly.

In practice, there are four basic approaches to standardisation and pre-assembly in construction:

- *Component manufacture and sub-assembly* – the basic level of standardisation and pre-assembly beyond the facility being completely one-off. This is widely used.
- *Non-volumetric pre-assembly* – typically, 'flat-pack' sub-assemblies which can be quickly erected on site. Such systems have been in use since at least the industrial revolution, providing easily erected homes and public buildings for the colonists of the nineteenth century, and housing for the working class after 1945. These systems are now widely used in various types of construction.
- *Volumetric pre-assembly* – where a whole enclosed space is built in the factory and installed on site complete. These methods are widely used by hotel chains and the like. For instance, a Hotel Ibis bedroom is identical in the Manchester

Panel 12.9 Pre-assembly for the main structure

The structural frame for the new Inland Revenue Centre in Nottingham relied heavily on pre-assembly under factory conditions. The two main components were brick-clad piers and standardised pre-cast concrete flooring units. The latter were 3.2 m wide and weighed 25 tonnes. The piers were produced in a factory, where teams of bricklayers worked indoors in a steady rhythm which improved process capability. The flooring units were cast using reusable moulds around pre-assembled reinforcement cages, and light fittings were then fixed. Both were then stored at the factory and delivered to the site around 4 km away as required – placement of a flooring unit typically took 15 min. While the capital cost of this approach was higher than a traditional approach, predictability of both schedule and conformance were improved. More generally, higher than normal process capability was achieved because of the control over the working conditions, and the opportunity to prototype and debug assemblies prior to volume production.

Source: *Construction Productivity Network* Workshop Report 723L, 1997.

and Lübeck branches of its chain. The classic example here is the bathroom pod, described in panel 12.10.

- *Modular buildings* – where the entire building is supplied on a volumetric basis, perhaps with only the addition of a brick cladding. Assembly times on a prepared site can be as low as 11 hours for a McDonald's drive-thru outlet[20].

The case for standardisation is often damaged by incautious rhetoric comparing the construction industry to the car industry, and reached its nemesis in industrialised building. It is worth quoting the leading historian of industrialised building:

> '. . . this analogy was a false one. Car prices initially were high, to cover high tooling costs and disproportionate overheads, while production steadily increased. But as a generic product the car was unique, and its manufacturers had a monopoly; one either paid the high price or did not acquire a car . . . . Industrialized housing did not produce a unique product, the competition of the traditionally built house was an ever-present factor, and the industry was denied that sheltered growth period it needed to reach the critical mass of production'[21].

Panel 12.10 Bathroom pods

E.J. Badekabiner A/S supplies bathroom pods on a mass customisation basis. The basic bathroom design is customised to meet the needs of particular clients, and then produced on an assembly line in Badekabiner's dedicated factory in Denmark. This allows uncertainties in task execution duration to be minimised, and therefore allows the complex parade of trades through a single bathroom to be redesigned in the most efficient and effective manner. The process is supported by extensive investment in IT. The pods are then sealed and shipped to site where they are connected to the external services fitted to the building structure. Seals need not be broken until commissioning. The benefit is much greater process capability, with predictability of schedule and conformance. The disadvantage is that the units are more expensive than their on-site equivalent because of the higher factory overheads, transportation costs and the necessary increase in floor-to-ceiling heights to accommodate the units. Bathroom pods therefore offer a trade-off between budget on the one hand, and schedule and conformance on the other.

Sources: http://www.ej-badekabiner.dk/, CIRIA (1999) and material supplied by Niels Albertsen.

The car offered such massive functional advantages over the horse and cart that affluent people were prepared to pay a major premium for those advantages; it was only later that prices started to come down to create the mass market. Interestingly, the housing sector does appear to follow the early phases of the car industry in take-up of prefabrication and pre-assembly. One of the most successful prefabricated housing systems in Europe is offered by the Huf Haus range of 11 model homes aimed at a high-end market niche[22]. The key to its success would appear to be the very high design values achieved by the product that make it instantly recognisable and stand out from the surrounding homes. The Huf Haus group also provides a range of associated services including finance, on-site assembly by Huf Haus employees, and refurbishment through life analogous to purchasing a high-end car.

Standardisation in construction simply does not offer the level of functional advantage offered by early cars – indeed, for many clients, it offers disadvantages – and so it is very difficult to get the virtuous cycle started[23]. However, there are many other industries from which construction can learn, and which can learn from construction – the so-called complex systems industries[24]. There is much that can be done to increase levels of standardisation and pre-assembly in the construction industry, but it requires the following:

- A willingness to invest in developing process capability on the basis illustrated in Fig. 10.5. The returns on that investment tend to come not through savings on budget but through greater process integrity, particularly with respect to schedule and conformance[25].
- A much greater willingness to define design problems as tame in the terms defined in Fig. 9.5, so that their solutions are capable of deploying solutions already tried and tested on other projects.
- The extension of the market through clients insisting that standard rather than bespoke solutions are chosen on their projects.
- A willingness to learn from other complex systems industries such as oil and gas, and aerospace, without being blinkered by inappropriate comparisons with the volume industries such as car manufacture.
- A willingness to accept that, at least in housing, prefabrication is probably most appropriate for high-end branded styles that cannot be achieved through

conventional construction methods and that prefabrication in volume housing will be based on the aspirations of less affluent householders making the popularisation of high-end styles viable[26].

12.11 Summary

The management of conformance is where information processing meets materials processing and lies at the heart of achieving process integrity. Conformance is about ensuring that the physical act of construction is closely related to intention – when it is not, loss occurs. Whether this is the loss to society embodied in the QLF or environmental damage, or the loss to individuals of life or limb, failures of conformance are about physical losses that frequently cannot be recovered. In order to avoid such losses, product integrity must be appropriately defined and process integrity appropriately developed. Failures of conformance – and much evidence for this has been presented through this part of the book – often occur when product integrity is poorly defined, or the definition is changed, when budgeting and scheduling are poorly managed. As the data presented in section 12.4 show, late delivery of design information can cause fatal accidents as well as building failures. Working 'out of process' will always make effective conformance management difficult. In the context of adversarial relations, discussed in section 6.7, budgeting and scheduling are compromised, and what gives is conformance.

Perhaps most importantly, the widespread abdication of managers of construction resource bases for the processing of materials on site – basic site operations – makes a culture of improvement almost impossible to develop. This point has been made in various ways throughout this book – in the discussion of sequential spot contracting in section 7.5; adversarial relations in section 6.7; the new scheduling techniques in section 11.5; and the management of risks in section 13.5. To put the point bluntly, unless the construction industry moves towards making a commitment to its most valuable resource – people – virtually nothing can be done to improve process capability. Without such improvement, the problems of budgeting and scheduling which derive from difficulties in predicting task execution cannot be addressed[27]. We will return to these organisational matters in Part V.

Case 12
From Navvies to White Van Man: Managing Conformance at T5

The new Terminal 5a/b (5c is presently under construction) at London's Heathrow airport is a major infrastructure investment by BAA plc, the UK's principal airport operator. The terminal cost £4.3b and opened on schedule and budget in March 2008. The initial days of operation were marred by lack of readiness on the part of the airline using the facility, BA plc, which took a 'calculated risk' on staff training and lost. However, T5's effective realisation by BAA remains a considerable achievement and the few systems failures on opening day were trivial. As the project moved through the life cycle from 'navvies' employed by large civil engineering contractors working on rail tunnels, diverting rivers, aircraft stands and the like through to small 'white van man' firms during fit-out of the terminals, it posed enormous challenges for managing conformance in safety, quality and environment in the context of a schedule fixed in 2001 and a budget fixed in 2003.

The project mission – honed through a regulatory process that took from February 1993 to November 2001 – set high standards in terms of specification and conception and aspired 'to deliver the world's most successful airport development'. Challenging civil engineering problems in a very tight site melded with the ambition to achieve an architectural statement for the principal international gateway to the UK. For instance, regulatory constraints meant that relatively simple solutions in concrete for the construction of the new air traffic control tower were not acceptable on aesthetic grounds and were replaced with a more complex cable-stayed steel design. The solution to the briefing problem matured through the work of Richard Rogers Partnership during 2000 as a '-loose-fit, flexible envelope' following iteration over the previous 10 years through three very different visions of what T5 could be like. Once regulatory consent was obtained the process could move onto addressing the design problem. Some 2000 designers employed by 19 firms were deployed at the peak of this stage of the project life cycle working to 300 separate functional briefs prepared by the BAA development team. These designers were managed by BAA's T5 Development and Design Director supported by what had become Rogers Stirk Harbour + Partners as concept architects.

The outputs from this design activity formed the solution to the design problem for 147 sub-projects grouped into 18 projects, the first of which moved into execution from September 2002. The BAA programme team consciously set out to improve on the performance benchmarks of major projects and the context for achieving this vision was the cultural change facilitated by the T5 Agreement. Born from the traumatic experience of the Heathrow Express project described in panel 16.5, BAA decided that as it effectively held the risk on the project in any case, it might as well formally take that risk through the contract, thereby motivating co-operative rather than adversarial behaviour from suppliers when the going inevitably became tough. The T5 Agreement was founded on a number of principles:

- All suppliers would receive a guaranteed margin of between 5 and 15% depending on the trade on incentive contracts for each package. Bonuses of up to one-third were available for exceeding package targets.
- Single project insurance of up to £2.4bn with a maximum payout of £500m for any one incident was negotiated, which took suppliers' professional indemnity and insurance costs out of the equation and paid on a 'no-fault basis'.
- Supply chain segmentation (see section 7.6), differentiating between those firms ready for a long-term relationship with BAA and those which were not. The first BAA Framework Agreement had been negotiated in 1993; the second was negotiated in 2000 and identified 750 first-tier suppliers (see Fig. 7.1). The T5 supply chain had 60 first-tier suppliers, 500 at the second tier and over 20 000 in lower tiers.

- Strong performance management of suppliers through quarterly supply chain reviews which mobilised peer pressure between suppliers to meet agreed targets. Rework was paid for the first time, and redone without profit payment if still unsatisfactory. Two first-tier suppliers and 12 at other tiers were removed from the project at various times.
- Collegial dispute resolution at team level. If this failed, a 'star chamber' of senior management was convened. The T5 Agreement then provided for an external third-party mediator before adjudication processes could start. By early 2008, no issue had gone as far as mediation.
- Progressive account settlement with the 60 first-tier suppliers with the aim of settling most accounts prior to opening.

In this context, the BAA team set ambitious conformance targets, using the earlier scheduling of the smaller T5b satellite terminal as a test-bed for methods on the much larger T5a main terminal building. Fundamental to achieving these targets was the establishment of a strong conformance culture on site. The BAA leadership team made it clear that safety came first, and then that quality and environment mattered as much as budget and schedule. These aspirations were broadcast to the 50 000 people who worked on the project through the life cycle by using various combinations of training, induction, publicity campaigns and *The Site* monthly tabloid-style newspaper, and 'on time, on budget, quality, safely with care to the environment was the T5 mantra for all' according to the BAA Construction Director. Strategically, this was supported by maximising the use of off-site prefabrication of components, and an off-site consolidation centre for logistics.

Taking safety first, the number of workers on site at any one time peaked at 8000 – a massive cultural challenge. While the role of the Planning Supervisor was held by Bovis, the Principal Contractor role under the regulations was split across five different people with a common approach across the site being managed by the BAA programme team. Industry benchmarks suggested that 2 people would be killed and 600 people would suffer serious injury, but BAA set out improve on this by setting a one-in-a-million rate of reportable accidents by working hours. The outturn on T5 was indeed 2 fatalities, but only ~200 serious injuries. While the T5 project tragically failed to achieve its targets on safety it did develop what was generally acknowledged to be a strong incident and injury free (IIF) safety culture where workers could stop the job if they believed it was unsafe and supervisors could stop the work of gangs outside their responsibility on the same grounds. Eight hundred senior managers from across BAA and the supply chain went through 2-day commitment workshops, 1200 supervisors attended a 1-day workshop, and all 50 000 workers heard about IIF at induction. This was supported by half-day sessions and briefings for the workforce and the development of 150 in-house trainers from the operative workforce.

The culture was sustained by monthly leadership forums where senior managers from the supply chain discussed quality of realisation issues, especially safety. These were supported by monthly health and safety forums where production leaders reviewed issues that had arisen over the previous month and identified the 'best-performing team'. Each of the 18 sub-projects was tracked on a number of benchmarks – both input such as training hours invested, and outputs such as reportable injuries and minor incidents. Senior BAA management also conducted safety walkabouts, and stopped the job in the winter of 2003 because of unsafe working on night shifts, thereby reinforcing its message.

The approach to quality was equally rigorous. A five-step procedure for conformance quality was defined:

(1) *Agree specification* – over 10 000 were agreed on the project in the design phase.
(2) *Method statement and inspection plan* – plan how to achieve consistently the specification and how you know that you have achieved it.
(3) *Start on site* – ensure that all workers are trained, have the right equipment, and have the benchmark standard example of what 'good' looks like which took the form of laminated sheet available in the workspace.
(4) *Benchmark check* – check work done visually against the 1400 benchmark standards previously agreed with suppliers.
(5) *Quality audits* – the BAA programme team audited 5% of completed work, a total of 700 audits.
(6) *Handover* – this happens after a final review.

This procedure was supported by training for supervisors in right-first-time, a DVD on quality at induction for all workers, and 10 000 quality walkabouts by senior BAA managers, building on the back of the development of the IIF safety culture. The result of this effort was a climb of 79% right first time in October 2004 to over 90% during 2006, despite the annual tightening of targets.

Airports in operation are very sensitive environmentally, and many had opposed the expansion of Heathrow through T5. The regulatory consents posed strict requirements in terms of traffic movements, working hours, noise, dust and air quality during construction. All these were monitored and managed carefully through the project life cycle. Early in construction work, a team of 80 archaeologists unearthed 58 000 artefacts dating back as far as 3000 BC on the T5 site, and the scheduling of the works had to be changed to accommodate this activity. The diversion of the twin rivers through the site also led to significant environment challenges, including the capture, breeding and subsequent release of water voles.

In all, 97% of site materials were recycled on site including 300,000 tonnes of aggregates. Imported aggregates were also recycled, crushed glass was used for road bases, and pulverised ash made up an average of 30% of the concrete mix. Most timber came from Forestry Stewardship Council approved resources.

Although the performance of the T5 project in conformance – at least on safety and quality – is by no definition perfect, the project is exemplary in the effort put into these issues by leadership from the client programme team to generate a high conformance culture across workers employed by around 21,000 firms. In this it has set new industry benchmarks and provided learning for the next phases of development at Heathrow. The key learning points for BAA from the experience of managing the T5 a/b project are as follows:

- Suppliers should take more of the risk on the project – on both Heathrow East and T5c BAA will hold less risk. This is a function of both the cultural change associated with the purchase of BAA by Ferrovial (a Spanish contractor and infrastructure operator) in 2006, and the smaller relative size of these projects to BAA's business as a whole.
- BAA should not pay for rework anywhere near as generously as it did on T5 a/b, and suppliers should pay much more of the costs associated with not achieving right first time.
- The first tier needs to be reduced in size, and many firms are simply not able to work collaboratively, so there should be more appropriate choice of incentives aligned to the particular needs of suppliers. For many global suppliers, T5 was just another job and they made profit from ensuring consistency of their management systems across all their projects rather than aligning them to the particular needs of one project, no matter how prestigious. Many smaller firms do not carry the managerial overhead to allow full participation in improvement schemes and the like – they make their money from having a clear objective in a lump sum and aggressively managing against that.

Despite these reservations, there is a widely shared view of those who worked on the project that T5 raised the best practice bar for the management of large construction projects and its effects will undoubtedly be felt throughout the UK and more widely as those who worked there move on to other projects in the future.

Sources: *Building Services Journal*, December 2006; *Building* 01/02/08; Doherty (2008); *Financial Times* 08/05/08; interview BAA project team member 11/09/08; Gil (2009).

Notes

1 Cited in Sabbagh (1989, p. 220).
2 These are identified by Dale (1994).
3 This and the subsequent quality management definitions are taken from ISO 9000:2005 *Quality Management Systems – Fundamentals and Vocabulary*.
4 Phil Crosby's famous dictum (1979, p. 1) has become a rallying cry for all those leading quality improvement programmes. Crosby could only make this bald claim because he used the same definition of quality as ISO 9000 – 'conformance to requirements' – and ignored the other three aspects of quality defined in Fig. 3.2.
5 Taguchi and Clausing (1990).
6 See Ishikawa (1990). Others have also identified seven new QC tools, for example Oakland (1993).
7 Ishikawa (1985, p. 197).
8 Arbulu *et al.* (2003) provide a detailed value-stream mapping for pipe supports; while Winch and Carr (2001a) use value-stream mapping as the basis for activity sampling of concreting works.
9 Wilkinson and Dale (1999).
10 OHSAS 18001, launched in 1999, was developed by an international consortium of standards and health and safety agencies. There is no ISO standard in the area.
11 Adams (1995) provides an extensive analysis of these issues. The classic case of risk transfer is the way in which much safer car designs which protect the car occupants have increased the risk to

pedestrians and cyclists, an issue which is belatedly being tackled by traffic-calming measures. The French appear to have done something similar with the development of integrated working platforms for the main trades, which are not then available for the finishing trades.

12 The proposition of evolutionary psychology is that our behaviour is, to an important extent, evolved to meet the challenges of being a hunter-gatherer. Hunters took physical risks or starved – see Nicholson (2000).

13 While working as a labourer on roofs, I saw little deliberately dangerous behaviour. However, once training as an estimator commenced, it was clear that I was expected to take serious physical risks. Most roofing work is refurbishment, and estimators do their own surveying. This can involve hair-raising ladder work and clambering across steeply pitched roofs in the ram to inspect valleys and obtain measurements – all in a day's work for four offices of two companies that I worked in. After four years of this, I was still stunned by the sight of a colleague 'walking the bolts' on a fragile factory roof. (The bolts indicate where the fixings to the underlying rafters are, and hence which bits of the roof can take a person's weight.)

14 See Kerr (1975) and Hauser and Katz (1998) for a discussion of these issues.

15 See Garvin (1991) for a succinct discussion of learning organisations.

16 See George *et al.* (2005) for an introduction.

17 See http://www.efqm.org.

18 This section is based largely on CIRIA (1999) and Gibb (2001); see also Gibb and Isack (2003).

19 The concept comes from Pine (1993).

20 Bennett *et al.* (1996).

21 Herbert (1984, p. 308).

22 http://www.huf-haus.com/de/ (accessed 30/09/08).

23 The AMPHION social housing consortium is one of the more recent to fail to achieve the volumes that capital investment in factory facilities requires – see Kaluarachchi and Jones (2007).

24 See Hobday (1998) and Winch (1998b) for concepts and analysis.

25 See *Financial Times* 22/11/01 for the case of speculative housing, where the same situation applies.

26 At the time of writing, the viability of any innovation strategy in housing was looking very bleak as a result of the credit crunch.

27 Whether such a commitment should be made through direct employment or self-employment is moot; the crucial point is that the people are recruited on the basis of a continuing commitment across multiple projects. The author's personal experience – having managed roofing gangs on both bases and having witnessed the leader of a self-employed gang being given a carriage clock for 20 years' service – is that direct employment has the advantage. This is supported by the evidence on the performance of the British and French construction industries – see Winch and Carr (2001a).

Further reading

Barrie G. Dale, Ton van der Wiele and Jos van Waarden (2007) *Managing Quality* (5th ed.). Oxford, Wiley Blackwell.

An authoritative text on all aspects of quality management.

Kaoru Ishikawa (1985) *What is Total Quality Control? The Japanese Way*. Englewood Cliffs, Prentice-Hall.

The first book which introduced western managers to the Japanese approach to managing quality.

Nigel Slack, Stuart Chambers and Robert Johnston (2007) *Operations Management* (5th ed.). Harlow, FT Prentice-Hall.

The leading text on managing manufacturing operations.

Chapter 13
Managing Uncertainty and Risk on the Project[1]

13.1 Introduction

> 'I compare fortune to one of those violent rivers which, when they are enraged, flood the plains, tear down trees and buildings, wash soil from one place to deposit in another. Everyone flees before them, everybody yields to their impetus, there is no possibility of resistance. Yet although such is their nature, it does not follow that when they are flowing quietly one cannot take precautions, constructing dykes and embankments so that when the river is in flood it runs in to a canal or else its impetus is less wild and dangerous. So it is with fortune. She shows her power where there is no force to hold her in check: and her impetus is felt where she knows there are no embankments and dykes built to restrain her'.

Niccolò Machiavelli's advice to a new prince[2] on management of risk in statecraft is equally applicable to construction project management, and usefully extends our river metaphor from section 1.3 to perhaps the most difficult area of construction project management – managing risk and uncertainty. If the project is the process of reduction of uncertainty through time, then, in a profound sense, managing risk and uncertainty is at the heart of the management of projects. Why, then, a chapter devoted to the topic? First, because managing risk and uncertainty is a peculiarly difficult topic, and many of those responsible for designing, or managing schedule and budget, prefer to think in deterministic terms of optimal solutions, rather than probabilistic terms of robust solutions. Second, it allows us to focus on the problems of decision-making while managing projects.

Perhaps only 'quality' is used in as many different ways as the term 'risk'. So, in developing this perspective on managing risk and uncertainty, the chapter will first explore what is meant by this much-abused term in some detail focusing on the distinction between risk and uncertainty, and the viability of eliciting subjective probabilities. It will then go on to discuss some of the best practice routines for managing risk and uncertainty before turning to the more strategic aspects of the subject.

13.2 Risk and uncertainty: a cognitive approach[3]

We offered a working definition of *uncertainty* in section 1.3 as the absence of information required for the decision that needs to be taken at a point in time. In order to understand the broader issues in managing project risk and uncertainty we need to unpack this definition a little more and show how this definition of uncertainty relates to the commonly used definitions of *risk*. Before we do this, however, it will be useful to identify explicitly the inherent time dimension in the concept of risk, because the presentation of risk management in the literature can sometimes imply timelessness to the decision-making problem.

Figure 13.1 presents a time-based framework for understanding risk and uncertainty. The *risk source* is the underlying condition that can generate a possible *risk event* at some time forward from the point of decision-making. For instance, unsafe working practices are an existing risk source while an accident is a risk event that could occur at some point in the future. Management can *respond* to a risk source by the actions identified in section 13.5.3 or it can *plan to respond* to a risk event by improving its capacity to handle the *impact* of the risk event occurring. This framework thus introduces the three dimensions of managing risk and uncertainty on projects – the risk source, the impact of the risk event and the extent to which management can effectively respond to the risk source and the risk event. Typically both responses are used; for instance the T5 project presented in Case 12 was very proactive in responding to the risk source by creating a safe working environment (pre-event response by mitigation of accident risk sources) and responding to the impact of the possible risk event by providing an emergency facility on site (post-event improvement of response) which is presented in panel 13.7.

A risk source is, therefore, an underlying state of affairs; a risk event is an event that can happen *given* that underlying state of affairs. Conceptually, the relationship

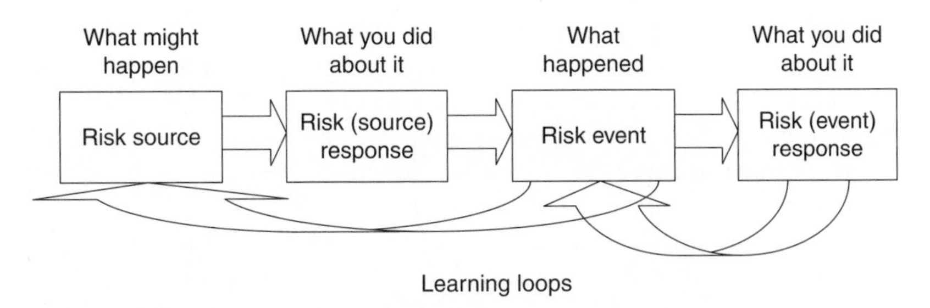

Fig. 13.1 Understanding 'risk' through time (source: developed from Dalton, 2007, Table 10.2).

between the risk source and the risk event is expressed in terms of the probability of its occurrence given the risk source; probability is, therefore, a property of the *event* and not the source. Broadly, there are four schools of thought on the definition of *risk* and its relationship to the concept of *probability*:

- The *objectivist* school argues that the probability of an event occurring in the future can be inferred from a sample of observations of previous occurrences drawn from a known population. This approach is inherently historical, and is associated with the science of statistics. Its most familiar implementation in managing projects is for conformance issues such as quality, particularly process capability discussed in panel 12.3. Its approach is essentially *predictive*[4] in that it attempts to predict future events from known data about risk sources.
- The *logical* school addresses the probability of a failure event in engineered systems. While there might not be a data set associated with failure – or only limited forensic data that does not meet the requirements for statistical inference – engineers' understanding of the design of the system and the scientific properties on which that design is based can be used to identify risk sources and hence the probable failure events in closed systems. This kind of analysis, for instance, is the basis of FMEA discussed in panel 12.4 and also aims to be *predictive*.
- The *subjectivist* school emphasises the degree of belief held by the decision-maker in the probability of a particular event and is the basis for the discipline of decision sciences where the elicitation of subjective probabilities extends the application of the tools associated with the objectivist school to future-orientated analysis. This approach is the intellectual underpinning of the toolbox of project risk management, although this is not always explicit in that practice. It is essentially *prescriptive*, in that it provides tools and techniques for how decisions ought to be made.
- The *behavioural* school focuses more on the actual behaviour associated with decision-making under uncertainty. The empirical research techniques associated with the behavioural school range from the ethnographic to the experimental in its ambition to be *descriptive* about how decisions are made in practice.

The *cognitive* approach to managing risk and uncertainty draws on important features of all these schools, but retains what we hold to be the vital distinction between uncertainty as defined in Chapter 1, and risk as the condition where a probability distribution can be applied to the occurrence of a risk event[5]. We call it cognitive because we accept the foundation insight of the subjectivist school that the probability of a future event occurring is a property of the decision-maker, and not the external world. The concept of subjective probability starts with a conundrum – is the 50% probability of a fair coin landing heads a property of the coin or the decision-maker? Clearly the coin is not making any decisions, but the (absolute) confidence of the decision-maker that there is a 50% chance of heads is apparently derived from empirical observations of actual events. Yet, the decision-maker is making a decision about something that does not yet exist – the orientation of the coin after the next toss – so it must be a property of the ability of individuals to mentally construct future states. A further twist to this conundrum is what if the fair coin has been tossed, but the tosser withholds the outcome information from the decision-maker? While the decision-maker is now being asked to identify a present state of the world, the information available to the decision-maker makes it appear that he or she is expected to predict a future state. Thus from the point of view of decision-making – as opposed to statistical inference – there is little significant difference between the *objectivis* and *subjectivist* views because the probability assessment in both cases is a property of the person and not the event – the consequences of any decision 'might appropriately be called states of the person, as opposed to states of the world'[6].

The cognitive model is presented in Fig. 13.2[7]. Logically, the occurrence of any future event is either certain, impossible or somewhere between the two. If an appropriate data set is available and a change in the underlying conditions of that data set is impossible, then we can infer an objectivist probability indicated as a point on the continuum between certainty and impossibility in the figure. Such desirable conditions are infrequent in general, and rare on construction projects – even if data are available, change in underlying conditions is usually possible – so we are usually in the area of the 'information space'[8] on the right of the figure. In the information space the perception by the project manager of the risk sources

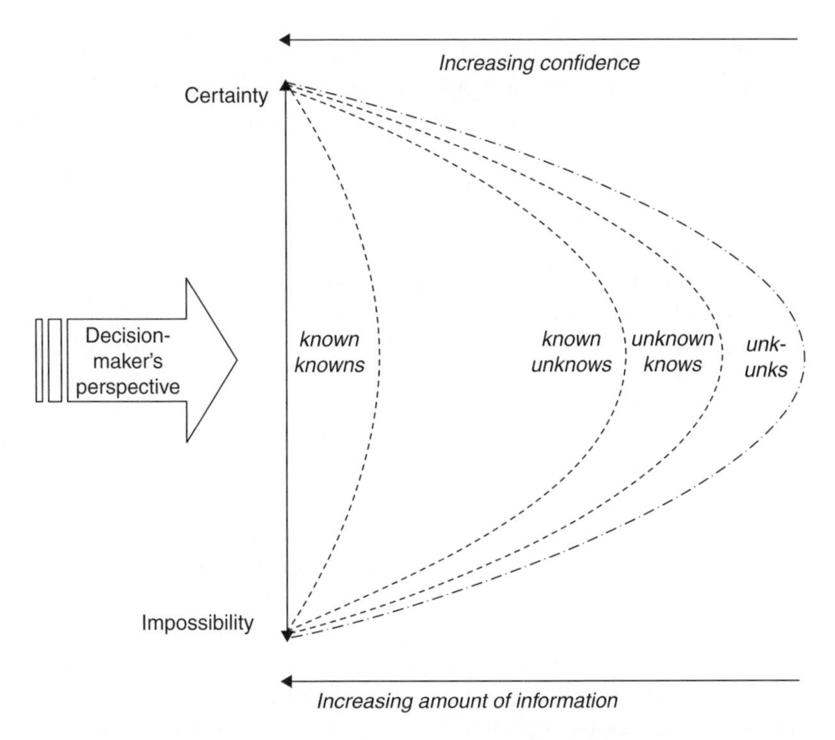

Fig. 13.2 A cognitive model of risk and uncertainty on projects.

in that space and the occurrence of risk events given those risk sources is fundamental to the sense made and hence decisions made by the project manager.

The cognitive model differs from the existing approaches to managing risk and uncertainty on the project in two ways:

- It makes explicit what is often only implicit in existing approaches that managing risk and uncertainty is fundamentally about the *perceptions* of risk events and their impact given risk sources. Risk is not a phenomenon 'out there' but a function of our perceptions given the information available at the time of decision.
- It makes clear the profound difference in those perceptions between the condition where a probability distribution can be assigned to the occurrence of an event by the decision-maker thanks to available data, and the condition where no such probability distribution can be assigned. Here we follow Knight and Keynes and define the former as the condition of *risk* and the latter the condition of *uncertainty*.

This multilayered definition of risk and uncertainty has become well known thanks to the award of the Plain English Campaign's Foot in Mouth trophy for mangling the English language to Donald Rumsfeld in 2003[9] and can be adapted from a cognitive standpoint as follows.

- *Known knowns* is the cognitive condition of *risk*, where the risk source has been identified and a probability can be assigned to the occurrence of a risk event given that risk source. This is the area of most contention, because many advocate the use of subjective probabilities which capture degrees of belief to include a much larger range of risk sources in the known knowns set, rather than leave them in the known unknown set – we will return to this in section 13.3.
- *Known unknowns* is the cognitive condition of *uncertainty* where a risk source has been identified, but a probability cannot be assigned to the occurrence of the risk event.
- *Unknown knowns* is the cognitive condition of *uncertainty* where somebody knows about the risk source and associated probabilities, but is keeping that information *private* – see section 9.3 – such as when the tosser conceals the tossed coin.
- *Unknown unknowns* (unk-unks) is the cognitive condition of *uncertainty* where the risk source has not been identified and therefore the risk event cannot be known – what has been called a 'black swan'[10].

Within the model in Fig. 13.2, the conceptual boundaries between the three main cognitive states are inevitably fuzzy, but valuable in deepening understanding of the uncertainties faced by the project manager. In particular it is important to understand whether there really is evidence to support placing an event in the known known category, or whether the project manager is being over-optimistic. Panel 13.1 presents a case where there clearly is good data available for decision-making from both an objectivist and a logical perspective, and can therefore be considered decision-making under risk, but it also indicates the limitations of such decision-making.

Panel 13.1 Known knowns in the North Sea

Working in demanding environments such as offshore oil and gas demands sophisticated management. The laying of pipe to offshore rigs is particularly difficult, combining risk sources associated with the weather and with the technology of pipe laying. The planned schedule identified the hook-up between the pipeline and the rig for August, when one risk source – weather conditions – in the North Sea would probably in the event be relatively good and a 1.6 m barge (that is one that could handle 1.6 m waves) could be specified. However, another risk source was possible schedule delays, pushing the planned hook-up to later in the year when in the event weather conditions would probably be worse and a 3 m barge be required. Alternatively, hook-up could be delayed to the following spring if the adverse weather risk event occurred, saving money on the barge, but delaying the start of the income stream from the facility. Thus, the impact of a schedule delay occurring was around up to £200m, yet the additional cost of the larger barge was £15m. The trade-offs here can be analysed using the concept of 'risk efficiency' derived from Markowitz' portfolio theory, and in this case, analysis clearly points to planning for the larger barge, even though in the event its capabilities may not be required.

This is an excellent example of the power of formal risk analysis and moves the decision from a judgement of an experienced project manager (who would typically opt for the larger barge in any case) to something that senior management can justify and defend. Its strengths are derived from working in the area of *known knowns*. Weather data for particular locations is readily available and can be effectively used to provide a probability distribution for a particular condition. The technologies for pipe laying – particularly those associated with pipe buckles which are a major risk source in this type of project – are logically knowable even if some engineering judgement is required and can also be treated probabilistically in the light of experience on similar projects as *known knowns*. Schedule delay is a *known unknown*, but can be treated on a present/absent basis for the analysis.

However, such analysis is limited to the information space of *known knowns*. As Taleb points out, portfolio analysis has proven to be counter-productive in the financial sphere – ironically the domain for which it was developed – by generating major problems for the integrity of the global financial system. This is because it was used in the analysis of expectations about future financial returns – the information space of *known unknowns* and *black swans* – and could not handle extreme events at the tails of the subjective probability distributions on which it relied.

Sources: Chapman and Ward (2003); Markowitz (1952); Taleb (2007).

In contemporary managerial decision-making, there is a strong bias towards over-optimism as discussed in section 10.9.

- Managers are easily 'fooled by randomness' and tend to credit lucky outcomes to their own skills; inversely they tend to credit unlucky outcomes to – bad luck![11]
- Contemporary management culture more readily accepts the statement 'the Monte Carlo analysis shows that there is a 5% chance of this project going over budget' than the statement 'in my judgement this project looks as if it will stay within budget' yet the former analysis is almost always based on the elicitation of subjective probabilities and therefore has no more empirical status than the latter statement.

A complementary dimension of a cognitive approach is the *confidence* in the level of information available[12]. For known knowns, this is a function of the probability distribution and can be assessed through the deployment of analytic techniques; for unknown knowns, this is a function of the level of trust in stakeholders associated with the project that they are disclosing all relevant information. For known unknowns and unknown unknowns, this is a function of judgement and experience. For instance, whether a regulatory authority will accept or reject project proposals is a known unknown – there is no data set available that can predict this stochastically. However, experience with working with a particular regulatory authority can generate a level of confidence as to what it will accept and what it will not, and effective advocacy can shift regulatory decisions. Case 9 presents a situation where lack of experience by a US-based client with UK regulatory authorities generated risks to schedule and budget that a UK-based client would probably not have experienced. Similarly, project managers working on an *incremental adaptation* as defined in Fig. 9.5 can have more confidence that there will be no unk-unks associated with the project than those working on *iconic architecture*.

The model is not dependent for its application on probabilistic thinking. Research within the behavioural school generally and on major construction projects in particular[13] has shown that managers do not think in probabilistic terms regarding the occurrence of an event, but simply in terms of impact should it occur and how capable they are of responding to the risk source or event. Manageability is particularly important for project managers – once a risk source has been identified, to what extent can it be managed either by reducing the likelihood of an associated risk event occurring, or by mitigating the impact of that event should it occur?

Before the reader is thrown into existential angst, it is worth remembering that most projects are delivered within narrow margins of performance on budget and schedule, and usually work well because of the following reasons:

- *Unk-unks* are remarkably rare in construction projects, and they are necessarily fatal to the project, as the Millennium Bridge case in panel 13.2 shows:

Panel 13.2 Unk-unks and opportunity on the Millennium Bridge

The Millennium Bridge that featured on the cover of the first edition of this book opened to the public on Saturday, 10 June 2000, and closed 2 days later when it generated a dangerous-feeling sway due to vibrations set up by pedestrian movement. This was entirely unexpected and prompted unwarranted fears of collapse. How could leading structural engineers like Arup make such a mistake – surely this was a known known? The problem lay in the combination of the lightness of the structure as a pedestrian bridge and the swaying movement of people as they walk; where the latter meets the resonant frequency of the former, movement starts. This is not the same effect as synchronised movements of marching soldiers; the sway synchronises the movement, not the other way round. The effect is not linear – for the Millennium Bridge it starts suddenly at around 166 people.

From a cognitive perspective, this is an *unk-unk*. This problem was unreported in the literature that bridge engineers read, and not mentioned in the codes they follow. The principal earlier publication on the topic was published in an earthquake engineering journal unlikely to be read by bridge engineers. Other cases observed only emerged in correspondence following the reporting of the Millennium Bridge problems and showed that it was independent of the bridge design – for instance, the Passerelle Léopold Sédar Senghor which opened across the Seine in Paris in December 1999 suffered similar problems. Arup were able to turn this risk event into an opportunity by taking full responsibility and developing an elegant damper solution which did not compromise the design intent and then allowed them to market their leading edge expertise in long-span foot bridges.

Sources: Sudjic (2001); Wikipédia (accessed 27/10/08).

- Failure to use appropriately readily available data can shift risk events from the *known known* to the *known unknown* category unnecessarily as is shown by the Challenger launch decision case in panel 13.3. There is a lot more data around than many realise, and one of the principal benefits of the lean construction initiatives presented in panel 12.8 is the encouragement of data logging and analysis at the process level;
- Many of the risk events in the *known unknown* category are internally generated and hence relatively manageable – see panel 13.4 for the research to support this claim.
- The new forms of collaborative working discussed in Part III are reducing, if not eliminating, the problems associated with *unknown knowns* as members of the project coalition are incentivised to release private information earlier rather than later;
- Many of the most spectacular 'project disasters' have not been the victims of external shocks but set up to fail through the strategic misrepresentation discussed in section 3.7 – see Cases 1 and 13 – although for obvious reasons project promoters focus on such shocks and try to blame the supply side and unmanageable *unk-unks*.

Panel 13.3 Misreading data at NASA

The space shuttle Challenger exploded shortly after launch at dawn on 28 January 1986 due to the failure of O-rings to seal properly in the rocket booster in unprecedented cold launch conditions. Seven crew members lost their lives. There are many important lessons that can be learned from this tragedy, but we will focus on one in particular – the failure to prove the importance of temperature in O-ring performance. Prior to the fatal launch, a pattern of O-ring failure had been identified – this was cognitively a known unknown – but its causes were not well understood. Some engineers believed the crucial variable to be temperature, but they could not produce objective data to support this contention. However, the data did exist; it simply had not been collected and analysed properly. The problem was that the engineers only plotted data from launches which had experienced some failure of O-rings which were small in number and yielded no obvious pattern. Investigators into the accident plotted (O-ring failure by temperature) data from all launches and found a clear patter of clustering of O-ring failures at the lower end of the temperature scale. The simple act of plotting the data properly would have shifted the risk of O-ring failure from a known unknown to a known known.

Sources: Vaughan (1996): see also Vick (2002).

Panel 13.4 The sources of risk on public projects

One of the best data sets for understanding the sources of risk on public sector construction projects is the Value for Money reports produced by the UK National Audit Office. These provide in-depth case study analysis of projects that have raised concern in government for one reason or another. A meta-analysis of 25 of these reports found that the problems they faced were overwhelmingly internally generated within the public sector client. The classic features of the risk register such as technological failures, incompetent contractors or unforeseen events rarely featured. The principal problems were to do with either the lack of appropriate routines to support project management by the client or the team's failure to follow the routines that were there – indeed only 11% of risk sources identified from the 25 reports could be attributed to factors external to the project organisation. There are many implications of these findings, but one is that the vast majority of risk sources on construction projects are manageable by the project team.

Source: Dalton (2007).

13.3 The elicitation of subjective probabilities

As argued in section 13.2, many of the more sophisticated techniques for risk analysis rely on quantitative methods. However, the reliable data sets that such methods rely on are not often available. The solution to this problem has been the development of routines for eliciting *subjective probabilities* from expert decision-makers[14] along the following lines:

- Identification and selection of the risk sources about which judgements are to be elicited with clear definition of variables and their measures. Each possible event and the scale by which it will be measured needs to be clearly and unambiguously defined, often through detailed decompositions.
- Identification and selection of experts who will provide the judgements who will be either interviewed individually or facilitated in a group. Where safety and similar risk sources are concerned, the independence of the experts from organisational pressures needs to be assured.
- Training for elicitation, so as to minimise the possibility of cognitive biases creeping into the judgements and to provide an understanding of the concept of probability.
- Elicitation, where distributions of the probability of risk events occurring given the identified risk sources and associated estimates of impacts are formally elicited through interviews. This can be done either *directly*, where the expert is asked by the analyst for the fractiles of the cumulative probability distribution or indirectly by asking the expert to a 'bet' on a specified event occurring using a tool such as a probability wheel. The most familiar implementation of this approach in construction is the simplified version known as three-point estimation of budgets and schedules. The three points can then be used to generate probability distributions[15].

- Analysis, aggregation and resolution of disagreements, so that all the participants have ownership of the resulting quantities.

An acknowledged difficulty with these commonly used routines is their handling of low-probability high-impact risks[16]. For instance, the probability wheel is claimed to be the most effective indirect method, yet it has difficulty at the tails of the distribution below 1.0 and above 9.0. This means that no probability can be elicited for 20% of the distribution. A more profound issue is raised by the work in the heuristics and biases line of enquiry which has identified some systematic biases in our perceptions of the probability of events[17].

- *Representativeness* captures our tendency to ignore the underlying base rate from which our sample of observations is drawn. A small sample will show more volatility than a large one according to statistical theory, yet decision-makers tend to ignore such issues. A construction project management example would be to accuse a project manager of incompetence if the project goes over schedule or budget. As discussed in section 10.2, the schedule and budget are the means of the distributions of the estimates, and projects will therefore overrun half the time in the absence of contingency buffers.
- *Availability* captures our tendency to base our perceptions of probability upon the most recent or high-profile events in our experience, rather than the whole distribution. A construction project management example would be to base actions on a subsequent project upon the experience of the last project without a reflexive evaluation of whether that experience was simply luck.
- *Anchoring* is the condition where our first estimate anchors our subsequent attempts to vary our estimate in the light of new information. This phenomenon is particularly challenging for the elicitation of subjective probabilities because many such routines rely on adjustments from first approximations, and the results are highly sensitive to the routine selected. A construction project management example would be the tendency to anchor on early estimates of budget and schedule despite the arrival of information that invalidates these figures.

This analysis of biases raises the question of whether experts can be debiased. Some advocates of methodologies for eliciting subject probabilities are confident that this can be done by confronting the expert with the data. But this is tautological – if the data are available we are in the world of objective probability as defined in section 13.2 and there is no need for an elicitation process. However, reviews of research on 'debiasing' indicate that this is no trivial problem[18]. In experiment after experiment, attempts to reduce bias in the elicitation of subjective probabilities have shown how intractable the problems are, and there is certainly no foolproof method of eliminating systematic biases in the probabilities elicited. Moreover, experts are found to be as equally prone to bias as the laity, as is shown in panel 13.5. The argument here is not that nothing can be done to reduce bias. Using frequencies rather than probabilities can lead to improved inferences; training can reduce biases; PowerPoint can be banned; and analytic models can be improved[19]. However, once the problem moves beyond low-impact/high-probability events the practical difficulties of calibration start to mount, and there is a growing danger that 'a debiasing procedure may be more trouble than it is worth if it increases people's faith in their judgmental abilities more than it improves the abilities themselves'[20].

Panel 13.5 The fallibility of experts

In a fascinating experiment on the judgement of experts, a group of geotechnical engineers attending a symposium at MIT was asked to predict the loading at which an existing embankment would fail. Seven experts were asked to give an estimate and a 50% probability distribution of the height of additional soil on embankment which would lead to failure. They had previously been given detailed geotechnical data on which to base their estimates. *None* of the experts encompassed the actual failure point in their confidence limits; five were pessimistic in that they expected failure at lower loadings than actual; two were optimistic. The *mean* of their estimates turned out to be the best predictor. The inference from this is that this particular group of engineers was very optimistic in its ability to predict embankment failure. The impact of the collapse was also greater than expected, and the organisers of the experiment lost their monitoring equipment in the collapse which was more extensive than expected. While geotechnical engineering is a more uncertain craft than most engineering, the results of this experiment do not instil confidence in the ability to elicit subject probabilities for logical risk sources – these engineers thought they were dealing with a *known known* which in the event was a *known unknown*.

Source: Vick (2002).

The partial exception to this statement is the use of 'calibration' or the confrontation of the decision-maker with the actual outcomes of the events predicted. This is, of course, only available with a large number of repeated elicitations of the probabilities of events with the outcomes of those events. Thus we know that the coin is true because our subjective probability of a heads of 0.5 is confirmed by repeated tosses of that same coin; if heads over a large number of tosses only turns up 0.3 times, we can recalibrate for further tosses of the coin. However, such calibration is typically unavailable for low-probability events because a very large number of outcomes would have to be plotted to allow calibration; even if the low-probability event occurs, this does not, of itself, change the prior probabilities because the occurrence could be due to chance. A moment's thought will show that if the average project lasts for 5 years, a construction project manager can only work on 8 projects in a 40-year career. This is nowhere near enough to generate any empirical insight into the factors that actually affect project performance at a level that would satisfy the requirements of the elicitation of subjective probabilities.

Is there, then, no role for subjective probabilities in project risk management? Clearly the expression of a range around an estimate derived from three-point estimation as a means of communicating the uncertainties associated with that estimate is helpful and is an advance on single-point estimates as discussed in section 10.2. It is a matter of taste as to whether these ranges are expressed as high or low chance of exceeding the stated range around the estimate, or a 90% or 10% chance of exceeding that estimate so long as it is clearly a judgement as part of a process of structured sensemaking. The problem comes when such estimates are combined through Monte Carlo analysis or other methods to produce quantitative measures of the project exceeding budget or schedule to satisfy a desire for *pseudocertainty*[21] regarding possible outcomes from a set of expert opinions.

From this analysis, we conclude that viability of the rigorous elicitation of subjective probabilities that meets the requirements of Savage and the other founders of the subjectivist school is in question. In terms of the model in Fig. 13.2 the space between the objectivist and the subjectivist boundaries is quite narrow when assessed against the overall information space of the project manager, and the role of uncertainty as defined by Knight – particularly in the information space of known unknowns – is a lot larger than the advocates of the techniques of project risk management typically allow. In the following sections we will start to develop a perspective on managing risk and uncertainty that does not rely upon the elicitation of subjective probabilities, but more on the processes of structured sensemaking under uncertainty[22].

13.4 Propensity for risk and uncertainty

So far the discussion has focused on what are sometimes called *state variables* – underlying conditions external to the project manager. However, project managers also have preferences which may be called their *propensity* or appetite for risk[23]. For a given perceived level of risk and uncertainty on a project as defined in Fig. 13.2, different decision-makers will have different appetites for the level of risk that they are willing to accept as a worthwhile proposition as illustrated in Fig. 13.3. This is a formal model of risk propensity for known knowns; the same principles, although not the formulae, will also apply to known unknowns, but not unknown unknowns – by definition decision-makers cannot have an appetite for something they do now know about. Figure 13.3 shows the relative probabilities associated with different decision-makers' risk preferences. The axes show the probability of a downside risk event on the x axis (e.g. loss due to an accident), and an upside risk event on the y axis (e.g. profit on a contract) occurring, for a constant monetary value. This definition allows us to identify some clear decision-making criteria in terms of risk profiles which are compared in a 'scratchcard' model of risk appetite in that it does not take into account any time dimension:

- Decision-makers are *risk-neutral* if they are indifferent between the chances of reward event y and risk event x occurring (i.e. $p[y] = p[x]$ where p is the probability of x or y occurring). This is indicated by a straight line at 45° in Fig. 13.3.
- Decision-makers are *risk averse* if they prefer situations where $p[y] > p[x]$. Such decision-makers either do not make the investment or are prepared to pay a premium to reduce either the probability of x occurring or the magnitude

of losses associated with x. This is indicated by the upper line in Fig. 13.3, with willingness to invest rapidly diminishing as x gets larger.

- Decision-makers are *risk-seeking* if they prefer situations where $p[y] < p[x]$. In other words, they are gamblers. This is indicated by the lower line in Fig. 13.3, with the willingness to gamble diminishing as x gets larger.
- In practice, transaction costs (e.g. dealing charges and stamp duty on equity purchases) mean that risk-neutral decision-makers have to be slightly risk averse in order to ensure that probable rewards are equivalent to probable losses associated with x plus transaction costs, t. This is illustrated by the dotted line in Fig. 13.3.

Transaction cost adjusted risk neutrality ($p[y] \star m[y] - t = p[x] \star m[x]$ where m is the magnitude of the loss or reward) is the robust position. Over repeated decision cycles, firms that are risk averse will fail to make viable business decisions, resulting in poorer relative performance. This happens in a number of ways:

- Decision-makers surround themselves with consultants and auditors, none of which add value to the process but are hired to reduce the probability of x occurring. Their fees are a cost which reduces the net value of y.
- Decision-makers take out insurance on the probable losses associated with x. Over repeated cycles, these insurance premia must be greater than the actual losses associated with x otherwise the insurers would not cover overheads and profit.
- Potentially viable investments will not be made, which a risk-neutral decision-maker would have made. Over repeated cycles, this means that the risk-neutral firm would either grow faster, be more profitable, or both.

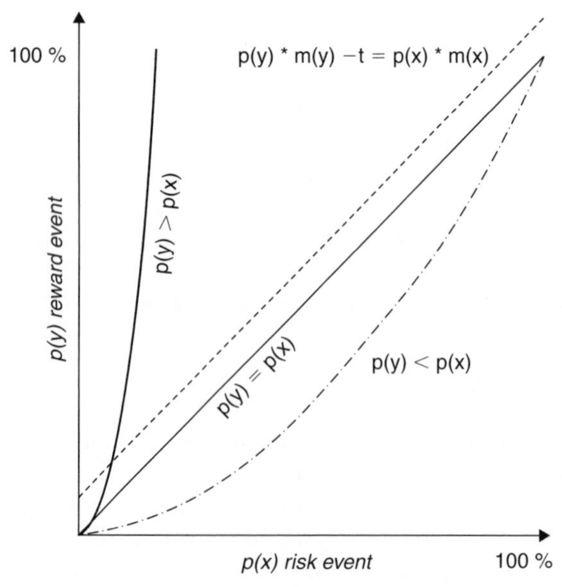

Fig. 13.3 Risk propensity.

Risk-seeking firms will tend to be less profitable than their risk-neutral competitors. Over repeated cycles, their losses will outweigh their rewards; just as gamblers as a group can never beat the house because the house would not be able to cover their overheads and profit, risk-seekers can never win in the long term. However, firms behave in risk-seeking ways for a number of reasons:

- An underperforming firm needs a quick success to remain viable – this is usually known as 'buying work' in the hope that something will turn up later to save the firm.
- Young architectural practices are trying to win a *concours*, and thereby get their big break on a nothing-ventured nothing-gained basis.
- Situations where the decision-makers are not risking their own assets. This is the classic moral hazard problem with highly leveraged firms where shareholders with limited liability risk mainly the bank's capital, yet reap all the rewards.

A large number of factors will affect the risk preference of decision-makers. Some of the more important are as follows:

- *The proportion of total assets at stake.* Where the possible loss is a high proportion of total assets, risk aversion will be higher. A decision-maker may be prepared to gamble on 5% of assets, but be completely risk averse on 50%.
- *The opportunity for the laws of chance to work.* An important assumption of the above-mentioned argument is that decision-makers have repeated opportunities to make decisions, so that outcomes approach their risk profile over time. Where decisions are unlikely to be repeated, risk aversion will tend to be higher, because bad luck is less likely to be balanced by rewards over time.
- *Sentiment.* If competitors are taking higher risk decisions, then this will tend to encourage risk-seeking by other decision-makers. This behaviour, known as herding by economists and more popularly as lemming behaviour, depends on the sense of security in numbers.
- *Organisational culture.* A bureaucratic culture in the organisation – see Table V.1 – punishes decision-makers for the costs associated with risk events occurring, but does not question excessive costs associated with avoiding or insuring against the risk.
- *Managerial capabilities.* As will be seen later, threats offer opportunities to be managed; those with greater confidence in their risk-management capabilities should be able to approach risk neutrality and hence be more effective in the management of their projects.
- *The human condition.* Our evolution has led us to fear loss more than we seek gain; humans are inherently risk averse[24].

13.5 The practice of managing risk and uncertainty

Effective routines for managing risk and uncertainty play an important part in achieving good sensemaking and reducing bad sensemaking, and can help project managers, and there is a number of protocols available[25]; the aim of this section is to take from them the basic principles of managing project risk and uncertainty.

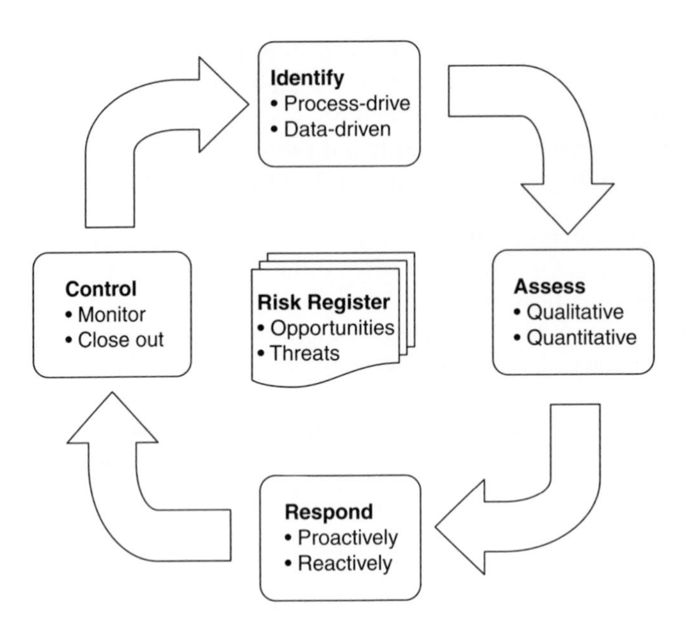

Fig. 13.4 The risk-management process.

A reading of the chapter so far may suggest that it is something of an oxymoron to talk about 'managing uncertainty' in the way we talk about managing risk. However, *known unknowns* can certainly be managed proactively as will be seen in Case 13 and the design of robust project coalitions can help to mitigate the impact of *unknown unknowns* reactively.

Figure 13.4 illustrates the four main elements of the project risk-management process[26]:

- identify and classify the risk sources in order to know what has to be managed;
- assess the risk sources so that they are fully understood both individually and in interaction with each other;
- respond to the risk sources in deciding what to do about each of them;
- control the risks while riding through the project life cycle.

The circularity of the model is designed to emphasise that risk management is, in essence, a learning process through time, and yet another application of basic information loop principles which were illustrated in Fig. IV.1.

13.5.1 *Identify and classify the risk sources*

Despite being the key to the whole process, this is one of the less formalised elements of risk-management practice[27]. Risk source identification is usually done through relying on the experience of older hands, or organising brainstorming sessions which identify the possible risks that might have an impact on the achievement of the project mission. The identification process produces a *risk register* containing all *knowns*, which is the baseline document of the process of managing risk and uncertainty and lies at the heart of the information loop shown in Fig. 13.4. It is the risk register which identifies what has to be managed from a risk and uncertainty perspective.

13.5.2 Assess the risk sources

Once risk sources have been identified and classified, there are a large number of tools available for assessing them, most of which have already been introduced in the context of managing schedule, budget and conformance. One of the most popular risk-management-specific tools is the probability/impact matrix shown in

> **Panel 13.6 Pascal's wager on the existence of God**
>
> The seventeenth-century French mathematician, Blaise Pascal – variously gambler and devout catholic – made a major contribution to the theory of probability and decision. He was the first to develop the binomial expansion which underlies the material in panel 10.2, and, perhaps most famously, he made *le pari de Pascal*. Pascal argued that either God exists or He does not, the question is how should one respond to these even odds in the absence of proof of his existence? If one acts as if He does not exist by not leading a pious life, and He turns out to exist, one risks eternal damnation. If one lives a pious life as if He exists, but He does not, only the possibility of salvation is lost. The impact of the horrors of eternal damnation is decisive, and the only rational course is to live a pious life.
>
> Keynes was scathing about this type of argument, arguing that 'no other formula in the alchemy of logic has exerted more astonishing powers. For it has established the existence of God from the premises of total ignorance'.
>
> Sources: Bernstein (1996); Keynes (1973, p. 89).

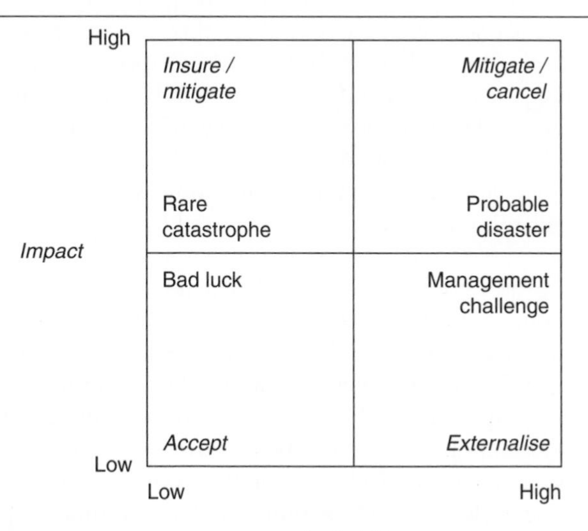

Fig. 13.5 The probability/impact matrix.

Fig. 13.5. The different risk sources identified are classified in terms of their probability of occurrence and the magnitude of their impact should they occur. Derived from Pascal's wager on the existence of God – see panel 13.6 – the probability/impact matrix can be used either with qualitative high to low scales for assessing *known unknowns*, or for eliciting subjective assessment of both probability and impact for assessing *known knowns*. It allows prioritisation of the risks on the project in terms of whether they are showstoppers or manageable within the NPV.

13.5.3 Respond to the risk sources

Having identified and analysed the implications of the risk sources, the next task is to decide what to do about each of them as shown in Fig. 13.2. In broad terms, the options are as follows.

- *Accept the risk* and plan to respond to the risk event – see section 13.5.4.
- *Externalise the risk* down the supply chain by subcontracting, as discussed in section 7.6. This should only be done when the agent is in a better position to manage the risk than the principal, because of having either more information or greater managerial capability. Externalising risk to an agent which does not have greater capabilities is folly and generates the secondary risk of the agent failing to meet commitments, and thereby effectively handing the risk back to the principal. The extreme case here is when the agent is bankrupted by the occurrence of the risk event.
- *Mitigate the risk* by changing the project mission or scope so as to minimise the probability of the risk event occurring. This is frequently the most appropriate response to identified risk sources, and a very good example of why risk management needs to start very early in the project life cycle.
- *Insure or hedge against the risk* where this is possible. With low-probability *rare catastrophes* beyond the control of the actors – such as a fire – insurance is usually possible. Where the risks are purely financial and spread across a large number of decisions, portfolio management techniques such as taking options to hedge losses are appropriate.
- *Delay the decision* until more information is available. This is frequently used, particularly in relation to risks generated by the regulatory system.

Which of the options is taken will depend very much on the classification of the risk source. With a *probable disaster*, the most appropriate action would be to mitigate the risk, if not avoid it altogether. Such risk sources are often associated with the choice of technology, in the product or process. Evaluating the options using a technique such as FMEA, presented in panel 12.4, would allow the identification of a more robust, and hence lower risk, choice of technology. If such alternatives are not available within the constraints of the NPV calculation, it would be sensible to cancel the project before too much capital is sunk.

A *rare catastrophe* is more likely to be insurable and, if so, this is the most appropriate approach; the classic example here is professional indemnity insurance.

If insurance is not available, then mitigation is preferable, before deciding to accept the risk. This choice will depend on the ratio of assets deployed to total assets. If the magnitude of the risk event could bring down the firm, then it would be better to cancel the project. This is why externalisation is not an option for large-impact risks, because clients are typically more financially capable than their suppliers of construction services, and principal contractors typically stronger than their trade contractors. If externalisation simply bankrupts the supplier firm, then the risk comes straight back to the externalising client or principal contractor. It was this understanding that led BAA to adopt the innovative project strategy of the T5 Agreement presented in Case 12.

The lower impact risk sources can be managed in a wider variety of ways. With *management challenges*, externalisation is an attractive and frequently taken option, because there should be suppliers who have better information or have developed superior management capabilities due to their exposure to similar risks on other projects. However, there will be associated transaction costs as the supplier allows a contingency for the risk. The general rule here is to allocate the risk to the actor closest to the sources of the risk and then to motivate them to manage them effectively. Where the risk is simply *bad luck* if it were to occur, the best approach is to accept it; any other approach is likely to involve the payment of excessive transaction costs.

With all types of risks, one of the most useful risk-management strategies is to delay the decision until more information is available, particularly for high-impact risks. This is an important benefit of the options thinking approach discussed in section 8.3 and last responsible moment in design task scheduling discussed in section 11.2. There is, therefore, a trade-off between risk and schedule on a project. Delay is particularly favoured where the risks derive from the regulatory context such as obtaining planning permission. It is for this reason that many clients prefer to procure design and site execution services separately. Most uncertainty reduction occurs through design, yet the costs of design are a relatively small proportion of total costs. As Fig. 5.8 shows, if the decision to procure the high-cost site execution services can be put off until more information has been obtained through design, then risks can be mitigated or even eliminated altogether. Where risks are more of the nature of management challenges, however, it may well be better to seek the early involvement of suppliers of site execution services to design out the risks.

13.5.4 Plan to respond to the risk event

If the risk has been accepted, either initially in assessment or as a residual of managerial responses to the risk source, plans will also need to be laid to respond to the risk event should it occur. The classic way of doing this is to provide a contingency in the budget or slack in the schedule. Slack and contingency are planned amounts which might be used – the problem is that knowledge of the availability of such resources can itself incentivise their use, so their effective management needs careful thought.

Contingency is not appropriate for physical risks associated with conformance. So far as quality of specification is concerned, the appropriate response is to leave an engineering margin in specifications so that performance is at the upper end of the distribution of expected use cases. So far as environmental and safety risks are concerned, plans for emergency response such as in panel 13.7 will need to be made.

Panel 13.7 Occupational health and safety at T5

The establishment of an occupational health and safety service on larger sites both responds to the threat of accidents and provides opportunities for health improvement. On the T5 project presented in Case 12, BAA paid for on-site facilities staffed by seven nurses complemented by a visiting general practitioner service. Drawing directly from experience on the Channel Fixed Link project presented in Case 1, the service provided:

- an emergency response to the occurrence of risk events causing accidents;
- routine medical services, thereby reducing lost working time caused by people visiting their own general practitioner, which meant the facility was cost-neutral;
- routine pre-screening for staff in safety-critical jobs such as crane operation – in 15% of cases this screening identified previously unknown medical conditions;
- the opportunity for lifestyle medicals for those who wanted them, which resulted in changes to the notoriously fat-laden menus on on-site canteens.

Source: Doherty (2008).

13.5.5 Control and monitor the risk source

The final phase is to monitor the risk source through the project life cycle so that as more information becomes available, the probability and impact can be reassessed, and once the point at which it could have become a risk event has been passed, it can be removed from the risk register. Monitoring risk sources requires that a *risk owner* be identified to do the monitoring. While this phase appears to be the least exciting of the process, it is no less important than the others and requires the full engagement of the project management team, not least because it is during this phase that awareness of possible unk-unk events will grow. Consummate control and monitoring of the risk sources on the project is a reflective practice, not a desk-bound manipulation of the risk register. Management techniques for control and monitoring include the following.

- 'Management by wandering around'[28] the site noticing what is going on and talking informally to the operatives and junior management of the trade contractors to get the story behind the progress, and identify potential conformance risks such as safety hazards and quality problems. This activity can identify both threats to schedule and budget, but also opportunities to do things better.

- Routine informal meetings such as those on the Waterloo International Project in Case 16 are used to freely exchange views without a record being taken, and hence liabilities incurred. These can be used to reveal potential problems and incentivise mutual problem solving.
- Spotting the weak trends and interpreting them correctly because 'Once is happenstance, twice is coincidence, the third time it's enemy action'[29] is one of the most challenging aspects of controlling and monitoring risk sources on a project – is a variance in performance noise around the mean or the first sign of a shift in the value of the mean? Certainly 'two points make a trend'[30], but it takes judgement to know whether this forewarns of 'enemy action' or is due to chance.

13.6 Managing opportunities and threats on projects

In formal probability theory, 'risk' refers to the whole range of the distribution whether the outcome is better or worse than the expected value. However, as a number of observers have noted[31], project managers tend to use the term 'risk' to refer to downside events that could affect the achievement of the project mission, particularly schedule or budget overruns. This approach has been criticised because the search for events to avoid means that events that might have a positive impact on project performance are ignored. In turn this has stimulated the advocacy of 'opportunity management' or 'uncertainty management' to complement 'risk management'[32]. We concur with the criticism, but agree with Knight that to use the term risk for a possible loss and uncertainty for a gain 'must be gotten rid of'[33]. We therefore propose that a more useful framework would be to refer to *threat* and *opportunity* in relation to potential downside and upside risk events respectively[34]. This has three advantages – it preserves the probabilistic meaning of risk; preserves the Knightian definition of uncertainty; and connects the project risk-management problem with the broader literature in strategic management and in particular with the useful SWOT tool.

At one level, the criticism of the focus on threats is misplaced. The seeking of opportunity is inherent in a project – it would not have been launched if there were no opportunity to create new value. This opportunity is enshrined in the NPV and permeates the project mission, so it is hardly surprising that the project managers focus more on the downside while riding the project life cycle. However, the focus on threats alone while riding the project life cycle can lead to emergent opportunities being missed, and so a reflective management of uncertainty and risk on projects could usefully include a special attention to opportunity.

Advocates of systematically addressing opportunity as a complement to risk – typically as part of a value management projects[35] – tend to rely upon the types of tools associated with managing design which were reviewed in section 9.7 and it follows that it is probably not appropriate to separate the search for opportunity from the design process in the earlier phases of the project. Similarly, during the later phases the search for opportunity is best placed with those responsible for execution. In both cases the options thinking presented in section 8.3 can provide a framework for assessment. Opportunity, it can be suggested, is most likely to arise from reflective practice in a collaborative project coalition and can be found in some surprising places. For instance, it is axiomatic for most project management teams that safety comes first because of the threat of accidents – a classic downside risk – yet responses to the safety risk can provide health improvement opportunities as shown in panel 13.7.

13.7 The strategic management of project risk and uncertainty[36]

The high levels of uncertainty during the early stages of the project life cycle mean that the greatest requirement for effectively managing risk and uncertainty occurs when there is the lowest level of reliable data for analysis. While there is a variety of quantitative tools available for analysis, they are usually starved of data when they are most needed. Yet many of the decisions taken early in the project life cycle will have important implications for the way risk sources and events are managed as they occur through that life cycle. The risk sources that need to be managed by project sponsors during mission definition can be grouped, broadly, into three categories:

- Market risk sources, such as whether the expected market – and hence income stream – for the facility will materialise, or whether capital funding will be available for the project.
- Completion risk sources, or whether the project life cycle can be ridden effectively.
- Institutional risk sources, or whether the project context in terms of external stakeholders or the regulatory environment can threaten the project.

There are no effective quantitative tools for assessing such risk sources, which do not rely largely on subjective probabilities – see section 13.3. While sensitivity analysis of a financial model will tell us the impact of a 20% drop in exploitation income streams for project viability, it will tell us nothing about how likely such an event is. Much of the problem is that our theories of risk are derived from research and practice in places such as Monte Carlo and the City of London. There, extensive data sets are available and repeated plays are the norm. Construction project managers cannot operate with the portfolio comfort of a win or lose situation on each throw of the dice or purchase of a share; projects are one-shot plays from the point of view of the project manager[37]. Because of this, construction project managers need to be much more proactive in managing risk and uncertainty than decision-makers who can rely on repeated plays – they need to rely on *project shaping*:

- *Information search* – including research studies, working parties, scenario planning and consultation of experts. Many of the cases and panels show how much investment needs to be made in early front-end of the project to confirm technological choices and identify risk sources. Boston CA/T in Case 13 made extensive use of expert and representative working parties to flush out potential problems and identify possible solutions, large numbers of different bridge/tunnel configurations were evaluated for the Channel Fixed Link in Case 1, while the Eden Project in Case 17 was able to persuade some of the most noted horticultural experts to work almost for nothing in the initial phases.

- *Network building and co-optation* – involving key players early, addressing external stakeholder concerns and partnering with key suppliers. CA/T promoters made extensive use of political networks both locally and nationally to obtain capital funding and minimise stakeholder opposition. Similarly, building on the credibility established through the earlier Lost Gardens of Heligan project, the champions of Eden were able to win over opposition and persuade potential sponsors to dig deep in their pockets. The support of political leaders was vital at key points in the Channel Fixed Link project.
- *Structures of incentives and contracts* – allocating risks to the most appropriate parties and creating appropriately aligned incentives was fundamental to the contracting strategies at Eden and on T5 in Case 12 to motivate collaborative working. The failure to achieve such alignment on the Channel Fixed Link and Boston CA/T led to extensive disputes which undermined the achievement of the project mission. Panel 6.5 compares the alignment of incentives on two stadium projects.
- *Project design and configuration* – avoiding locations and other features that will generate adverse stakeholder responses, designing flexible and modular solutions. The CA/T's Charles River Bridge was not the cheapest viable solution, but met the aesthetic criteria of the elites of Cambridge who would have it in their views across the river. Similarly, the architectural imagination behind the Eden Project appeased many sceptics who feared that a 'theme park' would be built.
- *Influence and mitigations* – seeking to change the regulatory context, compensating local losers, making symbolic gestures, exceeding minimum standards. The CA/T provided 11 ha of green space, among a large number of mitigations and BAA invested in extensive landscaping at T5. The clear economic regeneration affects of the Eden Project won over many local stakeholders who might otherwise have had a NIMBY reaction.

The shaping of the project requires explicit consideration of its *governability*, or the incentive structure that encourages project coalition members to respond positively and collaboratively to the (inevitable) occurrence of risk events. Some of the mechanisms for ensuring governability include:

- alignment of incentives through equity participation in the project such as on the Second Severn Bridge in panel 2.9;
- use of incentive contracts to motivate the search for cost-effective solutions, as discussed in section 6.5.3;
- flexibility and modularity in design so that the facility has alternative uses should the originally intended market shift or disappear – this led to the flexible open configuration of T5;

- deep pockets such as the lottery funding for the Sydney Opera House presented in panel 8.4 and London 2012 Olympics presented in panel 10.10, and a strategy for mobilising additional resources if required;
- long-term partnering agreements, as discussed in section 5.6.4.

In terms of the framework deployed in Fig. 5.7, mediated and unmediated project coalitions have high governability and integrated ones have low governability. Integrated project coalitions are particularly ungovernable if the unexpected happens, and they can easily degenerate into the cycle of adversarial relations identified in section 6.7, as the Channel Fixed Link described in Case 1 illustrates. In essence, project governability is achieved through sharing risks as appropriate through the project coalition, and not through shedding such risks to weaker parties. It is also enhanced if all the actors are engaged in a portfolio of projects, as illustrated in Fig. 1.4, rather than being project-dedicated entities, because they can benefit from portfolio effects and can afford to take a loss on any particular project as discussed in section 15.3. One of the major problems with the spread of concession contracting is that the benefit of the portfolio effect that public sector clients previously enjoyed is lost.

Institutional risks present a distinctive set of risks that can only be managed through networks of influence. Concession contracts often contain step-in clauses in favour of the public authorities should the terms of the agreement not be met; the interpretation of such clauses in the light of actual events is likely to be contentious. Where concession contracts – with the benefit of hindsight – appear to have been too loosely drawn, and the concessionaire is making greater profits than expected from exploitation, the renegotiation of the agreement or even expropriation of the asset is quite likely. On the other hand, where the exploitation turns out not to be viable, the public sector will always bail out the project as financier of last resort. This is because if the public authorities have a policy interest in the existence of the facility – and by definition they have such an interest in a concession – then they cannot afford to see it close if the concessionaire is failing to make a go of it[38].

One of the most important institutional risk-management strategies is to work in nations with stable and non-corrupt regulatory contexts – broadly, those scoring high on the transparency index presented in Table 5.2. Where this is not the case, risks mount and are largely incalculable. A sensitive ear to the political ground, and information garnered through well-connected networks, are essential risk-management strategies. The failure to do this can destroy coalition actors, as the case of Mitchell in panel 13.8 shows.

13.8 Summary

We started this chapter with an epigraph from Machiavelli, and he can provide further wisdom on the paradox of project risk management[39]:

> 'When trouble is sensed well in advance it can be easily remedied; if you wait for it to show itself any medicine will be too late because the disease will have

become incurable. As the doctors say of a wasting disease, to start with it is easy to cure but difficult to diagnose; after a time … it becomes easy to diagnose but difficult to cure'.

Panel 13.8 Institutional risk on the Kariba Dam North Power Station project

On 31 January 1973, Mitchell Construction Holdings, one of the leading UK civil engineering contractors of the day with extensive experience in hard rock tunnelling, was placed in receivership due to adverse cash flow on the project to build a new hydroelectric power station on the north bank of the Zambesi River in Zambia, powered by water from the Kariba Dam. This was to complement the one already constructed on the south bank in Rhodesia (now Zimbabwe). The works consisted mainly of an underground machine hall with associated intakes and outfalls. Mitchell had won the contract in competitive tendering in January 1971, and signed a standard international contract for work of this type for approximately £12.5m (1971 prices). The consultant engineers were a leading UK firm – Sir Alexander Gibb and Partners – and the client was a Zambian government-owned company, the Kariba North Bank Company (KNBC). The agent was the Central African Power Corporation (CAPCO) which operated the existing station on the south bank, and was based in the Rhodesian capital. The World Bank provided a large proportion of the costs, with the Zambians and other funders making up the rest of the capital. At the time, Rhodesia was a pariah state, subject to international sanctions having declared independence unilaterally from the UK in 1965.

It soon became clear that ground conditions were not as described in the tender documents – instead of high-grade gneiss, the rock was severely faulted with biotite schists. This rendered the agreed method of work slow and dangerous, and much more expensive. The tender documents had 'warranted' that the ground was of good-quality rock on the basis of cores taken in 1961. As work slowed to a snail's pace, and men started to be killed, Mitchell's high site overheads drained it of cash at a rate of some £200k each month. Yet Gibbs refused to certify any variations under the contract to take into account the extensive additional work required. In fact, unknown to Mitchell, Gibbs' terms of engagement did not allow it to certify anything more than £1.2k in additional costs without the agreement of KNBC and CAPCO. In any case, KNBC and the Zambian government did not have the resources to absorb any additional costs, as the market for copper – the country's principal export – dropped with the end of the long post-war boom. At the time it failed, Mitchell believed that it was owed £4.1m in under-certification of work completed. The project was finally completed by Energoprojekt of (the former) Yugoslavia for an additional £20m on a cost-plus basis.

The evidence that has come to light since 1973 suggests that both Mitchell and the Zambians were caught in a political exercise to circumvent the sanctions regime imposed on Rhodesia. The Zambians at the time were partially dependent for electrical power on the Rhodesians, and preferred to further exploit the hydroelectric potential of the Kafue Dam, entirely within their own territory. However, the claimed ease of constructing the north bank power station was used to persuade them – and perhaps more importantly, international funding bodies – to back the Kariba project first. CAPCO would then have a concession to operate the new power station before it reverted to the Zambians 35 years after completion. The ease of constructing the Kariba option turned on the quality of the rock, yet later inspection of the actual cores taken in 1961 show them to have been deliberately misrepresented in the tender documentation. In order to cover up this misrepresentation, it was necessary to blame the contractor for the delays to the works, for any certification for additional works that took into account the 'unexpected' ground conditions would have exposed the lie. Although a settlement was reached in May 1980 which gave very limited relief to both the Zambians and the creditors of Mitchell, paid from a fund created jointly by Mitchell's sureties for its performance bond and Gibb's professional indemnity insurers, aspects of the case were still the subject of litigation as late as 1995.

Sources: Morrell (1987); *Contract Journal* 06/07/95.

His advice to princes has become rightly famous, and much of what he had to say would be well taken by today's construction project managers. In particular he pinpointed the fundamental dilemma in managing risk and uncertainty – only if you can identify the risk sources before they become risk events can they be effectively managed.

The training of many professionals in the construction industry – particularly engineers – encourages the belief that there is one best way to solve a problem on a project. The training of many others – particularly architects – encourages the belief that qualitative judgement is the only way to solve project problems. The project risk manager requires the combination of the maturity to admit that there is not yet enough information to give the client a definitive answer, and the intellect to analyse rigorously the information that is available – to combine the intuition of the architect with the rigour of the engineer in managing project risks and uncertainties.

Construction projects – particularly major ones – are uncertain adventures. There will always be surprises as the unexpected happens. However, as Machiavelli appreciated five centuries ago, we can do much to prepare ourselves for those surprises and develop our ability to respond to them when they happen. Some of the most important project management skills lie not in quantitative simulation of dubious data sets, but in managing through networks, ensuring that stakeholders continue their support for the project, and lobbying those who might waver. Above all, the governability of the project coalition needs to be appropriate to the risks and uncertainties identified as present. Where they are perceived to be low, integrated routes which are relatively difficult to govern, but offer budget and schedule advantages, can be chosen. Where they are perceived to be higher, more governable project coalitions – particularly mediated ones – will be preferred, even if they do not offer the spurious comfort of having transferred risks.

Case 13
Managing Front-End Risks Through Networks: Boston Central Artery/Tunnel

As a result of the expansion of the US freeway system in the 1950s, downtown Boston was cut into two by a six-lane central artery, which had reached saturation capacity by the mid-1960s. However, the wholesale destruction of homes and neighbourhoods to build the original system had left a bitter legacy. Coupled with growing environmental awareness, this meant that solutions to Boston's traffic problems were very difficult to find. As a US Army Corps of Engineers official put it after a tunnel under the Hudson in New York was abandoned in 1985, 'You're not going to see the large project any more, because we've constructed so many hoops ... that it's almost impossible to get a large project through. There will be no more large projects'. It was against this background that the promoters and project managers of the Boston Central Artery/Tunnel worked to launch 'the largest, most complex highway design and construction project ever undertaken in the United States'. As with all projects, the greatest risk for these promoters and project managers was that it might be cancelled, and it is on the cancellation risk that this case focuses.

From the mid-1960s onwards, various ideas were developed for solving the limitations of the existing elevated freeway by the Massachusetts Department of Public Works (DPW). These inner belt schemes would have had a massive impact on Boston, such as the demolition of 3800 homes and loss of acres of open space. This planning process was described as 'pathological' because it was driven by the 10 cent dollar – for every dollar of funding raised locally, the federal government matched it with 9 dollars. Various study groups were convened, drawing heavily on the expertise of the local Harvard and MIT universities. During the 1970 gubernatorial race, both candidates expressed opposition to further expressway construction.

By 1972, the Boston Transport Planning Review, commissioned by the incoming Republican governor, had reviewed a complex set of projects which included placing the central artery underground, a third crossing of the Charles River – either by bridge or by tunnel – and investment in public transport. Dukakis, the new governor in 1975, was a Democrat noted as someone who 'hated [highways] as you would hate individual evil in people'. He appointed Salvucci – an Italian American with a reputation as something of a Machiavelli in his negotiating style – as his Secretary of Transportation. Salvucci had been an activist in the popular opposition to the DPW's schemes during the 1960s, and a member of the planning review study group. He promoted the underground central artery project because it would address the problem of the throttled expressway system and open up downtown Boston, while generating construction jobs and minimising the environmental impact.

The scope consists of 5.9km of tunnels, 3.7km of bridges and 2.4km of ground-level roads, including a third harbour tunnel and a new bridge crossing. Coupled with the requirement to keep Boston traffic moving during execution, their construction required some highly innovative civil engineering. The project scope also includes a sophisticated traffic management system to control traffic flows along the freeways which led to some problems in commissioning and late changes to the opening of particular tunnels. Unfortunately, the completed tunnels were plagued with hundreds of leaks costing some $10m to rectify, and one person was killed when a lining element fell on her car because of mis-specified adhesives.

In getting from a 1972 study group report to a working highway system in 2008, a large number of risk sources had to be addressed, any one of which could have killed the project and prevented it from ever being constructed. The cancellation risk was much higher than for a normal project because it is now clear that the business case was strategically misrepresented – see section 3.7 – in order to obtain highly favourable federal funding, and led to a series of interventions to stem the early release of information that would have shown this to be the case.

Risk 1: local loser opposition

The experience of the original construction programmes of the 1950s had been traumatic, and not forgotten, and the opposition mobilised against the schemes of the 1960s had led to the development of a number of vociferous campaign groups. The tenor of the times is indicated by Ed Logue, the Boston development director in 1966: 'In this business you've got to take some groups by the throat and say "Look, I'll do this or I'll break your neck". And they've got to believe you'll do it'. The most affected area was East Boston, where Salvucci had been a town hall official and campaigner against the DPW's proposals. He was able to work with his contacts from that earlier period to convince them that there would be no destruction of domestic property and minimal implications for commercial property. An important element in this was ensuring that the tunnel exit was on Logan Airport property, rather than in residential areas. To secure this, Dukakis had to wait until his appointees had gained enough influence on the airport management board. Salvucci's ethnic, working-class background gave him the credibility and networks to 'work' the East Boston community to gain support for the project. He was also ethnically linked to the Italian-American-dominated North End community which would benefit most from the project through the removal of the barrier effect of the existing elevated artery separating them from downtown. The 'mitigations' required to minimise local loser opposition included the understandable such as the $20m covering of the seaport access road that affected South Boston and an additional $450m of temporary works to ease the lives of Bostonians during construction and also included action to control the plague of rats released by the tunnelling. A lawsuit by the Conservation Law Association in 1990 led to action to improve public transport in the Boston area to offset the additional emissions from the traffic generated from the better road access. These mitigations totalled $2.8bn, adding around one-third to the total project budget.

Risk 2: federal funding

Ninety per cent of the funds were to come from the federal government, which meant scrutiny by the House of Representatives and Senate, and by the time the project came to this stage, the Republican Reagan was in the White House. A central element in the campaign for funding was the Democrat Speaker of the House of

Representatives, O'Neill, who represented an East Boston district. In this position, he was able to make political deals with Reagan on various matters, and was very keen to push for the funding of the project so long as he could be assured by Salvucci that there was no local loser opposition to the project – so much so that it became known in Washington as 'Tip's [O'Neill] tunnel'. This was officially recognised in 2006 when the Interstate 93 tunnels were dedicated as the Thomas 'Tip' O'Neill Tunnel. Suspicious of the merits of this and similar projects – once claiming that 'I haven't seen this much lard since I handed out blue ribbons at the Iowa State Fair' – Reagan prepared to veto the 1986 transportation bill of which this project formed a part. This threat motivated the Republican-dominated Senate to assert its authority against the President and to prevent the veto in 1986, the vote being won by offering to exchange support for tobacco subsidies with a North Carolina Democrat.

The second source of federal opposition was the Federal Highways Authority (FHWA). Here the principal weapon was the right to refuse to approve the environmental impact statement (EIS). The FHYA's head – a Reagan loyalist – hailed from Texas and was so frustrated by the anti-car lobbies in Boston that he was initially minded 'to let the bastards freeze in the dark'. However, having seen the levels of congestion on the existing artery for himself, and after further political manoeuvring by O'Neill and Salvucci, he was persuaded to approve the outline EIS in 1985. The compromise reached promised federal funding for specified parts of the system, leaving the central portion to be funded through other means. Federal funding was also capped at the budget approval level, leaving the citizens of Massachusetts, rather than Boston alone, liable for any budget overruns.

The threat of loss of federal funding encouraged deviousness on the part of the CA/T leadership team when it submitted a report in January 2000 to the FHWA which did not mention further budget overruns, despite knowledge from a review in late 1999 that the current $10.8bn budget would be significantly exceeded by at least $1.4bn. This stratagem backfired when it was inevitably revealed and was seen as a breach of trust with the FHWA, leading to a major overhaul of the FHWA's oversight arrangements.

Risk 3: management capabilities

There was considerable concern about the ability of the state agency designated to act as client – the DPW – which had little experience of this kind of project. The more experienced agency would have been the Massachusetts Bay Transportation Authority. However, there were fears that a disgruntled DPW 'could sabotage the project. Papers would get lost, that kind of stuff would happen'. The compromise was to appoint a very strong joint venture (JV) to design and manage the project – a 55:45 split between Bechtel and Parsons Brinckerhoff. The JV was appointed by DPW on a 1-year rolling contract in 1986 to be responsible for concept design and project management. The appointment of Bechtel also helped at the federal level, because two Bechtel principals held positions as Secretaries of State and Defense under Reagan's 'revolving door' with industry, and a third had been director of the

Central Intelligence Agency. In 1997, responsibility for the project was transferred to the Massachusetts Turnpike Authority (MTA). From 1998 on, the MTA and the JV merged into an Integrated Project Team. However, the JV was still obliged to repay $407m to the MTA for failures of oversight in January 2008.

Risk 4: environmental damage

The obligation to provide a project EIS under the National Environmental Policy Act of 1969 – see section 4.4.1 – created further risks and delays, because these statements required approval by a number of different authorities. This was mitigated by including Parsons in the JV, which had an unrivalled track record in preparing EIS. A 500-page supplementary EIS underwent public consultation in mid-1990. Later, the requirement for a third EIS also helped the project in that the more expensive option of tunnelling under, rather than bridging over, the Charles River favoured by NIMBY opposition was eliminated on environmental grounds.

Risk 5: NIMBY opposition

Many Bostonians, while welcoming the overall benefits of the project, objected to many detailed aspects of it. This problem was partially addressed by hiring a well-respected local architectural practice – Wallace, Floyd Associates – well networked among the city's cultural elites, to design vent shafts and similar installations. In particular, the Charles River Bridge caused a number of problems. Option Z – so called because it was the 26th to be developed – prompted a campaign of vilification in the local press, which grew in momentum. Stung by approval of Option Z by the Dukakis/Salvucci administration on its last day in office, the incoming Republican mayor in 1991 threw out Option Z and initiated the Bridge Design Review Committee, which included representatives of opponent stakeholders. Having been forced to reject the tunnel option on environmental grounds, this body chose a cable-stayed design by a Swiss architect, turning the bridge into an *ouvrage d'art* which was welcomed by Boston and Cambridge elites as a major statement and contribution to the urban scene. This issue delayed the project by 2 years and added an additional $1.3bn to the budget.

Risk 6: trade union opposition

In a heavily unionised industry such as construction on the north-eastern seaboard of the USA, union disruption during execution on site is always a major risk. Salvucci and O'Neill were both sons of Boston construction workers and were able to work these networks as well, promising thousands of construction jobs. Through the project, risks of disruption were minimised by ensuring that the package contractors agreed and adhered to a union labour only clause. However, the incoming Republican administration of Bush in 2001 made such clauses unlawful, and the Dewey Square package was the first to go out to tender without such a clause, in April. However, CA/T officials 'breathed a sigh of relief' when it was clarified that the new law did not apply to ongoing projects with existing union shop agreements.

Risk 7: budget and schedule overrun

The original project budget for completion in 1998 was $2.6bn (1982 dollars); it was finally handed over in December 2007 for £14.6bn ($8bn in 1982 prices). In 1991, the project budget was estimated at $5m. After scope changes associated with mitigations and other factors, the budget agreed for the programme with the FHWA was $8.6bn. By 1997, this had mounted to $10.8bn with inflation. However, it was announced in February 2000 that projected costs had mounted to $12.2bn, and $14.4bn in July 2001. These overruns were financed through additional bond issues by the Commonwealth of Massachusetts; it was estimated that the Commonwealth's tax and toll payers would be paying an additional $1500 each to allow this additional finance. State auditors and those politicians responsible for scrutiny of the FHWA in Washington were scathing about the ability of the FHWA and MTA to control the project budget.

In April 2001, it was revealed that in 1994 the JV had warned the governor and DPW that the budget would approach $14bn, but DPW officials had attempted to 'soften the sticker shock' by eliminating contingencies and reducing estimates of the costs of contracts still to be let and change orders. They did this by deploying federally sanctioned accounting assumptions. This move both kept the project from public scrutiny and implicated the FHWA in the suppression of the revised budget estimates. The programme manager of the JV publicly questioned the official DPW statement of budget in 1994, and was, apparently, sacked for his indiscretions.

The project has now been successfully delivered, but recriminations and litigation still shroud its technical achievement. While the benefits of the project largely flow to those who work and live in the Boston area, it is largely paid for by the citizens of the Commonwealth and federal funds. In many ways, the Boston political elites who promoted the project were remarkably skilful in their initial strategic misrepresentation and subsequent manoeuvres to address the cancellation risk. Unfortunately, the example they set in Boston is likely to mean that many other US cities which would benefit from the removal of elevated highways through their centres are unlikely to be able to launch similar projects.

Sources: Bushouse (2002); Committee (2003); Federal Task Force (2000); Hughes (1998); *Engineering News Record* (various dates); http://www.bigdig.com/ (various dates); Wikipedia (accessed 26/10/08).

Notes

1 I am grateful to Eunice Maytorena for her help with this chapter.
2 Machiavelli (1961, p. 130).

3 The ideas in this section building on section 13.2 of the first edition are work in progress in collaboration with Eunice Maytorena. They are particularly strongly influenced by the writing of Vick (2002).
4 These descriptors are adapted from Schoemaker (1982).
5 Knight (2002) has made this distinction most forcefully. The subjectivist school in particular conflates this important distinction.
6 Savage (1954, p. 12).
7 The initial shape of the model was derived from a reading of Keynes (1973), who also makes clear (1937) the distinction between risk and uncertainty.
8 The concept comes from Boisot (1995), although it is used in a rather different way.
9 Philip Stephens, 'The Unwitting Wisdom of Rumsfeld's Unknowns' *Financial Times* 12/12/03. The fourth category of 'unknown knowns' is Stephens' own contribution to project risk management. See De Meyer *et al.* (2002) for a similar categorisation.
10 Taleb (2007) defines a black swan as a highly improbably high impact event – an unk-unk with attitude.
11 The phrase comes from Taleb (2007); the more general point from Kahneman and Lovallo (1993).
12 Keynes (1973) refers to the weight of information available, while Vick (2002) distinguishes between weight and strength.
13 See March and Shapira (1987); Shapira (1997); and Shapira and Berndt (1997).
14 This section is based on Keeney and von Winterfeldt (1991) and Spetzler and Staël von Holstein (1975); see also Savage (1971).
15 See Lichtenberg (2000) for a discussion.
16 Schoemaker (1982).
17 This section is based on the classic review by Tversky and Kahneman (1982).
18 Fischoff (1982, 2002); Lichtenstein *et al.* (1982).
19 Gigerenzer (2002); Fischoff (2002); on PowerPoint and the visual display of quantitative data generally see the work of Edward Tufte at http://www.edwardtufte.com/tufte/index.
20 Fischoff (1992, p. 431).
21 Bazerman (2006).
22 See Winch and Maytorena (2009) for an initial attempt at grappling with the issues here.
23 Adams (1995).
24 See Nicholson (2000) for the reasons for this.
25 A useful overview is provided by Cooper *et al.* (2005).
26 The revisions in this section from the first edition are influenced by a reading of BP's *Risk Management Guidelines for Major Projects*, Sunbury 2005.
27 See Maytorena *et al.* (2007) for a review and critique.
28 The phrase comes from a Hewlitt Packard executive (cited Peters and Waterman, 1982, p. 289).
29 Auric Goldfinger to James Bond (Fleming, 2004, p. 1).
30 Jim Moore of BP.
31 Akintoye and MacLeod (1997); Shapira (1995).
32 For example, Chapman and Ward (2003); Hillson (2004).
33 2002, p. 233.
34 This is the terminology used in the BP Guidelines, and also by Chapman and Ward (2003).
35 For example, Dallas (2006); see also Kelly *et al.* (2004).
36 This section is largely based on my reading of the contributions to Miller and Lessard (2000), especially the contributions of Miller and Olleros, and Miller and Floricel.
37 Bernstein (1996) provides a comprehensive review of these developments; using tools derived from the analysis of repeated plays for one-shot risks is to fall into what Taleb (2008) calls the 'ludic fallacy'.
38 For example, the French *autoroute* system, built on a concession basis, has been through cycles of a greater and lesser role for the public authorities (Martinand, 1993).
39 Machiavelli (1961, p. 39).

Further reading

Peter L. Bernstein (1996) *Against the Gods: The Remarkable Story of Risk.* New York, Wiley.
A lively account by a Wall Street economist of the origin of the concepts of risk and probability, and the development of risk management.

Chapman, C. and Ward, S. (2003) *Project Risk Management: Processes, Techniques and Insights* (2nd ed.). Chichester, John Wiley.
The standard, authoritative, reference on the management of risks in a project context, although it does conflate uncertainty and risk.

Vick, S. G. (2002) *Degrees of Belief: Subjective Probability and Engineering Judgment.* Reston, VA, ASCE Press.
A profound and stimulating reflection on risk and uncertainty in construction projects which argues for the role of judgement rather than calculation in effective engineering.

Chapter 14
Managing the Project Information Flow

14.1 Introduction

> 'We are beginning to look at our industrial processes as complete, integrated systems, from the introduction of the raw material until the completion of the final product. This may be a physical product or it may be information. We look at this as an integrated system, and we try to weld together the parts of that system in order to optimise the use of our resources. It seems to me that this is basically a change in production philosophy. It is something analogous to Henry Ford's concept of the assembly line. It is a way of looking at, as much as a way of doing, technology'.

In a profound sense, the management of construction projects is about managing the project information flow. Why, then, a special chapter devoted to the topic? The reason is that specific tools and techniques have been developed for the manipulation and communication of information on construction projects, and it is these that are the focus of this chapter. Most importantly, this concerns the computerisation of the generation, storage and transmission of project information using what are commonly known as information and communication technologies (ICTs). This chapter will not be a comprehensive analysis of the application of ICTs to the construction process – many of these applications are the primary responsibility of the resource bases – but will focus on *project management information systems* (PMIS). The aim of a PMIS is to ensure that accurate and current project information is always available at the right time in the right format to the right person.

John Diebold, speaking before a US Congress subcommittee in October 1955[1], was articulating in the epigraph above the new theory of production based on the advances of Ford and others, and thereby laid the foundations for the management of production in the latter half of the twentieth century. There remain important challenges in applying Diebold's concept of information as system, which rests on the fundamental principle of *collect information once digitally*, to managing construction projects. The context of PMIS will be explored through discussion of the principles of integrated project information before turning to the discussion of ICT applications in construction project management. This is a rapidly moving area where extraordinary advances in technological capabilities are announced – so it seems – every month. There is no way a chapter such as this can be up to date even on the date it is published. The discussion, therefore, will focus on underlying principles of information management on construction projects, rather than reviewing the latest developments or discussing particular systems. In its very nature, this is an area where standards are important, so particular emphasis will be placed on the ISO standards that are applicable for information management.

The world of ICT is, perhaps inevitably, a world of acronyms. This can be difficult and confusing for both readers and writers. In an attempt to make this chapter easier to read, an appendix at the end of the chapter provides definitions of all the non-proprietary acronyms related to ICTs as they are used in this chapter and elsewhere in the book.

14.2 The principles of integrated project information

Vast amounts of information flow on a project, with the resolution of that information getting finer through the project life cycle in the rolling wave presented in Fig. IV.2. A basic strategy in developing the ability of the human mind to process large quantities of information is to structure that information. This is the fundamental principle behind traditional knowledge management systems such as libraries, and it remains fundamental to those that are currently being developed using ICTs. Knowledge management systems were discussed in section 8.9; here our focus will be on how classification underlies the management of information on the project.

Classification systems which attempt to order the knowledgebase of national construction industries have a long history. The Swedish SfB system has been under development since 1945[2] and although long superseded in Sweden itself, it remains the basis for many existing national knowledge classification systems such as CI/SfB which is widely used in the UK. CI/SfB has a number of weaknesses from a contemporary point of view:

- It applies only to building and not civil engineering.
- It does not contain classifications for process elements.
- Its coding system is inappropriate for computerisation.
- New facility types have developed, which are not included.

Awareness of these problems, growing experience with classification systems and the development of ICTs led to the ISO 12006 series aimed at establishing internationally recognised classification principles. Uniclass, published in 1997, is the UK replacement for CI/SfB which implements the principles of ISO 12006. This standard provides a classification framework in terms of knowledge related to:

- *Construction resources* – inputs to the construction process
- *Construction process* – processes and tasks associated with both the initial creation of the facility and maintenance during its life cycle

- *Construction result* – the outcomes of the construction process defined in terms of entities (complete facilities and complexes of facilities) and elements (components of those facilities).

These basic classifications, when broken down into detailed tables, provide a comprehensive classification system for knowledge of the construction process and constructed product which can be used for the storage of both physical media such as catalogues and drawings, and digital media in databases. International standards for the layering of CAD models – covered by the ISO 13567 series – also rely on ISO 12006. Uniclass incorporates the UK classification standards for the construction process – *Common Arrangement of Work Sections* (CAWS, 2nd ed.) of 1998 – and is therefore compatible with the current UK standard methods of measurement (SMM 7) for both building and civil engineering works[3].

14.3 The development of information and communication technologies

Complex technical systems do not evolve fully formed, but rather in fits and starts as the combination of technical possibility and economic advantage encourages localised developments. In the development of automated systems for the transfer and transformation of materials, this unbalanced evolution leads to the problem of 'islands of automation'[4] where highly automated material flows are mixed with completely manual ones. The same problem exists in the development of computerised information systems. The development of computing technology has meant that tools for analysis involving data manipulation have tended to develop earliest and in isolation. These tools play to the enormous strengths of computers in the rapid analysis of complex data sets – analysis that is frequently impossible manually. Thus, stand-alone applications dependent on numerical analysis, ranging from finite element analysis to critical path analysis, had been developed by the 1960s. Information flows between these types of application continued to use traditional information technologies such as the paper-based engineering drawing. During the 1970s, a new form of graphical manipulation developed to aid the creation of engineering drawings – computer-aided design (CAD). Again, the output from these systems relied largely on traditional technologies for communication between different applications. The construction industry was at the forefront of these developments – see panel 14.1. The 1980s saw the development of the personal computer (PC) which dramatically reduced the cost of computing power, and enabled a much wider diffusion of computers within the industry, while the processing power of computers continued to grow exponentially. Most importantly, site offices could now be equipped with computers.

The development of communication technologies has taken an independent path. In comparison to computer technologies, developments were earlier and more profound. The telegraph and, more importantly, telephone greatly improved communication capabilities. The fax and photocopier are more recent innovations

Panel 14.1 Information technologies in construction

The construction industry has long been at the forefront of the development of information technologies. By the fourteenth century, scaled technical drawings – probably the most important information technology of the last millennium after the printed book itself – were well established for use on religious and royal building projects. During the 1950s, Arup pioneered new computer-based structural analysis techniques for the design and testing of the distinctive shells of the Sydney Opera House roof which were also used to analyse structural deflection during construction. Without computers, the shell roof could not have been built. During the 1970s, large public sector projects – usually relying on extensive standardisation and prefabrication – offered the opportunity to develop CAD systems such as Harness and OXSYS. However, the demise of the large public sector construction programmes which have been essential to the development of ICT applications in all industrial sectors meant that this initial momentum was lost.

Sources: Howard (1998); Salzman (1952); Taffs (2006).

which have had a significant impact. However, these communication technologies did not allow any further manipulation of the data received. It was not until the 1970s that they began to be connected to computers to provide integrated systems for the direct communication of information between computer systems. The development of local and wide area networks (LANs and WANs) proceeded steadily, but interconnectivity between computers was transformed by the breathless diffusion of the Internet during the 1990s.

It is this rapid development of the interconnection between communication and information technologies over the past 20 years that has both opened up tremendous new opportunities and posed new technical challenges. When the interfaces between systems were paper based, it did not matter too much that different systems used different file formats; once computers directly communicated with each other this became a major problem. Many of the potential benefits were – and are still being – lost, because a system used by one resource base could not read files generated on the system used by another. This is the problem – rather inelegantly named – of *interoperability*.

There are a variety of different solutions to this problem, at varying levels of functionality and sophistication:

- *Neutral file formats* – the .dxf format for the exchange of CAD drawings is a well-known one. These neutral file formats suffer from the problem of being slow, and typically, some file formatting data are lost during the process. However, they are now well established and familiar to most users.
- *Metalanguages* are programming languages characterised by the ability both to define a new subset language and to dynamically extend their own functionality. In the context of the Internet, this usually involves the definition of new markup language 'tags' in order to handle data formats unique to a specific industry or application. XML (eXtensible Markup Language) was developed

by the World Wide Web Consortium (W3C) as a non-proprietary metalanguage. The basic XML schema was released in 1998; a generic construction schema – ifcXML – was released by the IAI (see panel 14.2) in 2004 with the latest release in 2006.
- *Industry Foundation Classes (IFCs)* facilitate the development of interoperability between object-oriented databases and are based on the development of the ISO 10303 series Standard for the Exchange of Product data (STEP). IFCs are a standardised framework for the definition of objects from which an object-oriented single product model can be generated for a particular project, and are developed under the auspices of the International Alliance for Interoperability presented in panel 14.2.

Panel 14.2 The International Alliance for Interoperability (IAI)

The IAI was founded at the instigation of Autodesk (suppliers of AutoCAD) in 1994, and became an international membership–based organisation in 1995. The IAI is a not-for-profit organisation. Its mission is to define, publish and promote specifications for IFCs as the basis for project information sharing on construction projects and through to facilities management. It is a 'fast-track' organisation, designed to move more quickly than is possible through the international standards setting processes associated with the ISO. In late 1999, the aecXML initiative sponsored by Autodesk's leading competitor, Bentley Systems (suppliers of Microstation), was also brought under the umbrella of IAI, which led to the launch of the ifcXML initiative in late 2000. This delivered the first schema in 2004.

Sources: http://iaiweb.lbl.gov (accessed 06/04/01); http://www.iai-international.org (accessed 03/10/08).

The continual attack on the problem of islands of computerisation by ICT suppliers has led to the development of two basic categories of information systems which are beginning to be deployed on construction projects:

- those orientated towards information about the product, increasingly called engineering information management systems (EIMS);
- those orientated towards information about the process, increasingly called enterprise resource management systems (ERMS).

Project management information systems lie at the interface between these two main categories of information systems used in the construction industry in the manner shown in Fig. 14.1 which also identifies some of the 'legacy' systems which make up EIMS and ERMS. The figure shows how the two main groupings of ICT systems in construction are presently being integrated through EIMS and ERMS, which will be discussed in turn. The figure also shows how the traditional project management software systems – notably for the critical path method – stand outside these two main integration movements.

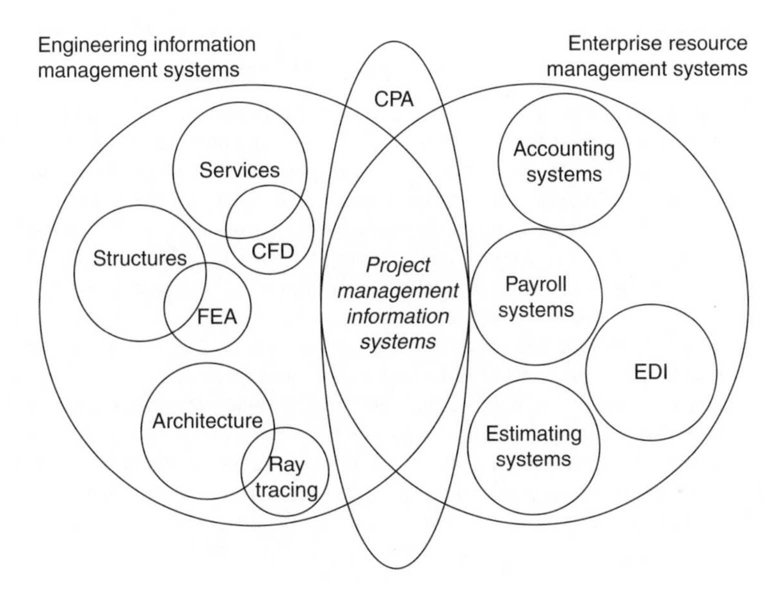

Fig. 14.1 Information and communication technology (ICT) systems for construction project management.

14.4 Engineering information management systems

Engineering information systems in construction can be broadly classified into two categories.

- *Engineering information creation systems (EICS)* – These are the basic building blocks of the project information system used for creating original input such as Bentley Microstation. They range from 2D CAD to sophisticated simulation systems using a virtual reality (VR) interface. EICS form an important type of *innovation technology*, defined as a technology designed to support more rapid and creative innovation[5] – panel 14.3 presents an innovative EICS for the simulation of building performance in fires.
- *Engineering information management systems (EIMS) themselves* – These interconnect EICS and allow the interchange of engineering information between the various resource base disciplines such as Bentley ProjectWise. Such systems are sometimes known as Product Lifecycle Management (PLM) systems when designed to support specific production processes.

The creation of construction project EIMS is presently the focus of intensive development, both commercially and in research laboratories. A typical contemporary implementation is built around an EICS consisting of a CAD system and suites of analysis programmes appropriate to the design discipline concerned. For instance, architects use ray tracing, services engineers use computational fluid dynamics (CFD) and structural engineers use finite element analysis (FEA). More

Panel 14.3 In case of fire use the lift

The development of sophisticated simulation technologies is rising to the challenge of ever-increasing demands from clients and regulators for building performance. For instance, achieving green building aspirations requires much greater modelling of the performance of the building using computational fluid dynamics (CFD) and other technologies. At Arup Fire, engineers have used bespoke programs linking CFD, finite element analysis (FEA) and pedestrian movement modelling software to simulate how buildings perform in a fire, in terms of both structural deformation and people evacuation. These simulations have challenged some well-accepted rules of thumb of fire engineering in the context of a shift towards performance-based regulation (see section 2.4). Better understanding of structural deformation has led to reduced requirements for fire protection of steel components, thereby reducing costs. Better understanding of evacuation movement patterns has led to the insight that in tall buildings lifts are the preferred method of emergency exit. Evacuation times can be almost halved, and risks for people of restricted mobility using stairs eliminated, by pressurising lift shafts so that smoke and fire do not enter them before evacuation is complete.

Source: Dodgson *et al.* (2007).

sophisticated users will be using full 3D models, perhaps visualised through a VR interface. However, these EICS typically remain discipline based; there is little potential for the electronic exchange of information between different resource bases unless they happen to have compatible EICS.

Electronic document management (EDM) systems are the basic level of EIMS, as illustrated in Fig. 14.2. Data from stand-alone EICS are either input from disk or scanned into the system. The EDM then provides a data storage and retrieval system with outputs in the form of hard-copy or computer files. EDM systems have the following advantages[6]:

- Generally efficient location and delivery of documentation;
- Ability to manage documents and data regardless of originating system or format;
- The ability to integrate computerised and paper-based systems;
- Control of access, distribution and modification of documents;
- Provision of document editing and mark-up tools.

However, EDMs also have the following limitations:

- Much effort is wasted in interfacing with non-compatible systems, particularly paper-based ones;
- Course granularity – information exchange is at the level of the drawing as a single unit of information, rather than the components depicted within the drawing;
- They do not allow concurrent working, where more than one designer works on the same drawing simultaneously.

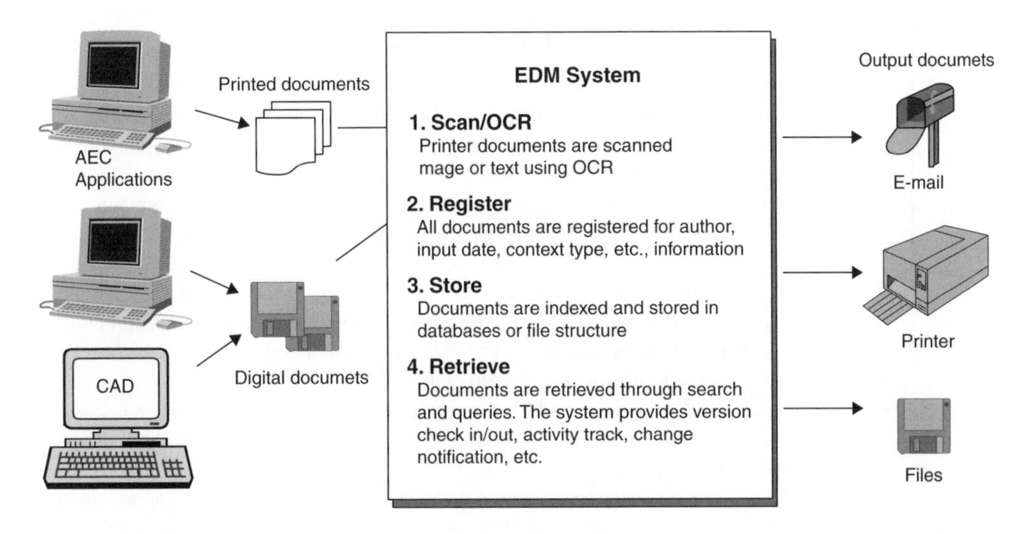

Fig. 14.2 Electronic document management system (source: Sun and Aouad, 1999).

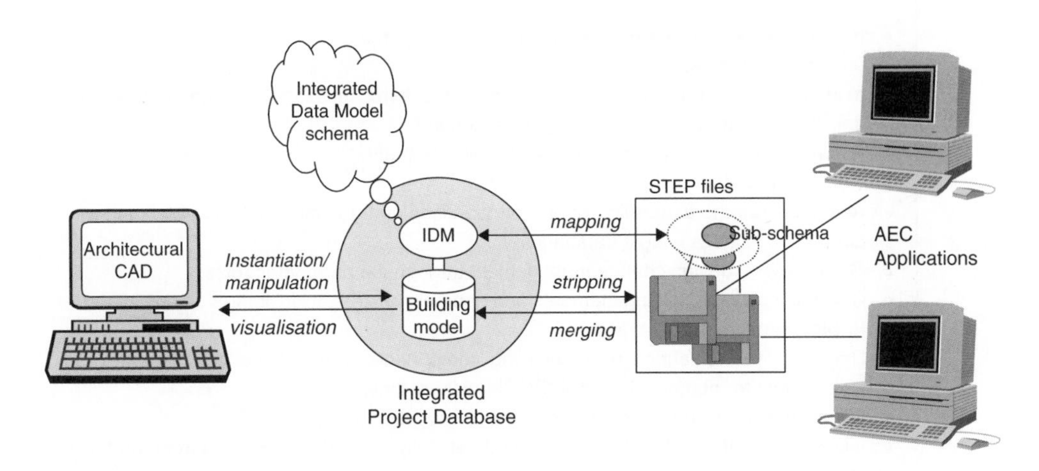

Fig. 14.3 A building information model (source: Sun and Aouad, 1999).

A more sophisticated approach is to set up a *Building Information Model* (BIM) – presented in panel 14.4 – of the proposed facility, with which all the EICS applications interact through schema-based data exchange. This development requires the use of a single project database which stores drawing files at the component level. Figure 14.3 provides an example of an EIMS organised around a BIM.

Panel 14.4 Building information models

Building information models (BIMs) represent a step change from drawing with CAD systems to modelling, and are now starting to diffuse within the construction sector – some 25% of US architectural practices reported using intelligent modelling techniques in 2007. Industry agreement on the term building information model emerged around 2003 to describe the state of the art in digital building and Case 14 describes one sophisticated implementation.

BIMs are characterised by the following features:

- Parametrically defined components that 'know' what they are and how they should interact with other building components. Parametric capability also allows design criteria to be input and the design to be generated from those criteria:
- Interoperability through the implementation of IFCs:
- Rich libraries of component details – both proprietary and component manufacturer supplied – which can be swiftly incorporated into the BIM, thereby supporting knowledge management:
- Consistent and non-redundant component data so that any change in the component is represented in all views of that component and all views of the model are represented in a consistent way:
- Virtual prototyping of both the final product and the process of its construction.

Once the BIM is developed it has an extensive range of uses, all of which are more effective the greater the interoperability of the various elements in the system:

- Rapid feasibility studies by linking parametrically generated building elements to standard cost databases;
- Incorporation of regulatory requirements so that error messages are generated when design solutions lie outside those requirements;
- Automation of the design of repetitive building elements;
- Quick visualisation of the design;
- Automation of the generation of bills of quantities and greatly eased cost planning with direct links to industry cost databases;
- Greatly eased design change management through the parametric rules;
- Clash detection of building elements and general constructability improvement;
- Facility management database;
- Direct link to the fabrication of building components where these are manufactured using computer numerical control (CNC) machine tools;
- 4D planning and virtual prototyping of the construction process;
- Incorporation of standard libraries of parametrically defined components from the firm's knowledge management processes or elsewhere such as component vendors' catalogue libraries;
- Improved quality control on site as coordinate data obtained from GPS and laser photogrammetry is fed back to the BIM to check for conformance in execution;
- Inputs to immersive VR for design reviews.

The promise first articulated in the 1970s of integrated building modelling is now being realised. The extraordinary organic forms created by architects such as Frank Gehry and Zaha Hadid – both users of the CATIA-based Digital Project – could not be built without the use of an integrated BIM. nD represents the next stage of development of BIM. While BIM is excellent for visualisation and simulation, it does not contain any decision support tools.

nD aims to supply such capability using tools such as analytic hierarchy process (AHP) to understand stakeholders preferences and to identify trade offs in the design to meet those preferences.

3D BIM technology is now well established, and starting to diffuse rapidly, but there remain a number of major challenges:

- Some BIMs, such as AutoDESK Revit, have low levels of interoperability; the tension between proprietary and open systems inherent in CAD is being reconstituted in BIM;
- Interfaces with other information systems in construction – notably geographical information systems (GIS) and enterprise resource management systems (ERMS) – remain rudimentary;
- The shift from 3D through 4D to nD incorporating analytic decision support capabilities is at an early stage;
- The potential for simulation using agent-based technologies is in its infancy;
- Debates are developing around which member of the project coalition should 'own' the BIM – the designers who generate it or the contractors who most benefit from it. In fact, it appears to be clients who are instigating the use of BIMs, probably with a view to the asset management benefits of BIMs through life, and the US General Services Administration is being proactive in ensuring that BIM technology meets its needs as a client;
- Implicit in the implementation of BIMs is a change in the relationships between the members of the project coalition, and a greater emphasis upon collaboration. It would appear that the kind of project coalition governance described in Case 7 is most appropriate to facilitate BIM implementation.

Sources: Aouad *et al.* (2008); Drogemuller (2008); Eastman (1999); Eastman *et al.* (2008); http://www.reedconstructiondata.com/bim/ (accessed 17/10/08); http://www.dte.co.uk (accessed 17/10/08); interview Ghassan Aouad (29/10/08).

14.5 Enterprise resource management systems

Running in parallel with the development of engineering information management systems during the 1990s has been the development of ERMS. These are widely known as enterprise resource planning systems (ERPS), but this is a legacy of their history and evolution from manufacturing resource planning (MRP II), and they are now used for much more than just planning. Some suggest that ERPS be called enterprise systems, but this is too broad. Most notably, the integrating scope of ERPS does not include those systems used for product development – what we have defined above as EIMS. Thus, the term ERMS is preferred here. By definition, ERMS are resource-base-orientated. Unlike EIMS, which are inherently project-orientated, ERMS are designed to support the resource base as a continuing business, rather than a one-off project.

ERMS typically integrate a number of different areas of application:

- *Manufacturing resource systems*. These are the heart of ITC systems within the manufacturing sector. The evolution from materials requirements planning (MRP), MRP II to ERPS has been a long and not altogether smooth one, but the vision is now one of an integrated information system for the management of manufacturing operations;

- *Financial systems.* These have evolved from traditional accounting systems, and provide for the financial management of the business;
- *Human resource systems.* These are the systems concerned with personnel administration – payroll, pensions and the like, and also provide a database of staff profiles ready for assignment to project teams;
- *Customer relationship systems.* These are the interface with the customer, providing marketing data on customer purchasing patterns and, with the arrival of B2C e-commerce, becoming a distribution channel in their own right;
- *Supply chain management systems.* Largely replacing electronic data interchange (EDI) applications, these are central to the development of B2B e-commerce systems. They largely automate the processes of ordering, logistics and invoicing between members of the supply chain.

ERMS provide the information backbone of firms. Where previously, information systems in the firm were either manual or consisted of a number of stand-alone systems with manual interfaces, ERMS provide 'seamless' integration between all the main functions of the firm required for continuing operations – Fig. 14.4

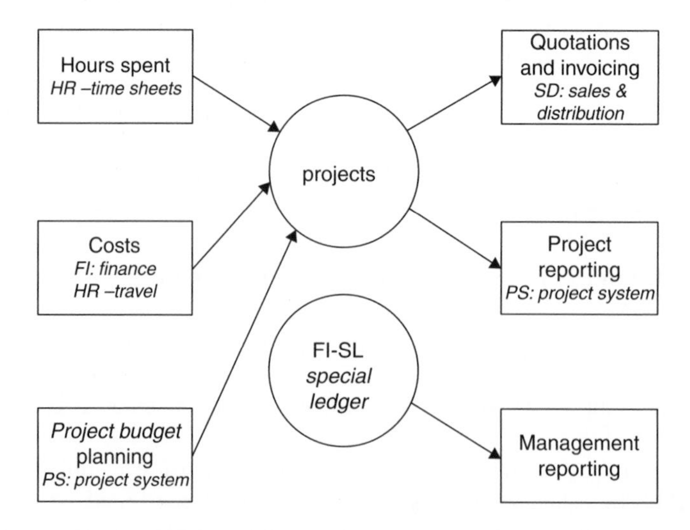

Fig. 14.4 SAP for project management at COWI (source: Interview, Jesper Hjerrild Rild, IT Systems Manager, COWI 10/04/01).

presents the SAP R3 implementation at COWI Consult showing the various SAP modules selected with their applications.

In technical terms, the main features of ERMS are[7] as follows;

- Modular construction, so that implementers need only purchase the modules that most suit their business needs. In addition, third parties supply bolt-on functionality, particularly in the area of supply chain management and professional services automation;

- A client–server architecture. Most ERMS evolved from mainframe computing, and require massive computing power for data storage and analysis. The heavyweight business is handled by the server while the user has a relatively thin or thick client depending on precise need;
- The ability to be configured to meet particular business process needs. In some industries, firms have worked collaboratively with ERMS suppliers to provide industry-specific configurations; see panel 14.5 for a construction example;
- A common central database. This is the whole point of an ERMS – a central database with common data standards imposed across the whole organisation. These are typically relational databases; the kind of object-oriented databases preferred for BIM have not found favour in the ERMS community;
- Web capability for the posting of information to intranets, extranets and the Internet as appropriate.

Panel 14.5 ERMS in construction

SAP, the German software company, is the global leader in ERPS, and claims some 1500 customers worldwide in the construction sector, although most of these are in the engineering construction and equipment manufacturing sub-sectors. Within the SAP Engineering, Construction and Operations industry solution there are three configurations focused on project management, procurement and facility management. These tend to be focused very much on estimating, cost planning, cost control and resource management. Interfaces are available with Primevera and BIM systems supplied by Autodesk and Bentley.

The high costs and generic nature of ERMS has left a niche market for construction-specific ERMS such as COINS (COnstruction INdustry Solutions) which provide integration modules for builders, civil engineering contractors, housebuilders and specialist contractors.

Sources: http://www.sap.com (accessed 03/10/08); http://www.coins-global.com (accessed 20/10/08).

Broadly there are four main modules available in most ERMS, which move beyond purely resource base management issues to address project management issues:

- *Supply chain management*. This is the horizontal dimension of the project coalition, as defined in section 7.2. In addition to considerably reducing the transaction costs associated with the administration of commercial relationships, ERMS supply chain applications can be used to meet the information requirements of lean production on a sell one, make one, buy one basis. Sophisticated optimisation algorithms can also be used to analyse the information generated by the system to improve decision-making.

- *Knowledge management.* ERMS generate vast amounts of information. By capturing, storing and analysing that information, ERMS can greatly enhance the ability of the resource bases to learn from the projects on which they mobilise.
- *Human resource management.* As well as integrating personnel services, ERMS store knowledge on the skills and capabilities of staff which can then be queried when projects need to be resourced.
- *Project management.* ERMS applications typically include project management modules, and interfaces with MS Project are common. Such applications tend, however, to be human resource management, rather than scheduling, orientated, although this is changing. Here, the suppliers or ERMS applications are starting to come into direct competition with the suppliers of traditional project scheduling software, which are also actively developing their resource scheduling offers.

ERMS are not easy to implement, and require significant changes to business processes – Atkins is a famous example of a fraught construction ERMS implementation in 2002[8]. In particular, most ERMS were not designed for project-based businesses, yet it is usually necessary to adapt to the system rather than adapt the system to the business because of the high costs of customisation. Fundamentally, the implementation of an ERMS is a major business change. However, a growing number of construction firms including COWI, Arup and Davis Langdon have implemented ERMS. Many contractors such as Balfour Beatty, however, prefer to implement COINS because it is specifically designed for their business processes.

Whatever the particular implementation, however, the more general and more fundamental point about the spread of ERMS in the construction industry is that they will become the principal source of knowledge about the construction process. Data on elemental costs and prices, task execution times, competent suppliers, resource availability and progress against schedule and budget will be increasingly stored in such systems – all the data required for full nD modelling. It will become increasingly difficult to establish accurate project budgets and schedules without accessing the ERMS used within the project coalition. Similarly, client ERMS will become the main repositories of data on the performance of buildings in use. The interfaces between ERMS and EIMS are, arguably, the central IT challenge facing the construction industry.

14.6 *e-construction*

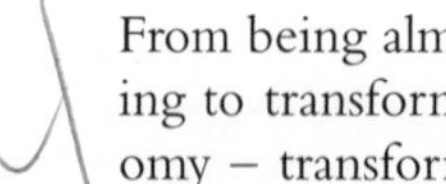

From being almost unknown in 1990, the Internet and World Wide Web are starting to transform the way in which business is done throughout the global economy – transformations which may usefully be grouped as e-business. E-business falls into two categories:

- B2C (business to consumer) covers the wide variety of new channels to consumer markets that the Internet allows, but these are of little concern to construction project managers.

- B2B (business to business) covers the use of the Internet for business relationships. The same web computing principles can be applied to *intranets*, which are internal networks within a single organisation, or *extranets*, which are extended networks between closed groups of companies.

In many industries, the potential of B2B is considerable; in a fragmented industry such as construction it is even greater. Fragmentation means that bespoke-integrated intra-firm networks are difficult to establish because of the relatively small size of firms, and where such networks are established, they are limited in their scope because of the necessity of interfacing on a temporary basis with a number of other firms within the project coalition. Similarly, the types of supply chain EDI networks that have become commonplace in many industries do not warrant the investment where relationships are one-off. In both cases, the number of repeated data transactions is simply not high enough to justify investment in bespoke networks. The Internet, supported by the diffusion of high-speed broadband, changes all this and offers considerable potential for transforming the ways in which information flows on construction projects.

Figure 14.5 tries to capture the range of applications of e-business in construction – let us call it *e-construction* – as a subset of what is more generally known as e-commerce or e-business. The applications fall into two broad areas:

- E-resources, which are essentially pre-contract and involve the search for resources such as people, components and suppliers;
- E-projects, which are post-contract and involve managing directly project information flows, or managing the supply chain.

Developments in e-procurement were discussed in section 5.7; the aim of e-construction more generally is to reduce transaction costs by cutting the information handling costs of the tendering process – most notably the printing and distribution of tender documents. While valuable savings can be made, such developments are unlikely to have a major impact on the management of construction projects. E-portals are similarly valuable; here the transaction cost savings principally come through reducing information search costs while opening up the possibility of finding better suppliers, particularly of idiosyncratic construction components and advertising for staff. Many of the traditional suppliers of construction industry information such as EMAP and Barbour[9] have now established portals through which decision-makers may search for component-specification information, search databases for the professional press and order materials. There are also a number of websites dedicated to diffusing industry best practice. E-auctions have received much attention, but it is unlikely that they will have much of an impact outside the supply of commodity materials and components such as cement and standard steel sections. Another interesting area of application is to the trade in used plant and machinery, where a broader market is of benefit to both buyer and seller, such as the site provided by Surplex[10].

Valuable as such initiatives are from the point of view of construction firms and their clients, it is our second e-construction category – e-projects – which is likely to have a much greater impact on the management of construction projects. There

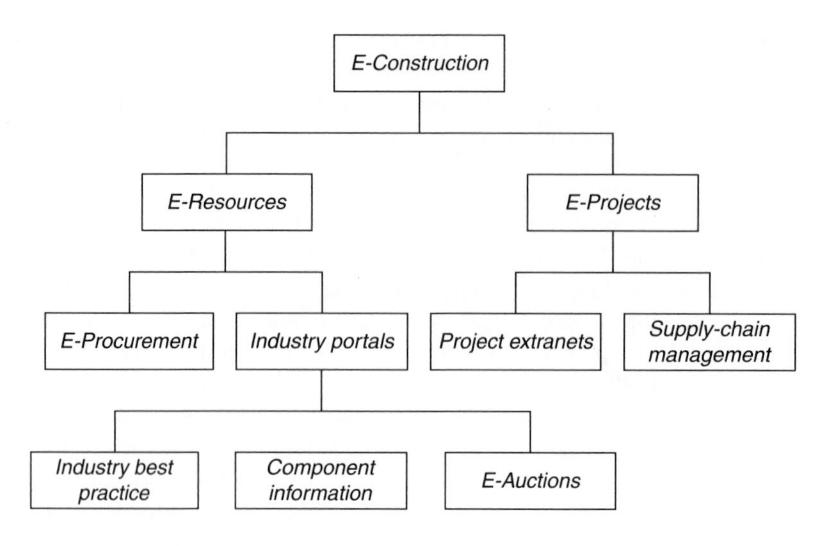

Fig. 14.5 The shape of e-construction.

are two broad categories of service here – project extranets, which will be discussed in the following section, and supply chain management. This latter category is essentially the Internet version of EDI. Developments such as XML, on-site bar coding, RFID tagging and ERMS are making these sophisticated supply chain management systems a possibility in construction. Their full development is likely to be dependent on the development of collaborative construction supply chains, moving towards what were defined in section 7.6 as construction quasi-firms.

14.7 Project extranets

Project extranets are a major new development, taking advantage of web technology to potentially transform the way in which the members of the project coalition interact with each other. In essence, they are web-enabled EDM systems which provide trackability and transparency in terms of information flows between the project coalition members without requiring individual systems to be interoperable, and offer claimed[11] savings of up to $70k on a medium-sized project and other benefits[12] including:

- average drawing approval times using extranets of 6.9 days, compared to 9.3 days for paper exchanges – a net improvement of 26%;
- audit trails of who did what and when to documents;
- 24/7 availability of documents;
- version control of project documents.

Over the past 10 years, the business model for the supply of project extranets has matured to the point where it is usually supplied on a Software as a service (SaaS)[13] basis by specialist suppliers who host the servers, although EIMS technologies such as ProjectWise do also have extranet capabilities. Many such suppliers came and went during the dot.com boom of the early 2000s, but one that has successfully survived and matured is BIW Technologies, presented in panel 14.6.

Panel 14.6 BIW Technologies

BIW Technologies originally started as a component supplier portal – the first in the UK – established in the mid-1990s. Strongly associated with the Process Protocol initiative, BIW Technologies developed the BIW Information Channel in collaboration with a number of leading retail clients and construction management companies and launched in 1998. It is currently one of the leading systems in the UK market.

BIW Technologies developed its own proprietary solution to the problem of how members of the project coalition could view construction drawings over the Web without themselves having applications such as AutoCAD and MicroStation. All functionality is achievable through a standard web browser and broadband connection.

The heart of the system is the collaboration platform which includes:

- document management, supported by audit trail of changes;
- integration with MS Outlook;
- a visual collaboration environment which allows documents to be viewed for comment without running the CAD application used for document creation;
- the ability to create an as-built archive for future asset management.

The document management system is complemented by project management applications such as:

- financial and other status reporting;
- integration with mobile technologies for logging site status such as snagging items;
- e-procurement capabilities;
- additional facilities available in support;
- contract administration, particularly the ECC discussed in section 6.4;
- incorporation of company and other standards such as standard details;
- archiving onto hard disk storage devices;
- creation of the obligatory (in the EU) health and safety file;
- interfaces with construction ERMS such as COINS.

Source: http://www.biwtech.com/ (accessed 22/09/01, 16/10/08).

Full-scope project extranets have the ability to allow the following from remote locations without the requirement for the user to have the application in which the original file was created:

- Evaluation of progress on site using webcams and digital photos
- Review of project documentation
- Review and marking up (redlining) of drawings without the need for them to be translated between CAD systems
- Issue and monitoring of the status of requests for information (RFIs)
- Management and tracing of correspondence, particularly e-mails

- Issue and monitoring of engineering change orders (ECOs)
- Sending and printing of drawings
- Data logging on site from mobile devices
- Provision of a Web presence to give the project a public face
- Support for contract administration.

Project extranets are very much a new technology and their potential has yet to be fully realised, but they are one of the most exciting developments in PMIS for a long time – by September 2008, BIW alone had a user base of nearly 11 000 projects.

14.8 The role of the project manager in managing project information

The pace of change in the application of ICTs to the construction industry is rapid, but their application is, primarily, the responsibility of the resource bases. The temporary nature of the project organisation, and hence the necessity to amortise any investments on a single project, means that investment in IT cannot easily be justified on a project basis. Moreover, the project manager is not responsible for either the generation of project information or the principal user of that information. The generation of project information – principally information regarding aspects of the product – is the responsibility of the designers, while the development and use of that information through the project life cycle is mainly by those responsible for execution on site.

What then is the responsibility of the construction project manager? Figure 14.1 helps us to identify this. In essence, just as the interfaces between the resource bases within the project coalition need to be managed, so do the interfaces between the ICT systems used by those resource bases. In other words, the construction project manager is responsible for the problem of interoperability at the project level in order to enable the most effective possible implementation of the PMIS. This will be limited by the current state of the technology and IT capabilities of the resource bases mobilised within the project coalition. It is likely that such capabilities will, increasingly, be selection criteria for membership of the project coalition. The project manager will also need to manage the 'wakes of innovation' presented in panel 14.7 that the new opportunities generated by the use of BIMs induce through the project coalition.

While the available solutions to this problem remained in paper form, and EICS were stand-alone, the responsibility was limited to ensuring the use of CI/SfB or similar classification system. As the Latham Report put it, co-ordinated project information 'is a technique which should have become normal practice years ago'[14]. The development of neutral file formats eased the problem but did not change its essence, and while benefits were gained in exchanges between particular EICS, relatively little progress was made at the project level. However, the development of the Internet and, most recently, XML has changed the situation dramatically. An unprecedentedly high level of interoperability is now attainable on most projects even if this does not achieve full BIM functionality. EIMS offer a considerable advance on previous capabilities, particularly when web

Panel 14.7 ITC induced wakes of innovation

The developments of BIMs is enabling architects such as Gehry Partners to design ever more complex buildings who then push the technology of construction creating 'wakes of innovation' through the project coalition. For instance, a number of trade contractors on the Peter B Lewis Building at Case Western Reserve University responded innovatively to Gehry's design:

- The structural engineers invented a new method for designing a steel roof with dramatic curved surfaces, winning an industry award.
- The drywall contractor licensed a Swiss soundproof plaster system and also developed a patentable method of framing undulating surfaces which opened up a new line of business.
- The local (Cleveland) fire authorities developed new techniques for modelling smoke evacuation which were presented at their national training academy
- The metal cladding contractor invented a waterproof shingling system that dramatically reduced the thickness, and hence cost, of roofing the building.
- The construction manager took responsibility for providing datum points on site, increasing its own risk but reducing the construction schedule and site errors.

Sources: Boland *et al.* (2007); see also Boland and Collopy (2004).

enabled through project extranets. Similarly, the limitations of EDI can be mitigated using XML-based supply chain management systems as part of ERMS.

Thus, the responsibility of the project manager lies at a number of interfaces, identified in Fig. 14.1:

- Within the EIMS domain, the interfaces between each of the EICS and their associated analysis programmes need to be co-ordinated by the project manager responsible for design.
- Within the ERMS domain, the interfaces along the supply chain need to be co-ordinated by the project manager responsible for execution on site.
- The interfaces between these two domains – the PMIS domain – need to be co-ordinated by the construction project manager and interfaced with the traditional project management tools such as CPM, in collaboration with the design and construction managers.

The term 'co-ordination' is here used to cover a number of responsibilities such as:

- ensuring that the resource bases are taking full advantage of the information management tools available for their domain;
- ensuring that the protocols for the interchange of information between the various systems deployed by the resource bases on the project are established and maintained – interface formats, password protection, marking-up conventions, RFI and ECO templates, security and the like;
- ensuring that a database of all communications – both electronic and paper – between resource bases is established and maintained;
- ensuring version control of files, plans and documents and the preparation of an as-built file for future asset management.

At the level of the conception and execution domains, these capabilities are available within EIMS and ERMS. However, at the interface between these two systems there is little available off the shelf to help the construction project manager. This is the potential of the project extranet. In combination with more traditional PM tools, it could become a very powerful means for the effective management of information flows on the construction project. It follows that the prime responsibility of the construction project manager in coming years may well be the effective implementation of the project extranet. In collaboration with the design and construction managers developing EIMS and ERMS applications, construction project managers will need to develop the capabilities of project extranets to meet the needs of all the members of the project coalition along the project chain. Integration with traditional PM tools will enable project progress to become visible to all, and will place a greater premium on the effective visualisation of the project management process.

14.9 Summary

The construction industry has long been at the forefront of the development of ICTs, but it has fallen back since the early 1980s as the investment cost of such technologies has risen while returns remain uncertain. Recent ICT developments could allow the construction industry to catch up, and it may be approaching a 'tipping point'[15] in its use of ITC. For some, such as Patrik Schumacher, a partner in Zaha Hadid Associates, parameter-driven design will stimulate a new architectural style – 'parametricism'[16] – that will finally replace modernism. More modestly, we can argue that the Internet is well adapted to the needs of an industry like construction where value is delivered to the client in shifting coalitions of relatively small, independent firms organised in projects. There are no common standards on the Internet, beyond minimal applications such as communications protocols and web browsers, and value is also delivered to customers through shifting coalitions of firms. New start-ups are easy, and there are no dominant firms. Earlier generations of integrated IT applications required high degrees of organisational centralisation; the Internet, on the other hand, is very decentralised with high levels of redundancy. Interoperability is largely achieved through the rapid negotiation of metalanguages, rather than through the *de jure* standardisation process. Perhaps most importantly, the Internet is – like construction – ideas led, where product integrity takes primacy over process integrity in delivering value for the client.

The Internet has implications for all the members of the resource base, but its implications are, arguably, greatest for the project managers at all levels as it transforms the management of project information flows. In particular, the development of project extranets and BIMs since 2000 is probably the single most important development in IT tools in the history of project management, for it places the construction project manager as the key node in the project information flows, and hence in a much easier position to exert project leadership, as will be discussed in Chapter 16. The fundamentals of construction project management still apply, but as time goes by, ICTs are opening new opportunities for the effective management of information flows on the project.

Appendix Acronyms for ICT applications in construction project management

Acronym	Full text	Definition
ICT	Information and communication technology	General term for interconnected systems of computers and telecommunications
PMIS	Project management information system	The subset of ICT which is explicitly orientated to supporting the project management function
ISO	International Organisation for Standardisation	See panel 12.6
CAD	Computer-aided design	General term for design and draughting support tools with a graphics interface
LAN	Local area network	Intraorganisational network of computers
WAN	Wide area network	Interorganisational network of computers
HTML	Hypertext markup language	Language for defining tags in Internet applications
XML	eXtensible markup language	A generic metalanguage for the exchange of information over the Internet
EIMS	Engineering information management system	Systems aimed at managing information generated during the product design process – typically in EICS
EICS	Engineering information creation systems	General term for systems used for creating and analysing design information – see panel 14.3
ERMS	Enterprise resource management systems	Systems aimed at integrating the diverse systems used within firms
VR	Virtual reality	A user interface with sophisticated graphical capabilities
CFD	Computational fluid dynamics	An engineering analysis technique for fluids
FEA	Finite element analysis	An engineering analysis technique for structures
CPM	Critical path method	See section 11.2
IAI	International Alliance for Interoperability	See panel 14.2
IFC	Industry foundation classes	Component-specific schemas for the exchange of data within BIMs
MRP	Material requirements planning	Planning system for scheduling the materials required for a manufacturing process
MRP II	Manufacturing resource planning	Planning system for scheduling all the resources required for a manufacturing process, incorporating MRP
B2C	Business to consumer	e-commerce marketing channel to individual consumers
B2B	Business to business	e-commerce marketing channel to other businesses
BIM	Building information model/modelling	See panel 14.4
EDI	Electronic data interchange	Dedicated WANs to enable electronic exchange of information in supply chains
SaaS	Software as a service	SaaS companies supply software applications to firms on a fee basis, thereby outsourcing IT capabilities from the resource bases

Case 14
Building Information Modelling at One Island East

港島東中心 (One Island East) opened in April 2008. Despite standing over 70 floors high at 308 m, it is not the tallest building in Hong Kong. However, it *is* notable for its leading edge use of building information modelling, as shown in Fig. 14.6 which shows the mechanical and electrical services for the whole building. The client, Swire Properties Limited, hired Gehry Technologies to set up a BIM project office adjacent to the construction site where the contractor (Gammon Construction) and key trade contractors were co-located with members of the design team to create and maintain the BIM from the scheme design stage onwards. Construction Process Simulation consulting services were provided separately by the Construction Virtual Prototyping laboratory of Hong Kong Polytechnic University.

Gehry Technologies was founded in 2002 to develop and market the innovative implementation of Dassault Système's CATIA 3D CAD system that had been behind the design of iconic buildings such as the Guggenheim Museum in Bilbao. Originally developed as a high-end surface modelling application for the aerospace industry and automotive industries, with the launch of Version 5 in 1998 CATIA was a fully functional CAD system implementable on PCs. Gehry Partners had originally adopted CATIA in order to meet the challenges of designing and constructing Frank Gehry's distinctively organic built forms during the 1990s. As their confidence in the technology grew, Gehry developed an engineering information management system called Digital Project which sits on top of CATIA v5 as a PLM implementation.

Following a presentation by Gehry Technologies at Hong Kong Polytechnic University in 2004, Swire adopted Digital Project as their company-wide BIM in early 2005 and instructed its implementation on One Island East. Until that time the architects – Wong and Ouyang – and the design team were working in 2D CAD. Following training by Gehry Technologies, they migrated to Digital Project working closely with other design consultants and Gammon. Some 500 person-years over 2 years were invested in the modelling with the BIM at a cost approaching 1% of total project cost. Although not confirmed by Swire, the industry target for cost savings using BIM is around 10% of project cost during execution. The One Island East project demonstrates many of the benefits of using a fully integrated BIM during the realisation phase of a construction project through detail design to execution on site.

- *Clash detection*. The enormous complexity of tall buildings generates the potential for large numbers of clashes which are typically resolved by change orders on site. Using Digital Project over 2000 clashes between services and structural building elements were identified prior to work on site as shown in Fig. 14.7. This work was particularly important for mechanical and electrical services, and the M&E contractor, Balfour Beatty worked particularly closely with the modelling team. Continual modelling during execution on side identified 150 potential clashes per week before they were 'discovered' on site.

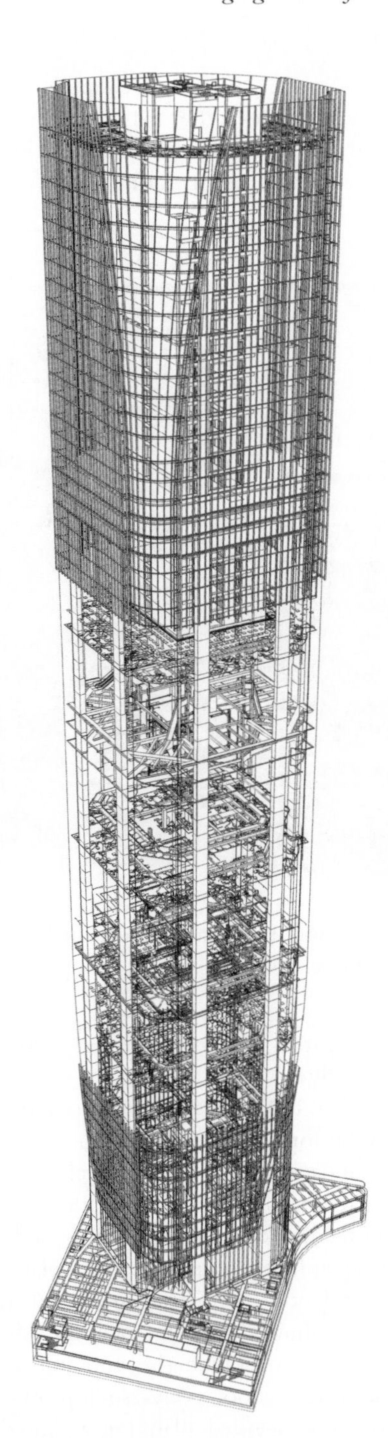

Fig. 14.6 One Island East: Entire M&E Services model (source: Gehry Technologies).

- *Parametric object creation*. Repeated elements of the design were created using a number of simple rules rather than being modelled one by one. For instance, steel structural outriggers were created in this way. These parametric objects were also stored for reuse on future projects, thereby supporting knowledge

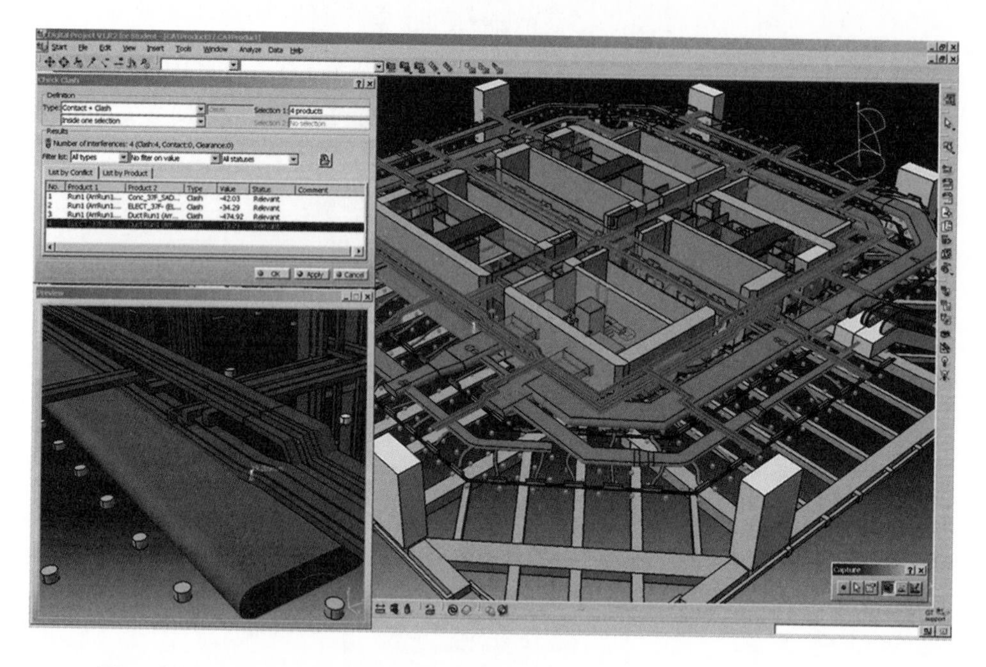

Fig. 14.7 One Island East clash detection between electrical and mechanical services (source: Riese, 2008, Fig. 5.5).

management. Figure 14.8 shows a repeated structural element that was generated parametrically.
- *Cost planning*. The BIM was used to automatically take off the quantities from the model in the format of the local standard method of measurement, which were quickly updated as the design developed. This facility allowed the quantity surveyors to focus on researching the market to establish costs and gave much quicker feedback on the developing budget.
- *Virtual prototyping*. In the context of 4D planning through an interface with Primavera, simulation of particularly challenging construction sequences during planning allowed, for instance, the identification of clashes between formwork elements. In a high-rise building where many tasks are inherently repetitive within clear floor cycles, such prototyping pays particular dividends and supported the detailed planning required to meet a 4-day floor construction cycle. This work was done in DELMIA (Digital Enterprise Lean Manufacturing Interactive Application), Dassault Système's virtual prototyping application for manufacturing processes, and involved the creation of

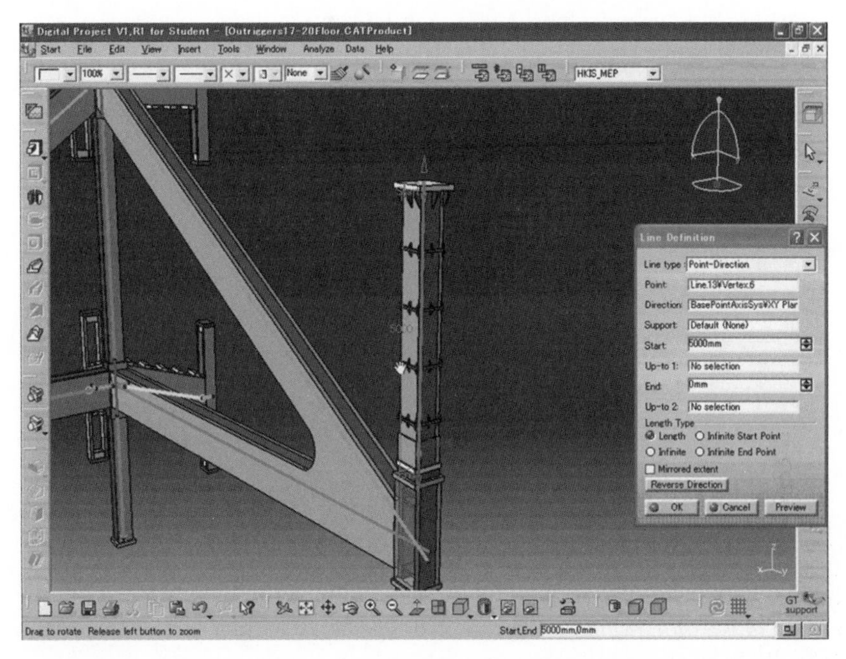

Fig. 14.8 Parametrically generated structural element (source: Riese, 2008, Fig. 5.6).

parametric models of temporary work elements for installation in the BIM. Figure 14.9 shows the results of the virtual prototyping of the 4-day floor cycle which gave considerable confidence to the contractor and client regarding the viability of the schedule. The construction of the outriggers halfway up the structure presented particular problems because of their irregular shape and so received particular modelling attention. This identified and resolved a 21-day 'mistake' in the schedule before site execution.

- *Safety.* High-rise construction is inherently dangerous, particularly on a very constrained site such as One Island East. Lay-down areas and lifting zones could be analysed prior so as to ensure safe working. Working spaces can also be visualised, as in Fig. 14.10.
- *Tendering.* All trade subcontractors invited to tender were provided with the BIM with the automatically generated bill of quantities held in an Excel spreadsheet. Tenderers were also able to use the 4D model to analyse alternative construction sequences. The greater confidence engendered by the integrity of the BIM allowed trade contractors to submit lower prices because of reduced requirements for contingency, particularly in services. This and the avoidance of claims was the source of many of the savings on the budget derived from the BIM.
- *Drawing co-ordination.* Co-ordination and version control of the paper 2D drawings issued to work teams was greatly facilitated by DP – paper was only an output from, not an input to, the BIM.

Fig. 14.9 One Island East virtual prototype of 4-day floor cycle (source: Riese, 2008, Fig. 5.11).

Fig. 14.10 One Island East: visualisation of working conditions (source: Baldwin *et al.*, 2008, Fig. 7.4).

- *Project extranet.* The BIM was complemented by a project extranet which was used by the design team when they were not co-located, and to interfaces with project information not contained within the BIM.

The implementation of Digital Project on One Island East was made part way through scheme design and so many of the advanced facilities that have made it such a versatile and attractive BIM system were not deployed. However, it undoubtedly supported the successful achievement of the One Island East project, and Swire intend to implement Digital Project at inception on their future projects. Gammon has also bought 10 seats based on their experience and are currently using Digital Project on other construction projects.

Sources: Baldwin *et al.* (2008); Eastman *et al.* (2008); Huang *et al.* (2007); Li *et al.* (2008); Riese (2008); telephone interview Martin Riese, Gehry Technologies (22/10/08). I am also grateful to Andrew Baldwin, Peter Brandon and Martin Riese for help with this case, and to Swire Properties Ltd. for their permission to use the case.

Notes

1 Diebold (1955).
2 See Giertz (1995) for the history, and Howard (2001) and Kang and Paulson (2000) for the critique.
3 These standards are currently being revised by the Construction Project Information Committee (CPIC), the body responsible for their co-ordination (http://www.productioninformation.org).
4 Bright (1958) was the first to use the term. Hughes (1983) provides a stimulating analysis of the evolution of complex systems focused on electrical power networks.
5 See Dodgson *et al.* (2005) for an extensive discussion of innovation technologies, drawing particularly on the work of Arup.
6 This analysis is largely based on Sun and Aouad (1999).
7 This analysis is largely based on Davenport (2000).
8 Yang *et al.* (2007) provide a case study of an ERMS implementation in a Taiwanese contractor.
9 http://www.constructionplus.co.uk and http://www.barbourexpert.com respectively.
10 http://www.surplex.com.
11 By Bentley Systems.
12 *Proving Collaboration Pays Report*, Network for Construction Collaboration Technologies Providers (2006).
13 Formerly known as Application Service Providers.
14 Latham (1994, s4.13).
15 Brandon *et al.* (2005).
16 See Smart Work: Patrik Schumacher on the Growing Importance of Parametrics. *RIBA Journal*, September 2008.

Further reading

Davenport, T. (2000) *Mission Critical: Realizing the Promise of Enterprise Systems.* Boston, Harvard Business School Press.
An excellent review and critique of the potential and implementation of enterprise management systems.

Dodgson, M., Gann, D., and Salter, A (2005) *Think, Play, Do: Technology, Innovation and Organization*. Oxford, Oxford University Press.
A thought-provoking book which explores the opportunities provided by innovation technologies, particularly in the construction sector.

Eastman, C., Teicholz, P., Sacks, R., and Liston, K. (2008) *BIM Handbook: A Guide to Building Information Modeling*. Hoboken, NJ, Wiley.
An authoritative account of the development and current state of the art of building information models in the construction industry.

Part V

Leading the Project Coalition

Modern societies rely extensively on complex organisations to achieve objectives. It is through complex organisations that firms implement their strategies and governments implement their policies. Complex private sector organisations are associated with the rise of capitalism; first the great trading companies of the seventeenth and eighteenth centuries, and then the factories of the eighteenth and nineteenth centuries, developed the capability to manage complex organisations. Meanwhile, the growing difficulties of governing in peacetime and campaigning in wartime forced the public sector to develop similar capabilities. By the twentieth century, many commentators were trenchantly analysing the central role of complex organisations in shaping the modern world.

Complex organisations are distinguished from simple ones when it no longer becomes possible for all of the members of an organisation to interact personally – typically above the size of a clan of around 150 members[1]. Complex organisations are characterised by:

- a purpose, without which there is no point in forming the organisation;
- a division of labour as different groups of organisational members carry out the different tasks which contribute to the purposes of the organisation;
- a structured hierarchy in order to co-ordinate the divided task groups;
- a set of processes by which inputs are transformed into the outputs which fulfil the organisation's purpose;
- a set of processes by which co-ordination is achieved.

Most commentators would add another feature of complex organisations – continuity through time. It is here that project organisations are distinguished from the larger class of complex organisations, for they are essentially temporary – formed when the client decides a new facility is required, and disbanded once that facility is completed. The distinctive challenges of managing construction project organisations derive from their typically being both complex and temporary[2]. How these challenges are met is the subject of this part.

The characteristics of complex organisations can be divided into two groups:

- structure, or the arrangements which relate the parts to the whole of the organisation – the banks in our river metaphor from section 1.3;
- process, or the flow of information and materials through that structure, and the processes by which that flow is controlled – the river in our metaphor – consisting of the interplay between tasks, teams, and routines shown in Fig. IV.3.

Structure and process are in continual tension in the organisation, with this tension between structure and process being transmitted through the teams and their interaction with the larger structure of the complex organisation. Just as the river erodes the banks and floods the plain, the demands of getting the work done can subvert the formal structure. This is what the classic Tavistock Institute research on the construction industry[3] called the informal organisation. Similarly, attempts to change the process are frequently made by changing the structure. Thus the implementation of PFI changes the structural relations within the project coalition, thereby significantly altering the project process. This continual tension and interplay between structure and process in complex organisations is called the tectonics of organisation[4] – tectonics being, according to the *Oxford English Dictionary* 'a series of arts which form and perfect vessels, implements, dwellings and places of assembly'.

Organisation design is the process by which choices regarding the appropriate configuration of structure and process for an organisation are made as a function of its purpose. These choices are difficult owing to the paradox of productivity and flexibility; organisations find it very difficult both to achieve high productivity in their use of resources and to respond flexibly to the needs of their customers and clients[5]. As discussed in section 7.6, construction firms have tended to opt for flexibility rather than productivity[6] in their organisation design choices; indeed, the construction industry is so well known for its flexibility that as customers became more demanding in terms of choice and delivery during the 1980s, many management analysts writing in journals such as *Harvard Business Review* turned to construction for a model of how to manage for flexibility rather than productivity[7].

These tensions between productivity and flexibility have led to the evolution of two broadly different types of organisation which are usually called the *bureaucratic* and *adhocratic* (or *organic*) types of organisation. The former, characterised by public sector organisations, mass service (e.g. retail banks) and mass production (e.g. car manufacturers) firms, emphasise order, procedure and hierarchy in ensuring that work is done as efficiently and consistently as possible. Much of the radical transformation of the productive capabilities of the Western industrial societies is a result of the development of bureaucracy, and its analysis is associated with the seminal work of the nineteenth-century German sociologist, Max Weber. The term '*adhocracy*' was coined by Alvin Toffler in 1970 to characterise a new type of organisation better adapted to the more turbulent economic environment that he (correctly) foresaw developing. Adhocracies emphasise interpersonal relations rather than procedure,

Table V.1 The two basic types of organisation

Bureaucratic organisation	Adhocratic organisation
Tall hierarchy	Flat hierarchy
Clear and precise specification of task responsibilities	Ambiguous and overlapping task responsibilities
Reliance on procedures for co-ordination	Reliance on leadership and teamworking for co-ordination
Search for technical fixes to co-ordination problems	Development of organisational capabilities to solve co-ordination problems
Production driven	Responsive to client needs
Strong emphasis on planning	Strong emphasis on learning by doing
Simple jobs in a complex organisation	Complex jobs in a simple organisation
Emphasis on productivity	Emphasis on flexibility

change rather than order, and process rather than hierarchy. Table V.1 summarises the principal differences between bureaucratic and adhocratic organisations.

A further difficulty is that complex organisations are rarely adapted to basic human needs. Just as the buildings and other facilities that construction projects deliver have to be designed within the basic constraints of how the human body has evolved, complex organisations need to be designed within the constraints of how the human mind has evolved. There is now a considerable body of evidence[8] that the mind is also an adapted organ, rather than a clean slate upon which socialisation and education can write. This is the central proposition of *evolutionary psychology*, which applies Darwin's analysis of the origin of the species to human behaviour. Evolutionary psychologists argue that our basic behaviour patterns are adapted to the needs of hunter-gatherers of the African savannah, and have not evolved since. Thus the remarkable continuities in organisations internationally, such as the prevalence of hierarchy, sexual divisions of labour, interdepartmental disputes, the need for leadership and poor decision-making as a result of difficulties in understanding risks, are signs of the difficulties of hunter-gatherers working in complex organisations. This is not to argue that effective management cannot mitigate the weaknesses of such behaviour and build on its strengths – for such evolutionary factors explain only around half of all human behaviour – but that they do place real constraints on organisation design.

The two chapters in this part will explore various aspects of the paradox of productivity and flexibility, identifying the extent to which project management can be considered to be an attempt to get the best of both worlds. It will thereby identify the key issues in leading the project coalition.

Notes

1 See Nicholson (2000).
2 For an account derived from organisation theory, see the paper by Bryman and his colleagues (1987).
3 Tavistock Institute (1966).
4 Winch (1994a).
5 Winch (1994a).
6 Ball (1988) argues this most strongly; see Eccles (1981a) for a different view.
7 See Winch (1994b) for a review of this literature.
8 See the collection edited by Barkow and his colleagues (1992) and Nicholson (2000) for an application to management.

Chapter 15
Designing Effective Project Organisations

15.1 Introduction

> '[Co-ordination] expresses the principles of organization *in toto*; nothing less. This does not mean that there are no subordinate principles: it simply means that all the others are contained in this one of co-ordination. The others are simply the principles through which co-ordination operates, and thus becomes effective'.

The vice-president of General Motors, James D. Mooney[1], reflecting in 1931 on the rise of the modern corporation, identified the essence of managing complex organisations. The resource bases providing the skills required for the execution of tasks can usually be organised simply; the complexity arises in the co-ordination of these resource bases so that they achieve the overall purpose of the organisation. This chapter will explore the distinctive organisational solution that has evolved to solve the co-ordination problem in construction – project management.

The chapter will start by briefly describing the evolution of project management as a response to the challenges of managing organisations that are both complex and temporary. It will then move on to its distinctive applications within the construction industry in co-ordinating the project coalition and to identify the responsibilities of the client for the governance of the project. Specific attention will be focused on the distinction between 'project managers' as an actor within the project coalition, and 'project managing' as the process of delivering the project mission for the client. The distinctions between managing projects, programmes and portfolios of projects will also be investigated. The argument will then move on to examine the concept and application of teamworking to the project management process. The emphasis in this chapter will be on the structural aspects of organisation, while Chapter 16 will explore in detail aspects of processes such as leadership and culture within the project coalition.

15.2 The rise of the project management concept

It is a truism that construction projects have always been 'managed' in some sense – under the craft system, trades followed one another under the more or less watchful eye of the master. However, the 'client' for virtually all major construction works prior to the modern period was either the church or the crown, with rich merchants becoming increasingly important during the Renaissance; at the heart of the project mission was the glorification of God or man, not an NPV calculation. Only when significant construction came to be undertaken for commercial reasons did the role of a specialist co-ordinator of the process start to evolve with the emergence of the general contractor during the industrial revolution, as discussed in Case 2. For the first time, the benefit for the client of a co-ordinator of the different trades on site was recognised, and was quickly appreciated as the railway construction boom began to take off in the middle part of the nineteenth century.

It was the challenges of constructing large, complex systems – most notably railways, and then electrical power distribution – that stimulated the formal articulation of the role of project management[2]. In 1909, an engineer with Stone and Webster drew on his experience of building the Boston Elevated Railway to argue that flexibility to allow deadlines to be met was more important than organising for productivity, and to argue for the role of a dedicated 'organisation for dispatch' co-ordinating the process as a whole. In 1915, the concept of 'construction management' was debated in the pages of the *Journal of the Western Society of Engineers*. Debate split along lines familiar today between the advocates of scientific management, with every aspect of the work tracked in detail and advocates of a more pragmatic approach adapted to the 'sporadic' nature of the construction project. The advocates of the more scientific approach had the better of the argument when it came to building the large dams in western USA during the 1930s. The contractors who successfully undertook these mammoth projects – most notably Bechtel – were able to apply their organisational and logistical skills to the building of ships and aircraft during the following war.

Although the construction industry was the first to formulate the project management concept, it was the aerospace industry which brought it to full maturity. By 1926, the role of 'project engineer' was established at the Naval Aircraft Factory in Philadelphia, although by 1939 Pratt and Whitney were having problems making the concept work, despite support from a punch-card based scheduling system[3]. It was during the Cold War that the concept reached full maturity for the large programmes to develop weapons systems and to carry men to the moon[4]. US defence programmes were, and are, the source of many of the tools and techniques that were discussed in Part IV, but as Peter Morris strongly argues, the core of project management lies not with these tools and techniques, but with the establishment of an organisational function concerned with the delivery of the system to the client, such as the Polaris Special Projects Office (SPO) and the Ramo–Wooldridge Corporation (now TRW) described in panel 15.1.

Paul Gaddis[5] was one of the first to identify the implications of this organisational innovation, characterising the project manager as the 'manager in the middle' between the client and the resource bases. He argues that the project manager can be neither an expert in the domain of the resource bases nor the client as such. The role of the project manager is to act as the interface between the client's desires and the capabilities of the resource bases – to sit at the interstices of the project coalition matrix. In this respect, the project manager performs a role in complex projects that neither the client's own functional managers nor the managers of the resource bases can achieve – the co-ordination of the project so that it fulfils the client's business needs.

A broad review of research on the application of the project management concept across a number of industries[6] has identified a range of applications, as presented in Fig. 15.1. The differences in application depend on the balance between the extent to which the manager of the resource bases has more control within the organisational matrix illustrated in Fig. 1.4, or the project manager has more control. Where resource-base managers are in complete control, and there is no attempt at lateral co-ordination, or the project manager is little more than a liaison role, this can be considered to be *functional* organisation. Where the project co-ordinator has clear responsibilities for overall co-ordination, monitors progress and brokers competition for resources, but is, in the end reliant on the resource-base managers to make appropriate allocations of resources to the project, this can be considered *lightweight* project management. Where the converse is true, and it is the project co-ordinator which has the stronger hand, overriding resource allocations made by resource base managers, then this is *heavyweight* project management[7]. Finally, where the project manager has complete autonomy – including the possibility of hiring staff directly – then this is *cell* organisation. Cell organisations

Panel 15.1 Project management as organisational innovation

When the US Navy wanted to develop a submarine-based nuclear missile capability – Polaris – it established an SPO to oversee system delivery in 1955. The actual work was executed by some 250 prime contractors and 9000 subcontractors. The job of the SPO as an elite function within the Navy was to provide systems integration and project co-ordination capabilities. In particular, it was to give Congress, which held the purse strings, the assurance that public money was being wisely spent. To achieve the latter it developed the Program Evaluation and Review Technique (PERT), hailed by the head of SPO as 'the first management tool of the computer and nuclear age' – see section 11.2.

The Air Force took a rather different approach with the development of the Atlas/Titan Intercontinental Ballistic Missiles (ICBM). It appointed an external company – the Ramo–Wooldridge Corporation – to manage the programme. Again, the project managers were responsible for co-ordination and systems integration, managing some 200 prime contractors and thousands of sub-contractors in the two parallel programmes. None of the prime contractors, who had aircraft manufacturing backgrounds, was deemed to have the technical skills to manage the ICBM programme. The intellectual contribution of this programme was systems engineering, and more broadly, systems thinking, which was then applied less successfully to the urban problems of the USA.

Sources: Hughes (1998); Morris (1994).

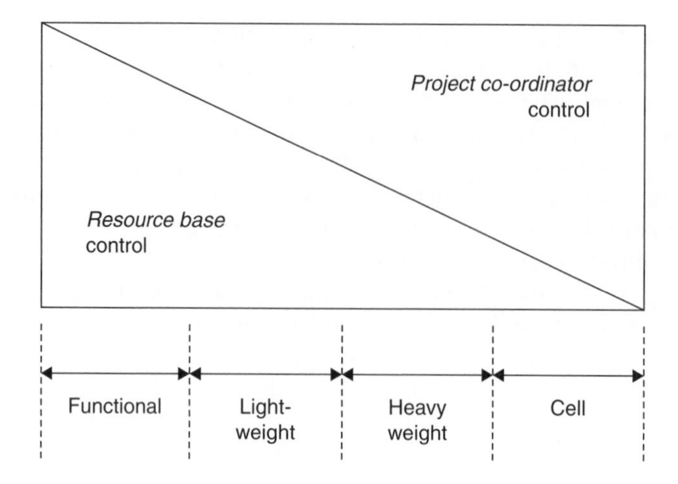

Fig. 15.1 The implementation of the project management concept.

often have a slightly covert aura; the famous 'skunk works' of Lockheed were project cells. The test of which form of project organisation is being implemented is who holds the project budget – the resource-base managers or the project manager.

As the project organisation matures, the distinctive roles of the resource-base and project managers become more clearly articulated and are summarised in Table 15.1. As is implicit in Fig. 15.1, it is never possible to fully distinguish these roles – some tension, and even conflict, is inevitable but not necessarily dysfunctional. Rather than this tension being solely negative, it can be a source of creativity in problem-solving. What is dysfunctional is poor conflict resolution as a result of suppressing or ignoring the tensions.

Table 15.1 The differing responsibilities of project and resource-base managers (source: developed from Cleland and King, 1983, Table 13.1).

Project manager	Resource-base manager
What is the task to be done?	How will the task be done?
When will the task be done?	Where will the task be done?
Why will the task be done?	Who will do the task?
What is the budget for the task?	What are the resources required for the task?
Quality of integration of task output into final product	Quality of task output

15.3 Projects, programmes and portfolios

As project organisations grow in size, hierarchies of project managers develop above the level of the resource-base managers. Although all of these managers are project managers in the sense of this book, the desire for organisations to define and reflect differing levels of responsibility has favoured the emergence of a more sophisticated terminology deploying the terms *programme* and *portfolio* to complement that of project. Debates around the definitions of these terms can be heated[8], but the following offers one view of these important distinctions which are illustrated in the three parts of Fig 15.2:

- *Portfolios* (of projects) share scarce resources but have different missions and are not sequentially dependent;
- *Programmes* (of projects) contribute towards a common mission while consisting of identifiably separate projects which may also be sequentially interdependent.

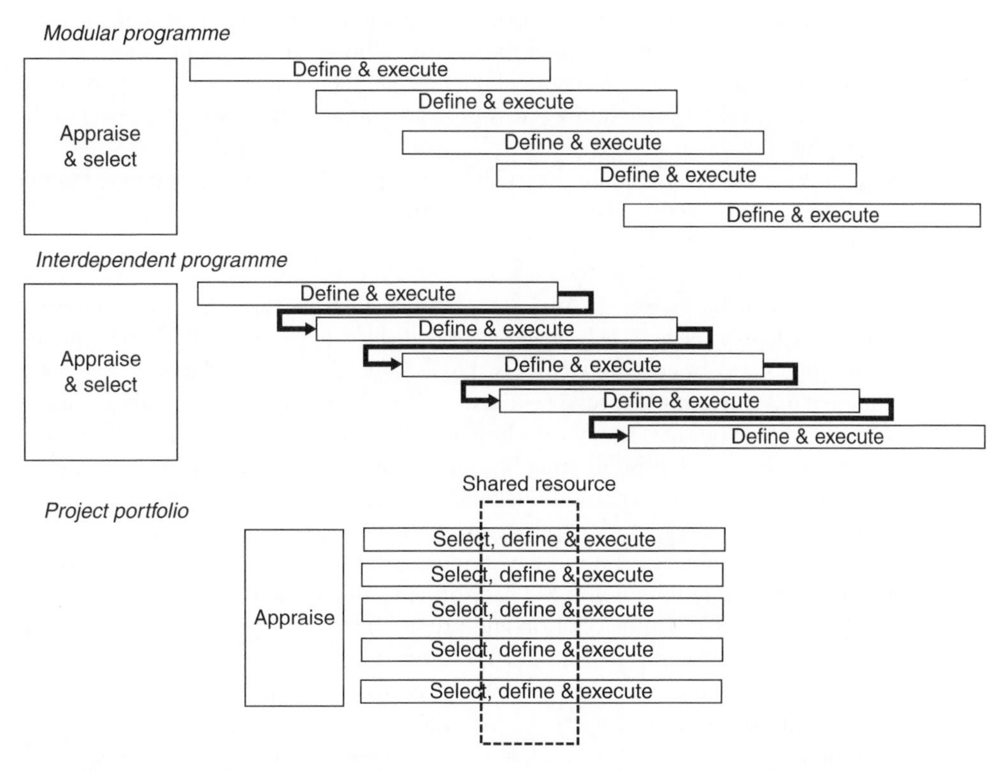

Fig. 15.2 Programme and portfolio management. NB the project life-cycle terminology refers to the BP Capital Value Process.

The most obvious resource that projects might share is capital. Organisations have to choose between different uses of their capital resources, and the tools for doing this were discussed in chapter 3. An important, but often overlooked, resource constraint is project management capability – just as the growth of the firm is limited by its ability to develop managerial capability rather than access to capital[9], the size of a client's project portfolio is limited by its ability to manage the projects within that portfolio. There are a number of advantages to managing projects as an explicit *portfolio.*

- Human or other scarce resources such as large cranes can be more effectively managed across projects[10]. One tool for managing human resources across a portfolio of projects is critical chain, discussed in section 11.5.1.
- Risk can be shared across the portfolio. This is of course the basic principle of the diversification of assets in investment management, but it also applies to projects. Project performance is typically evaluated project by project, and all projects are expected to perform within parameters. The effect of this is inherently conservative in terms of performance because if targets are always met, then they are not stretch targets. A performance policy that the portfolio of projects has to perform within parameters can facilitate innovation because the risk of a project failure is hedged against the successful projects.
- The costs of developing and maintaining project management skills can be shared across projects, and used to develop a Project/Programme Management Office or a Project Management Centre of Excellence to develop training and standardised procedures across projects and also to provide a project audit function[11].
- It is possible to use strategic partnering approaches using Integrated Supply Teams – see section 7.6 – moving from one project in the portfolio to another.

Programmes can be viewed in various ways, but all have the sense of an interdependent set of projects. *Modular programmes* do not have sequential interdependencies, but share important elements. For instance, the new Marks & Spencer's initiative to refurbish its stores, presented in Case 5, could be beneficially managed as a programme because the refurbishment will share a number of common design features to ensure consistent branding and customer experience, and a number of components and systems will be bulk ordered as done previously on Project Robin. On the other hand, site-specific work on particular stores will be managed as discrete projects, and most working on the project may not even be aware of the larger programme context. These provide an excellent context in which to justify investment in standardisation and pre-assembly – see section 12.10 – which is why these have been called modular. Projects may also be sequenced to smooth overall cash flow and enable learning from project to project regarding the technologies deployed or the construction methods selected.

Interdependent programmes consist of discrete projects which also have sequential interdependencies. For instance, the construction of the biomes for the Eden

Project in Case 17 and the acquisition, curating and growing of the plants that would go in them were organisationally very different projects, located in different places. Yet, a delay to one would have had a serious knock-on effect on the other and vice versa. Similarly, the West Coast Main Line renewal project in the UK has consisted of multiple blockades between 2000 and 2008. Each blockade needs to be managed as a separate project to very tight schedule constraints, but also as part of a programme implementing a common technology over 1031 route kilometres cumulatively building to a high-speed line by 2009.

A rather different definition of projects and programmes is that projects deliver *outputs* such as a new school and programmes deliver *outcomes* such as more effective learning for children[12]. In other words, programmes concern themselves with the value added by the facility, while projects do not. An advantage of this definition is that it puts the emphasis on the realisation of the benefits of asset creation and considerable stress on the role of the business change manager responsible for benefits management. However, it poses the curious challenge of trying to find a project where the asset created by it provides no benefits. Under the OGC definition, all projects must be programmes, as all are aimed at providing benefits for those that invest in them; it is a distinction without a difference. However, the perceived need to make the distinction shows how limited our notion of construction project management has typically been in the past and illustrates the problems associated with an execution orientated perspective on the process.

An important issue in the relationship between projects and programmes is the sets of skills required. In one perspective, programmes are aggregations of projects, and the skills required are a development of those developed on projects with a focused attention on process integrity. Programme managers tend, therefore, to be the more successful project managers. In a second perspective, programme managers are the implementers of corporate and business strategy, and the skills that are required are much closer to those required of senior managers in general, and people with a project management background are likely to flounder[13].

There would appear to be two elements to this debate which are confusing the issues. The first is that programme management is being defined against a restrictive view of project management as an execution-only discipline rather than one orientated to the management of projects as espoused in this text. The second is that, from a distance, any programme can be defined as a project, and, close up, any project can be defined as a programme. From a perspective of the management of projects, the real issue is whether managing a set of projects as a programme adds value. In the case of Marks & Spencer's store refurbishment programme, it clearly would, because there are synergies between the distinct projects. The case for bundling doctors' surgeries together simply to make the programme viable for private finance, as in the UK's Local Improvement Finance Trust (LIFT) programme, is less obvious because it can lead to large increases in costs and delivery times and facilities poorly adapted to users' needs[14]. Panel 15.2 compares two airport-construction projects where explicit programme management yielded benefits which were sorely missed on the other side of the Atlantic.

Panel 15.2 A tale of two airports.

Denver International Airport (DIA) and Heathrow Terminal 5 (T5) are both major international airline hubs; both have baggage-handling systems ranking high in the public perception of their success as projects, and both are strong architectural statements. The principal differences are that DIA is a completely new facility promoted by a public sector client, while T5 forms a self-contained part of an even larger complex operated by a private sector client. In this panel, we want to draw out the organisation aspects of the two projects, particularly with respect to programme management and client capabilities.

The DIA project started on site in 1989 and was handed over in 1995, 16 months behind schedule at a cost overrun of almost $2bn on a $4.8bn outturn cost. The baggage-handling system proved unreliable, was never fully used and was finally terminated in 2005 with a reversion to manual baggage handling. Heathrow T5 (T5a & T5b) was completed on schedule and within budget in 2008, having started on site in 2002 for a cost of £4.3bn. Although there were initial problems with baggage handling, these were more operational than technical and the system has now settled down.

Political pressures within Denver socio-economic elites stressed the urgency of the project, and it went ahead, in effect, as a 'build-design' project and an 'airport built by committee'. The project management team (PMT) was staffed by Denver city employees supported by project management consultants. It organised the project into five distinct and largely self-contained areas. However, a contract was signed in 1992 – well into the execution of the project – for an integrated state-of-the-art baggage-handling system that cut across all areas. This meant considerable rework on some structures already completed, and installation of the system required co-ordination with all five sub-projects. The baggage system contractor found that 'there was no one to tie it all together. . .it was pandemonium' and was obliged to work in very unsatisfactory conditions. Worse, the proactive DIA project manager died in late 1992, and was replaced by a functionary who also kept her operational roles within DIA and 'was overwhelmed'.

Work on T5 commenced after the longest planning enquiry in UK history – see panel 4.4 – so there was plenty of time to research the experience of other airports and develop the design concept. Execution was managed by a strong PMT organisationally distinct from the rest of BAA, which broke T5 down into 18 projects consisting of 147 sub-projects and used explicit programme management techniques to co-ordinate them. Baggage handling was a separate project of equal status to the construction works. The PMT controlled the project using earned value analysis – see section 10.8 – and an integrated schedule in Artemis. Integrated baseline reviews – see section 10.7 – were held every six months from July 2004 and used to slowly tighten up the programme management approach and initiate rescheduling and value engineering as required. An early change was the shift of £100m of contingency to the programme level rather than the project level facilitating trade-offs between projects.

Sources: Doherty (2008); Montealegre *et al.* (1996); Wikipedia (accessed 04/11/08).

15.4 The responsibilities of the client

The client mobilises a coalition of firms to deliver the project mission, and the contractual relationships between those firms define the project organisation structure and charter the project coalition. But the client retains important responsibilities towards the project coalition through the life cycle; indeed the US courts have found the client liable in some failed construction projects for not fulfilling their 'legal responsibility to participate actively in the project's management'[15]. These responsibilities are to act as:

- *Promoter* – defining the need for the project and ensuring that it meets that need;
- *Financier* – obtaining the capital required to finance the project;
- *Decision-maker* – making those decisions required to push the project through the life cycle;
- *Recruiter* – mobilising the most appropriate and capable firms to realise the project. This is a legal obligation under European health and safety legislation.

These are onerous responsibilities, and not all clients have the organisational capabilities to fulfil them. Many clients that sponsor a number of projects, such as BAA in Case 12, have developed substantial in-house managerial capabilities to fulfil these responsibilities . Many public sector organisations have also developed strong in-house project management capabilities, like the US Army Corps of Engineers presented in panel 5.2. However, many, if not most, clients for construction either do not wish for such a diversion from their core business, or do not have a large enough volume of construction to justify the development of such capabilities in-house.

This need by clients for organisational capabilities in relation to the projects they sponsor has led to the development of specialist providers of project management services. Confusingly, these firms are often called 'project managers', but in practice, they do not manage the project in terms of providing co-ordination between resource bases, but help the client fulfil its responsibilities towards the project coalition. These alternative arrangements are illustrated in Fig. 15.3, where the term *executive project manager* is used to denote these firms; this is the distinction identified in panel 15.1 between the US Navy's in-house Special Projects Office and the Air Force's outsourced Ramo–Wooldridge Corporation. Members of the project coalition are frequently unclear about the role of such executive project managers, particularly as they are not usually capable of making decisions themselves but have to refer back to the client for authorisation. As Morris argues, it is preferable for clients to have in-house project managers who can act in an authoritative way in relation to the project coalition. This is also the recommendation of the Latham Report. Where this is not possible, a good executive project manager can play an important role in achieving project success by managing the client rather than the project coalition. Whatever the final solution, it is vital that the client's project sponsor be engaged with the development of the project[16].

These issues of *project governance* have been of growing concern of late in both the public and private sectors. Concerns for corporate governance have led to the passing of the Sarbanes–Oxley Act in 2002 and subsequent issuance of internal control requirements in the USA, and the incorporation of the 1999 Turnbull Guidance on internal control into the Combined Code which regulates audit

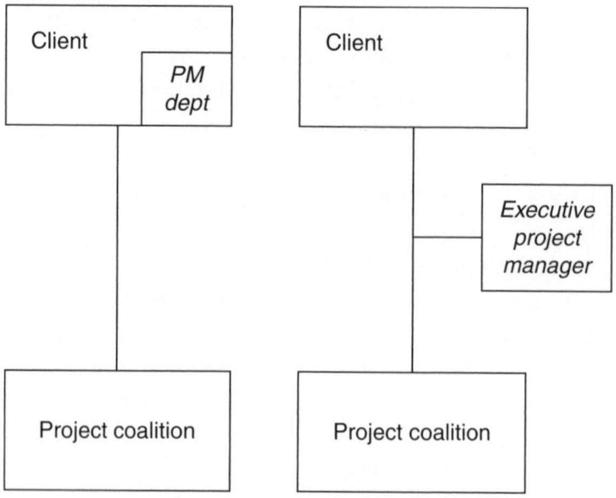

Client
PM dept
Project coalition

Client
Executive project manager
Project coalition

Fig. 15.3 Client project management options.

practice in the UK. Thus corporations quoted on the world's two largest stock exchanges – London and New York – are obliged to meet new and demanding requirements for internal control, which require their corporate boards to have much more knowledge of what is happening on the projects in which they are investing on a real-time basis. Similarly, the New Public Management[17], which emphasises performance and outcomes, has placed new obligations on the senior management of public sector organisations.

These new pressures for accountability have led to new roles emerging in project organisations. The first is the *project sponsor*, known within the UK public sector as the Senior Responsible Owner (SRO);[18]the second is the *project board*. There is rather limited research on the role of the project sponsor, but it is clear that the role involves working with the project or programme manager as the interface between the project organisation and the client organisation sponsoring the project – in an important sense, the project sponsor is the new 'man in the middle'. From the point of view of the project or programme manager, the ideal project sponsor:[19]

- is of appropriate seniority and power in the client organisation so that he or she can solve issues;
- has political knowledge of the client organisation;
- is able and willing to make connections between the project and the organisation;
- has courage and willingness to battle on behalf of the project;
- has the ability to motivate the project team;
- is willing to partner with the project team;
- has excellent communication skills;

- is personally compatible with other key players;
- has the ability and willingness to challenge the project team if necessary.

However, this is from the point of view of the project team, and the role is essentially about keeping the project going and on track – there is a risk here of the 'capture' of the project sponsor by the project team. A review of some 60 large-scale engineering projects[20] suggests that a project sponsor needs:

- an integrative business perspective;
- the ability to evaluate complex systems from multiple perspectives;
- relational and coalition-building competencies;
- political and negotiating skills;
- access to the resources necessary to support a long-term development process;
- the possibility of diversifying risk through a portfolio of projects;
- the will to abandon bad projects.

The second main organisational element in project accountability is the *project board*. This body is appointed from three different stakeholder groups – the client, the user groups and the suppliers. For instance, ProCure 21 presented in Case 6 provides for a Project Board if required. Within a PRINCE 2 environment, the Project Board has a more formal responsibility; it has representation from users and suppliers, and is chaired by the SRO with the project manager reporting to it on an exceptional basis. It is expected to be most involved at project initiation and closure, and at key stage gates in the project life cycle. Clearly, project boards can only work effectively if the project coalition is governed through an alliance relationship between client and suppliers.

According to the OGC[21], the role of the Project Board is to be responsible for directing the project, which includes:

- ensuring the ultimate success of the project;
- managing the risks identified on the project;
- ensuring effective management of the project;
- committing the required resources;
- making decisions on changes to project objectives when requested by the project manager;
- providing overall direction and guidance to the project;
- making decisions on exceptional situations;
- the project and its outputs/outcomes remain consistent with the business plan and the external environment;
- ensuring the necessary communication mechanisms are in place;
- sponsoring appropriate external communication and publicity about the project.

However, the reality of managing construction projects is a lot messier than these formal government guidelines would suggest.

15.5 Who is the project manager?

The project management function on construction projects is typically very diffused and lacks the clarity advocated in the previous section – probably much more diffused than in other project-orientated sectors. Panel 15.3 shows the different people who had the formal title of Project Director on the Tate Modern project, while Fig. 15.4 shows the allocation of project management responsibilities on the Boston Central Artery/Tunnel Project[22] described in Case 13. It shows how two project management functions faced each other from the client side in the Massachusetts Turnpike Authority and the supply side in the Bechtel/Parsons joint venture (JV). Within each member of the project coalition, the project management function was split between the more senior manager responsible for *external relations* (i.e. the issues discussed in Part II) and his direct report responsible for *internal effectiveness* (i.e. the issues discussed in Parts III and IV). Notably, only one of these four people had a background in a construction discipline – Bechtel's Project Manager was a civil engineer. The other three were a lawyer, a former army general and a former public official with an MBA.

	External relations	Internal effectiveness
Client side	Project Director	Deputy Project Director
Supply side	Program Manager	Project Manager

Fig. 15.4 The different project management roles on the Boston Central Artery/Tunnel.

15.6 Organising the project through the life cycle

Henry Mintzberg has characterised project management organisations as 'administrative adhocracies'[23], thereby suggesting that it is an adhocracy with bureaucratic tendencies, mixing characteristics from both columns of Table V.1. However, what this misses is that project organisation is not constant through the project life cycle, but dynamic with that life cycle. As analysed in section 8.5, levels of uncertainty are very high in the early stages of the life cycle, and they progressively reduce towards zero as the project approaches completion. A second dynamic is that the project organisation grows significantly in size through the life cycle. The initial work is typically done by a few very highly skilled people, while the project on site mobilises much larger numbers of less skilled people. On the largest projects, thousands of people may be working on the project simultaneously both on site and off site in design and the supply of components. These shifts in both size and the level of uncertainty with which the project organisation needs to cope lead to important changes in organisation as it moves from divergent to convergent iterative information processing, and on through reciprocal flows to much more of a pooled co-ordination with tasks running in parallel through the life cycle.

Panel 15.3 Who was the Tate Modern's Project Manager?

Three different people from three different organisations working on the Tate Modern project had the title of Project Director and, in effect, shared the project management role between them. These three were supported by a number of people with the title of *Project Manager*.

- The first *Project Director* was employed by the Tate. She was the formal interface between the client and the project coalition, who saw her role as being able to 'make sure that in 2000 or thereabouts the Tate has a new gallery, a new organisation and a new institution at Bankside'. Her role was wide ranging, including fund-raising and relations with the local community as well as monitoring project progress. She chaired many of the meetings held by the Tate with its principal advisors.
- The second *Project Director* was employed by Stanhope and saw his role as 'to help the Tate deliver the project . . . a question of managing the design process, and going and buying the construction and making sure the project actually gets built'. The Stanhope role also included advice on property acquisition and the appointment of the design team. He chaired many of the project progress meetings, and was supported by a *Project Manager*.
- The third *Project Director* was employed by Schal, the construction managers. He argued that 'we like to be involved in the development of the design . . . I do think that it is important for us to understand their way of thinking, but equally they need to understand that there is a budget for this job, there is a programme for this job, and there are client brief requirements that have to be maintained, and that's where our role really starts to come into its own'. He chaired many of the meetings with trade contractors and had a number of *Project Managers* reporting to him.

Source: Sabbagh (2000).

A basic tenet of designing effective organisations is that organisations facing high levels of uncertainty need to be relatively adhocratic in organisation and cannot be large, while organisations facing lower levels of uncertainty can be both relatively bureaucratic and large. This shift in project organisation is known as *matrix swing*[24], typically occurring around 15% of the way through the project. While there is more research required on appropriate organisation for the early stages of the project life cycle, there is general agreement in the literature that project organisation during the realisation phases should be clearly structured into independent packages, with managerial responsibility devolved down to the project managers of those packages. In this latter context, project management is very clearly interface management. Panel 15.4 shows how the organisational structure was changed as the Boston Central/Artery Tunnel project described in Case 13 evolved.

Panel 15.4 Matrix swing on the Boston Central Artery/Tunnel Project

As the C/AT project moved into the construction phase, an external review of the design of the project management organisation was undertaken during 1994 and 1995. The consultants were Peterson Consulting of Boston, Lemley and Associates (the consultancy founded by the former chief executive of TML – see Case 1 – who later took over for a short while as Chair of the London 2012 Olympic Delivery Authority) and Professor Jay Lorch, one of the most distinguished experts on organisation design from Harvard Business School. There were two aspects to the report:

- The relationship between the Bechtel/Parsons JV and the client
- The organisation of the 1000-strong JV itself.

The principal recommendation on organisation design was that the JV reorganise itself from a functionally structured one, with different disciplines working separately (e.g. design, procurement, construction) leading to a fragmentation of managerial responsibility, to one structured by geographic area (e.g. East Boston) with multi-disciplinary teams. Within each area, project managers would manage the separate work packages under way in that area, giving much clearer lines of responsibility for budget and schedule.
As the client's Project Director characterised the new JV organisation: 'someone should go home at night with a stomach ache about making sure something occurs. Senior management always has stomach aches, but must parcel out some of these to other people.'

Source: Hughes (1998).

15.7 Project organisation in construction

The relationship between the firms or departments providing the resources to the project and the mobilisation of those resources to deliver the project mission creates a classic matrix management situation. The resource-base teams deployed on the project are responsible to both the managers of the firm which employs them and has contracted to supply them as a resource to the project and to the managers of the project on behalf of the client. Different firms within the project coalition may commit their resources in different ways to the project. Figure 15.5 shows how different firms make differing commitments to the project. At one extreme the firm and the project may be co-extensive, such as teams E and D – a small architectural practice or a subbie bricklaying gang may only be working on one project. At another extreme, a larger firm may commit a dedicated team to the project, such as in cases A and C. Teams B and F, on the other hand, are only partially committed to the project – either because they are overstretched and the firm is trying to serve more than one client with a single team or because the level of resources required for the project does not warrant the commitment of a whole team.

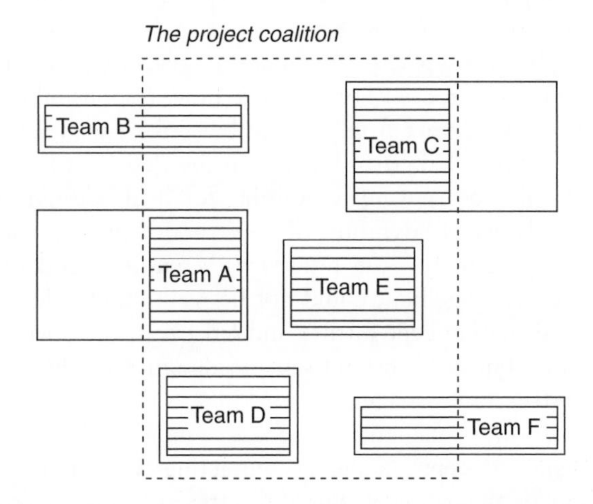

Fig. 15.5 Resource-base teams and commitment to the project coalition.

The first line of management on the job is the resource-base team leader – frequently known by titles such as job architect or foreman. The management of a single resource-base team is, by definition, not a project management job. It is at the level of co-ordination between different resource-base teams that the project management role emerges. Where the resource commitment of a single firm is large, or involves more than one set of skills, there may be internal project management between the in-house teams to ensure the meeting of the firm's commitments to the client; however, this is not always the case. More frequently, the first project management roles begin to emerge in construction at levels above the individual firm. A common example of this role is the general contractor taking responsibility for the co-ordination of all the different trades deployed on site. In parallel, the architect traditionally takes responsibility for the co-ordination of the different resource bases involved in design.

This traditional arrangement for the management of construction projects has a number of important limitations.

- There is no co-ordination at the level of the process as a whole – the project management of the design process and the project management of execution on site are completely separate processes. This is one reason why the design/construction interface has become such a contentious one on many projects.
- Those responsible for co-ordination also have resource-base responsibilities. This is particularly true of architects, whose main responsibilities are delivering the architectural design, not the co-ordination of the other members of the project coalition.
- Those charged with actually doing the co-ordination are rarely well trained in project management tools and techniques.

The development of mediated coalition structures such as construction management in the 1960s – see section 5.4.3 – was a major attempt to apply the principle of a separate organisation specialising in project co-ordination to the construction process. For the first time, responsibility for co-ordination was organisationally distinct from responsibility for execution. However, this responsibility rarely included co-ordination of the design process – except to the extent that trade contractors were responsible for detail design. Indeed, one of the frequent complaints of advocates of construction management was that they were not being brought into the process early enough, and therefore full benefits of construction management could not be realised. Developments during the 1990s have pushed the concept further and led to various forms of design and manage organisations, but concept and scheme design typically remain outside the remit of the construction manager.

Construction management, then, is a lightweight application of the project management concept. General contracting was traditionally a heavier-weight application of the concept, but the retreat of general contractors from actually performing site operations to simply co-ordinating others doing it, coupled with the expanding role of the specialist trade contractors, has lightened the role with respect to site execution. During the design stage, project management is typically even lighter in weight – on most construction projects it is largely functional. This is in marked contrast to the situation on new product development in the car industry, where heavyweight project managers expend most of their effort during the design stage, ensuring that the project stays within budget and schedule while safeguarding the integrity of the product concept.

An important implication of the above argument is that there are project managers at all levels in the construction project organisation, particularly if it is a complex one. A number of levels can be identified above the manager of the resource-base team:

- Project managers internal to a single organisation responsible for providing more than one set of resources – the classic example of this is the general contractor's site agent or contracts manager, co-ordinating the trades on site;

- Project managers who co-ordinate different firms providing sub-groupings of the resources required, such as construction managers or design managers;
- Project managers responsible for the entire process on behalf of the client – the executive project manager or in-house team such as the BAA T5 team.

The greater the breadth of responsibility of the construction project manager for the process, the lighter the weight of the responsibilities for that process. Many project managers working directly for clients have little more than a monitoring and advisory role, and little or no executive responsibility. Yet, it is these managers, whether they work directly for the client or provide their services as independent consultants, who carry the title 'project manager'. Thus, one of the main peculiarities of the construction industry compared with other sectors is that those who carry the title 'project manager' frequently have little responsibility for actually realising the project.

15.8 Determining the organisation breakdown structure

As in all organisations, the allocation of responsibilities within the project coalition for task execution is central to effective project organisation, and this allocation creates the OBS introduced in section 10.3. Because of the temporary nature of the organisation, this process needs to be explicitly tackled very early in the life cycle, rather than being allowed to evolve naturally. While the precise nature of responsibilities will change through the life cycle, the principles of allocating responsibilities are valid both before and after matrix swing. The method used for identifying task execution responsibilities and thereby creating the OBS is responsibility charting[25], which combines the functional organisation of resource bases in the traditional organisation chart, with the horizontal identification of project responsibilities. A responsibility chart (or matrix) consists of:

- an X axis of the array of resource bases mobilised on the project forming the OBS;
- a Y axis of the tasks to be executed on the project derived from the WBS;
- a system of symbols to identify different types of responsibility for a specified task.

The X axis may initially be at the level of the resource bases themselves, but soon needs to be developed to identify the person responsible for that resource base. Similarly, the Y axis also becomes progressively more detailed as the project moves ahead. The responsibility matrix is thereby built up from modules of the type illustrated in Fig. 15.6. Symbol systems vary and can be adapted to the precise requirements of the project; a useful one is DECA[26]:

- **D**ecide – the person with overall accountability for the task including deciding to start the task and declaring it completed;

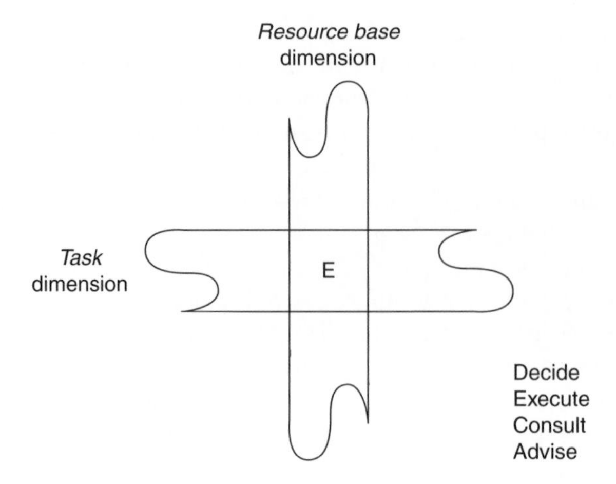

Fig. 15.6 The basic module of a responsibility chart.

- **E**xecute – the team actually responsible for executing the task;
- **C**onsult – those people who need to be consulted regarding aspects of task execution, and their opinion sought;
- **A**dvise – those people who need to know that the task has started, is in progress or finished, but without further input.

Bundles of tasks allocated to particular resource bases create the OBS and identify budgetary responsibility for task execution. However, the responsibility chart, through its symbol system, also identifies important reporting and liaison responsibilities beyond direct budgetary responsibilities.

15.9 Project teamworking

Good teamworking is an axiom of contemporary management – a group is merely a number of people engaged in a common activity, while a team adds value to the group because it is a team. Teams are the living stuff of the routines/tasks/teams dynamic, but the application of the teamworking concept in the design of project organisations is not straightforward. Firstly, project teams are very different from other types of management teams because they are temporary. The top management team in a firm, while not unchanging, does not have a determinate life cycle. Similarly, the crew of a fire engine would expect to work together through a number of emergencies. Project teams know that they will disband in the near future. Effective teamworking has been the subject of a large amount of research[27]; the results are clear and can be simply stated in four propositions.

Firstly, there is a finite maximum size for a team – most research places this at between five and seven members. Beyond this the quality of interpersonal relationships starts to break down, and the team starts to fragment. Teamworking is very dependent on high-quality interpersonal relationships, and this can only be fostered through interpersonal contact. This is why many team-building programmes involve the participants in getting cold, wet and muddy. The sense of struggling through together creates these personal bonds as well as individual self-awareness. A very important implication of this is that teams are not appropriate forms of organisation on their own for large-scale undertakings such as construction projects. Such undertakings are achieved through project managing coalitions of teams, not though teamworking alone.

Secondly, the team needs a mix of members with complementary skills. Obviously, these skills are first and foremost the technical skills required for task execution – if the services design team needs an electrical engineer, a mechanical engineer will not do. However, a mix of teamworking skills is also required. Research has identified the critical team roles shown in panel 15.5. While the analysis is, perhaps, overelaborate, the fundamental point is that teams need to be heterogeneous but balanced on both the technical and team skills dimensions. As illustrated in Fig. 15.7, the team needs to avoid both too homogenous a perspective on how the task should be tackled – known as 'groupthink' in policy analysis circles, the 'Nut Island effect' in management[28] – and unmanageable conflict through having little common ground in approaching the task.

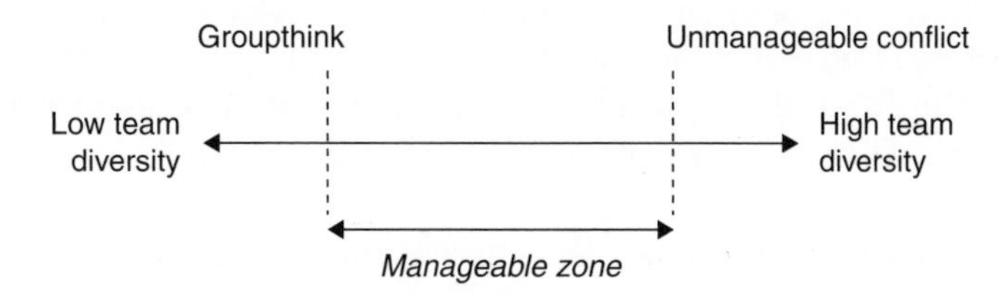

Fig. 15.7 The manageable zone in teamworking.

Thus any effective team needs:

- a *leader* to ensure that it stays focused on its goals, and differences between team members are handled sensitively;
- a *generator* of solutions to the team task;
- an *evaluator* of those solutions;
- a *finisher* who ensures all the details are handled.

Thirdly, the team needs clear goals and appropriate incentives in the context of mutual accountability in order to focus its efforts. Teams are no different from individuals in terms of what motivates them, except that all team members need to share the same goals and incentives. In particular, rewards such as bonuses and

Panel 15.5 Team roles

On the basis of extensive experimental research with management teams, Belbin and his colleagues identified eight basic team roles that need to be fulfilled for high team performance. The people most able to fill these roles were identified through psychometric tests. The roles are:

Chair	Co-ordinates and facilitates team process
Shaper	Drives the team forward, articulating task objectives
Plant	Generates solutions to the team task
Monitor–evaluator	Analyses proposals and interrogates them for flaws
Resource-investigator	Keeps the team in contact with other teams, and identifies the sources of additional resources that may be required
Company worker	Keeps the team on track, reminding them of their goals
Team worker	Keeps the team together and smoothes over differences
Completer-finisher	Ensures that the task is 100% complete

These are roles, and as the ideal team size is less than eight, each member may need to play more than one role.

Source: Belbin (1981).

the like need to be on the basis of group output, not individual contributions. The team-building process is very much one of the group mutually identifying and owning common goals; this will always be threatened if the members have different underlying interests.

Fourthly, teams go though an evolutionary cycle as they learn to work as a team. This is usually presented as a four-phase cycle.

- *Forming*, where some individuals come together self-consciously as a group with a task to do – the *working group*.
- *Storming*, where the group members negotiate their positions with each other. Personal agendas are revealed, and power struggles are worked through. Sometimes a false consensus is achieved for external purposes, hiding conflicts from the outside – the *pseudo team*.
- *Norming*, where the norms of group behaviour are established and trust begins to build up – the *potential team*.
- *Performing*, where the whole is greater than the sum of the parts – teamworking is adding value to the efforts of the individual members of the group – the *real team*.

An important implication of this cycle is that investing in team building is a cost – initially performance drops below what could be achieved by the group members individually. It is only once the team has passed through the storming phase that value starts to be added. Thus pseudo teams can be dangerous for management, because the false appearance of teamworking can hide dysfunctional behaviour. This cycle is illustrated in Fig. 15.8.

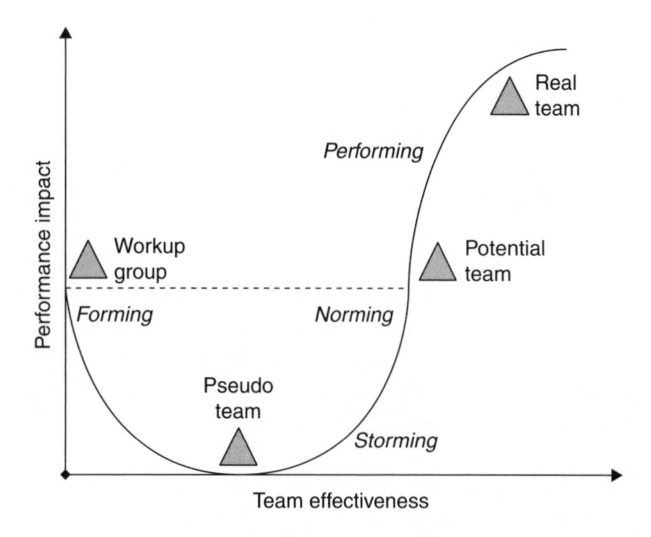

Fig. 15.8 The team development cycle (source: developed from Katzenbach and Smith, 1993, Fig. 11.1 and Tuckman, 1965).

15.10 Constructing the team

There are three very different senses in which the project team concept is used in construction:

- the resource base team charged with executing a particular set of tasks;
- the team supporting the project manager;
- the coalition of firms deployed to realise the project mission.

It is this third sense of team that the Latham Report deployed, which became widely used in the industry during the 1990s, sustained by the advocacy by the Strategic Forum of Integrated Project Teams (IPTs) of the kind presented in Cases 6, 7 and 12[29]. The Construction Industry Board's[30] definition of 'the project team' is:

> all the consultants, contractors, specialists, and the others who come together to design, manage, and construct a project.

However, it is clear from the above discussion that this is not a very appropriate use of the concept; it is more a euphemism for partnering. At best, it can refer to the group of representatives of the various resource base firms realising the project mission, but this is an unlikely team because:

- Many of these representatives will not actually meet each other – the contracts manager of the roofers is unlikely to meet the project engineer from the services engineers;

- The group will consist of representatives of around 20 to 30 firms on a typical medium-sized project; more on the larger ones;
- The representative nature of the group means that they do not have mutual accountability. The representatives are accountable to the client on the one hand and their line manager in the home firm on the other, not to each other;
- They will rarely have had the opportunity to move through the team-building cycle.

This is not an argument that teams cannot draw members from more than one firm in the coalition. However, unless the team building takes place in the context of a partnering type arrangement, there is a danger that all that will result is a pseudo team which performs less well than if they had been honest about being a representative group. Management development techniques can have an enormous impact on project performance, as the case of the Heathrow Express in panel 16.5 shows, but it needs to be in the context of awareness of the particular dynamics of project coalitions.

In the first sense, team performance is the responsibility of the resource-base managers, not the project managers. However, project team performance in the second sense is a primary responsibility of the project manager.

Few projects are small enough to be managed by one person, and large projects will be managed by a number of teams within the programme management function. While there is very little research on the particular dynamics of project management teams – particularly their temporary nature – it would appear that the general principles of teamworking described above apply equally well to project management teams, as shown in panel 15.6.

Panel 15.6 High-performance project team characteristics

Data were collected by a number of different means on the experiences of 16 project teams at NASA and its contractors. Analysis of the results shows that high-performing teams are characterised by:

- a clear focus on the project mission;
- an open and honest exchange of information within the team;
- members with appropriate skills, who can use that competence to convince other members of their position;
- appropriate technical skills;
- willingness to draw on the skills of other members;
- a strong team identity which is inclusive where members socialise together;
- willingness to make personal sacrifices for the project;
- diversity in terms of gender, culture and age;
- clear roles with overlapping responsibilities;
- mutual recognition of milestones passed, celebrated by project outings and socials.

Source: Hoffman *et al.* (2002).

15.11 Summary: project organisation design

Project organisations are in no less need of design in order to enable effective co-ordination than their counterparts with an indeterminate life; indeed, the peculiar position of the project management team as the team in the middle between the client and the resource bases means that particular attention has to be given to the project organisation design problem. The evidence from other industries suggests that the heavyweight project management form is the most effective, yet this is rarely found in construction, particularly since the demise of the general contractor with in-house capabilities. However, new forms of procurement such as prime contracting and concession contracting would appear to be shifting the practice of construction project management to the right in Fig. 15.1, and experienced clients using unmediated procurement routes, such as Slough Estates, Stanhope and BAA described in section 5.4.4, are taking heavyweight project management to levels approaching manufacturing sectors.

More generally, project management organisation is growing as the need of society to create large, complex systems increases. The construction industry was the first to develop project organisation, but since at least 1945, it has fallen behind sectors such as aerospace in the development of its project management capabilities. It now lags behind many other industrial sectors in the sophistication of the design of its project management organisations, preferring a rhetoric of teamworking to seriously addressing the problems of project organisation design and leadership.

Case 15
Glaxo Project Organisation

The Glaxo Group Research (GGR) campus project – presented from a process perspective in Case 8 – had two distinct phases. The first phase was abandoned by Glaxo in 1989 – construction was to start in 1990 – because of escalating costs and the realisation that the overall design was of a poor quality owing to the use of integrated coalitions as defined in section 5.4.2 for each of the separate buildings on the campus. The original scheme had been intended to provide similar facilities to the successful project but was organised in a fundamentally different way.

The second phase of the GGR project began with an 'internal Value Engineering exercise' in late 1989, which resulted in the suspension of the first phase and the determining of all existing contracts. Glaxo appointed the US-based Kling Lindquist Partnership as their master planner (MP) and concept designers, and to assist in the early data-gathering exercise because they had provided a similar service on the Glaxo's Research Triangle Park project in the USA. It was decided that master planning and concept development were to be carried out in the USA, and a UK-based firm was to be appointed to participate with them and complete the engineering and construction document preparation.

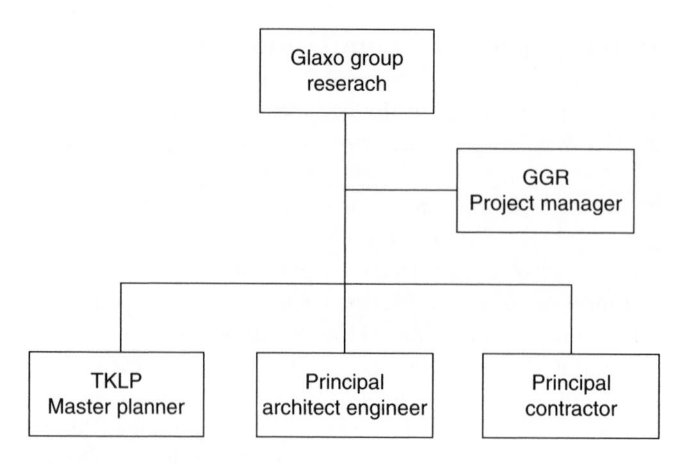

Fig. 15.9 Glaxo project organisation.

By early 1990, Ove Arup and Partners had been appointed as principal architect-engineer (PAE) with Sheppard Robson as architects and Davis, Langdon & Everest as quantity surveyors in a sub-contract relationship. By August 1990, Laing had been appointed as the principal contractor (PC), and formed the consortium of Laing Morrison Knudson (LMK) with MK Ferguson (a division of the then US company Morrison Knudson). Glaxo co-located their key project management staff in the same building with the rest of the PAE team. The engineering manager, representing the client, his six project managers and four support staff were available for participation in the development of designs and to resolve any problems. This ensured complete client involvement at each stage of the process. The structure of the project coalition is illustrated in Fig. 15.9. As can be seen, they opted for a mediated coalition because they were not able to specify precisely enough their requirements to allow a fixed-price contract. As a result, Glaxo appointed the PAE team on a fee-based contract, as defined in section 6.5.1, in order to be more closely involved in the design process.

The GGR project was organised so that the three main teams were involved in overlapping phases of the project. As Fig. 8.6 illustrates, all three principal teams were involved in the earlier stages, before the project was handed over to them. The MP was responsible for the master plan and concept design, but the PAE was also involved in the development of concept design. Similarly, the PC was involved in the process from the scheme design stage. In order to achieve the required project realisation criteria, strict management procedures were developed for the control of the budget and the progress of the project. One of the most important measures taken for effective planning and control of all dimensions of performance was to ensure a high level of communication between all the participants in the project through structured meetings. This was accomplished in two ways.

- Through the organisation of *structured meetings* on a weekly, fortnightly and monthly basis for each level of hierarchy, as shown in Table 15.2. The objective of these meetings was to review performance against budget and schedule targets; they did not cover quality, which was handled at the design reviews, as discussed in Case 8. Such meetings provided an internal discipline to the PAE, facilitated liaison between the three main members of the project coalition (MP, PAE and PC) and enabled the line of visibility between GGR and the project coalition to be pushed back.
- Through *co-location*. Representatives from the PAE team were located with the MP during the early stages of the project to ensure effective co-ordination between the master planners and the architect-engineers. Perhaps more importantly, the GGR project management team and all three project teams from the PAE were housed together in specially rented offices in Central London. Staff from LMK and trade contractors contributing to the design were also co-located there before the project organisation relocated to the Stevenage site. This greatly facilitated informal interaction and enhanced the visibility of the project process to the client.

Table 15.2 Glaxo project management meeting structure.

Meeting (mtg)	Frequency	Participants
Principal's meeting	quarterly	GGR/PAE/TKLP principals and project directors
Project progress mtg	monthly	GGR/PAE/PC principals and project directors
Design progress mtg	monthly	GGR/PAE/PC directors and project managers
Contract and finance mtg	monthly	GGR/PAE/PC directors and project managers
Design review meeting	monthly	GGR/PAE project directors and accountants
Subcontract parcel mtg	fortnightly	GGR/PAE/PC controls staff
PAE executive mtg	fortnightly	Directors
PAE management mtg	fortnightly	Directors and project managers
PAE team meeting	fortnightly	Project managers and design leaders
Administrators' mtg	fortnightly	Project administrators
PAE design executive	weekly	Directors
PAE design leaders' mtg	weekly	Directors and design leaders
PAE team meeting	fortnightly	Project managers and design leaders

The organisation structure of the PAE is illustrated in Fig. 15.10, which should be compared with Fig. 1.4. It shows that the project was headed by the PAE executive, which consisted of the project director and five directors drawn from the three participating firms. Reporting to the PAE executive were five project managers, each responsible for the design of a specific set of buildings and facilities as shown on the horizontal axis. They reported to the project director. Each member of the PAE executive was responsible for a particular aspect of the design work as shown on the vertical axis. The project teams were supported by the project control group which included a project controller, an administrator assisted by a document controller and a technical co-ordinator assisted by a drawings controller. These co-ordination and control services were provided for the whole project by this central group to achieve consistency of approach across the different building sub-projects.

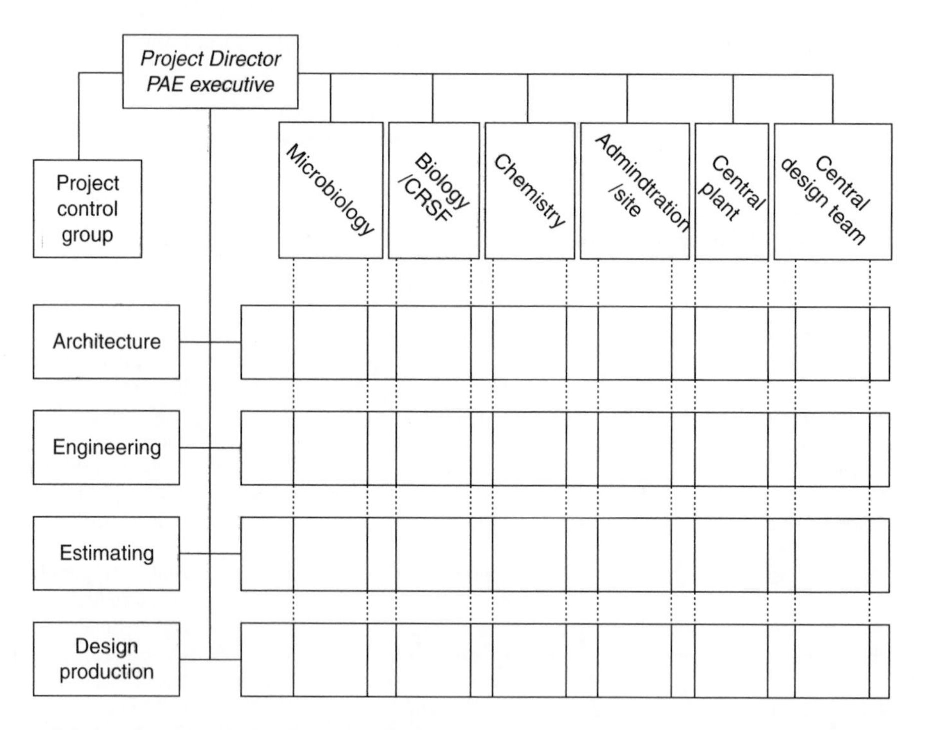

Fig. 15.10 Project team organisation.

A separate manager co-ordinated design production – the production and storage of drawings prepared by each team. The team project managers were responsible for mobilising the discipline resources on each of the five parts of the project – microbiology; biology/central research and service facilities; chemistry; administration and site services; and central plant. There was also a central design team for common design elements.

The matrix organisation of disciplines and projects was divided into six integrated, multi-disciplinary design teams responsible for the design of specific buildings and facilities. A typical design team structure is shown in Fig. 15.11. Each team was run by an individual project manager supported by a team administrator. The team administrators acted as design team managers and were responsible for monitoring the schedule, resources and design costs within the cost account – see section 10.7 – on a day-to-day basis. The project managers and administrators were selected from all three firms forming the PAE organisation. Each discipline within a design team had a group of professionals, assisted by technicians, working on a specific building. These disciplines included architecture, civil/structural engineering, mechanical engineering, electrical engineering, and cost planning.

The use of such integrated design teams ensured that each team not only monitored its own progress but also liaised with other design teams. A group of professionals from each discipline met across teams to ensure that there was consistency of standards. In addition to these professionals from each design team, representatives from the project control group also participated in liaison meetings. These horizontal co-ordination meetings were held fortnightly. The purpose was to monitor standards across all teams, co-ordinate individual progress and raise matters of concern for discussion. Cross-functional meetings were necessary because with five teams working in parallel on the same campus, not only was careful planning required, but effective communication across all teams was also important for achieving the required objectives.

In addition to these teams working together on individual buildings and then meeting across disciplines, there was a central design team which dealt with

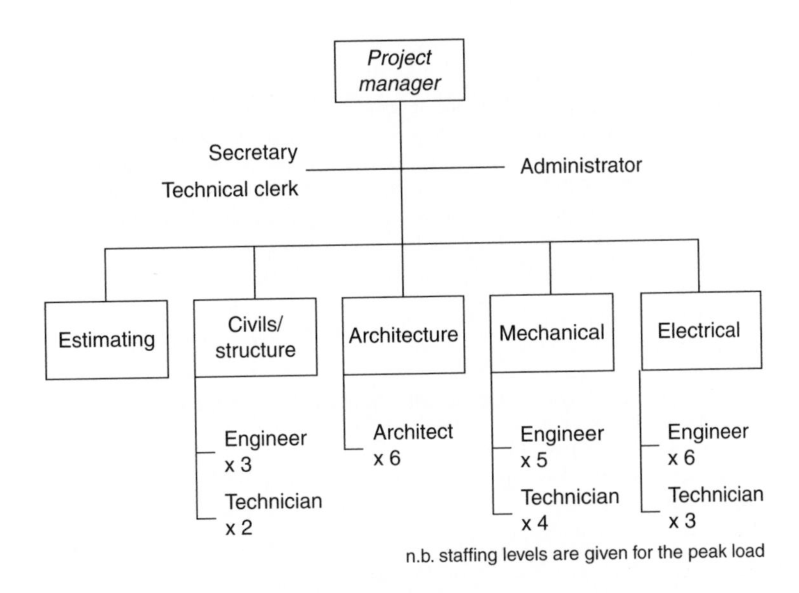

Fig. 15.11 Central plant design team organisation.

planning and standardisation of certain details such as cladding, toilets and staircases. The purpose of this central team was to develop design solutions to ensure continuity of design. This meant that each team was not 'reinventing the wheel'. This system results in two types of 'parcel': one set of parcels was specific to individual buildings, and the other was a set of site-wide parcels designed by the central team. The use of such site-wide parcels had two advantages; it made procurement more time- and cost-effective and led to a consistent design approach.

Thus co-ordination within the project coalition was given a high priority and took effect at five levels:

- within project teams;
- across disciplines;
- with the central design team;
- with Glaxo's management team;
- between the members of the project coalition.

Source: Usmani and Winch (1993).

Notes

1 Cited in Urwick (1937, p. 49).
2 Pinney (2001).
3 See Pinney (2001) and Womack and Jones (1995) for accounts of this experience.
4 See Hughes (1998), Morris (1994) and Sayles and Chandler (1971/1993).
5 His seminal article was published in 1959.
6 Winch (2000a).
7 The term comes from the car industry – see Clark and Fujimoto (1991) and Cusumano and Nobeoka (1998).
8 See, for instance, *Project* April 2008.
9 Penrose (1995).
10 The shortage of catenary operatives and supervisors across a number of rail projects over the holiday shut-down period of Christmas and New Year 2007/8 in the UK is a classic failure of portfolio management. A number of renewal projects across the country had no dependencies, but in total simultaneously required more specifically skilled operatives than were available and the delegation of responsibility for resourcing meant that the client, Network Rail, lacked oversight of capacity until things started to go wrong (Office of Rail Regulation, 2008).
11 See Powell and Young (2004).
12 OGC (2003, Appendix E).
13 Pellegrinelli *et al.* (2007).
14 Holmes *et al.* (2006).
15 Cited in Morris (1994, p. 252).
16 Thurm (2005) provides an account of his experiences as project sponsor of the *New York Times* HQ building.
17 Barberis (1998); guidance on the implications of these new obligations for the management of projects is available from the Association for Project Management.
18 OGC (2003).
19 Helm and Remington (2005).
20 Miller and Hobbs (2005).
21 OGC (2005a).
22 See Hughes (1998).

23 Mintzberg (1979, Chapter 21).
24 Morris and Hough (1987).
25 See Cleland and King (1983).
26 An alternative one used in BP is Responsible; Accountable; Consult; Inform (RACI).
27 This is well reviewed in Handy (1993) and Katzenbach and Smith (1993).
28 Groupthink derives from an analysis of policy disasters such as the Bay of Pigs invasion of Cuba by the USA, while the Nut Island effect explains how pollution in Boston Bay was made worse by those trying to make it better. See Janis (1972) and Levy (2001).
29 Strategic Forum (2002).
30 CIB Working Group 12 (1997).

Further reading

Katzenbach, J. and Smith, D. (1994) *The Wisdom of Teams: Creating the High Performance Organization*. New York, Harper Business.
Valuable synthesis of how to create high-performance business teams.

Mintzberg, H. (1992) *Structure in Fives: Designing Effective Organizations*. Englewood Cliffs, Prentice-Hall.
Synthesising his research, this is a standard reference on the design of organisations.

Morris, P.W.G. (1998) *The Management of Projects* (2nd ed.). London, Thomas Telford.
An inspiring account of the development and contemporary relevance of organising by projects.

Chapter 16
Infusing the Project Mission

16.1 Introduction

> 'Leadership does not annul the laws of nature, nor is it a substitute for the elements essential to co-operative effort; but it is the indispensable social essence that gives common meaning to common purpose, that creates the incentive that makes other incentives effective, that infuses the subjective aspect of countless decisions with consistency in a changing environment, that inspires the personal conviction that produces the vital cohesiveness without which co-operation is impossible'.

Chester Barnard, reflecting on the functions of the executive towards the end of his distinguished executive career in the 1930s, identifies the essential role of the leader as defining the social essence that infuses the organisation, giving it an organic cohesiveness[1]. In combination with Gaddis' identification of the project manager as mediator between the client's desire for a new facility and the abilities of the resource bases to deliver that facility, this identifies the distinctive challenge of construction project leadership, and it is a challenge at two very different levels.

Firstly, it is the challenge of leading the project management team, but secondly and more importantly, it is the challenge of leading the entire project coalition. On a large project, this can be an enormous challenge, demanding both the ability to articulate clearly and consistently the project mission across an organisation consisting of thousands of people and the ability to mobilise the project management team to ensure that the rhetoric of project realisation is supported by the reality. This is a process of infusing the project mission throughout the project organisation.

In Chapter 15 we investigated how to design the project organisation most effectively. In this chapter, we investigate how to ensure that this organisation achieves its mission. The argument will be around the two basic processes of leadership – facilitating the definition of the organisational mission and ensuring the delivery of that mission. It will start by reviewing some basics of leadership, before applying them to the two challenges of leading the project management team and leading the project coalition. An important aspect of this leadership role is conflict resolution, so this will receive attention. The argument will then turn to look in more detail at the levers of power available to a project manager as he or she ensures that the project mission is defined and delivered. A more contemporary term for 'social essence' is culture, so the chapter will also examine the role of project manager in establishing the project culture. The chapter will close by an integrative model of construction project leadership pulling together many of the key themes from the earlier chapters.

16.2 Appropriate leadership

The challenge of leadership derives, in essence, from the problem of uncertainty[2]. Absent uncertainty, and the organisation can be programmed to deliver; accept uncertainty, and a continual process of interpretation of the organisational context and capabilities is required in a manner that the organisation's members find credible. Leadership in a business organisation is required at three distinctive levels:

- Leadership of the organisation overall – the chief executive role
- Leadership of the principal divisions of the organisation – the senior management role
- Leadership of the various units which make up the organisation – the team leadership role.

Our focus here will not be on team leadership – leading a small team is a qualitatively different process from leading a whole organisation[3]; those that are good at the latter are not necessarily good at the former and vice versa. Our focus will be on the upper two levels and the challenge of facilitating the definition and delivery of the project mission as a whole, and of its principal phases – the four identified in the gap analysis model presented in section 8.5.

The source of the challenge of leadership in the level of uncertainty has meant that the level of agreement about what constitutes good leadership is not the same as there is regarding teamworking. Different traditions of leadership research tend to look at different aspects, principally:

- the capabilities of the leader;
- the task facing the organisation;
- the expectations of the led.

The appropriate manner of leading and motivating the project coalition depends upon the interaction of these three factors to generate an appropriate leadership style, as illustrated in Fig. 16.1.

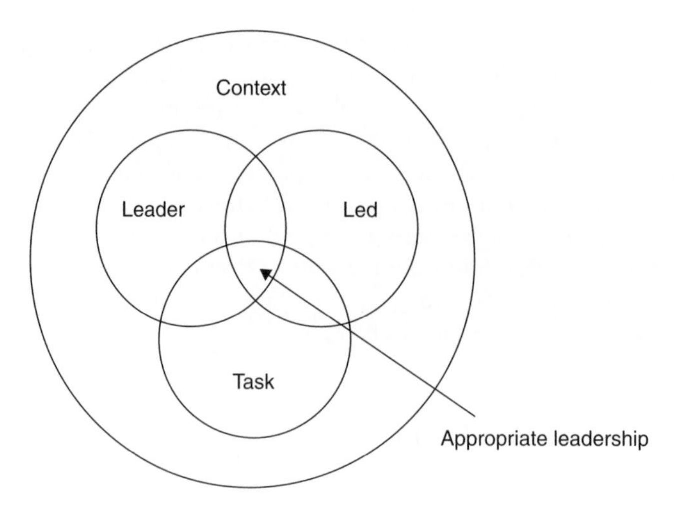

Fig. 16.1 Appropriate project leadership.

16.2.1 The capabilities of the leader

There is now considerable research[4] to suggest that leaders, in comparison to the people they are leading, usually have the following:

- Greater analytic and problem solving skills, and the ability to set the position of the organisation in a broader context. In other words, they can articulate a vision of where the organisation is going.
- More energy, being physically able to set the pace and show by example what needs to be done.
- Consummate technical skills, showing to both their peers and the led that they are worthy of their position.

However, more recent research has identified a further set of capabilities relative to the led, which can be broadly grouped under the heading of *emotional intelligence*. This addresses the softer issues of how the leader not only sets a direction and marches off quickly but ensures that the led will follow; the concept is summarised in Table 16.1. Of course, the table paints the picture of a paragon, and it should always be borne in mind that the characteristics of the leader are *relative* to those of the led, and not absolute requirements. Many successful leaders have been woefully inadequate on a number of these criteria, but all successful leaders are likely to have some of them.

16.2.2 The characteristics of the led

Just as important as the characteristics of the leader are the expectations that the led have of that leader. Different groups have different expectations of their leaders; factors which affect such expectations include the following.

Table 16.1 The five components of emotional intelligence at work (source: Goleman, 1998).

Component	Definition	Hallmarks
Self-awareness	The ability to recognise and understand one's moods, emotions and drives, as well as their effect on others	Self-confidence Realistic self-assessment Self-deprecating sense of humour
Self-regulation	The ability to control or redirect disruptive impulses and moods. The propensity to think before acting	Trustworthiness and integrity Comfort with ambiguity Openness to change
Motivation	A passion for work for reasons that go beyond money or status	Strong drive to achieve
	A propensity to pursue goals with energy and persistence	Optimism, even in the face of failure Organisational commitment
Empathy	The ability to understand the emotional make-up of other people	Expertise in building and retaining talent
	Skill in treating people according to their emotional reactions	Cross-cultural sensitivity Service to client and customers
Social skill	Proficiency in managing relationships and building networks	Effectiveness in leading change
	An ability to find common ground and build rapport	Persuasiveness Expertise in building and leading teams

- *Level of education* – the more educated tend to expect greater deference and sensitivity from their leaders. Thus professionals with graduate qualifications who trained to the highest level are likely to expect a leadership style different from that expected by operatives who left school at 16.
- *Cultural factors* – different organisational and national cultures expect different styles in their leaders. For instance, Geert Hofstede[5] has argued that Anglo-Saxon cultures value much less 'power distance' in their leaders than Latin cultures.
- *Previous experience* – the styles of leaders perceived to have been successful in the past are more likely to find favour in the future.

- *Perception of the situation* – if there is a widely perceived crisis, then different actions will be expected from the leader than if the situation is perceived to be stable.

16.2.3 The nature of the mission

Extraordinary times require extraordinary measures; the converse is also true. The appropriate way of leading will depend considerably on whether the mission is to solve a crisis or to manage steady growth. Different leadership capabilities will be required in each case, and those to be led will expect different things. The 'change agent' is a type of leader who is charged with turning the situation round through redefining the organisational mission; the 'safe pair of hands' is charged with ensuring that the organisation stays on course for the previously defined mission. From our point of view, the key variable is the level of uncertainty – leadership in the early phases of the project where dynamic uncertainty is high will require a style different from the one required by leadership in the later phases where dynamic uncertainty has been reduced. Similarly, projects with high mission uncertainty will require leadership styles different from those where it is lower – especially in the early phases of the project.

16.3 Leadership style

The combination of these three elements identifies the appropriate leadership style. There are different classifications of leadership style; we will deploy here an early one which has stood the test of time, has intuitively meaningful categories and is broadly supported by a number of subsequent qualitative and quantitative studies which identified four different styles[6]:

- *Autocratic*, where the leader takes decisions alone or relies on an exclusive circle of advisors and expects the led to implement them without further discussion. The leader *tells* the followers what to do, is in control and takes full responsibility for the decision.
- *Paternalistic*, where the leader takes the decision alone, but takes pains to explain the reasons for the decision to the led, aiming to convince them of the case by *selling* it to them.
- *Consultative*, where the leader discusses the issues with the led prior to taking the decision, listens to and considers their views, but takes the decision alone.
- *Participative*, where the decision is discussed by the led as a whole, and the majority view is taken, giving the led ownership of the decision. In professional consultancies, this style is typically called *collegial*.

Many people do not wish to take responsibility for a difficult decision – leadership in such a context is inherently lonely, and a participative style is not possible.

Where there are well-established factions within the organisation, such an approach can also lead to stalemate. However, where the decision is complex, and large amounts of, perhaps conflicting, information have to be processed to reach a decision, the participative approach can yield the best results. However, some leaders may find such participation difficult; it requires considerable confidence in oneself and trust in one's team to be fully participative in decision-making.

The autocratic style is not typically favoured by the led, and places an enormous strain on the shoulders of the leader, but where there is considerable urgency or uncertainty as to the best way forward, it may be the most appropriate. The crisis situation – particularly where that crisis has been worsened by inertia or gridlock in decision-making – may require someone to cut to the quick and articulate a clear vision and direction out of the mess. Autocrats may leave blood on the carpet, but they get things done quickly. However, such leaders can be disastrous in situations where the led do not share the sense of crisis, or once the crisis is over and a steadier approach is required.

The styles of most leaders in contemporary organisations are either paternalistic, or consultative. *Paternalism* can work well in stable situations, where the led do not have a strong desire to participate in decisions, but a *consultative* approach is more appropriate and welcomed by most of the led. It also significantly reduces the chances of making a poor decision because of the broader search for possible solutions, and difficulties with the ones proposed. It also avoids the risks of the decision-deadlock that the *participative* approach runs.

One of the continuing debates is whether individual leaders can effectively operate in more than one style as the task at hand demands. Some argue that they can, advocating that effective leaders select appropriate styles like they select golf clubs[7]. However, the leader who can actually do this is rare; although they may experiment with other styles, when hard decisions have to be made, they tend to revert to their habitual style. Others argue that individuals are usually only capable of adopting a narrow range of leadership styles[8]; the key to effective leadership is to match the person to the situation.

The question of whether leaders are born or made is similarly contentious. Emphasis on the characteristics of leaders tends to invoke the argument that leadership is not a capability that can be developed; others argue that leadership training is both possible and desirable. As usual, there is truth in both sides – nature and nurture are mutually reinforcing. Analytic skills, technical skills and physical capabilities are both innate and can be developed, while emotional intelligence has its roots in experiences during early development as much as in experience at work. However, the relative nature of leadership suggests that most people have the potential for leadership in the right situation, and it is well demonstrated that personal development can bring out the potential for leadership.

16.4 Construction project leadership

Leading project organisations requires some of the highest skills of leadership. Construction project leadership is a distinctive management task, with similarities to, but also important differences from, senior management in permanent organisations. The traditional view of management being about the processes of planning, organising, commanding, co-ordinating and controlling has withered under empirical investigation by a number of researchers since it was first articulated by Henri Fayol in 1916, but it does describe the distinctive features of the construction project management job rather well[9].

More broadly, the distinctive challenges of construction project leadership derive from three particular features of project organisations:

- They are typically large and dispersed, and so leadership has to be 'broadcast' over a wide area.
- They are typically diverse in that the levels of education and organisational cultures of the different resource bases vary enormously.
- As discussed in section 8.5, the nature of the task changes significantly over the project life cycle.

There would appear to be a particular personality type that is associated with being a construction project manager. The skills are not primarily technical, simply because the whole point of construction project management is to co-ordinate and integrate across a number of skill sets provided by the resource bases. However, analytic capabilities are important – the ability to quickly grasp what the client, or the architect or the operative faced with an unbuildable detail, is saying is vital, as is the ability to put those individual contributions in the broader context of the project mission.

It would appear that other sectors have found the same thing. Research at NASA has found that its successful project managers are mature, intuitive and outgoing, with high emotional intelligence scores, as shown in panel 16.1. Similar research among project managers in US military procurement has shown that the high performers are strongly committed to a clear mission and thrive on relationships and influence[10]. Programme managers in the car industry are tireless in talking to all the representatives of the project coalition about their ability to meet their commitments, continually articulating and re-articulating the project mission, and what it means for that particular resource base[11]. Much of this networking is done with the project sponsors and the external stakeholders, ensuring that they remain committed to the project and, in particular, keep the finance flowing that is the project's life-blood. Thus much of the project manager's role is external to the project, managing its context to ensure smooth delivery – a role consummately played by Alastair Morton on the Channel Fixed Link, as shown in panel 16.2. Even at the level of the management of the principal project phases, considerable external liaison is now expected; the construction manager for the Waterloo International Terminal did around 100 presentations on the project to external bodies ranging from the Institution of Civil Engineers to a local school, as shown in Case 16.

As the project is ridden through its life cycle, the most appropriate leadership style tends to change. In the early phases of solving the briefing and design

Panel 16.1 The characteristics of NASA project managers

Ten NASA project managers were interviewed and set standard psychometric tests. The Myers–Briggs test measures personality type drawing on Jungian psychology, while ER89 measures emotional intelligence. The project managers tested showed a clear tendency to be

1. mature (mean age 50 years);
2. well educated (over half with masters degree);
3. extrovert rather than introvert;
4. intuitive rather than sensing;
5. thinking rather than feeling;
6. high in emotional intelligence.

Source: Mulenberg (2000).

Panel 16.2 Leading the Channel Fixed Link project

The Channel Fixed Link – see Case 1 – demanded strong and decisive leadership. On such a sensitive project, the effective management of the external stakeholders was vital. In particular, financiers had to be wooed as construction costs mounted. A major barrier to securing additional finance was the strong suspicion among the financial community that the principal contractor – Transmanche–Link (TML) – had signed a contract with itself when the original contract was negotiated. This was because its member firms had constituted the majority of the shareholders in the special project vehicle (Eurotunnel) at the time of signing. Therefore, the principal leadership task on the project was to convince the financial community that no excess profits were being earned by TML, so as to convince them to advance additional funds. The chief executive of the SPV (Sir Alastair Morton) was, therefore, publicly very tough on TML, putting on theatrical displays of righteous anger at their performance. The results were very successful – the finance was raised, the project performed better than most large infrastructure projects in terms of schedule and budget, and no member of TML lost money.

Source: Winch (1996b).

problems, a participative style is most likely to be appropriate to ensure that all the issues and options are brought out. Time is not of the essence here; quality of decision-making is. As the project moves towards the planning and execution problems, uncertainty is reduced and schedule issues press harder – see Case 16. A switch to a more consultative mode is therefore appropriate to speed up decision-making. An autocratic approach is unlikely to be the most appropriate at any stage of the life cycle, although the evidence is that construction managers – who are typically focused on the later realisation phases of the project – tend to be at the autocratic end of the leadership style spectrum, as shown in panel 16.3.

Panel 16.3 Autocracy in construction project leadership

The French and British managers of TML were administered a questionnaire derived from the work of Hofstede (2001) which measured their perception of their bosses' management style.

There was no difference between the two groups which were both at the autocratic end of the spectrum compared to IBM managers. More generally, although important differences in cultural values were identified between the two groups, these differences in values could not be related to differences in managerial behaviour.

Sources: unpublished research; Winch *et al.* (1997).

16.5 Resolving conflict on the project

As argued in section 15.2, conflict is inherent within the project coalition matrix; the challenge for the project manager is to manage that conflict so that it becomes a source of creativity, rather than a sink of energy. Conflict within project organisations is mainly, but not entirely, around schedules, priorities for the use of resources and the allocation of human resources to the project[12]. These issues go to the heart of the relationship between the project management function and the resource bases; direct conflict between the resource bases themselves is unusual. The essence of the problem is that the client's interest, as articulated by the construction project manager, is to maximise the allocation of resources to its project, while the interest of the resource bases is to maximise the utilisation of resources across the portfolio of projects. Thus the matrix presented in Fig. 1.4 is in continual tension.

One solution to conflict is to ignore it, or to smooth it over with a rhetoric of teamworking. This is very much the contemporary tenor of the debate within the UK construction industry, where conflict is seen as dysfunctional and its occurrence a sign of failure. More seasoned observers see conflict as something to be acknowledged and managed so that its positive aspects are realised for the benefit of all. There are various ways of managing conflict within the project coalition before it reaches outside the coalition and moves through the familiar escalator of adjudication, arbitration and litigation. These are as follows:

- *Articulate a clear and coherent project mission* – much of the research on conflict has shown that it is more likely to occur where there is no clear point around which to rally.
- *Argue about facts, not opinions* – in a context where technical people predominate, respect is more likely to be given to positions backed by data. It is the responsibility of the project manager to ensure that appropriate data are collected, collated and deployed in dispute resolution.
- *Use third-party experts to break deadlocks* – this was successfully used a number of times on the Sheffield Arena project, as shown in Case 9.
- *Use decision-making tools that allow both the expression of difference and the identification of commonality* – such as stakeholder mapping presented in section 4.3.

5. *Enter into a partnering relationship* – the need to work together in the future greatly encourages the spirit of compromise in the present.

The research on the evolution of co-operation places the greatest emphasis on the importance of creating a sense of common objectives between the parties that are at a higher level than the particular issues in dispute – see panel 16.4. Similarly, research derived from game theory has shown that co-operation between the parties is strongest when they know that they are going to be working together again – see panel 5.5.

16.6 The levers of power

The exercise of power and influence are an inevitable part of organisational life. This is not to argue that all project managers are students of Niccolò Machiavelli – although that is not unknown – but to suggest that the exercise of power is essential for the definition and delivery of the project mission. Power is essentially the ability of A to persuade B to do what A wants – in other words, it is a relationship. Just as there are no leaders without the led, power is only manifest in its exercise – it is not something that can be stored. There are three dimensions to this exercise of power[13], as shown in Fig. 16.2:

- the overt power of A to directly influence B to choose one option rather than another.
- the power of A to set the agenda so that B's preferred option is not 'on the table'.
- the hegemonic power of A to set the rules of the game so that B cannot conceive of options other than the set acceptable to A.

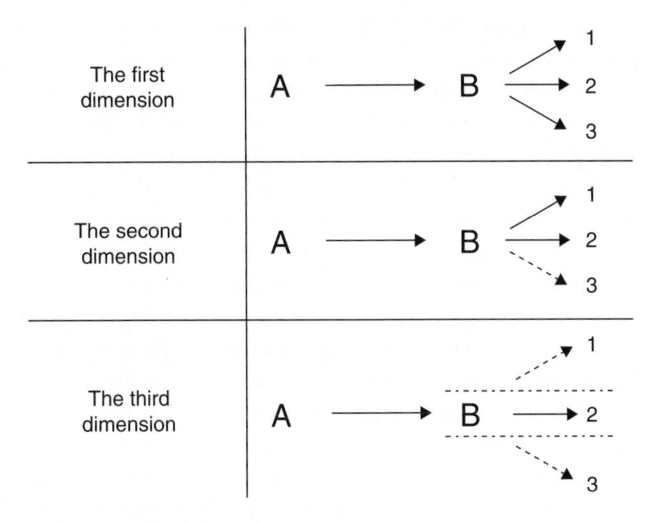

Fig. 16.2 The three dimensions of power.

Panel 16.4 Common objectives and conflict resolution

In one of the most famous series of experiments in social psychology, groups of boys in summer camps were manipulated to both generate and reduce inter-group conflicts. In the first stage of the 1954 experiment, boys were encouraged to form cohesive groups by giving each group a goal that could only be attained through collective activity. At the next stage, competition between groups was generated by offering prizes that could only be achieved at the expense of other groups. Over a period of time negative stereotypes and unfavourable attitudes towards other groups developed, reinforced by significantly greater in-group cohesion. The final stage was to break down this hostility and generate positive intergroup relations. Various strategies were tried to achieve this. Simply providing common activities, such as showing a movie, did not work and, indeed, provided opportunities for aggressive action between groups; neither did 'educational' measures aimed at reversing negative stereotypes. The most effective way of reducing – even eliminating – inter-group conflicts was to introduce common goals that could only be achieved through different groups collaborating. These included combating water shortage and repairing the only vehicle that could take them to food supplies located at a distance. Through tackling these tasks, interpersonal relations between members of different groups improved significantly, and negative stereotyping diminished. These changes did not appear to diminish intra-group cohesion.

Source: Sherif (1958).

The five levers of overt power are as follows[14]:

- *Physical power* – is only occasionally used in a project management context in modern organisations. Its most obvious manifestation is physical force, but it also has its negative side in the ability of those who feel aggrieved with the situation to commit *sabotage* covertly or openly refuse to work. Even the most powerless can throw sand in the lubricants of the organisation.
- *Reward power* – is the ability of A to favour B with something if B does A's bidding. It could be a bonus, promotion or some other incentive, or it may be more negative *sanctions* such as the ability of A to terminate B's employment or to shift B to a less favourable position. The locations of the holders of formal reward power frequently follow the hierarchy of the organisation.
- *Positional power* – derives directly from the position of A in the organisation, be it the security guard who insists on seeing identification, or the auditor who inspects the books. It is rarely effective on its own and is typically backed with sanctions rooted in reward power should B not see A's exercise of power as legitimate. However, positional power carries with it one extremely important attribute – the right to be part of the networks through which information flows. Knowing what is going on, and being able to fit the elements of the picture together, is in itself a crucial lever of power[15].

- *Expert power* – derives from the need of B for A's expertise, be it knowledge of the thinking of the local planning committee, or how to design a bridge. B is often tempted to resist this power, and A's strength may well depend on how much A is trusted by B. The widespread importance of the exercise of this power in construction project coalitions, and the weak position of B in the face of A's expert power, is indicated by the fact that the A's failure to properly exercise expertise on B's behalf can lead to litigation on the grounds of negligence.
- *Personal power* – is rooted in a much more emotional relationship between A and B. Where relationships are purely interpersonal, this can reside in the trading of favours, but at the larger level this is known as *charisma* – A's power to swing B's opinion through force of personality and argument and to communicate effectively to a larger audience.

All five of these sources of power are governed socially through both the legal framework and cultural values regarding acceptable behaviour at work. This is most obviously true of physical power, but there are also formal limitations on the exercise of reward and positional power, while the inappropriate use of expert power is also subject to legal regulation, albeit in a more complex way.

16.7 Project culture and leadership

That the creation and maintenance of the organisation's culture are the most important aspects of leadership is now widely accepted. So, too, is the definition of the project culture. Organisational cultures can be defined at three levels[16] as shown in Fig. 16.3.

- *Artefacts and creations* – these are the visible symbols of organisational culture, such as the way people dress, the amount of marble in the head office foyer and the layout of the offices.
- *Values* – these are the espoused values in terms of what people say and how they justify decisions – the *vocabularies of motive* that people deploy in managerial debates.
- *Basic assumptions* – these are largely the product of socialisation and education, and may not be explicit, even to the individual who holds them. These are the taken-for-granted values-in-use which actually drive decision-making; these are not necessarily completely congruent with the espoused values.

The problem for the project manager is that the client and the resource bases making up the project coalition each will have its own organisational culture. These organisational cultures also vary systematically between the different types of resource bases typically found deployed on a project. Thus managing construction projects is multi-cultural management, as a glance round the table

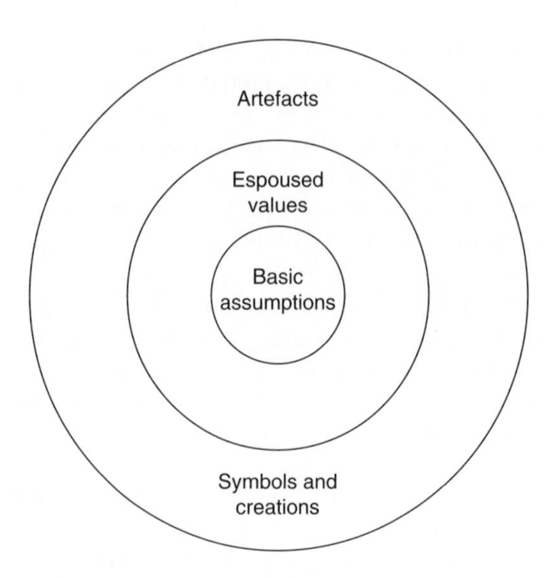

Fig. 16.3 The levels of organisational culture.

at any site meeting at the way people are dressed and present their arguments will confirm. On top of this *potpourri* of organisational cultures, the project manager must try and develop a distinctive project culture.

There are a number of complementary ways in which an organisational culture is infused[16,17].

- *What is noticed and measured* – only a limited number of aspects of organisational performance can be actively monitored by the project management team. Clear messages by the team regarding those they are measuring will help coalition members identify what is important on this particular project. The types of performance that stimulate compliments, rather than being taken for granted, help coalition members focus their efforts.
- *The project manager's response to critical incidents* – emotional outbursts help coalition members identify what is really driving the project manager. If the project manager is passionate about an issue, then failures on that issue will stimulate a public dressing down for those that fail.
- *Deliberate coaching* – again, project managers cannot provide a role model in all areas of performance. Those in which they choose to be supportive and provide coaching will send messages regarding what is important.
- *Explicit and public criteria for the selection of resource bases* – these will send messages regarding what is important. It will also be necessary to sack individuals or firms that fail to perform on the most valued criteria.

Clearly there is little possibility of affecting basic assumptions within the life cycle of a project organisation; the project manager needs to work on artefacts and creations, and values. Particular areas for action here include the following:

- Monitoring the way in which project review meetings are organised. Is open and honest debate encouraged, or are such meetings the arena for finding who is to blame for the latest disruption to schedule? Some project managers have found that organising special off-the-record meetings, where nobody is held to account for what they say, is an important way of generating a co-operative culture within the project coalition – see Case 16.
- Demonstrating that the workforce is a valued contributor to the project by providing high standard site installations – see Case 5.
- Demonstrating that safety is a priority by allowing no exceptions to rules on conduct on site – see Case 12.
- Refusing to procure on the basis of price alone if quality and schedule are important elements of the project mission and being *seen* to reject the lowest tender.
- Engaging in a formal management development programme such as that described in panel 16.5.

One of the most important initiatives in attempting to change the culture of projects is the development of partnering arrangements. These aim to change the culture of the project from an 'adversarial' one to a more collegial set of values. One of the most important symbolic artefacts of this set of values is the alliancing charter that is typically signed by the main representatives of each of the members of the project coalition; panel 16.6 provides the text of such a charter for the project presented in panel 6.4.

Panel 16.5 Organisational transformation on the Heathrow Express

In October 1994, a section of the tunnel being built under London's Heathrow Airport for the express train link to Central London using the innovative New Austrian Tunnelling Method collapsed – luckily without loss of life. This technique relies strongly on teamworking. This was a major, and very public, setback to the project. Instead of taking an adversarial approach of seeking to identify and blame the guilty party through legal action, the client, BAA, decided to take an entirely different approach to building an effective project organisation. Organisational development consultants were hired to create a culture of trust and 'a seamless team', thereby overcoming the very low morale within the project organisation following the collapse. The management development consultants intervened at a number of levels. For instance, they

- worked with BAA's Construction Director to ensure that his language and behaviour continually supported the desired cultural change;
- ran a 'values exercise' to help the senior project management team to align their expectations from the project;
- worked with supervisors to help them to improve their motivation and delegation skills;
- organised special personal development sessions for a group of middle management 'dinosaurs', using the Construction Director as a moderator;
- facilitated joint supplier working groups to enable them to solve problems together, rather than to pass the buck to each other;

- responded to requests from individual managers who felt that they had difficult personal or interpersonal issues to handle with one-to-one and small group sessions;
- ran 'Managing the Future' workshops to help staff learn from this project and focus on what they wanted to achieve on their future projects.

As a result, the new rail link was successfully opened in June 1998, having been completed well within the revised schedule and budget, and without further incident. The number of quantity surveyors working on the project was halved, and staff turnover was a quarter of expectations. This experience then fed forward into the thinking behind the T5 Agreement presented in Case 12.

Sources: Lownds 1998; *Construction Productivity Network Report*, Technical Day 1, 1999.

16.8 Leading the construction project

We presented a contingency framework for construction project leadership in section 16.2 and looked at different leadership styles in section 16.3. We close the chapter by presenting an integrative model of the process of leading a construction project to address the question of how effective construction project managers infuse the project mission to deepen understanding of the practice of leading presented by Patrick Crotty in Case 16. Recent research at MIT has developed a four-dimensional framework for understanding the process of leading, which has become known as the Sloan Leadership Model (SLM) illustrated in Fig. 16.4. The SLM[17] asserts the following.

- Leadership is pervasive – it is not merely the activity of the project director at the most senior level of the project but leading takes place at all levels of the project organisation identified in section 15.6 from team leader upwards.
- Leadership is personal and developmental as we learn by doing – we all have our own leadership 'signature' rooted in our capabilities explored in section 16.2.1 as matured through experience.
- Leadership is incomplete – no one person can excel at all aspects of leadership, and each leader has a preferred style as discussed in section 16.3; each leader needs a strong team to complement their weaknesses.

Within the SLM there are four distinctive and mutually reinforcing leadership processes which all effective leaders deploy whatever their capabilities and styles.

- *Visioning* – the process of creating a compelling vision of the future. It is essential to the effective definition of the project mission discussed in Chapter 3 supported by future-perfect thinking discussed in section 8.2.
- *Inventing* – the process of creating the means to deliver the vision. In a temporary project organisation this is even more essential than in a permanent one. While routines can be adopted from previous organisations, they will always need to be configured to meet the needs of the tasks and teams involved as Chapters 8 and 15 show.

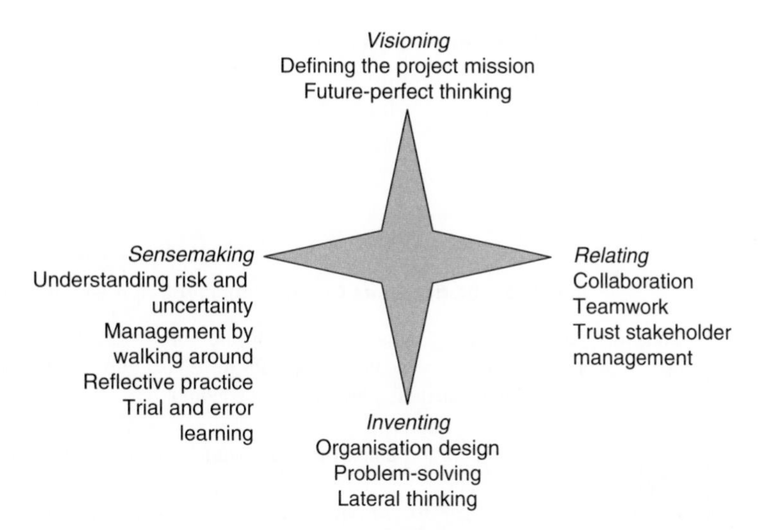

Fig. 16.4 The Sloan Leadership Model applied to managing projects (source: adapted from Ancona *et al.*, 2007).

- *Relating* – the process of mobilising the resources required to achieve the vision through processes of enquiry, advocacy and connecting. These are essentially collaborative processes of understanding the other's point of view and generating trust through relationships rather than contracts with internal and external stakeholders, as discussed in Chapter 4 and explored more deeply with respect to internal stakeholders in Part III.
- *Sensemaking*[18] – the cognitive process of riding the project life cycle adapting as new information arrives while keeping the vision clear, spotting weak signals that hint at underlying problems. While the routines that make up the toolbox of riding the project life cycle discussed throughout Part IV play a vital role in structuring sensemaking, effectiveness in leading the project coalition derives from an engagement with the process in collaboration with the members of the project coalition rather than following prescribed templates.

We will see how this works in practice in Case 17.

16.9 Summary: infusing the project mission

Infusing the project mission is about leadership, but about leadership in a very distinctive organisational context. Project organisations are temporary coalitions of firms that need to be motivated towards a common purpose – the project mission. The task of the construction project manager is to facilitate both the definition and realisation of this mission. Yet, the project manager does not control directly the resources to do this definition and realisation – those are within the resource bases. As argued in section 15.2, the project manager is the person in between the resource bases and the client. At one level this is a position without power, at another it is one which is central to the project process and hence strategically important.

Panel 16.6 Staffordshire County Council/Birse Alliancing Charter

Construct Tunstall Western bypass Phase 2 to the appropriate quality and to the mutual benefit of all the parties. In so doing our joint objective is to free the potential of the team to achieve the following goals during the life of the scheme:

- Complete the works for the employer at an outturn cost within the contract budget and with budgetary certainty;
- Enable the contractor to maintain a positive cashflow and achieve a reasonable profitability;
- Complete on time;
- Maintain an impeccable safety record;
- Build and maintain good relationships and reputation with the community;
- Make Tunstall a model project;
- Create an environment based on respect, trust and fairness that promotes honest, open communication between all;
- Make the partnering arrangement work by breaking down the chains of convention and tradition to promote efficiency, imagination and innovation;
- Create an environment and atmosphere in which all the team members are happy to come to work;
- Encourage alternatives and share in cost savings;
- Implement a procedure to jointly solve problems.

Source: CIB Working Group 12, 1997.

In fulfilling this strategically important role, the project manager has few direct levers of power which are not counterbalanced by the possession of strong levers within the resource bases. The levers are mainly positional and, for the fortunate few such as Alistair Morton in Case 1 and Tim Smit in Case 17, personal. However, that positional power needs the continued backing of the reward power of the client for its credible deployment. Within a project coalition, the position of project manager is the central node in the project information flows. The project manager's power largely lies in this nodal position in the project information flows and places the resource-base managers at a relative disadvantage. Thus the construction project manager is in the best position to articulate the 'big picture' – to define and diffuse the project mission – as a direct result of being in the middle of the relationships between the client and the resource bases within the project coalition.

These information flows are in two directions: information flows towards the project manager by virtue of position, but as a result flows from the project manager who is in the best position to communicate the project mission in whichever terms are most appropriate for a particular resource base or group of resource bases. It is this process of articulating and re-articulating the project mission which lies at the heart of project leadership, and which we have called *infusing* the project mission captured in the SLM. This is done through repeated one-to-one interactions with representatives of the client, the external stakeholders and the resource bases, through project newsletters and other media, through articulating a clear project culture and most obviously at formal meetings which the project manager typically chairs. Sometimes more informal means are used. We have called the process one of *infusion* because the aim is to ensure that the project mission pervades the project coalition and its meaning has been internalised by the key decision-makers representing the coalition members. If this can be achieved, then conflict within the coalition will be about means rather than ends, and a shared understanding can enable the development of trust between the coalition members.

Case 16
Patrick Crotty: Project Director on the Waterloo International Terminal

'What is distinctive about project management is the time span. It is not unending like normal management. In project work, you bring people together for a specific job and when it is finished they all disband and go elsewhere. Then I move to another project'.

When you are working forwards towards the completion of a project, it is like riding on the Cresta Run – it feels as though you are out of control, there is absolutely no way of stopping and you have got to steer it to stay in that narrow groove all the way to the end. And, almost always, I have done. But it has definitely got no brakes. Nothing stops time.

Another thing that is peculiar to project work is that at different stages of the process, the kinds of problems and the kinds of questions that emerge and that need to be dealt with are different. Yet you need to work with the group of people that you assembled at the start, whose skills may be more suited to one stage than another. You cannot just keep taking people out and putting in more appropriate people because then you lose the benefit of continuity. So I have to be able to judge, more than other types of managers, the significance and the reliability of the advice that I get, depending on who gives it. A lot of that judgement comes down to perception and past experience.

You do have to get up in your helicopter, because it is very easy to get wrapped up in the day-to-day problems. There is so much detail in a vast project like this, you have to come out of that mass of detail occasionally. . . .

16.10 The project life cycle

Through the life of a project there are a number of quite distinct stages, each with its own challenges and pitfalls, and each requiring different skills from me. To give you an idea of the time frame, I was appointed to the project in the middle of 1989, and I expect to hand over the finished station to the railway operators in May 1993. But within that period there are probably five distinct phases.

Stage 1

During this stage I was based away from the site, at 100 Piccadilly. There were a number of distinct tasks within this stage. In the first fortnight after being appointed, three of us chopped the job into a series of components each one of which would become an enquiry, a quotation and an awarded contract. We produced a book called the *Element Scope Definition* which listed each package of tasks which we grouped together to send out, each as one enquiry. We put together some packages and rapidly awarded them, to do with relocations, cutting and diverting services and pipes.

A major task in 1990 was building my team. I interviewed everybody who came to work here, even people who already worked for Bovis elsewhere, whereas in the past they would just have arrived and started. I only rejected one or two people, who were inappropriate, but I used the opportunity of the interview to give each person a little inspirational pep-talk.

At that stage of any construction project, you have to make sure your growing team gets to know each other and identify whether they can work together. Generally you can make people work together, but inevitably there are stresses and you have to find out what they are, and to make sure everyone is communicating with one another. I had to build my own team, make them pleased to be here and keep them inspired.

You have to make sure that everyone understands the systems that are job-specific. There are lots of systems in big companies like Bovis that are normal and standard, but on every construction project, and probably more on this one than on most, there are job-specific differences. So I had to make that work. We were heavily involved in work with the designers to create enquiries, quotations and the logistics to plan the awards for the main bulk of the project. That was when we concluded the big packages for work to be put out to contract, sent them out as individual enquiries, got back quotes, sifted them, analysed them all and recommended to British Rail with whom they should place their orders for each package.

We were also working quite closely with the people that the project would affect. This is a very big city-centre project with operating railways above, six underground platforms below, which has meant liaising with the London Borough of Lambeth, ambulance, fire and police.

Another key task at this stage was to build up personal credibility with the client and with the designers. Bovis is slightly unusual in not being a builder. Not many people are used to managing contractors, so it was necessary for me to establish credibility and recognition within the design teams. . . .

In the first stage, we were physically remote from the site and doing lots of work with the designers. It would have looked very small as a project; you would hardly have seen us. There was lots of staff involvement and nothing much to show for it.

Stage 2

In the first stage I was mostly in an office-type environment; my suit stayed fairly clean. In the second stage we physically moved in and took possession of the footprint of the terminal. The trains stopped running where we were to build and we put hard fences round a much bigger zone of the project. We had progressively to close off some of the routes into the London Underground because we were shifting the underground station to tuck it away under the existing station. . . .

We did a lot of buying while we had not very serious construction to manage the invitations for contractors to come in, to be interviewed and vetted. In stage two we were doing construction of the shell and we were buying the big mechanical and electrical packages. It is not until you have bought those packages that you can move on from the bidding, competition, handing out of awards and handshaking stage.

Alongside the buying process, the real construction got under way. So my job effectively doubled. I still had the responsibility for what was now a more established team. I got involved in a lot of dialogue with contractors. We started our trade contractors' directors meetings, where I met monthly (now six-weekly) with a main board director for every single contractor – presently this is 30 individuals. Gradually, the physical work on-site gathered momentum, which brought with it the new challenges of safety and operational logistics. When you have a project like that going on in the field, you have contractors with hundreds of men and the surprises and little crises that come along and need handling. We also had to continue our relations with the people who would be affected. We brought in all those other people, like the Highways Authority, the local police and New Scotland Yard, London Underground, Network South East. We had, for example, to satisfy the Health and Safety Executive that our station roof would be safe. We built a mock-up of it because it was unique and we needed to try out the erection methodology.

Stage 3

We are now in the third stage where the structure is in place. The big building construction is done and so is the buying. We are now in the business of letting contractors do their designs and then working to co-ordinate one contractor with another. The designer input to the project involves taking the contractor's design and checking it, making sure it complies with specifications. So we are heavily into contractor response and into co-ordination in terms of making things fit – pipes, cables and wires are notoriously designed by different practices to go through spaces that are never big enough or the right shape to take them. So the co-ordination process is heavily reinforced by a Bovis team involved in making sure that pipes are of a certain size. They sometimes have to be made smaller, but with higher pressures and higher volumes, it means the pumps have to be of higher specification – and you have the problem of the money saved by having smaller pipes being used up on making the pumps a higher capacity. There is also the problem of whose responsibility that is. Did the designer design an impossible system or should the contractor be liable for the cost of the solution? So there is a complex operational and highly technical process going on now.

If you prioritise the process of procurement buying, you buy the structure first and the innards second. It is not until you have bought the innards that you can move on to co-ordination. So typically high-tech co-ordination dominates this third stage. The outcome of this co-ordination process consists of production drawings and the manufacture of things like escalators and lifts off-site. The high-tech end is not in manufacture yet. So far we have only built one of the fifteen motor control centres. They are the complicated brains of these intelligent building systems. The other fourteen are waiting for the outcome of the co-ordination process which will affect them.

The third stage, then, is all about co-ordination. My role in this third stage is to arrange and chair meetings between the contractors and the consultants. We create the forums for the co-ordination process to happen. It is all very well writing into contracts, 'You are responsible for co-ordinating with other contractors', but if you simply beat people over the head as they sit in their separate work-places, that does not get you anywhere. The idea is to make sure that when the contractors go away and draw it and submit it, it all fits together and you can get at it to maintain it. So we have a group of managers whose job is to facilitate that, and they have to be able to spot the difference between what is technically difficult for a contractor to change and what is simply costly to change. We have to be able to cut across that commercialism.

I have a sort of arbitration role at this stage because it is then that you find out whether the logistics plans are working. Inevitably, you find that some things do not work and some things are missing. There is a lot of potential for quarrels and arguments about what people in other practices and other firms did that was not right. That is normal in the installation phase of construction. There is a tendency for people to blame everyone but themselves and to come back at any criticism with things the other party may have omitted.

One of the things British Rail have set up is a regular risk review, where the leader of each one of the separate practices has a meeting away from the project to 'admit things' with no minutes, no notes, no biros or tape recorders. That is helping us a lot. It allows one person to say, 'This happened, that happened. I really do not know why'. And somebody else will say, 'It is because your man did not do that'. Some people are responding better and others are not relaxing into it; they are not really participating. We are hoping that the ones who are willing to admit problems in-house without recrimination will encourage others to come out, because we are in that stage where, admit or not, the warts-and-all are going to show.

Stage 4

During the fourth stage the project will look like a finished building. There will still be contractors doing work in outlying zones, but the big activity will be commissioning, the testing and the setting-to-work of all the electrical and electronic systems. It is all about making intelligent systems talk to each other, proving that they work within their parameters.

In parallel there will be all the things one has not thought of: the mistakes, the defects, remedial work, training the new incoming staff who will operate the terminal. It is rounding out all the accounts. We will be dealing with any financial claims, where contractors think they have done more work than they were contracted to do.

The project can be 99% complete, but in the last three months there is a lot of checking and tying up of loose ends. It is about checking both physically and checking the listings of commissioning schedules. It is about gaining the approval of the Health and Safety Executive, the railway operators, Her Majesty's Customs, the police and fire services. All the incoming occupiers have individually to accept their components: the tram crew, the package handling crew, the Network South East staff. It is all outlined in the Plan for Completion.

Stage 5

The test of whether you have achieved the fourth stage or not is the architect's Certificate of Practical Completion. That triggers a whole lot of things: for instance, the insurance risk for the project moves; half of the money retained from every payment made to every contractor is released; the station legally begins to be operated by the users; and the one-year defects-liability period begins.

At the end of that one-year period there is a revisiting to inspect and discover whether every defect noticed has been remedied. Generally that period is fairly low-key and would involve only small numbers of Bovis people. My personal input to the defects-liability stage should be quite small, but if the whole thing became fundamentally unworkable or if there were court cases I would be called back. So I do not personally expect to be very heavily involved in that final stage. I would expect to be released from full project duty around June 1993 . . .

16.11 Keeping control

We have two important planning documents. The first is the Planned Access and Logistics book. It specifies level by level, stage by stage, every little bit of separation of the public from the workplace, traffic flows, how the work would affect the walking public through the station and how we would affect London Underground. We did lots of plans like that which took an iterative approach. You would put together a plan that suited nine out of the ten groups and the tenth one would come up with a basic reason why that whole thing had to go back – and we did a lot of work on that which cost us about £50 000, though initially there was very little to show for it, and I am glad we did it because it is now working very well and it is recognised by lots of people. I have never used such a process before. We have not stopped updating it. Some of the drawings are on their fifth incarnation.

The other control tool is the Scope Definition book. That book is intended to make sure that there are no gaps or overlaps between packages, to make the designer concentrate on who is responsible, say, for the channel in the floor for the glass wall to come out of, between where one bit of floor ends and another begins; who is responsible for the software that drives the electronic doors – is it the door contractor or the building management system contractor? It is in the later stages of the project process that you find inevitable areas where the system did not work, but the intention is to limit the problem.

We are audited and you must be able to prove to people how you made decisions and why you spent the money. Having said that, when you work for a big organisation, you can go overboard on systems and frameworks. I think you have to be careful of that in project management. You must not let that framework rule everything. You must let the people make the project happen and not have 'the system' dominate. There is a job out there that is happening and it is people who are building it.

I have also tried to keep an operational grasp on the project by making myself an essential signatory on the final buying of every single package. I make it my business to be the person who signs for Bovis so that I can see who the winning bidder is, and so I know what is going on. That way I am not going to be embarrassed by somebody saying, 'I see you have got so-and-so caterers on your project', and for me to say, 'Have we?'. Also I have some things about which I am a stickler and I sometimes make new members of my team go through hoops before I will sign the buying report – especially if I think they have skipped an important stage in the process that I believe is important.

Increasingly, I am finding that it is more important to keep tabs on people than on systems and programmes, which is where I am more naturally drawn. I am changing during this job in a way that I like – if you make a change and it works it reinforces it in you. I hope that I am moving towards being more flexible and having a more humanist approach.

We have used computers more here than on previous projects to help architects to see what things would look like and to produce working drawings with dimensions and calculations of the stresses and strains. We also used computer graphics to create a series of images showing what the terminal would be like to walk through.

It is useful for me to keep an image in my mind of how the project is supposed to go and to work hard against any big deflection. It is easy to be too receptive to frightening stories, because lots of people have vested interests in coming to me directly or in a fairly direct route with a smokescreen. People bring you tales of woe. So you have to ask them what makes them think is so different about this project that a particular thing should throw us off track. I will constantly remind people of the programme in quite high-powered-level talks. I try to resist alarmism. I am fairly relaxed about relying on my own judgement. When I get disturbing advice, I am not easily thrown off the track.

I am very prepared to set a framework and to make sure that things happen in accordance with it. The framework in this project is primarily guided by safety and time, with cost and quality as controlling influences. Because it is such a highly visible project and because my company knows me, they know that I am prepared to put a lot of pressure on everyone including myself, to stick to that framework. That is both a positive quality and something I have to be careful of: it must not blind me to real issues. But I think that aspect of my contribution is quite important and it is a key to why I finish jobs on time.

I do not leave anything alone. I am always looking into what my people are doing, and not just by meeting them or by reading the correspondence. I go in from different angles. I walk the project, I talk to people, I compare what one person tells me with what I have read in a report. I cross-check all the time and if I find something that does not look quite right, I meddle. I hope I am learning to meddle in a way that supports the people and that does not undermine the lines of authority and their pride in the job. . . . '

Source: Stewart and Barsoux (1994, Chapter 4).

Notes

1 Barnard (1968, p. 283).
2 Barnard (1938/1968); Katz and Kahn (1978).
3 Katz and Kahn (1978); Katzenbach and Smith (1993).
4 Handy (1993).
5 The new edition of his classic based on research in IBM was published in 2001.
6 Developed by Harbison and Myers (1959) from their survey of management across a number of different countries.
7 For example, Goleman (2000).
8 For example, Nicholson (2000).
9 Fayol (1999); Stewart and Barsoux (1994).
10 Gadekin (2002).
11 See Walton (1997).
12 Thamhaim and Wilemon (1975).
13 Lukes (1974).
14 In identifying the levers of power, the framework established by French and Raven (1960) has proved robust, although the terminology used by Handy (1993) is preferred here because it is more intuitive.
15 Pettigrew (1972) provides the classic analysis.
16 Schein (1992).
17 Ancona *et al.* (2007).
18 Weick (1995).

Further reading

Handy, C. (1993) *Understanding Organizations* (4th ed.). Harmondsworth, Penguin.
Well-written review of the research in management and organisation – a classic.

Nicholson, N. (2000) *Managing the Human Animal*. London, Texere.
The first sustained application of the concepts of evolutionary psychology to management.

Schein, E.H. (1992) *Organizational Culture and Leadership* (2nd ed.). San Francisco, Jossey-Bass.
A comprehensive statement of the role of culture in the effectiveness of organisations.

Chapter 17
Conclusions: Managing Construction Projects Consummately

> 'I love construction. There are two reasons. One is because we're very much a people-orientated business – there'll never be robots climbing around scaffolds laying bricks, so we'll never be automated to the extent that industry is. And the second thing is, you start with a concept and a big hole in the ground, and one day you walk away and you've actually built something. I still get a buzz out of walking around London, Saudi Arabia, Burma, wherever I've built buildings, and you look there, and there's something left, not so much a monument to yourself but just that you've built something and you can see something for it. That's a big buzz, it really is'.

> 'After all it's gone, you forget all the problems. You just look up in the sky and say 'Wow, that's a beautiful building'. That's the reward. You feel that great sense of satisfaction, and that makes everything else go away'.

Driving together through Colwyn Bay – a small North Wales resort – one of my slaters suddenly pointed up to a hotel on a corner site with its riot of mitred hips and valleys in Penrhyn slate and exclaimed, 'I did that'. He was proud of something he could show me – his new manager – as well as his children. Mick O'Rorke and Dominic Fonti – the straight-talking construction managers of the Tate Modern and Worldwide Plaza projects[1] – capture in the epigraphs above exactly why construction people, from the architects to the operatives, love working in construction when they could be earning better money with better working conditions in other industries. The satisfaction is in the product, in having crafted something that is useful. This is the same point that was put more conceptually in Chapter 1 – construction projects are, fundamentally, about adding value, so it is to the revaluing of construction that we turn first. We will then review what this means for the consummate achievement of product and process integrity, before reviewing the keys to managing construction projects – systems thinking, professionalism and judgement.

17.1 Revaluing construction[2]

There has been much debate around the world about finding news ways of managing construction projects – Case 2 reviews developments in the UK. However, such debates have tended to be process orientated and to ignore the product, particularly the way in which it generates value for the client and instils pride in the people who created it. Perhaps a broader ambition can be visualised around the idea of *revaluing construction* on three dimensions:

- Generating a much better understanding of how constructed assets add value for the clients that finance them and the people that use them – these are the issues that were addressed in Part II. It is still the case that constructed products are seen as *artefacts* – as things that cost money, rather than *assets* that provide a return on investment as discussed in section 3.2.
- Developing a much more effective capture of the value generated by riding the project life cycle – in terms of both profits and learning – in the manner shown in Fig. 1.1. These issues were discussed in depth in the introduction to Part IV.
- As a result of these two revaluations, revaluating the image of the industry and the way it is perceived by those outside it who equate construction with disruption, which is reinforced by the perceived environmental impact of the creation and operation of the built environment.

We discussed the many facets of the concept of 'value' in section 3.2, focusing on the process of value creation. To understand more deeply the relevance of the revaluing construction perspective, it will be helpful to consider the relation of the value-creation process to the stock of value, or what economists call *capital*.

In pre-industrial economies, land was the most important type of wealth, typically owned by a very small minority of the population. Capital – traditionally held as gold – provided a fund which could be used to exploit land by paying labour. Again, it was typically held by a relatively small minority of the population. Labour was the only type of capital that could create value and was ubiquitously held, but was useless without land or money. Thus land, labour and capital were defined as the *factors of production*. As modern economies have developed through the industrial revolution and moved towards what many see as a post-industrial era, economists have reviewed these classical definitions and developed more sophisticated definitions where capital is seen as a multi-faceted entity rather than a factor of production.

Capital is very much the focus of this new analysis – indeed, we often talk of the modern economic system as the capitalist system – but capital is now seen as having a number of different, and not completely substitutable, dimensions[3]. We will see shortly that construction is an important element of all these forms of capital.

- *Physical capital* can be held in a liquid form such as money, or, more widely, as investments in productive assets such as machines and equipment.

- *Human capital* is held by people in the form of their skills and capabilities.
- *Natural capital* is the environment around us. It includes land that can be used for farming and amenity purposes, but also wilderness that has a purely ecological value.
- *Social capital* is the latest addition to the forms of capital and aims to capture the intangible social elements that hold society together.

Constructed facilities – housing, buildings, infrastructure – account for around three-quarters of the total *physical capital* of a modern society and around half of all the new capital created (defined as gross fixed capital formation) each year. The stock figure is significantly larger than the flow figure because constructed facilities are relatively long-lasting compared with other physical capital. Thus, collectively, constructed facilities are our most valuable assets, and we can also see this individually because, for home-owners at least, the family house is far and away their most important physical asset.

According to the World Bank, around three-quarters of the total wealth of an advanced economy is *human capital* – people are certainly our most important assets. Human capital includes the capability for hard physical work – the traditional view of labour – but much more importantly it includes the skills and capabilities developed through education, training and learning on the job. In comparison with many other sectors, construction is certainly a skilled occupational sector. Relatively large numbers of technical and managerial staff have graduate-level qualifications in their chosen professions, and much of the manual workforce has relatively high levels of apprenticeship-based craft qualifications. This human capital represents the basic capacity to create the built environment in any national economy, and it is relatively mobile. Where it is in excess and cannot be immediately deployed, it is mobile to places where it is in demand. Construction boom economies such as those of Dubai thereby become melting pots of internationally mobile human resources.

The relationship between *natural capital* and construction is a difficult one. Construction projects inherently convert the natural environment into the built environment, and, by any calculation, their impact on the natural environment is a negative one. The building of roads and the conversion of farmland to housing estates turns natural capital into physical capital. The basic tool for evaluating the advantages of such a conversion is cost-benefit analysis, but as we saw in section 3.6, the pricing of natural capital is fraught in comparison to the pricing of physical and human capital, and so there is considerable room for debate around the value of natural capital such as a bucolic view or a nature reserve. As a result, most nations have developed more or less democratic procedures for evaluating the inevitable trade-offs between investment in constructed physical capital and the loss of natural capital that those investments entail, which were explored in section 4.4.

The awareness of the importance of *social capital* has been growing in recent years. One definition is that 'social capital is a capability that arises from the prevalence of trust in a society'[4]. Construction can affect the development of both the positive and negative aspects of social capital through the design of its products and the conduct of its processes. It can affect it positively by designs that facilitate the development of social capital, such as creating urban spaces that facilitate interaction within the community – one model here is the classic European city square, *piazza, platz* or *place*. It can affect the negative aspects by designing urban environments that discourage crime as discussed in section 3.3.

A further aspect of social capital which extends the accepted definition but is highly relevant to the construction sector is the way in which physical capital can become social capital through perception as *heritage*. As we explored in section 3.5, constructed facilities have important symbolic functions, and these symbolic aspects tend to increase with facility age. The process is obvious with buildings such as cathedrals and other major public buildings such as the Sydney Opera House in panel 8.4, but it can also transform the most utilitarian of buildings such as factories, power stations and warehouses into symbols of an industrial heritage in a post-industrial society, as shown in section IV.3. Through this process, physical capital takes on a social dimension that leads it to be perceived in the same way as natural capital, as something that can be threatened by further investment in physical capital.

Capital as a stock of value, as we can see, has four different aspects, and in all of them the construction sector plays a major social and economic role in both creating and destroying value. The net effect is usually positive, but the vehement opposition to some construction projects suggests that not everybody is convinced that it is always positive. The issues around the different perceptions of the stakeholders in a project regarding the net benefit it is creating were explored in section 4.2 and are fundamental to the definition of product integrity. The different forms of capital interact – only human capital can create new capital, yet it is useless without access to physical capital. Social capital facilitates the development of human capital and the creation of physical capital, yet at the same time natural capital can be destroyed. Physical capital can take on social aspects and thereby come to be seen as similar to natural capital.

These considerations suggest value creation is a complex trade-off negotiated by stakeholders[5] and therefore at the heart of revaluation lies the *alignment of incentives*, upon which Part III focused, where social capital plays a major role. Construction project coalitions are typically complex and difficult to manage; the task is made much easier if the incentives that motivate coalition members are aligned, so as to enhance competitive collaboration. The individual members of the project coalition as firms in a market economy have the obligation to their shareholders to maximise their share of the value generated in the realisation of the project; what the project manager needs to do is to ensure that what Adam Smith called 'enlightened self-interest' prevails, and that the achievement of these objectives is done jointly rather than severally by finding the largest area of overlap of interests between the actors in the project coalition. Such alignment is required at a number of levels:

- Between the stakeholder groups making up the client, particularly between the promoter, operator and financiers of the proposed facility, as discussed in sections 4.3 and 13.7.

- Between the designers and the client – the use of third-party value management is a symptom of poor alignment here. These issues are discussed in sections 9.3 and 10.5.
- Between the client and the principal contractor, through the use of incentive contracts, often in a partnering or alliancing context – the problem of vertical project governance discussed in sections 5.6 and 6.8.
- Through the supply chain – the problem of horizontal project governance discussed in section 7.6.
- Between management and operatives to allow the removal of padding from task execution times and release cycles of continuous improvement in process capability – see sections 7.4, 11.5 and 12.7.

The role of the construction project manager is to facilitate the achievement of these alignments, and to act as the first point of call in dispute resolution before the parties enter the adjudication, arbitration, litigation escalator. Once incentives are aligned, there are two symbiotic dimensions to managing construction projects consummately:

- Managing for product integrity
- Managing for process integrity.

Thus the task of managing construction projects is firstly to enable the design of facilities with integrity – in other words, facilities which match externally with client and user needs, and internally as engineered systems. Secondly, it is to develop process integrity so that products with integrity can be effectively and efficiently realised.

17.2 Managing for product integrity

Under the separated routes discussed in section 5.4.1 that traditionally predominated in construction procurement, the project coalition was led by the architect or consulting engineer as both design professions and, often reluctantly, project manager. The performance of the design professional – whose heart lay in the product rather than the process – in managing the process was frequently poor. However, an obsession with process is to put means before ends; the end must determine the means, not the other way round. The end is the asset that the client can exploit for its own business processes – be they driven by profit, community or a personal dream. How can we articulate a design-led process that puts the process at the service of the product, without crippling the efficiency and effectiveness of the process in favour of the product? Table 17.1 captures some of the changes in the role of the professional (traditional fee based) designer in the UK as the professional system identified in Case 2 evolved over the last quarter of the twentieth century.

In enabling product integrity as defined in section 3.8, the construction project manager plays a number of vital roles. As much of the evidence presented

Table 17.1 The changing role of the design professional (source: developed from Gray *et al.*, 1994, Fig. 1).

Yesterday	Today
Designers dominate the market, offering a professional service led by the architect or consulting engineer	New designers, e.g. interiors, services and other engineers, are emerging to fragment the established design professions
Designers hold the dominant position of authority in the design process	Designers are losing position and authority within the design team to project managers
Professionalism is based on clear disciplinary boundaries formed in independent schools	Professional designers are required to become generalists with less control over details, which are now dealt with by specialist designers from a wide variety of organisations
Designers are natural leaders of the process	Multiple control of the design and construction process
Fee agreements are simple and loose	Fee agreements are complex, restrictive and competitive
Designers determine the client's 'real' problem, and hence the definition of the project mission	Expert clients now dominate project mission definition
Design issues establish the quality of intention, and consequently the cost	Value management through design quality/cost/ time trade-offs
Self-governing design professionals are relied upon to deliver a competent service	Project managers manage designers
The professional designer has an overall responsibility for the management of the whole process	A wide variety of sophisticated procurement techniques are used to coordinate design and execution

in this book has demonstrated – and the gap analysis model in Fig. 8.2 captures conceptually – defining the project mission is the key to the minimisation of client surprise. If this is not done well, little can be done to rescue the situation. Project managers do not define the mission – that is the responsibility of the client working with the designers – but they can facilitate that definition process so that the brief is complete and appropriate. In many larger projects – such as Heathrow Terminal 5 presented in panel 4.4 and Case 12 – winning regulatory consent becomes a project in its own right. During the resolution of the briefing problem, construction project managers, among other roles, can:

- ensure that realisation considerations are fully articulated in the definition process, as illustrated in Fig. 3.2;
- ensure that stakeholders' voices are adequately and appropriately heard, thereby minimising the chance of the expression of such interests disrupting project execution, as discussed in section 4.5;
- define the ICT strategy of the project in terms of project management information systems, as defined in Fig. 14.1;

- develop a risk management strategy – particularly along the lines discussed in section 13.7 – ensuring that the project coalition is governable;
- develop the strategy for the procurement of the early-phase resource bases required, as discussed in sections 5.3 and 6.5;
- start articulating and infusing the project culture, as discussed in section 16.7.

As the project moves through the life cycle from addressing the briefing problem to the design problem, the construction project manager can:

- identify and implement the appropriate design management tools such as those introduced in section 9.7;
- define and implement the rhythm of stage gates within the project life cycle – Case 8 provides one way of doing this;
- ensure that the ICT strategy is implemented, and that protocols for the exchange of information between project coalition members are maintained, and a project extranet is established as discussed in section 14.7.

Moving on through the planning and execution problems, the challenge shifts from design-led construction project management focused on product integrity, to performance-driven construction project management focused on process integrity.

17.3 Managing for process integrity

Process integrity is achieved through riding consummately the project life cycle through the realisation process. Process integrity – in essence – is about the ability of the project coalition to keep its promises. During mission definition, objectives for the quality of specification and conception will have been set, and the associated schedules and budgets established. Realisation of this mission needs to be predictable, so as to minimise client surprise about the project performance gap. This realisation requires all the issues addressed in Parts III, IV and V to be addressed by the construction project manager, ranging from the soft skills of managing teams, discussed in section 15.9, and conflict resolution, discussed in section 16.5, through to the rigours of controlling budget and schedule against plan, as discussed in sections 10.7 and 11.2.

At the heart of process integrity is *process capability* as defined in panel 12.3 – in essence, process capability is the measurable dimension of process integrity. Process capability is about ensuring that the discrete tasks that make up the material flows during project execution are earned out as predictably as possible. The more immediate benefits of enhancing process capability include:

- less disruption to dependent tasks – see section 11.5;
- reduced rework and attendant waste of resources – see section 12.9;
- more efficient allocation of resources – see section 11.3;
- better fit, finish and durability in the completed facility – see section 12.2;
- fewer accidents – see section 12.7.

A basic prerequisite for enhancing process capability is the performance measurement and benchmarking of discrete processes. This provides the basis for achieving some of the larger benefits that will justify the return on the investment that will be required to improve process capability, and hence product integrity.

- Cost planning will be easier, and value engineering more viable if task execution is more predictable, because costs can then be known more precisely, as discussed in section 10.2.
- The impact of different design options on process performance can be determined more accurately in constructability, as discussed in section 10.6.
- Project schedules can be planned with less contingency, as discussed in section 11.5.
- Levels of conformance can be improved and a culture of improvement developed, as discussed in section 12.7.

Overall, it will be easier for the supply side in construction to keep its promises, and there will therefore be fewer surprised clients, thereby closing the *project performance gap* identified in Fig. 8.2.

The site-specific and one-off nature of construction means that process capability will never achieve the same levels as the much more predictable mass and lean production environments. However, if the dispersion around the mean level of performance can be reduced, this will be of benefit to a proportion of projects, and if the learning thereby acquired can be deployed to incrementally improve mean levels of performance, then all projects will benefit.

Part IV explored some of the ways in which process capability can be enhanced in construction, and it is worth pulling a few of them together here:

- Using new scheduling tools to ensure that only quality assignments are allocated – section 11.5
- Using TQM techniques to tackle poor conformance quality – see section 12.7
- Using standardisation and pre-assembly of components and component sub-systems – see section 12.10
- Using integrated project management information systems – see section 14.8
- Investing in constructability on the basis described in section 10.6.

How, then, can managing for product and process integrity be developed? We will first explore an answer to the question that is given by many, but, we submit, does not really address the scale of the challenge.

17.4 Construction as a manufacturing process

Throughout the nineteenth century the construction sector was associated with advanced technology[6]. The extraordinary engineering achievements in bridging and tunnelling that associated the construction of railways throughout the world; the advanced materials used in Paxton's Crystal Palace; the development of the skyscraper in Chicago requiring innovation in both structures and internal vertical transportation all testify to this claim. During the latter part of the nineteenth century the manufacturing sector began to innovate rapidly – the development of interchangeable parts, scientific management, the production line and automation in the context of rapid organisational innovation led to the remarkable gains in productivity and reductions in costs that laid the foundations of the consumer society. Construction did not appear to share in this innovation and thereby reap the resulting benefits, and so the argument that construction was in some way 'backward' compared with the manufacturing sector gained ground. In particular, housing production became an area of urgent concern as pressure grew to tackle the woefully inadequate housing produced during the rapid urbanisation of the nineteenth century[7].

Some of the first to address this issue were the architects associated with the Modern Movement in Europe after the end of World War I. Their vision of unadorned buildings, where form followed function, being mass produced from standardised modules was both seductive and very influential. While its successes in the area of housing have been mixed, it rapidly diffused as the international style in commercial and public building. Reactions to it such as the Art Deco of 1930s skyscrapers and the post-modernism of the late twentieth century have done little to mitigate its pervasive influence. In practice, though, the Modern Movement was more of a stylistic influence than an innovation which transformed the construction process. While there have been major innovations in building sub-systems over the last 100 years which have transformed their performance and the functionality delivered for clients by the modern constructed product is much superior to that of 100 years ago[8], in terms of process, most buildings are still crafted on site in a way that would be basically familiar to building workers from the nineteenth century. Thus the debate on the extent to which the construction sector can emulate the manufacturing sector continues.

Over the past 20 years, volume manufacturing processes have been transformed by the success of Japanese companies in international competition. What has become known as *lean production* is now established best practice in the car industry and a number of other sectors[9]. While the concept of lean production has a number of variants[10], its principal features are the following.

- The production flow is paramount – the flow of components through a factory, or the flow of passengers through an airport, should be maximised and the old concept of batch and queue to maximise capital utilisation is outmoded.
- The production process should be pulled by customer demand rather than be pushed by production scheduling which requires flexibility in production processes.
- Suppliers should be tiered in proactively managed and partnered supply chains.
- The elimination of in-process and finished inventory by the focus on flow and pull-scheduling leads to reduced working capital requirements for production.

- Continuous improvement of the production process takes place through team-oriented activities such as total quality management.
- The challenges in improving performance are largely organisational and do not depend on high levels of technology in the production process.

The undoubted success of lean production concepts in improving performance in the car industry and a number of other sectors has provided an inspirational model for those wanting to improve the integrity of the construction process. These concepts underlay the Egan report published in the UK in 1999 – the chair of the Commission had an extensive background in the car industry and one of the commission members had been a member of the original MIT research team. They are also the focus for the work of the US-based Lean Construction Institute[11]. However, we submit that this approach is essentially neo-bureaucratic[12] rather than professional because of its emphasis on the standardisation of process and product and its intellectual roots in high-volume lean manufacturing.

The lean construction perspective emulates the broader lean production perspective in that it focuses on materials flows – the transformation of materials into components and the assembly of those components into completed systems. It is, essentially, about what happens in the factory or on the construction site, treated as a problem in production engineering. This is, of course, enormously important, but it can be argued that this is not the only, or even the principal, problem in managing construction projects at this stage of the development of the art. A fundamental feature of the type of volume manufacturing firms analysed by the advocates of lean production is that design is largely separated from manufacture. While the design of a product and its sub-systems is only done once for each product launch, repeated examples of that product, customised as appropriate, are manufactured. Most construction remains in design-once/produce-once mode, as opposed to design-once/produce-many. In volume manufacturing the design project finishes at 'job 1' and the design, often together with a design for the manufacturing system, is handed over to one or more factories so that manufacturing can commence[13]. While variants on the design can be manufactured, they are drawn from a pre-planned menu in what is known as mass customisation. In construction, designs are usually specific to the site where the building is to be constructed and to the requirements of the client that commissioned it.

The crucial question in assessing the potential for the application of the lean approach to construction is the extent to which the one-off nature of the construction process can be changed. To answer this question, it is useful to separate the new-build construction sector – lean concepts as a whole are unlikely to be applicable to refurbishment projects, let alone repair and maintenance, because of their inherent specificity to the building being worked upon – into four broad sub-sectors:

- large infrastructure works, typically the preserve of civil engineering;
- prestige building projects, where the architectural distinctiveness of the project is part of the brief;

- 'routine' building projects that provide the bulk of new buildings;
- housing.

There have been numerous attempts to build houses on a volume scale. Ever since the Modern Movement turned its attention to the construction of social housing, there has been the ambition to move to volume production. In countries such as the UK where land and house are bundled together at purchase, a major constraint on tackling the issues is the profits to be made by the builder from gains in the value of the land compared with reductions in the cost of construction[14]. Where land and house can be separated and sold separately, as in Japan, there is considerable potential to reach the same annual volumes of some of the more specialist cars, and there is a significant potential for the application of lean concepts because builders can only make profits from building, with the gains from land speculation going to the house-owner[15]. Indeed, the construction example used by Womack and Jones is from the volume house-building sector.

However, there are difficulties which cause pressures to reduce levels of standardisation at the level of the product as a whole. In countries such as the UK, construction on brownfield as opposed to greenfield sites is government policy; many of these are constrained in ways that require high levels of design input to make the best use of the site. Further, there are pressures from those concerned with housing design quality to reduce repeatability. For instance, the Commission for Architecture and the Built Environment's (CABE) Building for Life criteria include the requirement[16] that the design be specific to the scheme being built on the grounds that a housing development should contribute to its local community by engaging rather than ignoring its context to create a 'sense of place'.

Routine buildings include several in which many of us spend our working lives – schools, clinics, offices. The extent to which they add value for the clients that commission them in the sense analysed in section 3.2 is vital for the performance of the economy overall, and the extent to which they provide a pleasing environment in which to work, learn or play is vital to our overall quality of life. On the one hand, there have been many experiments in gaining volume for routine buildings ranging from the prefabricated churches exported to the British colonies in the nineteenth century to the large-panel prefabrication of schools during the 1960s. On the other hand, CABE has been promoting distinctively designed public buildings appropriately designed and consummately delivered.

If there are major difficulties in achieving volume in routine public buildings, it is unlikely that it can be achieved in symbolic public buildings and large infrastructure. Arguably, for the most part outside housing, construction projects will remain a one-off production process where design and execution on site are intimately linked. Can the challenges really be considered to be merely a problem in production engineering?

An alternative view suggests that the way forward is principally one of stressing the importance of managing projects, and in particular, how the relationships between the members of the project coalition are governed[17]. The lean-inspired improvement activities presented in Chapter 12 and Case 7 certainly have their place in the effectively managing construction projects because they provide a valuable toolset for improving process capability, but we submit that their effective deployment relies on three complementary managerial processes:

- Systems thinking rather than lean thinking;
- Professionalism rather than neo-bureaucracy;
- Judgement rather than pseudo-certainty.

17.5 Systems thinking and managing projects

The argument in this book has placed considerable emphasis on the role of uncertainty in creating bounded rationality in decision-making, which is the central paradox in the management of construction projects. Where a number of decision-makers – responsible, say, for the execution of sequentially dependent tasks in a parade of trades – act independently of each other with incomplete information about the implications of their decisions on those dependent on them, perverse dynamics can be generated. This was particularly important for the discussions in section 7.8 on the dynamics of the supply chain, and in section 6.7 on the dynamics of adversarial relations. Figure 6.7, in effect, presents a systems view of the vicious circle that constrains the performance of construction projects. Those that have addressed these issues have advocated the use of systems thinking – understanding the parts in the context of the whole, and the whole as greater than the sum of the parts, as presented in panel 7.5. The explicit use of systems dynamics modelling – an application of systems thinking to schedule delay and disruption – was also presented in section 11.6.

Intellectually, this view of the system as something both greater than the parts and constraining the performance of the parts has considerable parallels with the sociological approach to understanding the individual and society in terms of actors and systems, which underlies the tectonic approach developed in this book and encapsulated in Fig. 1.5. At one level, systems thinking is common sense; at another level it is very difficult to carry out because it requires the ability to take a view of the whole while keeping the parts in focus. It is for good reason that both Eli Goldratt and Peter Senge write of the *discipline* of systems thinking. A prerequisite for systems thinking is to be in a position to have information about both the whole and the parts, and as discussed in section 15.2, systems thinking and project management evolved together from meeting the same challenges during the 1950s. On a construction project, the project management function is best placed to have access to both these categories of information. This link between systems thinking and managing projects is not original[18], but that is no reason not to reiterate the point. Traditionally, the construction project process was fragmented along the lines of professional disciplines; as these rifts are being overcome, the process has become fragmented along the lines of management fad. Valuable competencies tend to become bundled up and sold as stand-alone services to the client. Suppliers of these services inevitably start to compete with each other, so we see value management separated from the supply of design services, risk management from the supply of construction services and supply-chain management separated from responsibility for execution on site.

17.6 Professionalism and managing construction projects

We argued in Chapter 1 that construction projects are one of the most important ways of creating the modern world. Before the emergence of the Renaissance merchant, the central mission of most construction projects was to express power – be it the power of God or the throne. Even the merchants of Florence, Amsterdam and London built mainly to express their wealth[19]. It was not until the industrial revolution that construction began to create facilities principally for wealth creation rather than wealth consumption. We live with many of the achievements of the nineteenth century today – the railways and urban infrastructure of our cities were created by entrepreneurs such as Eiffel, the Brunels and Aird[20]. It was from the challenges of constructing such massive facilities that the role of the project manager began to emerge. As Prometheus was unbound[21], project managers increasingly played their role in developing and applying his arts.

Perhaps the most notable feature of the modern world compared to earlier eras is the existence of complex organisations. Max Weber argued that the growth of *bureaucracy* as the epitome of rational organisation was central to modernisation. Although the term now tends to have a pejorative tone, Weber saw bureaucracies as the central way of managing a modern economy and society. Talcott Parsons criticised Weber[22]. He argued that there was an alternative to bureaucracy – professionalism. Where needs are predictable and can be standardised, bureaucracy remains the most effective way to meet them. For this reason many services provided by the state are organised as bureaucracies. Where needs are more complex or dynamic, professionalism defined as the ability to configure established expertise to solve novel problems is more appropriate. Arthur Stinchcombe's classic[23] comparison of craft production in construction with bureaucratic production in cars poses exactly this issue.

One of the most penetrating criticisms of the approach to managing projects that evolved with the arming of the USA in the period after 1945 is that it is, at heart, a bureaucratic approach[24]. While internally, such projects met and matched enormous engineering and scientific challenges, they operated in very stable external environments. So long as the threat of the USSR could be mobilised, funding was assured, opposition was minimal and the project managers could get on with the job of realising the project mission undisturbed. One does not have to accept Thomas Hughes' argument that the Boston Central Artery/Tunnel project discussed in Case 13 is a post-modern project, to appreciate that something has fundamentally changed in the context of the management of construction projects over the last 20 years or so.

Project managers now face much more complex environments where stakeholders are voicing their opposition and have to be accommodated; funding can no longer be assured as state budgets are cut, and clients face more turbulent economic environments. At the same time, constructed facilities are becoming more complex as technologies change and performance expectations are raised. In response, new approaches to managing projects are being developed which, we submit, can address these issues and consummately deliver facilities with high product and process integrity[25]. This text is one attempt to present such an approach, and professionalism[26] is central to its effectiveness.

The application of expertise to novel problems creates considerable expert power as defined in section 16.6. For this reason, professionals and the institutions that organise them have developed the ethical principle of disinterested advice with regard to both the client who is paying their fees and the wider society which charters them[27]. While traditionally, the 'professional' approach has been used to distinguish consultants from the 'commercial' approach of contractors, developments in the commercialisation of consultancy and the professionalisation of construction management make this distinction less relevant. However, what remains completely relevant is the ethical stance implied. In deploying expert power under uncertainty on the project there is an obligation to make decisions based on ethical principles rather than on self-serving expediency. This can be dangerous because project mangers have been sacked for exposing strategic misrepresentation – see section 3.7 – by their clients. But it remains central to the professionalism of project managers[28].

17.7 Judgement in managing construction projects

Judgement is the essential complement to decision-making under certainty. Making the call in the information space identified in Fig. 13.2 is a judgement call – the project manager can only do his or her best to make the decision at the last responsible moment with the available information. In this definition[29], judgement is about *what* should be done rather than *why* it should be done. Table 17.2 attempts to understand more deeply the nature of judgement and to relate it to Weick's model of sensemaking in organisations. In this perspective, judgement has three elements:

- *Intuition* to generate hypotheses for solutions by defining the problem through forensic analysis of solutions to earlier problems and reflection upon their implications for the current problem complemented by broad search around the problem. This process generates an *enactment* which orientates further action and stimulates the garnering of more information.
- *Induction* to assess the available evidence and stimulate the search for more evidence, thereby testing the hypotheses generated and *selecting* the most plausible as the basis for further action.
- *Interpretation* to review the assessed information distilling clarity from equivocality *retaining* the most plausible for future information processing.

This sensemaking process of judgement is a continuous one as the information flows through the information loops shown in Fig. IV.1. The outputs from judgement processes are, in effect, mental models of the situation – what Weick calls an *enacted environment*. However, they have real consequences because they allocate resources – skills, effort, plant – in one way rather than another[30]. Thus at some

Table 17.2 Judgement and sensemaking (source: adapted from Vick, 2002, Table 3.1; Weick, 1979, Chapter 5).

Aspect	Use	Requirements
Intuitive (Enactment)	Hypothesis formation Problem definition Guiding what to look for Identifying predictions to be made	Forensic skills Reflection Understanding Visualisation
Inductive (Selection)	Synthesis of evidence, information, and underlying knowledge from different sources Assessing probable truth of hypotheses Generalising from specific cases	Recognition of evidence Awareness of signs Observational abilities Analogy
Interpretive (Retention)	Critical review Evaluation Establishing meaning and context	Pattern recognition • Patterns of consistency and anomaly • Spatial relationships • Correlative relationships (from data) • Causal relationships (from theory)

point the enacted environment meets the real environment – what Weick calls *ecology* – and the outcome of the project is known. While each of the chapters in this text has introduced routines which support effective sensemaking, the presence of uncertainty means that making good sense is a matter of judgement rather than sticking to those routines. While there is nothing wrong in principle in expressing those judgements quantitatively[31], it should always remain clear that the numbers embody professional judgements about the future, not facts.

What role then for neo-bureaucratic methods? As might be expected by the reader who has reached this point in the book, the answer to that question is that it all depends on the project mission. Taking some of the concepts from Chapter 9, we can suggest that where the project mission is a tame one, a bureaucratic approach is appropriate; where it is more wicked, a professional one is more likely to be successful. It is entirely appropriate for many project missions – broadly those described as incremental in Fig. 9.5 – which create standardised building types such as housing, schools and retail units, and form the majority of construction projects by number, if not by value. It is less clear that it is appropriate for more complex projects – those which push the envelope or create iconic architecture.

17.8 Summary of the book

This book has been about the process of managing construction projects, but it has never lost sight of the fact that it is the product that gives meaning to that process. It has articulated a tectonic approach to the structure and process of construction projects. The river metaphor, introduced in Chapter 1, shows how the information flows which initiate and control materials flows are at the centre of construction project management, yet cannot be managed directly. It is through the structure of the project coalition that these flows are governed, as uncertainty is reduced through time. The rest of the argument took various aspects of the construction project management problem from this perspective.

In Part II, the complex, messy, inherently political process of defining the project mission was addressed as a problem in strategic decision-making. The difficulties clients have in deciding what they want, and then negotiating that vision with the various project stakeholders, was analysed, and the importance of product integrity identified. In Part III, the difficulties that asymmetries of information between principal and agent pose for the effective mobilisation of the resource base were identified and analysed. Part IV turned to the problem that is at the heart of construction project management – riding the project life cycle through the various stages from brief to execution on site. Here the toolbox of construction project management techniques was presented and assessed, showing how and when each tool could be most appropriately deployed to achieve process integrity. Part V turned more towards organisational issues, investigating the importance of leading the project coalition through the life cycle.

This concluding chapter has developed the concept of revaluing construction as a much broader concept than re-engineering or rethinking, encompassing client interests, project coalition interests and the image of the construction industry more broadly. The construction industries in many countries are presently in a state of flux, and ways of doing things differently are widely sought. This book has offered frameworks for thinking about how to manage construction projects in different ways so that they deliver value for both clients and the actors in the project coalition, and thereby encourage a revaluation of the tarnished image of the industry. We submit, therefore, that good judgement, supported by professionalism and systems thinking, in the context of aligned incentives is the fundamental attribute of effectively managing construction projects, and that this is how the common causes of project failure summarised in section 1.9 can be most consummately addressed as can be seen from Case 17.

17.9 A concluding thought

This book started by citing the poetry of T.S. Eliot; it is perhaps fitting that it should end by citing a Project Director:

'Building should be fun after all'.[32]

Case 17
Tinker Bell Theory in Practice

The Eden Project in Cornwall is one of the most successful UK Millennium projects; images of Grimshaw's structural drawing for the biomes and the completed facility are on the cover of this book. The facility provides an environmental visitor experience of over 1m plants from a variety of climes. A large covered biome provides a humid tropical environment, while a smaller one provides a warm temperate environment totalling 2.1 hectares. The cool temperate environment is in the uncovered outdoor biome. Constructed in a redundant south-facing china clay pit, the project presented an enormous range of challenges and provides a wonderful example of the power of future-perfect thinking presented in section 8.2, or what the project champion, Tim Smit, calls Tinker Bell theory – the project only exists if you believe enough in it.

The project architects, Grimshaw Architects working in collaboration with structural engineers Anthony Hunt and structural contractor Mero, designed biome covers constructed from a tubular steel space-frame (tri-hex-net) to form a geodesic spherical network creating very wide span free-standing spaces for the plants up to 125m in diameter and 55m high. This steel frame was clad with lightweight hexagonal panels made from three layers of thin UV-transparent ETFE film which are sealed around their perimeter and inflated to create large thermally efficient cushions. The panels vary in size up to 11 m across, with the largest at the top of the structure. The erection of the structure on the 858-m-long ground beam required the largest free-standing scaffold in the world, followed by installation of the cladding panels by abseilers. Civil engineering works included moving 800 000 m^3 of fill and extensive drainage systems, by the construction manager McAlpine JV. A visitor centre was built which opened over a year before the completion of the facility so that tourists could view the construction works, generating much needed income. An education centre – The Core – opened in 2005. Eden is currently seeking funding for a third covered biome – The Edge – to provide an arid environment.

The inspiration behind the Eden Project is Tim Smit – currently Chief Executive – who had rescued and opened to the public the Lost Gardens of Heligan in 1992. The idea for Eden was prompted by the garden festivals of the early 1990s which attempted to regenerate run-down urban areas and distilled from a conversation over a bottle of whisky in May 1994. Funded by pump-priming money from the local authority, a mix of Smit, architects and horticulturalists energetically developed their idea. It moved from fantasy to future-perfect thinking thanks to the launch of the Millennium Commission with a brief to fund capital projects to celebrate the new millennium – the Millennium Bridge (panel 13.2), which featured on the cover of the first edition, and the Tate Modern (section IV.3) were also funded in this way. An initial bid – based on 'back of fag packet' budgeting – was submitted in April 1995, but was turned down. Here, a little strategic misrepresentation came into play, because Smit decided to withhold this information from his growing team so as not to discourage them!

Smit managed to convince some of the leading consultancies in their respective fields (including Ove Arup on services and Davis Langdon as project managers and cost consultants) to work for free to develop the design while Smit and the team worked on the Millennium Commission. The Commission did not fund development work before bids, and so it was not obvious anything was amiss and the team struggled on private donations and small grants. By mid 1996, the lobbying achieved results and Eden was back in the competition with a submission due in

December. The construction budget was £74.3m reached after aggressive value engineering through which project lost a biome and an oceanic feature. The news that Eden had been successful was announced in May 1997, and the McAlpine JV was notified as preferred bidder in June 1997. The relationship was reinforced by appointing a Director of Sir Robert McAlpine to the Eden Board in 1998. This relationship would be of enormous benefit later during construction when the project nearly ran out of cash owing the JV millions and the Director, Cullum McAlpine, steadied the boat by saying 'we're still here'.

With the funding announcement, the project reached a turning point; as Smit put it (p. 117):

> 'There comes a time in all great ventures when the talking has to stop. We'd created the constituencies, we'd talked the hind legs off donkeys, we'd been snake-oil salesmen with attitude and a dream to peddle, but turning a dream into a reality needs iron in the soul, money in the bank, and military organization'.

As the design developed it became clear that it would not work – basically the proposal was for a Waterloo International Terminal (see Case 16) propped against the side of the pit, but the structure was too heavy for the span and the ground too uneven and changing because of continued working of the pit for clay. The inspiration from Grimshaw was a soap bubble which can mould itself to whatever surface it alights upon, the solution a geodesic dome and so Mero – a German specialist in this kind of structure – joined the coalition. Further value engineering was required, so Grimshaw halved the cost of the visitor centre, completely redesigning it in two days and bonding with McAlpine in the process. Finally, the clay pit was purchased in October 1998, and the ECC contract (see section 6.4) was signed in January 1999 as a target cost contract with a guaranteed maximum price – McAlpine had worked for nearly two years without a contract, as had most of the consultants. Intensive construction on site started in February 1999, and the complete facility opened in March 2001 before schedule and to budget. In the meantime, Mero was obliged to take over the supplier of the ETFE cushions because it was too small to deliver on a project of this scale. Alongside the construction, a second project involved the construction of greenhouses in the Eden nursery a few miles away, selection and purchase of plant specimens, growing them on and planting them in the biomes in the different types of soil manufactured by the project.

Eden was a remarkably successful project, and now draws double the number of visitors annually that was envisaged in its 1997 business plan. It is worthwhile, therefore, revisiting the OGC's common causes of project failure to help understand how it overcame them to be so successful.

- *Aligning with strategic priorities* – as the client was a special purpose vehicle established to deliver the project, this was not a problem.
- *Clear senior management leadership* – the leadership of Smit as project champion, then Project Director, then Chief Executive was unconventional but

extraordinary. He effectively mortgaged Heligan to the hilt and persuaded all the consultants and the construction managers to work virtually for free in the early phases on the promise of reimbursement when the funding started to come through.

- *Engagement with stakeholders* – Smit's ability to network both locally within Cornwall and nationally garnering enthusiastic commitment was impressive, mobilising the right people to solve difficult problems – particularly those associated with finding the other half of the funding for the project according to Millennium Commission rules. These skills encouraged the head of a neighbouring county, Somerset, to back publicly the Cornwall project for European funding at his own county's loss.
- *Lack of skills and proven approach* – Smit had the ability to surround himself with a good team, including a highly professional client-side project manager, world-class horticulturalists and a dedicated administrative team complemented by a network of wise counselors. The top firms in the country were selected to join the project coalition and committed wholeheartedly to Smit's vision.
- *Breaking down into manageable steps* – the effectiveness of the McAlpine JV which was involved early enough to have a significant influence on the design, and was incentivised to solve problems rather than make claims, was important here, although there was no stage-gate process. This was supported by active risk management.
- *Emphasis on initial cost rather than on whole-life value* – the effective cap on costs forced by the Millennium Commission rules meant a strong emphasis on cost control and effective value management, while the Eden vision was essentially an operational one and the values expressed in that operation were continually reiterated as the design developed. This extended to refusing to hire a specialist facility operating company for fears of losing control.
- *Lack of client engagement with the supply chain* – Eden worked actively to engage with the JV, selecting them because the JV considered it 'the ultimate construction project' and inviting a representative onto the board. The Eden project managment team then left relationships with the second tier to McAlpine.
- *Poor project team integration* – the first tier of the project coalition was not integrated formally, but as it learned to work together – virtually unpaid – it generated high levels of trust, and the long gestation period of the design thanks to the time taken to raise funds enabled relationships to be built and risk taken out. This was reinforced by co-location. All acknowledged that this was a design-led project, and it was the job of the JV to deliver on that design.

One factor that is not in the OGC's list, but as we have seen in this book is vital to many projects, is faith – faith that the project will be realised. This faith inspired Smit and the initial champions of the project, it inspired the design team and the construction managers, and it even inspired some of the financiers. As Smit says, it was Tinker Bell – the fairy who only exists if you believe in her – who really built Eden.

Sources: Barrie (1995); Rawlinson (2006); Smit (2001); http://www.mero-tsk.de; http://www.wikipedia.com; http://www.edenproject.com/ (all accessed 06/11/08).

Notes

1 Cited in Sabbagh (2000, p. 177) and Sabbagh (1989, p. 63).

2 This concept is derived from discussion at an invited seminar sponsored by CIB and hosted by VTT in Helsinki in December 2001. The contribution of David Hall of BAA was particularly valuable here. It has been developed further in a special issue of *Building Research and Information* (**31**, 2, 2003) and contributions, among others, from Barrett (2007; 2008) and Winch (2008).

3 This discussion was developed from Pearce (2003).

4 Fukuyama (1995, p. 26).

5 Winch (2008).

6 See Briggs (1979); Peters (1996).

7 Woudhuysen and Abley (2004) provide a contemporary restatement of this position.

8 Gann (2000).

9 Womack *et al.* (1990); Womack and Jones (1996).

10 Winch (1994a).

11 http://www.leanconstruction.org/

12 Readers may be surprised to see lean manufacturing described as bureaucratic in the non-pejorative sense. Yet lean manufacturing has its roots deep in the bureaucratic approach to management of Taylor and Ford (see Littler 1982), and, with its reliance on stable gross output, is more an intensification of mass production than a replacement for it. Perhaps most tellingly, it is forthright in its rejection of such developments as autonomous group working (Womack *et al.*, 1990). Much of the debate on lean construction has ignored the alternative models of manufacturing that are available, particularly from the capital goods sector (Winch 1994a).

13 Winch (2003).

14 Ball (1983).

15 Gann (1996); Barlow *et al.* (2003).

16 www.cabe.org.uk/buildingforlife (accessed 12/04/07).

17 Winch (2003).

18 It is central to Cleland and King's (1968/1983) classic, Walker's theory (1996) and most recently Blockley and Godfrey's (2000) ways of doing it differently.

19 This is well captured in two very different books – Goldthwaite (1980) on the construction process in Florence in the fourteenth and fifteenth centuries and Jardine (1996) on the worldliness of the Renaissance merchant and his family. Alberti, the original Renaissance man, took considerable interest in architecture, as well as mathematics and science – see Gadol (1969).

20 The first two are well known; Aird constructed much of the water and gas infrastructure of the UK from the 1850s on, and the Millwall Docks described in panel 2.8. See Middlemas (1963).

21 The reference is to David Landes' 1969 book on the industrial revolution. *The Unbound Prometheus*. According to Greek mythology, Prometheus was taught the useful arts by Athene and passed them on to mankind, most notably the secret of fire. As punishment for this act of treachery, Zeus had him bound naked to a pillar in the Caucasian mountains. See also Thomas Hughes' *Rescuing Prometheus*.

22 See the selections of his translated works edited by Gerth and Mills (Weber 1948), and Parsons (Weber 1947).

23 Stinchcombe (1959).

24 This point has been made by many; see, especially, the contributions from Peter Morris (1994) and Irving Horwitch (1987). This is also the thrust of the critique of project management from the advocates of lean construction – see Koskela and Howell (2008).

25 See, for instance, the Rethinking Project Management initiative, *International Journal of Project Management* 2006 **24** (8).

26 This argument does *not* automatically lead to support for the professionalisation of project management – see, for instance, Hodgson and Cicmil (2007) for a critique of this professionalisation process.

27 Perkin (1989).

28 Vick (2002); Wachs (1990).

29 Derived from Simon (1976).

30. Winch and Maytorena (2009).

31. Schutz (1967) argues that they are a useful component in future-perfect thinking.

32. Ian Blake of Schal, cited in Sabbagh (2000, p. 153).

References

Abbott, A. (1988) *The System of Professions: An Essay on the Expert Division of Labour*. Chicago, University of Chicago Press.

Adams, J. (1995) *Risk*. London, UCL Press.

Adamson, D.M. and Pollington, T. (2006) *Change in the Construction Industry: An Account of the UK Construction Industry Reform Movement 1993–2003*. London, Routledge.

Adler, P.S. and Obstfeld, D. (2007) The Role of Affect in Creative Projects and Exploratory Search. *Industrial and Corporate Change* **16**: 19–50.

Akerlof, G.A. (1970) The Market for 'Lemons': Quality, Uncertainty and the Market Mechanism. *Quarterly Journal of Economics* **84**: 488–500.

Akintoye, A. and MacLeod, M. (1997) Risk Analysis and Management in Construction. *International Journal of Project Management* **15**: 31–38.

Allen, T.J. (1977) *Managing the Flow of Technology: Technology Transfer and the Dissemination of Technological Information within the R&D Organization*. Cambridge, MIT Press.

Ancona, D., Malone, T.W., Orlikowski, W.J., and Senge, P.M. (2007) In Praise of the Incomplete Leader. *Harvard Business Review* **February**, pp. 92–100.

Anguera, R. (2006) The Channel Tunnel – An Ex Post Economic Evaluation. *Transportation Research Part A* **40**: 291–315.

Ansoff, I.H. (1968) *Corporate Strategy*. Harmondsworth, Penguin.

Aouad, G., Wu, S., and Lee, A. (2008) nD Modelling, Present and Future. In: P. Brandon and T. Kocatürk (eds.) *Virtual Futures for Design, Construction and Procurement*. Oxford, Blackwell

Arbulu, R.J., Tommelein, I.D., Walsh, K.D., and Hershauer, J.C. (2003) Value Stream Analysis of a Re-Engineered Construction Supply Chain. *Building Research & Information* **31**: 161–172.

Armor, D.A. and Taylor, S.E. (2002) When Predictions Fail: The Dilemma of Unrealistic Optimism. In: T. Gilovich, D. Griffin, and D. Kahneman (eds.) *Heuristics and Biases: The Psychology of Intuitive Judgment*. Cambridge, Cambridge University Press.

Austin, S.A., Baldwin, A.N., Li, B., and Waskett, P. (1999) Analytical Design Planning Technique: A Model of the Detailed Building Design Process. *Design Studies* **20**: 279–296.

Axelrod, R. (1990) *The Evolution of Co-operation*. Harmondsworth, Penguin.

Ayas, K. and Zeniuk, N. (2001) Project-based Learning: Building Communities of Reflective Practitioners. *Management Learning* **32**: 61–76.

Bain, S. (2005) *Holyrood: The Inside Story*. Edinburgh, Edinburgh University Press.

Bajari, P. and Tadelis, S. (2001) Incentives versus Transaction Costs: A Theory of Procurement Contracts. *RAND Journal of Economics* **32**: 387–407.

Baldwin, A., Kong, C.W., Huang, T., Guo, H.L., Wong, K.D., and Li, H. (2008) Planning and Scheduling in a Virtual Prototyping Environment. In: P. Brandon and T. Kocatürk (eds.) *Virtual Futures for Design, Construction and Procurement*. Oxford, Blackwell.

Baldwin, A.N., Austin, S.A., Hassan, T.M., and Thorpe, A. (1999) Modelling Information Flow During the Conceptual and Schematic Stages of Building Design. *Construction Management and Economics* **17**: 155–167. Ball, M. (1983) *Housing Policy and Economic Power: The Political Economy of Owner Occupation*. London, Methuen.

Ball, M. (1988) *Rebuilding Construction: Economic Change in the British Construction Industry*. London, Routledge.

Ballard, G. and Howell, G. (1998) Shielding Production: Essential Step in Production Control. *Journal of Construction Engineering and Management* **124**: 11–17.

Ballard, G. and Howell, G. (2003) Lean Project Management. *Building Research and Innovation* **31**: 119–133.

Ballard, G., Tommelein, I., Koskela, L., and Howell, G. (2002) Lean Construction Tools and Techniques. In: R. Best and G. de Valence (eds.) *Design and Construction: Building in Value*. Oxford, Butterworth-Heinemann.

Barber, P., Tomkins, C., and Graves, A. (1999) Decentralised Site Management – A Case Study. *International Journal of Project Management* **17**: 113–120.

Barber, W.J. (1967) *A History of Economic Thought*. Penguin, Harmondsworth.

Barberis, P. (1998) The New Public Management and a New Accountability. *Public Administration* **76**: 451–470.

Barkow, J.H., Cosmides, L., and Tooby, J. (1992) *The Adapted Mind: Evolutionary Psychology and the Generation of Culture*. New York, Oxford University Press.

Barlow, J. (2000) Innovation and Learning in Complex Off-shore Construction Projects. *Research Policy* **29**: 973–989.

Barlow, J., Childerhouse, P., Gann, D., Hong-Minh, S., Naim, M., and Ozaki, R. (2003) Choice and Delivery in Housebuilding: Lessons from Japan for UK Housebuilders. *Building Research and Information* **31**: 134–145.

Barlow, J., Jashapara, M., Cohen, M., and Simpson, Y. (1997) *Towards Positive Partnering: Revealing the Realities in the Construction Industry*. Bristol, Policy Press.

Barnard, C.I. (1938/1968) *The Functions of the Executive*. Cambridge, Harvard University Press.

Barnes, M. (1988) Construction Project Management. *International Journal of Project Management* **6**: 69–79.

Barrett, P. (2007) Revaluing Construction: A Holistic Model. *Building Research and Information* **35**: 268–286.

Barrett, P. (2008) A Global Agenda for Revaluing Construction: The Client's Role. In: P. Brandon and S.-L. Lu (eds.). *Clients Driving Innovation*. Oxford, Wiley-Blackwell.

Barrett, P. and Stanley, C. (1999) *Better Construction Briefing*. Oxford, Blackwell Science.

Barrie, J.M. (1995) *Peter Pan*. London, Penguin Books.

Bazerman, M.H. (2006) *Judgment in Managerial Decision Making* (6th ed.). New York, John Wiley.

Bazin, M. (1993) *Analyse Stratégique en Science et Technologie*. Cahiers du CSTB Rapport 2643.

Becker, M.C. (2004) Organizational Routines: A Review of the Literature. *Industrial and Corporate Change* **13**: 643–667.

Belbin, R.M. (1981) *Management Teams: Why they Succeed or Fail*. Oxford, Heinemann Professional.

Bennett, J. and Jayes, S. (1998) *The Seven Pillars of Partnering; A Guide to Second Generation Partnering*. Reading, Centre for Strategic Studies in Construction.

Bennett, J., Pothecary, E., and Robinson, G. (1996) *Designing and Building a World Class Industry*. Reading, Centre for Strategic Studies in Construction.

Berger, P.L. and Luckmann, T. (1970) *The Social Construction of Reality*. Harmondsworth, Penguin.

Bernstein, P.L. (1996) *Against the Gods: The Remarkable Story of Risk*. New York, Wiley.

Biau, V. (1998) *Les Concours de Maîtrise d'Oeuvre dans L'Union Europeéenee*. Paris, Centre de Recherche sur L'Habitat.

Blockley, D. and Godfrey, G. (2000) *Doing it Differently; Systems for Rethinking Construction*. London, Thomas Telford.

Blyth, A. and Worthington, J. (2001) *Managing the Brief for Better Design*. London, Spon Press.

Boisot, M. (1995) *Information Space: A Framework for Learning in Organizations, Institutions and Culture*. London, Routledge.

Boland, R.J. and Collopy, F. (eds.) (2004) *Managing as Designing*. Stanford, Stanford Business Books.

Boland, R.J., Lyytinen, K., and Yoo, Y. (2007) Wakes of Innovation in Project Networks: The Case of Digital 3-D Representations in Architecture, Engineering and Construction. *Organization Science* **18**: 631–647.

Bonke, S. (1996) *Technology Management on Large Construction Projects*. London, Le Groupe Bagnolet Working Paper 4.

Bonke, S. (1998) *The Storebælt Fixed Link: The Fixing of Multiplicity*. London, Le Groupe Bagnolet Working Paper 14.

Bourdieu, P. (1989) *La Noblesse d'Etat: Grandes Ecoles et Esprit de Corps*. Paris, Editions de Minuit.

Boussabaine, A. and Kirkham, R. (2004) *Whole Life-cycle Costing: Risk and Responses*. Oxford, Blackwell.

Bowley, M. (1966) *The British Building Industry: Four Studies in Response and Resistance to Change*. London, Cambridge University Press.

Brand, S. (1994) *How Buildings Learn: What Happens After They're Built*. New York, Viking.
Brandon, P., and Kocatürk, T. (eds.) (2008) *Virtual Futures for Design, Construction and Procurement*. Oxford, Blackwell.
Brandon, P., Li, H., and Shen, Q. (2005) Construction IT and the "Tipping Point". *Automation in Construction* **14**: 281–296.
Braverman, H. (1974) *Labor and Monopoly Capital: The Degradation of Work in the Twentieth Century*. New York, Monthly Review Press.
Bremer, W. and Kok, K. (2000) The Dutch Construction Industry: A Combination of Competition and Corporatism. *Building Research and Information* **28**: 98–108.
Bresnen, M. (2007) Deconstructing Partnering in Project-based organizations: Seven pillars, Seven Paradoxes, and Seven Deadly Sins. *International Journal of Project Management* **25**: 365–374.
Bresnen, M., Goussevskaia, A., and Swan, J. (2004) Embedding New Management Knowledge in Project-based Organizations. *Organization Studies* **25**: 1535–1555.
Bresnen, M., Goussevskaia, A., and Swan, J. (2005) Organizational Routines, Situated Learning and Processes of Change in Project-based Organizations. *Project Management Journal* **36**: 27–41.
Bresnen, M. and Marshall, N. (2000a) Partnering in Construction: A Critical Review of the Issues, Problems and Dilemmas. *Construction Management and Economics* **18**: 227–237.
Bresnen, M. and Marshall, N. (2000b) Motivation, Commitment and the use of Incentives in Partnerships and Alliances. *Construction Management and Economics* **18**: 587–598.
Briggs, A. (1979) *Iron Bridge to Crystal Palace: Impact and Images of the Industrial Revolution*. London, Thames and Hudson.
Bright, J.R. (1958) *Automation and Management*. Boston, Harvard University Press.
Brooks, F.P. (1995) *The Mythical Man-month: Essays in Software Engineering* (2nd ed.). Reading, Addison-Wesley.
Bryman, A., Bresnen, M., Beardsworth, A.D., Ford, J., and Keil, E.T. (1987) The Concept of the Temporary System: The Case of the Construction Project. *Research in the Sociology of Organizations* **5**: 253–283.
Buchanan, R. (1995) Wicked Problems in Design Thinking. In: V. Margolin and R. Buchanan (eds.) *The Idea of Design*. Cambridge, MIT Press.
Buehler, R., Griffin, D., and Ross, M. (2002) Inside the Planning Fallacy. In: T. Gilovich, D. Griffin, and D. Kahneman (eds.) *Heuristics and Biases: The Psychology of Intuitive Judgment*. Cambridge, Cambridge University Press.
Buehler, R., Messervey, D., and Griffin, D. (2005) Collaborative Planning and Prediction: Does Group Discussion Affect Optimistic Biases in Time Estimation? *Organizational Behavior and Human Decision Processes* **97**: 47–63.
Building EDC. (1987) *Achieving Quality on Building Sites*. London, NEDO.
Burckhardt, J. (1990) *The Civilization of the Renaissance in Italy*. London, Penguin.
Burningham, D. (1995) Environmental Impact Analysis. In: J.R. Turner (ed.) *The Commercial Project Manager*. London, McGraw-Hill.
Burrell, G. (1988) Modernism, Post Modernism, and Organizational Analysis 2: The Contribution of Michel Foucault. *Organization Studies* **9**: 221–235.

Bushouse, B. (2002) Changes in Mitigation: Comparing Boston's Big Dig and 1950s Urban Renewal. *Public Works Management and Policy* **7**: 52–62.
Cacciatori, E. and Jacobides, M.G. (2005) The Dynamic Limits of Specialization: Vertical Integration Reconsidered. *Organization Studies* **26**: 1851–1883.
Cain, C.T. (2003) *Building Down Barriers: A Guide to Construction Best Practice*. New York, Taylor and Francis.
Cameron, K. and Lavine, M. (2006) *Making the Impossible Possible: Leading Extraordinary Performance – The Rocky Flats Story*. San Francisco, Berrett-Koehler.
Cammock, P. (1987) Industrial Relations at Motonui – An Examination of a Successful Industrial Relations Strategy on a Large Construction Project. *New Zealand Journal of Industrial Relations* **12**: 71–80.
Campagnac, E. (1996) *La Maîtrise du Risque entre Différences et Cooperation: le Cas du Severn Bridge*. London, le Groupe Bagnolet, Working Paper 12.
Campagnac, E. and Winch, G.M. (1997) The Social Regulation of Technical Expertise: The Corps and the Profession in France and Great Britain. In: R. Whitley and P.H. Kristensen (eds.) *Governance and Work: The Social Regulation of Economic Relations in Europe*. Oxford, Oxford University Press.
Campinos-Dubernet, M. and Grando, J.-M. (1988) Formation Professionnelle Ouvrière: Trois Modèles Européennes. *Formation Emploi* **22**: 5–29.
Carlile, P.R. (2002) A Pragmatic View of Knowledge and Boundaries: Boundary Objects in New Product Development. *Organization Science* **13**: 442–455.
Carlson, S. (2001) Dazzling Designs, At a Price. *The Chronicle of Higher Education* 26/01/01.
Carmona, M. (2004) Adding value through better urban design. In: S. Macmillan (ed.) *Designing Better Buildings*. London, Spon.
Carr, B. and Winch, G.M. (1999) *Mapping the Construction Process in Britain and France*. London, Bartlett, Research Paper 8.
Chandler, A.D. (1977) *The Visible Hand: The Managerial Revolution in American Business*. Cambridge, The Belknap Press.
Chang, C.-Y. and Ive, G. (2007a) The Hold-up Problem in the Management of Construction Projects: A Case Study of the Channel Tunnel. *International Journal of Project Management* **25**: 394–404.
Chang, C.-Y. and Ive, G. (2007b), Reversal of Bargaining Power in Construction Projects: Meaning, Existence and Implications. *Construction Management and Economics* **25**: 845–856.
Chapman, C. and Ward, S. (2003) *Project Risk Management: Processes, Techniques and Insights* (2nd ed.). Chichester, John Wiley.
Chaslin, F. (1985) *Les Paris de François Mitterand*. Paris, Gallimard.
Cherns, A.B. and Bryant, D.T. (1984) Studying the Client's Role in Construction Management. *Construction Management and Economics* **2**: 177–184.
CIB Working Group 3. (1997) *Constructing Success: Code of Practice for Clients of the Construction Industry*. London, Thomas Telford.
CIB Working Group 12. (1997) *Partnering in the Team*. London, Thomas Telford.
CIRIA. (1999) *Standardisation and Pre-assembly: Adding Value to Construction Projects*. London, CIRIA.

Clark, I. and Ball, D. (1991) Transaction Cost Economics Applied? Consortia within Process Plant Contracting. *International Review of Applied Economics* **5**: 341–357.

Clark, K. and Fujimoto, T. (1991) *Product Development Performance: Strategy Organization and Management in the World Auto Industry*. Boston, Harvard Business School Press.

Cleland, D.I. (1998) Stakeholder Management. In: J. Pinto (ed.) *Project Management Handbook*. San Francisco, Jossey-Bass.

Cleland, D.I. and King, W.R. (1968/1983) *Systems Analysis and Project Management* (3rd ed.). New York, McGraw-Hill.

Clegg, S.R., Pitsis, T.S. Marosszeky, M., and Rura-Polley, T. (2006) Making the Future Perfect: Constructing the Olympic Dream. In: D. Hodgson and S. Cicmil (eds.) *Making Projects Critical*. Basingstoke, Palgrave Macmillan.

Clegg, S.R., Pitsis, T.S., Rura-Polley, T., and Marosszeky, M. (2002) Governmentality Matters: Designing an Alliance Culture of Inter-Organizational Collaboration for Managing Projects. *Organization Studies* **23**: 317–337.

Cohen, R., Standeven, M., Bordass, B., and Leaman, A. (2001) Assessing Building Performance in Use 1: The Probe Process. *Building Research and Information* **29**: 85–102.

Committee for Review of the Project Management Practices Employed on the Boston Central Artery/Tunnel ("Big Dig") Project. (2003) *Completing The "Big Dig": Managing The Final Stages Of Boston's Central Artery/Tunnel Project*. Washington, National Academies Press.

Conlon, D.E. and Garland, H. (1993) The Role of Project Completion Information in Resource Allocation Decisions. *Academy of Management Journal* **36**: 402–413.

Construction Task Force. (1998) *Rethinking Construction*. London, Department of the Environment, Transport and the Regions.

Cooper, D., Grey, S., Raymond, G., and Wlaker, P. (2005) *Managing Risk in Large Projects and Complex Procurements*. Chichester, Wiley.

Cooper, K.G. (1994) The $2000 Hour: How Managers Influence Project Performance through the Rework Cycle. *Project Management Journal* **25**: 11–24.

Cooper, R.G. (1993) *Winning at New Products: Accelerating the Process from Idea to Launch* (2nd ed.). Reading, Perseus Books.

Cousin, V. (1998) Innovation Awards: A Case Study. *Building Research and Information* **26**: 302–310.

Cox, A. and Townsend, M. (1998) *Strategic Procurement in Construction*. London, Thomas Telford.

Crosby, P.B. (1979) *Quality is Free*. New York, McGraw-Hill.

Cross, N. (1984) *Developments in Design Methodology*. New York, Wiley.

Cuff, D. (1991) *Architecture: The Story of Practice*. Cambridge, MIT Press.

Curtis, B., Ward, S., and Chapman, C. (1989) *Roles, Responsibilities and Risks in Management Contracting*. London, CIRIA.

Cusumano, M.A. and Nobeoka, K. (1998) *Thinking Beyond Lean: How Multi-Project Management is Transforming Product Development at Toyota and Other Companies*. New York, Free Press.

Cyert, R.M. and March, J.G. (1992) *A Behavioral Theory of the Firm* (2nd ed.). Oxford, Blackwell.

Dale, B.G. (1994) *Managing Quality* (2nd ed.). London, Prentice-Hall.

Dallas, M.F. (2006) *Value and Risk Management: A Guide to Best Practice*. Oxford, Blackwell.

Dalton, J. and Kenward, K. (1975) Economics of Dimensional Accuracy. In: D. Turin (ed.) *Aspects of the Economics of Construction*. London, Godwin.

Dalton, M. (2007) *A Risk Breakdown Structure for Public Sector Projects*. PhD Thesis, University of Manchester.

Davenport, T.H. (1993) *Process Innovation: Reengineering Work Through Information Technology*. Boston, Harvard Business School Press.

Davenport, T.H. (2000) *Mission Critical: Realizing the Promise of Enterprise Systems*. Boston, Harvard Business School Press.

De Meyer, A., Loch, C.H., and Pich, M.T. (2002) Managing Project Uncertainty: From Variation to Chaos. *MIT Sloan Management Review* **43**: 60–67.

Delgado-Hernandez, D.J., Bampton, C.E., and Aspinwall, E. (2007) Quality Function Deployment in Construction. *Construction Management and Economics* **25**: 597–609.

Department for Transport. (2004) *Procedures for Dealing with Optimism Bias in Transport Planning*. London, Department for Transport.

Dewey, J. (2002) *Human Nature and Conduct: An Introduction to Social Psychology*. Amherst, Prometheus Books.

Dickson, M. (2004) Achieving quality in building design by intention. In: S. Macmillan (ed.) *Designing Better Buildings*. London, Spon.

Diebold, J. (1955) *Automation and Technological Change*. Washington, Government Printing Office.

DiMaggio, P.J. and Powell, W.W. (1983) The Iron Cage Revisited: Institutional Isomorphism and Collective Rationality in Organizational Fields. *American Sociological Review* **48**: 147–160.

Dixit, A.K. and Pindyck, R.S. (1995) The Options Approach to Capital Investment. *Harvard Business Review* **May–June**, pp. 105–115.

Dodgson, M., Gann, D.M., and Salter, A. (2005) *Think, Play, Do: Technology, Innovation and Organization*. Oxford, Oxford University Press.

Dodgson, M., Gann, D.M., and Salter, A. (2007) "In Case of Fire, Please Use the Elevator": Simulation Technology and Organization in Fire Engineering. *Organization Science* **18**: 849–864.

Doherty, S. (2008) *Heathrow's Terminal 5: History in the Making*. Chichester, Wiley.

Dorée, A.G. (2004) Collusion in the Dutch Construction Industry: An Industrial Organization Perspective. *Building Research and Innovation* **32**: 146–156.

Drogemuller, R. (2008) Virtual Protyping from need to Pre-construction. In: P. Brandon and T. Kocatürk (eds.) *Virtual Futures for Design, Construction and Procurement*. Oxford, Blackwell.

Druker, J. and White, G. (1996) *Managing People in Construction*. London, Institute of Personnel and Development.

Dyer, J.H., Cho, D.S., and Chu, W. (1998) Strategic Supplier Segmentation: The Next 'Best Practice' in Supply Chain Management. *California Management Review* **40**: 57–77.

Eastman, C., Teicholz, P., Sacks, R., and Liston, K. (2008) *BIM Handbook: A Guide to Building Information Modeling*. Hoboken, John Wiley.

Eastman, C.M. (1999) *Building Product Models: Computer Environments Supporting Design and Construction*. Boca Raton, CRC Press.

Eccles, R.G. (1981a) Bureaucratic versus Craft Administration: The Relationship of Market Structure to the Construction Firm. *Administrative Science Quarterly* **26**: 449–469.

Eccles, R.G. (1981b) The Quasi-Firm in the Construction Industry. *Journal of Economic Behavior and Organization* **2**: 335–357.

Eden, C., Williams, T., Ackermann, F., and Howick, S. (2000) The Role of Feedback Dynamics in Disruption and Delay on the Nature of Disruption and Delay (D&D) in Major Projects. *Journal of the Operational Research Society* **51**: 291–300.

Edkins, A.J. (1998) *Managing the Design Process in Construction: A Cognitive Approach*. PhD Thesis, University of London.

Edkins, A.J. and Winch, G.M. (1999) *Project Performance in Britain and France: The Case of Euroscan*. London, Bartlett Research Paper 7.

Eggleston, B. (2006) *The NEC3 Engineering and Construction Contract; A Commentary*. Oxford, Blackwell Science.

Ekins, R. (2006) Delivery of Works – Safely. *Proceedings: Railway Bridges Today and Tomorrow*. London, Network Rail.

Eliot, T.S. (1974) *Collected Poems 1909–1962*. London, Faber & Faber.

Emery, A. (1989) Kirby Muxloe Castle. *Archaeological Journal* **146** (Suppl.): 71–77.

Engwall, M. (2003) No Project is an Island: Linking Projects to History and Context. *Research Policy* **32**: 789–808.

Evans, R., Haryott, R., Haste, N., and Jones, A. (1998) *The Long Term Costs of Owning and Using Buildings*. London, Royal Academy of Engineering.

Ewenstein. B. and Whyte, J.K. (2007) Beyond Words: Aesthetic Knowledge and Knowing in Organizations. *Organization Studies* **28**: 689–708.

Fayol, H. (1999) *Administration Industrielle et Générale*. Paris, Dunod.

Federal Task Force on the Boston Central Artery Tunnel Project. (2000) *Review of Project Oversight & Costs*. Washington, Federal Highways Administration.

Feldman, M.S. and Pentland, B.T. (2003) Reconceptualizing Organizational Routines as a Source of Flexibility and Change. *Administrative Science Quarterly* **48**: 94–118.

Fellows, R., Langford, D., Newcombe, R., and Urry, S. (1983) *Construction Management in Practice*. London, Construction Press.

Fetherston, D. (1997) *Chunnel: The Amazing Story of the Undersea Crossing of the English Channel*. New York, Times Books.

Fichman, R.G., Keil, M., and Tiwana, A. (2005) Beyond Valuation: "Options Thinking" in IT Project Management. *California Management Review* **47**: 74–96.

Field, C., Gamble, M., and Karakashian, M. (2000) Design and Construction of London Bridge Station on the Jubilee Line Extension. *Civil Engineering* **138**: 26–39.

Fine, B. (1975) Tendering Strategy. In: D. Turin (ed.) *Aspects of the Economics of Construction*. London, Godwin.

Fischer, M.A. (2008) Reshaping the Life Cycle Process with Virtual Design and Construction Methods. In: P. Brandon and T. Kocatürk (eds.) *Virtual Futures for Design, Construction and Procurement*. Oxford, Blackwell.

Fischoff, B. (1982) Debiasing. In: D. Kahneman, P. Slovic, and A. Tversky (eds.) *Judgement under Uncertainty: Heuristics and Biases*. Cambridge, Cambridge University Press.

Fischoff, B. (2002) Heuristics and Biases in Application. In: T. Gilovich, D. Griffin, and D. Kahneman (eds.) *Heuristics and Biases: The Psychology of Intuitive Judgment*. Cambridge, Cambridge University Press.

Fisher, N, and Morledge, R. (2002) Supply Chain Management. In: J. Kelly, R. Morledge, and S. Wilkinson (eds.) *Best Value in Construction*. Oxford, Blackwell.

Flanders, A. (1970) *Management and Unions: The Theory and Reform of Industrial Relations*. London, Faber and Faber.

Fleming, I. (2004) *Goldfinger*. London, Penguin.

Fleming, Q.W. and Koppelman, J.M. (2000) *Earned Value Project Management* (2nd ed.). Newtown Square, Project Management Institute.

Flyvbjerg, B. (2005) Design by Deception: The Politics of Megaproject Approval. *Harvard Design Magazine* **Spring/Summer**, pp. 50–59.

Flyvbjerg, B. (2006) From the Nobel Prize to Project Management: Getting Risks Right. *Project Management Journal* **37**: 5–15.

Flyvbjerg, B., Bruzelius, N., and Rothengatter, W. (2003) *Megaprojects and Risk : An Anatomy of Ambition*. Cambridge, Cambridge University Press.

Flyvbjerg, B., Holm, M.S.K., and Buhl, S. (2002) Underestimating Costs in Public Works Projects: Error or Lie? *Journal of the American Planning Association* **68**: 279–295.

Ford, D.N., Lander, D.M., and Voyer, J.J. (2002) A Real Options Approach to Valuing Strategic Flexibility in Uncertain Construction Projects. *Construction Management and Economics* **20**: 343–351.

Fortune, C. and Weight, D. (2002) Building Project Price Forecasting. In: J. Kelly, R. Morledge, and S. Wilkinson, S. (eds.) *Best Value in Construction*. Oxford, Blackwell.

Fraser, R. (1969) *Work 2; Twenty Personal Accounts*. Harmondsworth, Penguin.

French, J.R.P. and Raven, B.H. (1960) The Bases of Social Power. In: D. Cartwright and A. Zander (eds.) *Group Dynamics: Theory and Research* (2nd ed.). New York, Row Peterson, pp. 607–623.

Frenkel, S. and Martin, G. (1986) Managing Labour on a Large Construction Site. *Industrial Relations Journal* **16**: 141–157.

Fukuyama, F. (1995) *Trust: The Social Virtues and the Creation of Prosperity*. London, Hamish Hamilton.

Gaddis, P. (1959) The Project Manager. *Harvard Business Review* **May–June**, pp. 89–97.

Gadekin, O.C. (2002) What the Defense Systems Management College has Learned from Ten Years of Project Leadership Research. In: D.P. Slevin, D.I. Cleland, and J.K. Pinto (eds.) *The Frontiers of Project Management Research*. Newtown Square, PA, Project Management Institute.

Gadol, J. (1969) *Leon Battista Alberti: Universal Man of the Renaissance*. Chicago, University of Chicago Press.

Galbraith, J.R. (1977) *Organization Design*. Reading, Addison-Wesley.

Gann, D. (1996) Construction as a Manufacturing Process? Similarities and Differences between Industrialised Housing and Car Production in Japan. *Construction Management and Economics* **14**: 437–450.

Gann, D. (2000) *Building Innovation: Complex Constructs in a Changing World*. London, Thomas Telford.

Gann, D., Salter, A.J., and Whyte, J.K. (2003) The Design Quality Indicator as a Tool for Thinking. *Building Research and Information* **31**: 318–333.

Garvin, D.A. (1991) Building a Learning Organization. *Harvard Business Review* **July–August**, pp. 78–91.

Gehry, F.O. (2004) Reflections on Designing and Architectural Practice. In: R.J. Boland and F. Collopy (eds.) *Managing as Designing*. Stanford, Stanford Business Books.

George, M.L., Rowlands, D., Price, M., and Maxey, J. (2005) *Lean Six Sigma Pocket Toolbook*. New York, McGraw-Hill.

Giard, V. and Midler, C. (1993) *Pilotages de Projet et Entreprises: Diversités et Convergences*. Paris, Economica.

Gibb, A.G.F. (2001) Standardization and Pre-assembly: Distinguishing Myth from Reality Using Case Study Research. *Construction Management and Economics* **19**: 307–315.

Gibb, A.G.F. and Isack, F. (2003) Re-engineering through Pre-assembly: Client Expectations and Drivers. *Building Research and Information* **31**: 146–160.

Giddens, A. (1984) *The Constitution of Society*. Cambridge, Polity Press.

Giertz, L.M. (1995) Integrated Construction Information Efforts since 1945. In: P. Brandon and M. Betts (eds.) *Integrated Construction Information*. London, E & FN Spon, pp. 101–116.

Gifford, R. Hine, D.W. Muller-Clem, W., and Shaw, K.T. (2002) Why Architects and Laypersons Judge Buildings Differently: Cognitive Properties and Physical Bases. *Journal of Architectural and Planning Research* **19**: 131–148.

Gigerenzer, G. (2002) *Reckoning with Risk: Learning to Live with Uncertainty*. London, Penguin.

Gil, N. (2007) On the Value of Project Safeguards: Embedding Real Options in Complex Products and Systems. *Research Policy* **36**: 980–999.

Gil, N. (2009) Developing Cooperative Project Client-Supplier Relationships: How Much to Expect from Relational Contracts? *California Management Review* **51**: 144–169.

Gleick, J. (1988) *Chaos: Making a New Science*. London, Cardinal.

Goffee, R. and Scase, R. (1982) Fraternalism and Paternalism as Employer Strategies in Small Firms. In: G. Day (ed.) *Diversity and Decomposition in the Labour Market*. Aldershot, Gower.

Goldratt, E.M. (1997) *Critical Chain*. Great Barrington, The North River Press.

Goldratt, E.M. and Cox, J. (1993) *The Goal* (2nd ed.). Aldershot, Gower.

Goldthwaite, R.A. (1980) *The Building of Renaissance Florence: An Economic and Social History*. Baltimore, John Hopkins University Press.

Goleman, D. (1998) What Makes a Leader? *Harvard Business Review* **November–December**, pp. 93–102.

Goleman, D. (2000) Leadership that Gets Results. *Harvard Business Review* **March–April**, pp. 78–90.

Goodpasture, J.C. (2004) *Quantitative Methods in Project Management.* Boca Raton, J. Ross Publishing.

Gray, C. (1996) *Value for Money: Helping the UK Afford the Buildings it Likes.* Reading, Reading Construction Forum.

Gray, C., Hughes, W., and Bennett, J. (1994) *The Successful Management of Design.* Reading, Centre for Strategic Studies in Construction.

Green, S.D. (1992) *A SMART Methodology for Value Management.* Ascot, CIOB Occasional Paper 52.

Green, S.D. (2006) The Management of Projects in the Construction Industry: Context, Discourse and Self-identity. In: D. Hodgson and S. Cicmil (eds.) *Making Projects Critical.* Basingstoke, Palgrave Macmillan.

Guildford, J.P. (1959) The Three Faces of Intellect. *American Psychologist* **14**: 469–479.

Guillery, P. (1990) Building the Millwall Docks. *Construction History* **6**: 3–21.

Hale, J. (1993) *The Civilization of Europe in the Renaissance.* London, Harper Collins.

Hall, P. (1982) *Great Planning Disasters.* Berkeley, CA, University of California Press.

Handy, C. (1993) *Understanding Organizations* (4th ed.). Harmondsworth, Penguin.

Hansen, M.T., Nohria, N., and Tierney, T. (1999) What's Your Strategy for Managing Knowledge? *Harvard Business Review* **March–April**.

Harbison, F. and Myers, C.A. (1959) *Management in the Industrial World.* New York, McGraw-Hill.

Harvey, J. (1972) *The Mediaeval Architect.* London, Wayland.

Hauser, J. and Katz, G. (1998) Metrics: You are What You Measure. *European Management Journal* **16**: 517–528.

Haymaker, J. and Fischer, M. (2001) *Challenges and Benefits of 4D Modeling on the Walt Disney Concert Hall Project.* CIFE Working Paper 64, Stanford University.

Health and Safety Executive. (1991) *Workplace Health and Safety in Europe.* London, HMSO.

Helm, J. and Remington, K. (2005) Executive Project Sponsorship: An Evaluation of the Role of the Executive Sponsor in Complex Infrastructure Projects by Senior Managers. *Project Management Journal* **36**: 51–61.

Henderson, N. (1987) *Channels and Tunnels.* London, Weidenfield and Nicholson.

Herbert, G. (1984) *The Dream of the Factory-made House: Walter Gropius and Konrad Wachsmann.* Cambridge, MIT Press.

Herrigel, G. (1996) *Industrial Constructions: The Sources of German Industrial Power.* Cambridge, Cambridge University Press.

Hillier, B. (1996) *Space is the Machine.* Cambridge, Cambridge University Press.

Hillier, B. and Hanson, J. (1984) *The Social Logic of Space.* Cambridge, Cambridge University Press.

Hillier, B. and Shu, S.C.F. (2000) Crime and Urban Layout: The Need for Evidence. In: S. Ballintyre, K. Pense, and V. McLaren (eds.) *Secure Foundations: Key issues in Crime Prevention, Crime Reduction, and Community Safety.* London, Institute for Public Policy Research.

Hillson, D. (2004) *Effective Opportunity Management for Projects: Exploiting Positive Risk*. New York, Marcel-Dekker.

HM Government. (2000) *Better Public Buildings: A Proud Legacy for the Future*. London, Department for Culture, Media and Sport.

HM Government. (2006) *Better Public Building*. London, Department for Culture, Media and Sport and Commission for Architecture and the Built Environment.

HM Government. (2008) *Strategy for Sustainable Construction*. London, Department for Business Enterprise and Regulatory Reform and Strategic Forum.

HM Treasury. (2002) *Review of Large Public Procurement in the UK*. London, HM Treasury.

HM Treasury. (2003) *The Green Book: Appraisal and Evaluation in Central Government*. London, HM Treasury.

HM Treasury. (2004a) *Supplementary Green Book Guidance: Optimism Bias*. London, HM Treasury.

HM Treasury. (2004b) *The Orange Book: Management of Risk – Principles and Concepts*. London, HM Treasury.

Hoballah, K. (1998) *The Redevelopment of Beirut Central District Through Public/Private Partnership: A Study of SOLIDERE*. MSc Construction Economics and Management Report, University College London.

Hobday, M. (1998) Product Complexity, Innovation and Industrial Organisation. *Research Policy* **26**: 689–710.

Hobhouse, H. (1971) *Thomas Cubitt: Master Builder*. London, Macmillan.

Hodgson, D. and Cicmil, S. (2007) The Politics of Standards in Modern Management: Making "The Project" a Reality. *Journal of Management Studies* **44**: 431–450.

Hoffman, E.J., Kinlaw, C.S., and Kinlaw, D.C. (2002) Developing Superior Project Teams: A Study of the Characteristics of High Performance in Project Teams. In: D.P. Slevin, D.I. Cleland, and J.K. Pinto (eds.) *The Frontiers of Project Management Research*. Newtown Square, PA, Project Management Institute.

Hofstede, G. (2001) *Culture's Consequences: Comparing Values, Behaviours, Institutions, and Organizations Across Nations* (2nd ed.). London, Sage.

Holmes, J., Capper, G., and Hudson, G. (2006) LIFT: 21st Century Health Care Centres in the United Kingdom. *Journal of Facilities Management* **4**: 99–109.

Holti, R., Nicolini, D., and Smalley, M. (2000) *The Handbook of Supply Chain Management: The Essentials*. London, CIRIA.

Horner, R.M.W. and Talhouni, B.T. (1995) *Effects of Accelerated Working, Delays & Disruption on Labour Productivity*. Ascot, CIOB.

Horwitch, M. (1987) Grands Programmes: l'Expérience Américaine. *Révue Française de Gestion* **Mars–Avril–Mai**, pp. 54–69.

Hounshell, D.A. (1984) *From the American System to Mass Production, 1800–1932*. Baltimore, John Hopkins University Press.

Howard, R. (1998) *Computing in Construction: Pioneers and the Future*. Oxford, Butterworth-Heinemann.

Howard, R. (2001) *Byggeklassiikation: International Practice and Experience*. Taastrup, Teknologisk Institut.

Howell, S., Stark, A., Newton, D., Paxson, D., Cavus, M., Pereira, J., and Patel, K. (2001) *Real Options: Evaluating Corporate Investment Opportunities in a Dynamic World.* London, Prentice-Hall.

Huang, T., Kong, C.W., Guo, H.L., Baldwin, A., and Li, H. (2007) A Virtual Prototype System for Simulating Construction Processes. *Automation in Construction* **16**: 576–585.

Hughes, T.P. (1983) *Networks of Power; Electrification in Western Society 1880–1930.* Baltimore, Johns Hopkins University Press.

Hughes, T.P. (1998) *Rescuing Prometheus: Four Monumental Projects that Changed the Modern World.* New York, Vintage Books.

Husted, B.W. and Folger, R. (2004). Fairness and Transaction Costs: The Contribution of Organizational Justice Theory to an Integrative Model of Economic Organization. *Organization Science* **15**: 719–729.

Hunt, V.D. (1996) *Process Mapping: How to Reengineer your Business Processes.* New York, Wiley.

Ishikawa, K. (1985) *What is Total Quality Control? The Japanese Way.* Englewood-Cliffs, Prentice-Hall.

Ishikawa, K. (1990) *Introduction to Quality Control.* Tokyo, 3A Corporation.

Ive, G. (2006) Re-examining the cost and value ratios of owning and occupying buildings. *Building Research and Information* **34**, 230–245.

Ive, G. and Gruneberg, S. (2000) *The Economics of the Modern Construction Sector.* Basingstoke, Macmillan.

Ivory, C., Alderman, N., McLouglin, I., and Vaughan, R. (2006) Sense-Making as a Process within Complex Projects. In D. Hodgson and S. Cicmil (eds.) *Making Projects Critical* Basingstoke, Palgrave Macmillan.

Janis, I.L. (1972) *Victims of Groupthink.* Boston, Houghton Mifflin.

Jardine, L. (1996) *Worldly Goods: A New History of the Renaissance.* London, Macmillan.

Johnson, G. and Scholes, K. (2002) *Exploring Corporate Strategy* (6th ed.). Harlow, FT Prentice Hall.

Kahneman, D. and Lovallo, D. (1993) Timid Choices and Bold Forecasts: A Cognitive Perspective on Risk Taking. *Management Science* **39**: 17–31.

Kaluarachchi, Y.D. and Jones, K. (2007) Monitoring of a Strategic Partnering Process: The Amphion Experience. *Construction Management and Economics* **25**: 1053–1061.

Kang, L.S. and Paulson, B.C. (2000) Information Classification for Civil Engineering Projects by Uniclass. *Journal of Construction Engineering and Management* **126**: 158–167.

Kaplan, R.S. and Norton, D.P. (1996) *The Balanced Scorecard.* Boston, Harvard Business School Press.

Katz, D. and Kahn, R.L. (1978) *The Social Psychology of Organizations* (2nd ed.). New York, John Wiley.

Katzenbach, J.R. and Smith, D.K. (1993) *The Wisdom of Teams: Creating the High Performance Organization.* Boston, Harvard Business School Press.

Keegan, A. and Turner, J.R. (2001) Quantity versus Quality in Project-based Learning Practices. *Management Learning* **32**: 77–88.

Keeney, R.L. and von Winterfeldt, D. (1991) Eliciting Probabilities from Experts in Complex Technical Problems. *IEEE Transactions on Engineering Management* **38**: 191–201.

Kelly, J., Male, S., and Graham, D. (2004) *Value Management of Construction Projects*. Oxford, Blackwell.

Kerr, S. (1975) On the Folly of Rewarding A, While Hoping for B. *Academy of Management Journal* **18**: 769–783.

Keynes, J.M. (1937) The General Theory of Employment. *The Quarterly Journal of Economics* **51**: 209–223.

Keynes, J.M. (1973) *A Treatise on Probability: The Collected Writings of John Maynard Keynes VIII*. London, Macmillan.

Kibert, C.J. (2005) *Sustainable Construction: Green Building Design and Delivery*. Hoboken, John Wiley.

Kirkham. R (2007) *Cost Planning of Buildings* (8th ed.). Oxford, Blackwell.

Knight, F.H. (2002) *Risk, Uncertainty and Profit*. Washington, Beard Books.

Knott, T. (1996) *No Business as Usual: An Extraordinary North Sea Result*. London, British Petroleum Company.

Koskela, L. and Ballard, G. (2006) Should Project Management be Based on Theories of Economics or Production? *Building Research and Economics* **34** (2): 154–163.

Koskela, L. and Howell, G. (2008) The Underlying Theory of Project Management is Obsolete. *IEEE Engineering Management Review* **36**: 22–34.

Kristensen, P.H. (1996) On the Constitution of Economic Actors in Denmark: Interacting Skill Containers and Project Coordinators. In: R. Whitley and P.H. Kristensen (eds.) *The Changing European Firm: Limits to Convergence*. London, Routledge, pp. 118–158.

Lahdenperä, P. (1995) *Reorganizing the Building Process: A Holistic Approach*. Espoo, VTT Publications 258.

Landes, D. (1969) *The Unbound Prometheus: Technological Change and Industrial Development in Western Europe from 1750 to the Present*. London, Cambridge University Press.

Landes, D. (1983) *Revolution in Time: Clocks and the Making of the Modern World*. Cambridge, Belknap Press.

Latham, M. (1994) *Constructing the Team*. London, The Stationery Office.

Layzell, J. and Ledbetter, S. (1998) FMEA Applied to Cladding Systems – Reducing the Risk of Failure. *Building Research and Information* **26**: 351–357.

Lee, S.-H. and Peña-Mora, F. (2007) Understanding and Managing Iterative Error and Change Cycles in Construction. *Systems Dynamics Review* **23**: 35–60.

Levy, P.F. (2001) The Nut Island Effect: When Good Teams Go Wrong. *Harvard Business Review* **March**, pp. 51–59.

Li, H., Huang, T., Kong, C.W., Guo, H.L., Baldwin, A., Chan, N., and Wong, J. (2008) Integrating Design and Construction through Virtual Prototyping. *Automation in Construction* **17**: 915–922.

Lichtenberg, S. (2000) *Proactive Management of Uncertainty Using the Successive Principle*. Copenhagen, Polyteknisk Press.

Lichtenstein, S., Fischoff, B., and Phillips, L.D. (1982) Calibration of Probabilities: The State of the Art to 1980. In: Kahneman *et al.* (eds.). Kahneman, D., Slovic, P., and Tversky, A. Judgement under Uncertainty: Heuristics and Biases. Cambridge, Cambridge University Press.

Littler, C.R. (1982) *The Development of the Labour Process in Capitalist Societies*. London, Heinemann.

Locke, D. (1996) *Project Management* (6th ed.). Aldershot, Gower.

Lockyer, K.G. (1963) *An Introduction to Critical Path Analysis*. Course Notes, Department of Management Studies, The Polytechnic.

Lookman, A.A. (1994) *Project Management of the Sheffield Arena: A Case Study*. MSc Architecture Report, University College London.

Lowe, D. and Leiringer, R. (eds.) (2006) *Commercial Management of Complex Projects: Defining the Discipline*. Oxford, Blackwell.

Lowe, D. and Skitmore, M. (2006) Bidding: In: D. Lowe and R. Leiringer (eds.) *Commercial Management of Complex Projects: Defining the Discipline*. Oxford, Blackwell.

Lownds, S. (1998) *Fastrack to Change on the Heathrow Express*. London, IPD Books.

Luck, R. (2007) Using Artefacts to Mediate Understanding in Design Conversations. *Building Research and Information* **35**: 28–41.

Lukes, S. (1974) *Power: A Radical View*. London, Macmillan.

Lumley, R. (1980) Industrial Relations on a Large Industrial Construction Site: The Influence of Security of Employment. *Journal of Management Studies* **17**: 68–81.

Lundin, R.A. and Söderholm, A. (1995) A Theory of the Temporary Organization. *Scandinavian Journal of Management* **11**: 437–455.

Lyons, B. and Mehta, J. (1997) Contracts. Opportunism, and Trust: Self-Interest and Social Orientation. *Cambridge Journal of Economics* **21**: 239–257.

MacCrimmon, K.R. and Ryavec, C.A. (1964) An Analytic Study of the PERT Assumptions. *Operations Research* **12**: 16–37.

Machiavelli, N. (1961) *The Prince*. Harmondsworth, Penguin.

Macmillan, S. (2006) Added Value of Good Design. *Building Research and Information* **34**: 257–271.

Magnussen, O.M. and Olsson, N.O.E. (2006) Comparative Analysis of Cost Estimates of Major Public Investment Projects. *International Journal of Project Management* **24**: 281–288.

Manning, S. (2008) Embedding Projects in Multiple Contexts – A Structuration Perspective. *International Journal of Project Management* **26**: 30–37.

March, J.G. and Shapira, Z. (1987) Managerial Perspectives on Risk and Risk Taking. *Management Science* **33**: 1410–1418.

March, J. and Simon, H.A. (1993) *Organizations* (2nd ed.). Oxford, Blackwell.

Marcou, G., Vickerman, R., and Luchaire, Y. (1992) *Le Tunnel sous la Manche: Entre Etats et Marchés*. Lille, Presses Universitaires de Lille.

Markowitz, H. (1952) Portfolio Selection. *Journal of Finance* **1**: 77–91.

Marsh, P.D.V. (1994) *Comparative Contract Law: England, France, Germany*. Aldershot, Gower.

Martinand, C. (ed.) (1993) *Le Financement Privé des Équipements Publics: l'Expérience Française*. Paris, Economica.

Marx, K. (1968) The Eighteenth Brumaire of Louis Bonaparte. In: K. Marx and F. Engels (eds.). *Selected Works*. London, Lawrence and Wishart.

Marx, K. (1976) *Capital* (vol. 1). Harmondsworth, Penguin.

Masterman, J.W.E. (2001) *An Introduction to Building Procurement Systems* (2nd ed.). London, E & FN Spon.

Maytorena, E., Winch, G.M., Freeman, J., and Kiely, T. (2007) The Influence of Experience and Information Search Styles on Project Risk Identification Performance. *IEEE Transactions on Engineering Management* **54**: 315–326.

McAfee, P.R and McMillan, J. (1986) Bidding for Contracts: A Principal-agent Analysis. *Rand Journal of Economics* **17**: 326–338.

McAfee, P.R and McMillan, J. (1987) Auctions and Bidding. *Journal of Economic Literature* **25**: 699–738.

McMeeken, R. (2008) Egan 10 Years On. *Building* 09/05/08, pp. 30–33.

Meredith, J.R. and Mantel, S.J. (2000) *Project Management: A Managerial Approach* (4th ed.). New York, John Wiley.

Merrow, E.W. (1988) *Understanding the Outcomes of Megaprojects: A Quantitative Analysis of Very Large Civilian Projects*. Santa Monica, RAND Corporation.

Merrow, E.W., Phillips, K.E., and Myers, C.W. (1981) *Understanding Cost Growth and Performance Shortfalls in Pioneer Process Plants*. Santa Monica, RAND Corporation.

Micelli, E. (2000) Mobilising the Skills of Specialist Firms to Reduce Costs and Enhance Performance in the European Construction Industry: Two Case Studies. *Construction Management and Economics* **18**: 651–656.

Middlemas, R.K. (1963) *The Master Builders*. London, Hutchinson.

Milgrom, P. and Roberts, J. (1992) *Economics, Organization and Management*. Upper Saddle River, Prentice Hall.

Miller, J.B. (2000) *Principles of Public and Private Infrastructure Delivery*. Boston, Kluwer Academic Publishers.

Miller, R. and Hobbs, B. (2005) Governance Regimes for Large Complex Projects. *Project Management Journal* **36**: 42–50.

Miller, R. and Lessard, D.R. (2000) *The Strategic Management of Large Engineering Projects: Shaping Institutions, Risks, and Governance*. Cambridge, MIT Press.

Ministry of Public Building and Works (MPBW). (1970) *The Building Process: A Case Study from Marks and Spencer Limited*. London, HMSO.

Mintzberg, H. (1979) *The Structuring of Organizations*. Englewood Cliffs, Prentice-Hall.

Mintzberg, H. (1987) The Strategy Concept 1: Five Ps for Strategy. *California Management Review* **Fall**, pp. 11–24.

Mintzberg, H. (1994) *The Rise and Fall of Strategic Planning*. New York, Prentice Hall.

Moore, R. and Ryan, R. (2000) *Building Tate Modern: Herzog & De Meuron Transforming Giles Gilbert Scott*. London, Tate Gallery Publishing.

Montealegre, R., Nelson, H.J., Knoop, C.I., and Applegate, L.M. (1996) *BAE Automated Systems (A & B): Denver International Airport Baggage-handling System*. Boston, Harvard Business School Publishing.

Morrell, D. (1987) *Indictment: Power and Politics in the Construction Industry*. London, Faber and Faber.

Morris, P.W.G. (1994) *The Management of Projects.* London, Thomas Telford.

Morris, P.W.G. (2002) Research Trends in the 1990s: The Need Now to Focus on the Business Benefit of Project Management. In: D.P. Slevin, D.I. Cleland, and J.K. Pinto (eds.) *The Frontiers of Project Management Research.* Newtown Square, PA, Project Management Institute.

Morris, P.W.G. and Hough, G.H. (1987) *The Anatomy of Major Projects: A Study of the Reality of Project Management.* Chichester, John Wiley.

Mudarri, D.H. (2000) The Economics of Enhanced Environmental Services in Buildings. In: D. Clements-Croome (ed.) *Creating the Productive Workplace.* London, Spon.

Mulenberg, G.M. (2000) Report of Research Examining the Characteristics of Managers of Complex Contemporary Projects in the National Aeronautics and Space Administration. *Project Management Research at the Turn of the Millennium: Proceedings PMI Research Conference,* Paris, pp. 385–399.

Murray, P. (2004) *The Saga of the Sydney Opera House.* London, Spon.

Murray, M. and Langford, D. (eds.) (2003) *Construction Reports 1944–98.* Oxford, Blackwell Science.

National Audit Office. (2001) *Modernising Construction.* London, NAO.

National Audit Office. (2003) *The English National Stadium Project at Wembley.* London, NAO.

National Audit Office. (2005a) *Improving Public Services Through Better Construction.* London, NAO.

National Audit Office. (2005b) *Improving Public Services Through Better Construction: Case Studies.* London, NAO.

National Audit Office. (2007a) *Improving the PFI Tendering Process.* London, NAO.

National Audit Office. (2007b) *The Budget for the London 2012 Olympic and Paralympic Games.* London, NAO.

Nelson, R.R. and Winter, S.G. (1982) *An Evolutionary Theory of Economic Change.* Cambridge, Belknap Press.

Nicholson, N. (2000) *Managing the Human Animal.* London, Texere.

Nicolini, D., Holti, R., and Smalley, M. (2001) Integrating Project Activities: The Theory and Practice of Managing the Supply Chain Through Clusters. *Construction Management and Economics* **19**: 37–47.

Nicolini, D., Tomkins, C., Holti, R., Oldman, A., and Smalley, M. (2000) Can Target Costing be Applied in the Construction Industry – Evidence from Two Case Studies. *British Journal of Management* **11**: 303–324.

Nonaka, I. (1994) A Dynamic Theory of Organizational Knowledge Creation. *Organization Science* **5**: 14–37.

Normann, R. and Ramirez, R. (1993) From Value Chain to Value Constellation: Designing Interactive Strategy. *Harvard Business Review* **July–August**, pp. 65–77.

North, D.C. (1990) *Institutions, Institutional Change and Economic Performance.* Cambridge, Cambridge University Press.

Northcraft, G.B. and Wolf, G. (1984) Dollars, Sense, and Sunk Costs: A Life-Cycle Model of Resource Allocation Decisions. *Academy of Management Review* **9**: 225–234.

Nutt, B. (1993) The Strategic Brief. *Facilities* **11**: 28–32.

Nutt, B. and McLennan, P. (2000) *Facility Management: Risks and Opportunities.* Oxford, Blackwell Science.

Oakland, J.S. (1993) *Total Quality Control: The Route to Improving Performance* (2nd ed.). Oxford, Butterworth-Heinemann.

Office of Government Commerce. (2003) *Managing Successful Programmes.* London, TSO.

Office of Government Commerce. (2005a) *Managing Successful Projects with PRINCE 2.* London, TSO.

Office of Government Commerce. (2005b) *Common Causes of Project Failure.* London, OGC.

Office of Government Commerce/Commission for Architecture and the Built Environment (2002) *Improving Standards of Design in the Procurement of Public Buildings.* London, OGC/CABE.

Office of Rail Regulation (2008) *Report of ORR's Investigation into Engineering Overruns* London, ORR.

Parasuraman, A., Zeithaml, V.A., and Berry, L.L. (1985) A Conceptual Model of Service Quality and its Implications for Future Research. *Journal of Marketing* **49**: 41–50.

Patel, K., Paxson, D., and Sing, T.F. (2005) *A Review of the Practical Uses of Real Property Options.* London, RICS.

Pearce, D. (2003) *The Social and Economic Value of Construction.* London, Construction Industry Research and Innovation Strategy Panel.

Pellegrinelli, S., Partington, D., Hemingway, C., Mohdzain, Z., and Shah, M. (2007) The Importance of Context in Programme Management: An Empirical Review of Programme Practices. *International Journal of Project Management* **25**: 41–55.

Penn, A., Desyllas, J., and Vaughan, L. (1999) The Space of Innovation: Interaction and Communication in the Work Environment. *Environment and Planning B* **26**: 193–218.

Penrose, E. (1995) *The Theory of the Growth of the Firm* (2nd ed.). Oxford, Oxford University Press.

Perkins, H. (1989). The Rise of Professional Society: England Since 1880, London, Routledge.

Peters, T.F. (1996) *Building the Nineteenth Century.* Cambridge, MIT Press.

Peters, T.J. and Waterman, R.H. (1982) *In Search of Excellence: Lessons from America's Best Run Companies.* New York, Harper & Row.

Pettigrew, A.M. (1972) Information Control as Power Resource. *Sociology* **6**: 187–204.

Pfeffer, J. and Sutton, R.I. (2006) Evidence-based Management *Harvard Business Review* **January**, pp. 63–74.

Phiri, M. (1999) *Briefing a Continuous Building Programme.* York, Institute of Advanced Architectural Studies.

Pietroforte, R. (1997) *Building International Construction Alliances: Successful Partnering for Construction Firms.* London, E & FN Spon.

Pinch, T.J. and Bijker, W.E. (1987) The Social Construction of Facts and Artifacts: Or How the Sociology of Science and the Sociology of Technology Might

Benefit Each Other. In: W.E. Bijker, T.P. Hughes, and T.J. Pinch (eds.) *The Social Construction of Technological Systems: New Directions in the Sociology and History of Technology*. Cambridge, MIT Press.

Pine, B.J. (1993) *Mass Customization: The New Frontier in Business Competition*. Boston, Harvard Business School Press.

Pinney, B.W. (2001) *Projects, Management and Protean Times: Engineering Enterprise in the United States 1870–1960*. PhD Thesis, Massachusetts Institute of Technology.

Pitsis, T.S., Clegg, S.R., Rura-Polley, T., and Marosszeky, M. (2003) Constructing the Olympic Dream: A Future Perfect Strategy of Project Management. *Organization Science* **14**: 574–590.

Ponthier, P. (1993) L'Insertion des Concepteurs dans une Système d'Acteurs Etrangers. *Recherches sur le Projet et Les Concepteurs: Actes du Séminaire Euro-Conception*. Paris, Plan Construction et Architecture.

Popper, K.R. (1969) *Conjectures and Refutations: The Growth of Scientific Knowledge*. London, Routledge and Kegan Paul.

Poppo, L. and Zenger, T. (2002) Do Formal Contracts and Relational Governance Function as Substitutes or Complements? *Strategic Management Journal* **23**: 707–725.

Porter, M.E. (1985) *Competitive Advantage: Creating and Sustaining Superior Performance*. New York, Free Press.

Porter, M.E. (1990) *The Competitive Advantage of Nations*. New York, Macmillan.

Powell, J.A. and Brandon, P.S. (1984) An Editorial Conjecture Concerning Building Design, Quality, Cost and Profit. In: P.S. Brandon and J.A. Powell (eds.) *Quality and Profit in Building Design*. London, Chapman and Hall.

Powell, W. and DiMaggio, P.J. (eds.) (1991) *The New Institutionalism in Organizational Analysis*. Chicago, IL: University of Chicago Press.

Powell, M. and Young, J. (2004) The Project Management Support Office. In: P.W.G. Morris and J.K. Pinto (eds.) *The Wiley Guide to Managing Projects*. New York, Wiley.

Prasad, S. (2004) Inclusive maps. In: S. Macmillan (ed.). *Designing Better Buildings*. London, Spon.

Prebble, J. (1979) *The High Girders: The Story of the Tay Bridge Disaster*. Harmondsworth, Penguin Books.

Price, R. (1980) *Masters, Unions and Men: Work Control in Building and the Rise of Labour 1830–1914*. Cambridge, Cambridge University Press.

Priemus, H. (2004) Dutch Contracting Fraud and Governance Issues. *Building Research and Information* **32**: 306–312.

Pryke, S.D. (2004) Analysing Construction Project Coalitions: Exploring the Application of Social Network Analysis. *Construction Management and Economics* **22**: 787–797.

Pryke, S.D. (2005) Towards a Social Network Theory of Project Governance. *Construction Management and Economics* **23**: 927–939.

Pryke, S.D. and Pearson, S. (2006) Project Governance: Case Studies on Financial Incentives. *Building Research and Information* **34**: 534–545.

Pugh, S. (1990) *Total Design: Integrated Methods for Successful Product Engineering*. Wokingham, Addison-Wesley.

Ramirez, R. (1999) Value Co-production: Intellectual Origins and Implications for Practice and Research. *Strategic Management Journal* **20**: 49–65.

Rawlinson, S. (2006) Successful Projects. *Building* 10/02/06.

Reitsma, D. (1995) Major Public Works: Cultural Differences and Decision-Making Procedures. *Tijdschrift voor Economische Geografie* **86**: 186–190.

Riese, M. (2008) One Island East, Hong Kong: A Case Study in Construction Prototyping. In: P. Brandon and T. Kocatürk (eds.) *Virtual Futures for Design, Construction and Procurement*. Oxford, Blackwell.

Riley, D.R. and Sanvido, V.E. (1995) Patterns of Construction-Space Use in Multi-storey Buildings. *Journal of Construction Engineering and Management* **121**: 464–473.

Rittel, H.W.J. and Webber, M.W. (1973) Dilemmas in a General Theory of Planning. *Policy Sciences* **4**: 155–169.

Rosenhead, J. and Mingers, J. (eds.) (2001) *Rational Analysis for Problematic World Revisited*. Chichester, John Wiley.

Rouse, J. (2004) Measuring Value or Only Cost: The Need for New Valuation Methods. In: Macmillan (ed.). *Designing Better Buildings*. London, Spon.

Royer, I. (2003) Why Bad Projects Are so Hard to Kill. *Harvard Business Review* **February**, pp. 48–56.

Rummler, G.A. and Brache, A.P. (1995) *Improving Performance: How to Manage the White Space on the Organization Chart* (2nd. ed.). San Francisco, Jossey-Bass.

Sabbagh, K. (1989) *Skyscraper*. London, Macmillan.

Sabbagh, K. (2000) *Power into Art: Creating the Tate Modern, Bankside*. London, Allen Lane.

Sadeh, A., Dvir, D., and Shenhar, A. (2000) The Role of Contract Type in the Success of R&D Defense Projects under Increasing Uncertainty. *Project Management Journal* **31**: 14–22.

Salzman, L.F. (1952) *Building in England Down to 1540: A Documentary History*. Oxford, Oxford University Press.

Samuelson, W. (1986) Bidding for Contracts. *Management Science* **32**: 1533–1550.

Savage, L.J. (1954) *The Foundations of Statistics*. New York, John Wiley.

Savage, L.J. (1971) Elicitation of Personal Probabilities and Expectations. *Journal of the American Statistical Association* **66**: 783–801.

Sayles, L.R. and Chandler, M.K. (1993) *Managing Large Systems: Organizations for the Future*. New Brunswick, Transaction Publishers.

Schein, E.H. (1992) *Organizational Culture and Leadership* (2nd ed.). San Francisco, Jossey-Bass.

Schoemaker, P.J.H. (1982) The Expected Utility Model: Its Variants, Purposes, Evidence and Limitations. *Journal of Economic Literature* **20**: 529–563.

Schön, D.A. (1983) *The Reflective Practitioner*. New York, Basic Books.

Schutz, A. (1967) *The Phenomenology of the Social World*. Evanston, Northwestern University Press.

Schutz, A. (1973) *Collected Papers I: The Problem of Social Reality*. The Hague, Martinus Nijhoff.

Scott, B. (2001) *Partnering in Europe: Incentive-based Alliancing for Projects*. London, Thomas Telford.

Senge, P.M. (1990) *The Fifth Discipline: The Art and Practice of the Learning Organization*. London, Century Business.

Sha, K. (2004) Construction Business System in China: An Institutional Transformation Perspective. *Building Research and Information* **32**: 529–537.

Shapira, Z. (1997) *Risk Taking: A Managerial Perspective*. New York, Russell Sage Foundation.

Shapira, Z. and Berndt, D.J. (1997) Managing Grand-scale Construction Projects: A Risk-taking Perspective. *Research in Organizational Behavior* **19**: 303–360.

Shenhar, A. and Dvir, D. (2007) *Reinventing Project Management: The Diamond Approach to Successful Growth and Innovation*. Boston, Harvard Business School Press.

Sherif, M. (1958) Superordinate Goals in the Reduction of Intergroup Conflict. *American Journal of Sociology* **63**: 349–356.

Sidwell, A.C. (1990) Project Management: Dynamics and Performance. *Construction Management and Economics* **8**: 159–178.

Sieff, M. (1990) *On Management*. London, Weidenfeld and Nicholson.

Simon, H.A. (1955) A Behavioral Model of Rational Choice. *The Quarterly Journal of Economics* **69**: 99–117.

Simon, H.A. (1973) The Structure of Ill-structured Problems. *Artificial Intelligence* **4**: 181–202.

Simon, H.A. (1976) *Administrative Behavior: A Study of Decision-Making Processes in Administrative Organization* (3rd ed.). New York, Free Press.

Simon, H.A. (1977) *The New Science of Management Decision* (revised edition). Englewood Cliffs, Prentice-Hall.

Slack, N., Chambers, S., and Johnston, R. (2007) *Operations Management* (5th ed.). Harlow, FT Prentice-Hall.

Smit, T. (2001) *Eden*. London, Corgi.

Smith, A. (1970) *An Inquiry into the Nature and Causes of the Wealth of Nations*. Harmondsworth, Pelican.

Sommer, S.C. and Loch, C.H. (2004) Selectionism and Learning in Projects with Complexity and Unforeseeable Uncertainty. *Management Science* **50**: 1334–1347.

Spender, J.-C. (1989) *Industry Recipes: The Nature and Sources of Managerial Judgement*. Oxford, Blackwell.

Spencer, N.C. and Winch, G.M. (2002) *How Buildings Add Value for Clients*. London, Thomas Telford.

Spetzler, C.S. and Staël von Holstein, C.-A.S. (1975) Probability Encoding in Decision Analysis. *Management Science* **22**: 340–358.

Spring, M. (2007) Top Drawer. *Building* 13/07/07.

Standish Group. (1995) *Chaos*. Boston, Standish Group International Inc.

Stanghellini, S. (1996) *Le Système Contractuel Italien en Phase de Changement*. London, Le Groupe Bagnolet, Working Paper 9.

Stanwick, S. and Fowlow, L. (2006) *Wine by Design*. Chichester, Wiley.

Star, S.L. and Griesemer, J.R. (1989) Institutional Ecology, 'Translations' and Boundary Objects: Amateurs and Professionals in Berkeley's Museum of Vertebrate Zoology 1907–39. *Social Studies of Science* **19**: 387–420.

Staw, B.M. (1997) The Escalation of Commitment: An Update and Appraisal. In: Z. Shapira (ed.) *Organizational Decision Making*. Cambridge, CUP.

Stern, N. (2007) *The Economics of Climate Change: The Stern Review*. Cambridge, Cambridge University Press.

Steward, R.A. and Spencer, C.A. (2006) Six-sigma as a Strategy for Process Improvement on Construction Projects: A Case Study. *Construction Management and Economics* **24**: 339–348.

Stewart, R. and Barsoux, J.-L. (1994) *The Diversity of Management: Twelve Managers Talking*. London, Macmillan.

Stinchcombe, A.L. (1959) Bureaucratic and Craft Administration of Production: A Comparative Study. *Administrative Science Quarterly* **4**: 168–187.

Stinchcombe, A.L. (1985) Contracts as Hierarchical Documents. In: A.L. Stinchcombe and C.A. Heimer (eds.) *Organization Theory and Project Management: Administering Uncertainty in Norwegian Offshore Oil*. Oslo, Norwegian University Press.

Strauss, A., Schatzman, L., Ehrlich, D., Bucher, R., and Shabshin, M. (1971) The Hospital and its Negotiated Order. In: F.G. Castles, D.J. Murray, and D.C. Potter (eds.) *Decisions, Organizations and Society: Selected Readings*. Harmondsworth, Penguin.

Strategic Forum. (2002) *Accelerating Change*. London, Rethinking Construction.

Stringer, J. (1995) The Planning Enquiry Process. In: J.R. Turner (ed.) *The Commercial Project Manager*. London, McGraw Hill.

Sudjic, D. (2001) *Blade of Light: The Story of the Millennium Bridge*. Harmondsworth, Penguin.

Sun, M. and Aouad, G. (1999) Control Mechanism for Information Sharing in an Integrated Construction Environment. *Proceedings of the Second International Conference on Concurrent Engineering in Construction – CEC '99*. Espoo, Finland, 25–27 August 1999, pp. 121–130.

Syben, G. (1996) *Learning the Rules of the Game Abroad: The Case of Friedrichstadtpassagen* 207. London, Le Groupe Bagnolet Working Paper 15.

Syben, G. (2000) *Die Baustelle der Bauwirtschaft: Unternehmesentwicklung und Arbeitskräftepolitik auf dem Weg ins 21. Jahrhundert*. Berlin, Edition Sigma.

Sydow, J. (2006) Managing Project in Network Contexts: A Structuration Perspective. In: D. Hodgson and S. Cicmil (eds.) *Making Projects Critical*. Basingstoke, Palgrave Macmillan, pp. 252–264.

Sykes, A.M. (1969) Navvies: Their Work Attitudes. *Sociology* **3**: 21–35.

Taffs, D. (2006) Computers and the Opera House: Pioneering a New Technology In: A. Watson (ed.) *Building a Masterpiece: The Sydney Opera House*. Sydney, Powerhouse Publishing.

Taguchi, T. and Clausing, D. (1990) Robust Quality. *Harvard Business Review* **January–February**, pp. 65–75.

Taleb, N.N. (2007) *Fooled by Randomness: The Hidden Role of Chance in Life and the Markets* (2nd ed.). London, Penguin.

Taleb, N.N. (2008) *Black Swan: The Impact of the Highly Improbable*. London, Penguin.

Tassoul, M. (1998) Making Sense with Backcasting: The Future Perfect. *Creativity and Innovation Management* **7**: 32–45.

Tavistock Institute. (1966) *Interdependence and Uncertainty: A Study of the Building Industry*. London, Tavistock Publications.

Taylor, T. and Ford, D.N. (2006) Tipping Point Failure and Robustness in Single Development Projects. *System Dynamics Review* **22**: 51–71.

Thabet, W.Y. and Beliveau, Y.J. (1994) Modeling Work Space to Schedule Repetitive Floors in Multistory Buildings. *Journal of Construction Engineering and Management* **120**: 96–116.

Thamhaim, H.J. and Wilemon, D.L. (1975) Conflict Management in Project Life-cycles. *Sloan Management Review* **16**: 31–50.

Thoenig, J.-C. (1987) *L'Ere des Technocrats* (2nd ed.). Paris, L'Harmattan.

Thomas, H.R. and Smith, G.R. (1990) *Loss of Construction Labour Productivity due to Inefficiencies and Disruptions: The Weight of Expert Opinion.* PTI Report 9019, Pennsylvania Transportation Institute, Pennsylvania State University.

Thomas, J. (1972) *The Tay Bridge Disaster.* Newton Abbott, David and Charles.

Thomas, J.L. (2000) Making Sense of Project Management. In: R.A. Lundin and F. Hartman (eds.) *Projects as Business Constituents and Guiding Motives.* Dordrecht, Kluwer.

Thompson, D.S., Austin, S.A., Devine-Wright, H., and Mills, G. (2003) Managing Quality and Value in Design. *Building Research and Information* **31**: 334–345.

Thompson, J.D. (1967) *Organizations in Action: Social Science Bases of Administrative Theory.* New York, McGraw-Hill.

Thurm, D. (2005) Master of the House: Why a Company Should Take Control of its Building Projects. *Harvard Business Review* **October**, pp. 120–129.

Tommelein, I.D., Levitt, R.E., Hayes-Roth, B., and Confrey, T. (1991) SightPlan Experiments: Alternative Strategies for Site Layout Design. *Journal of Computing in Civil Engineering* **5**: 42–63.

Tommelein, I.D., Riley, D.R., and Howell, G.A. (1999) Parade Game: Impact of Work Flow Variability on Trade Performance. *Journal of Construction Engineering and Management* **125**: 304–310.

Tuckman, B.W. (1965) Developmental Sequence in Small Groups. *Psychological Bulletin* **63**: 384–399.

Turner, J.R. (1999) *The Handbook of Project-based Management* (2nd ed.). London, McGraw-Hill.

Turner, J.R. and Cochrane, R.A. (1993) Goals-and-Methods Matrix: Coping with Projects with Ill-defined Goals and/or Methods of Achieving Them. *International Journal of Project Management* **11**: 93–102.

Tversky, A. and Kahneman, D. (1982) Judgment under Uncertainty; Heuristics and Biases. In: D. Kahneman, P. Slovic, and A. Tversky (eds.) *Judgement under Uncertainty: Heuristics and Biases.* Cambridge, Cambridge University Press.

Ulrich, K.T. and Eppinger, S.D. (2008) *Product Design and Development* (4th ed.). London, McGraw-Hill.

Ulirch, R., Quan, X., Zimring, C., Joseph, A., and Choudhary, R. (2004) *The Role of the Physical Environment in the Hospital of the 21st Century: A Once in-a-Lifetime Opportunity.* Concord, Center for Health Design.

Urwick, L. (1937) Organization as a Technical Problem. In: L. Gulick and L. Urwick (eds.) *Papers on the Science of Administration.* New York, Institute of Public Administration.

Usmani, A. and Winch, G.M. (1993) *The Management of the Design Process: The Case of Architectural and Urban Projects.* London, Bartlett, Research Paper No. 1.

Van de Ven, A.H. and Ferry, D.L. (1980) *Measuring and Assessing Organizations*. New York, Wiley.

Vaughan, D. (1996) *The Challenger Launch Decision: Risky Technology, Culture and Deviance at NASA*. Chicago, University of Chicago Press.

Vick, S.G. (2002) *Degrees of Belief: Subjective Probability and Engineering Judgment*. Reston, VA, ASCE Press.

Wachs, M. (1990) Ethics and Advocacy in Forecasting for Public Policy Business and Professional. *Ethics Journal* **9**: 141–157.

Walker, A. (1996) *Project Management in Construction* (3rd ed.). Oxford, Blackwell Science.

Walsh, J.P. (1995) Managerial and Organizational Cognition: Notes from a Trip down Memory Lane. *Organization Science* **6**: 280–320.

Walton, M. (1997) *Car: A Drama of the American Workplace*. New York, W.W. Norton.

Watson, A. (ed.) (2006) *Building a Masterpiece: The Sydney Opera House*. Sydney, Powerhouse Publishing.

Way, M. and Bordass, B. (2005). Making Feedback and Post-occupancy Evaluation Routine 2: Soft Landings – Involving Design and Building Teams in Improving Performance. *Building Research and Information* **33**: 353–360.

Weber, M. (1947/1964) *The Theory of Social and Economic Organization*. New York, Free Press.

Weber, M. (1948) *From Max Weber: Essays in Sociology*. London, Routledge and Kegan Paul.

Weick, K.E. (1979) *The Social Psychology of Organizing* (2nd ed.). New York, McGraw-Hill.

Weick, K.E. (1995) *Sensemaking in Organizations*. Thousand Oaks, Sage.

Wheelwright, S.C. and Clark, K.B. (1992) *Revolutionizing Product Development: Quantum Leaps in Speed, Efficiency, and Quality*. New York, Free Press.

Whitley, R. (1992) *Business Systems in East Asia: Firms, Markets, and Societies*. London, Sage.

Whitley, R. (1999) *Divergent Capitalisms*. Oxford, Oxford University Press.

Wikinson, G. and Dale, B.G. (1999) Integrated Management Systems: An Examination of the Concept and Theory. *The TQM Magazine* **11**, pp. 95–104.

Williamson, O.E. (1975) *Markets and Hierarchies: Analysis and Anti-Trust Implications*. New York, Free Press.

Williamson, O.E. (1985) *The Economic Institutions of Capitalism*. New York, The Free Press.

Winch, G.M. (1986) The Labour Process and Labour Market in Construction. *International Journal of Sociology & Social Policy* **6**: 103–116.

Winch, G.M. (1994a) *Managing Production: Engineering Change and Stability*. Oxford, Oxford University Press.

Winch, G.M. (1994b) The Search for Flexibility: The Case of the Construction Industry. *Work Employment and Society* **8**: 593–606.

Winch, G.M. (1996a) *The Contracting System in British Construction: The Rigidities of Flexibility*. London, Le Groupe Bagnolet, Working Paper 6.

Winch, G.M. (1996b) *The Channel Tunnel: Le Projet du Siècle*. London, Le Groupe Bagnolet, Working Paper 11.

Winch, G.M. (1998a) The Growth of Self-employment in British Construction. *Construction Management and Economics* **16**: 531–542.
Winch, G.M. (1998b) Zephyrs of Creative Destruction: Understanding the Management of Innovation in Construction. *Building Research and Innovation* **26**: 268–279.
Winch, G.M. (1999) Project Management. In: C. Clegg, K. Legge, and S. Walsh (eds.) *The Experience of Managing: A Skills Guide.* Basingstoke, Macmillan.
Winch, G.M. (2000a) The Management of Projects as a Generic Business Process. In: R.A. Lundin and F. Hartman (eds.). *Projects as Business Constituents and Guiding Motives*. Dordrecht, Kluwer, pp. 117–130.
Winch, G.M. (2000b) Innovativeness in British and French Construction: The Evidence from Transmanche-Link. *Construction Management and Economics* **18**: 807–817.
Winch, G.M. (2000c) Institutional Reform in British Construction: Partnering and Private Finance. *Building Research and Information* **28** (2): 141–155.
Winch, G.M. (2001) Governing the Project Process: A Conceptual Framework. *Construction Management and Economics* **19**: 799–808.
Winch, G.M. (2003) How Innovative is Construction? Comparing Aggregated Data on Construction Innovation and Other Sectors – A Case of Apples and Pears. *Construction Management and Economics* **21**: 651–654.
Winch, G.M. (2004) Managing Project Stakeholders. In: P.W.G. Morris and J.K. Pinto (eds.) *The Wiley Guide to Managing Projects*. New York, Wiley, pp. 321–339.
Winch, G.M. (2005) Rethinking Project Management: Project Organizations as Information Processing Systems? In: D.P. Slevin, D.I. Cleland, and J.K. Pinto (eds.) *Innovations: Project Management Research 2004*. Newtown Square, Project Management Institute, pp. 41–55.
Winch, G.M. (2006a) The Governance of Project Coalitions: Towards a Research Agenda. In: D. Lowe and R. Leiringer (eds.) *Commercial Management of Complex Projects: Defining the Discipline*. Oxford, Blackwell, pp. 324–343.
Winch, G.M. (2006b) Towards a Theory of Construction as Production by Projects. *Building Research and Information* **34** (2): 164–174.
Winch, G.M. (2008) Revaluing Construction: Implications for the Construction Process. In: P. Brandon and S.-L. Lu (eds.). *Clients Driving Innovation* Oxford, Wiley.
Winch, G.M. and Bonke, S. (2002) Project Stakeholder Mapping: Analyzing the Interests of Project Stakeholders. In: D.P. Slevin, D.I. Cleland, and J.K. Pinto (eds.) *The Frontiers of Project Management Research*. Newtown Square, PA, Project Management Institute.
Winch, G.M. and Campagnac, E. (1995) The Organization of Building Projects: An Anglo/French Comparison. *Construction Management and Economics* **13**: 3–14.
Winch, G.M. and Carr, B. (2001a) Benchmarking On-Site Productivity in France and Great Britain: A CALIBRE Approach. *Construction Management and Economics* **19**: 577–590.
Winch, G.M. and Carr, B. (2001b) Processes Maps and Protocols: Understanding the Shape of the Construction Process. *Construction Management and Economics* **19**: 519–531.

Winch, G.M., Clifton, N., and Millar, C.J.M. (2000) Organisation and Management in an Anglo-French Consortium: The Case of Transmanche Link. *Journal of Management Studies* **37**: 663–685.

Winch, G.M., Grèzes, D., and Carr, B. (2002) Exporting Architectural Services: The English and French Experiences. *Journal of Architectural and Planning Research* **19**: 165–175.

Winch, G.M. and Kelsey, J. (2005) What do Construction Project Planners Do? *International Journal of Project Management* **23**: 141–149.

Winch, G.M. and Maytorena, E. (2009) Making Good Sense: Assessing the Quality of Risky Decision-making. *Organization Studies* **30**: 181–203.

Winch, G.M., Millar, C.J.M., and Clifton, N. (1997) Culture and Organisation: The Case of Transmanche Link. *British Journal of Management* **8**: 237–249.

Winch, G.M. and North, S. (2006) Critical Space Analysis. *ASCE Journal of Construction Engineering and Management* **132**: 473–481.

Winch, G.M. and Schneider, E. (1993) Managing the Knowledge-Based Organisation; the Case of Architectural Practice. *Journal of Management Studies* **30**: 923–937.

Winch, G.M., Usmani, A., and Edkins, A. (1998) Towards Total Project Quality: A Gap Analysis Approach. *Construction Management and Economics* **16**: 193–207.

Winter, M. (2006) Problem Structuring in Project Management: An Application of Soft Systems Methodology (SSM). *Journal of the Operational Research Society* **57**: 802–812.

Winter, M. and Szszepanek, T. (2008) Projects and Programmes as Value Creation Processes: A New Perspective and some Practical Implications. *International Journal of Project Management* **26**: 95–103.

Womack, J.P. and Jones, D.T. (1996) *Lean Thinking: Banish Waste and Create Wealth in your Corporation*. New York, Simon and Schuster.

Womack, J.P., Jones, D.T., and Roos, D. (1990) *The Machine that Changed the World*. New York, Rawson Associates.

Woudhuysen, J. and Abley, I. (2004) *Why is Construction so Backward?* Chichester, Wiley.

Wright, R.N., Rosenfeld, A.H., and Fowell, A.J. (1995) *National Planning for Construction and Building R&D*. Gaithersburg, National Institute of Standards and Technology.

Yang, J.-B., Wu, C.-T., and Tsai, C.-H. (2007) Selection of an ERP System for a Construction Firm in Taiwan: A Case Study. *Automation in Construction* **16**: 787–796.

Yeomans, D.T. (1988) Managing Eighteenth-Century Building. *Construction History* **4:** 3–19.

Zeithaml, V.A., Parasuraman, A., and Berry, L.L. (1990) *Delivering Quality Service: Balancing Customer Perceptions and Expectations*. New York, Free Press.

Zouein, P.P. and Tommelein, I.D. (1999) Dynamic Layout Planning Using a Hybrid Incremental Solution Method. *Journal of Construction Engineering and Management* **125**: 400–408.

Zweigert, K. and Kötz, H. (1998) *An Introduction to Comparative Law* (3rd ed.). Oxford, Oxford University Press.

People Index

Note: references to numbered panels are shown in bold; case references are shown in italics; notes are shown within square bracket

Project Index

Note: references to numbered panels are shown in bold; case references are shown in italics; notes are shown within square bracket

Subject Index

Note: references to numbered panels are shown in bold; case references are shown in italics; notes are shown within square bracket